NOBEL LECTURES IN MOLECULAR BIOLOGY

NOBEL LECTURES IN MOLECULAR BIOLOGY 1933-1975

with a Foreword by
David Baltimore

ELSEVIER · NEW YORK
NEW YORK · AMSTERDAM · OXFORD

Elsevier North-Holland, Inc.
52 Vanderbilt Avenue, New York, New York 10017

Elsevier Scientific Publishing Company
Jan Van Galenstraat 335, P.O. Box 211
Amsterdam, The Netherlands

Library of Congress Cataloging in Publication Data

Main entry under title:

Nobel lectures in molecular biology, 1933-1975.

Includes index.
1. Molecular biology—Addresses, essays, lectures.
2. Molecular genetics—Addresses, essays, lectures.
I. Baltimore, David.
QH506.N58 574.8'8'08 77-22255
ISBN 0-444-00236-7

Manufactured in the United States of America

Contents

Foreword

Alfred Nobel's will stipulated that all who come to Stockholm to be honored must give a lecture to their Swedish colleagues about their work. The recipients of his Prize take this as a special task—to produce a lecture worthy of the occasion. But they write their lectures in an odd and unique moment of their life, in the interregnum between the announcement of the award in October and its presentation in December. The heightened awareness of their life in science during this time makes this lecture more personal than any other they are likely to give. The special insights into the scientists and their science found in these lectures make them a unique resource.

The Nobel Lectures presented in this compendium are many of those that relate to molecular biology. Trying to see in them the underlying themes and contrasts offers many rewards. For one thing, these lectures almost define the imprecise but universally recognized concept of molecular biology. The Lectures also let us see the close connections between the most highly honored practitioners in the field: so many of them worked together, thought about similar problems, built on each others ideas and experiments that the unity of the quest of science is a continuing theme. And certain experimental threads emerge: the seminal and continual importance of radiation experiments in studying genes, the necessity of detailed structural knowledge in the understanding of functional problems, and the central role of a small number of simple systems like bacteriophages. Finally, the concern of many Nobelists about social issues comes through in these Lectures: from Muller through Dulbecco many of the Laureates used this occasion to voice their concern about the relation of science and society.

The first Lectures reprinted here represented a significant departure in the awarding of the Nobel Prize in Physiology or Medicine. Morgan and Muller dealt with biology at a much more fundamental level than any of their predecessors. They were the first to be honored for trying to comprehend the nature of the gene. They were the first scientific descendants of Darwin to win the Prize and they set the stage for many further awards in the continuing attempt to provide a chemical explanation for both the stability of inheritance and the variability underlying evolution. Morgan and Muller were the fathers of molecular biology.

Morgan saw clearly that his genetics was a formal science and noted that "at the level at which the genetic experiments lie, it does not make the slightest difference whether the gene is a hypothetical unit or whether the gene is a material particle." But he wondered if "now that we located them in the chromosomes are we justified in regarding them as material units . . .".

Following Morgan and Muller, the development of molecular biology can be seen as the "materialization of the gene." The first notable result in this development never was signaled by a Nobel award: the proof by Avery,

McLeod and McCarthy that genes are DNA. Avery, unfortunately, died before their contribution was recognized as the great event it was. The Prizes began in 1958 to reflect the progress in molecular biology by honoring Beadle, Tatum and Lederberg. This trio was responsible for two major advances. The first is the concept that a gene, whatever it may be physically, has as its primary function the synthesis of a protein. Second, they introduced microorganisms as the appropriate systems for studying what genes are and how they work, setting the stage for many future advances.

There were many questions that needed answers before the link between DNA and the synthesis of proteins could be understood. Kornberg and Ochoa, honored in 1959, took one step by elucidating the enzymatic machinery that makes DNA and RNA. But the prize that reflects best the revolution that is molecular biology was that to Crick, Watson and Wilkins in 1962. By describing the structure of DNA they showed what really underlies genetics. The road from Darwin and Mendel reached its end with the memorable line that starts Wilkins' lecture, "Nucleic acids are basically simple." Molecular biology is the working out of the implications inherent in the "simple" double helix model of DNA.

Molecular biology, then, is the study of how DNA, RNA and protein are interrelated. A look at the prizes following 1962 shows how the concepts of molecular biology developed and were proved. Jacob, Lwoff and Monod showed how protein regulates synthesis of RNA from DNA. Holley, Khorana, and Nirenberg showed how RNA can specify the structure of proteins. Anfinsen and Moore and Stein showed that proteins have a definable linear structure and that their three-dimensional structure is a result of the play of thermodynamic principles on linear amino acid sequences. And finally in 1969 a trio who had been the prophets, conscience and experimental innovators of molecular biology and who had exploited the genetically most manipulable of all organisms we know, the bacteriophage, were finally honored. Their Prize was so long in coming that all three wrote about their present concerns. Delbrück's made his philosophical parenthood to molecular biology especially evident. Sounding very much like Morgan had when he looked forward, Delbrück made the connection between chemistry and genetics by looking back. "Genes [in Muller's day] were algebraic units of the combinational science of genetics and it was anything but clear that these units were molecules analyzable in the terms of structure chemistry."

Since the broad outlines of molecular biology became clear, application of its principles to specific systems has been a major activity in biology. The systems of widest concern have been those that relate to major human diseases and the Nobel Prize has recently begun to reflect the new perspective. In 1972, Edelman and Porter were honored for elucidating the structure of the central molecule of the immune system, the antibody. They brought immunology into the fold of molecular biology by their simple but elegant dissection of elements that make up the antibody molecule. In 1975, Dulbecco, Temin and I were honored for bringing molecular biology to bear on the awful and awesome problem of cancer. We had provided crucial evidence that viruses able to cause cancer do it by adding new genes to the cell. Two of us had also added a footnote to what has been called the central dogma of molecular biology, that information flows from nucleic acids to proteins and never in reverse. We found that information flowed from RNA to DNA in the genetic system of certain viruses although in all other known situations it transferred undirectionally from DNA to RNA.

Molecular biology is now a science in its own right with departments, journals, courses, study sections and the rest of the paraphernalia of respectability. But in spite of its acceptance into the establishment, it is still a vibrant and continually surprising field of endeavor from which we can expect more Prize-worthy discoveries and hopefully major contributions to human welfare. The recent advent of recombinant DNA methodology, a breakthrough certain to generate Prizes of its own, should allow a new push into the molecular biology of higher cells, an area too complicated to be readily accessible to the tools of the past. From the puzzles of higher cell function, and especially neural function, will flow the surprises of the future.

David Baltimore

1933

Thomas H. Morgan

Thomas H. Morgan

The Relation of Genetics to Physiology and Medicine

June 4, 1934

The study of heredity, now called genetics, has undergone such an extraordinary development in the present century, both in theory and in practice, that it is not possible in a short address to review even briefly all of its outstanding achievements. At most I can do no more than take up a few outstanding topics for discussion.

Since the group of men with whom I have worked for twenty years has been interested for the most part in the chromosome mechanism of heredity, I shall first briefly describe the relation between the facts of heredity and the theory of the gene. Then I should like to discuss one of the physiological problems implied in the theory of the gene; and finally, I hope to say a few words about the applications of genetics to medicine.

The modern theory of genetics dates from the opening years of the present century, with the discovery of Mendel's long-lost paper that had been overlooked for thirty-five years. The data obtained by De Vries in Holland, Correns in Germany, and Tschermak in Austria showed that Mendel's laws are not confined to garden peas, but apply to other plants. A year or two later the work of Bateson and Punnett in England, and Cuénot in France, made it evident that the same laws apply to animals.

In 1902 a young student, William Sutton, working in the laboratory of E. B. Wilson, pointed out clearly and completely that the known behavior of the chromosomes at the time of maturation of the germ cells furnishes us with a mechanism that accounts for the kind of separation of the hereditary units postulated in Mendel's theory.

The discovery of a mechanism, that suffices to explain both the first and the second law of Mendel, has had far-reaching consequences for genetic theory, especially in relation to the discovery of additional laws; because, the recognition of a mechanism that can be seen and followed demands that any extension of Mendel's theories must conform to such a recognized mechanism; and also because the apparent exceptions to Mendel's laws, that came to light before long, might, in the absence of a known mechanism, have called forth purely fictitious modifications of Mendel's laws, or even seemed to invalidate their generality. We now know that some of these «exceptions» are due to newly discovered and demonstrable properties of the chromosome mechanism, and others to recognizable irregularities in the machine.

Mendel knew of no processes taking place in the formation of pollen and egg cell that could furnish a basis for his primary assumption that the hereditary elements separate in the germ cells in such a way that each ripe germ cell comes to contain only one of each kind of element: but he justified the validity of this assumption by putting it to a crucial test. His analysis was a wonderful feat of reasoning. He verified his reasoning by the recognized experimental procedure of science.

As a matter of fact it would not have been possible in Mendel's time to give an objective demonstration of the basic mechanism involved in the separation of the hereditary elements in the germ cells. The preparation for this demonstration took all of the thirty-five years between Mendel's paper in 1865, and 1900. It is here that the names of the most prominent European cytologists stand out as the discoverers of the role of the chromosomes in the maturation of the germ cells. It is largely a result of their work that it was possible in 1902 to relate the well-known cytological evidence to Mendel's laws. So much in retrospect.

The most significant additions that have been made to Mendel's two laws may be called linkage and crossing-over. In 1906 Bateson and Punnett reported a two-factor case in sweet peas that did not give the expected ratio for two pairs of characters entering the cross at the same time.

By 1911 two genes had been found in Drosophila that gave sex-linked inheritance. It had earlier been shown that such genes lie in the X-chromosomes. Ratios were found in the second generation that did not conform to Mendel's second law when these two pairs of characters are present, and the suggestion was made that the ratios in such cases could be explained on the basis of interchange between the two X-chromosomes in the female. It was also pointed out that the further apart the genes for such characters happen to lie in the chromosome, the greater the chance for interchange to take place. This would give the approximate location of the genes with respect to other genes. By further extension and clarification of this idea it became possible, as more evidence accumulated, to demonstrate that the genes lie in a single line in each chromosome.

Two years previously (1909) a Belgian investigator, Janssens, had described a phenomenon in the conjugating chromosomes of a salamander,

Batracoseps, which he interpreted to mean that interchanges take place between homologous chromosomes. This he called chiasmatypie – a phenomenon that has occupied the attention of cytologists down to the present day. Janssens' observations were destined shortly to supply an objective support to the demonstration of genetic interchange between linked genes carried in the sex chromosomes of the female Drosophila.

Today we arrange the genes in a chart or map. The numbers attached express the distance of each gene from some arbitrary point taken as zero. These numbers make it possible to foretell how any new character that may appear will be inherited with respect to all other characters, as soon as its crossing-over value with respect to any other two characters is determined. This ability to predict would in itself justify the construction of such maps, even if there were no other facts concerning the location of the genes; but there is today direct evidence in support of the view that the genes lie in a serial order in the chromosomes.

What are the genes?

What is the nature of the elements of heredity that Mendel postulated as purely theoretical units? What are genes? Now that we locate them in the chromosomes are we justified in regarding them as material units; as chemical bodies of a higher order than molecules? Frankly, these are questions with which the working geneticist has not much concern himself, except now and then to speculate as to the nature of the postulated elements. There is no consensus of opinion amongst geneticists as to what the genes are – whether they are real or purely fictitious – because at the level at which the genetic experiments lie, it does not make the slightest difference whether the gene is a hypothetical unit, or whether the gene is a material particle. In either case the unit is associated with a specific chromosome, and can be localized there by purely genetic analysis. Hence, if the gene is a material unit, it is a piece of a chromosome; if it is a fictitious unit, it must be referred to a definite location in a chromosome – the same place as on the other hypothesis. Therefore, it makes no difference in the actual work in genetics which point of view is taken.

Between the characters that are used by the geneticist and the genes that his theory postulates lies the whole field of embryonic development, where the properties implicit in the genes become explicit in the protoplasm of the cells. Here we appear to approach a physiological problem, but one that is new and strange to the classical physiology of the schools.

We ascribe certain general properties to the genes, in part from genetic evidence and in part from microscopical observations. These properties we may next consider.

Since chromosomes divide in such a way that the line of genes is split (each daughter chromosome receiving exactly half of the original line) we can scarcely avoid the inference that the genes divide into exactly equal parts; but just how this takes place is not known. The analogy of cell division creates a presumption that the gene divides in the same way, but we should not forget that the relatively gross process involved in cell division may seem quite inadequate to cover the refined separation of the gene into equal halves. As we do not know of any comparable division phenomena in organic molecules, we must also be careful in ascribing a simple molecular constitution to the gene. On the other hand, the elaborate chains of molecules built up in organic material may give us, some day, a better opportunity to picture the molecular or aggregate structure of the gene and furnish a clue concerning its mode of division.

Since by infinite subdivisions the genes do not diminish in size or alter as to their properties, they must, in some sense, compensate by growing between successive divisions. We might call this property autocatalysis, but, since we do not know how the gene grows, it is somewhat hazardous to assume that its property of growth after division is the same process that the chemist calls autocatalytic. The comparison is at present too vague to be reliable.

The relative stability of the gene is an inference from genetic evidence. For thousands–perhaps many millions–of subdivisions of its material it remains constant. Nevertheless, on rare occasions, it may change. We call this change a *mutation*, following De Vries' terminology. The point to emphasize here is that the mutated gene retains, in the great majority of cases studied, the property of growth and division, and more important still the property of stability. It is, however, not necessary to assume, either for the original genes or for the mutated genes, that they are all equally stable. In fact, there is a good deal of evidence for the view that some genes mutate oftener than others, and in a few cases the phenomenon is not infrequent, both in the germ cells and in somatic tissues. Here the significant fact is that these repetitional changes are in definite and specific directions.

The constancy of position of genes with respect to other genes in linear order in the chromosomes is deducible, both from genetic evidence and from cytological observations. Whether the relative position is no more than a *historical accident*, or whether it is due to some relation between each gene and its neighbors, can not be definitely stated. But the evidence from the dislocation of a fragment of the chromosome, and its reattachment to another one indicates that accident rather than mutual interaction has determined their present location: for, when a piece of one chromosome becomes attached to the end of a chain of genes of another chromosome or when a section of a chromosome becomes inverted, the genes in the new position hold as fast together as they do in the normal chromosome.

There is one point of great interest. So far as we can judge from the action of mutated genes, the kind of effect produced has as a rule no relation to location of the gene in the chromosome. A gene may produce its chief effect on the eye color, while one nearby may affect the wing structure, and a third, in the same region, the fertility of the male or of the female. Moreover, genes in different chromosomes may produce almost identical effects on the same organs. One may say, then, that the position of the genes in the hereditary material is inconsequential in relation to the effects that they produce. This leads to a consideration which is more directly significant for the physiology of development.

In the earlier days of genetics it was customary to speak of unit characters in heredity, because certain contrasted characters, rather clearly defined, furnished the data for the Mendelian ratios. Certain students of genetics inferred that the Mendelian units responsible for the selected character were genes producing only a single effect. This was careless logic. It took a good deal of hammering to get rid of this erroneous idea. As facts accumulated, it became evident that each gene produces not a single effect, but in some cases a multitude of effects on the characters of the individual. It is true that in most genetic work only one of these character effects is selected for study – the one that is most sharply defined and separable from its contrasted character – but in most cases minor differences are also recognizable that are just as much the product of the same gene as is the major effect. In fact, the major difference selected for classification of the contrasted character-pairs may be of small importance for the welfare of the individual, while some of the concomitant effects may be of vital importance for the economy of the individual, affecting its vitality, its length of life, or its fertility. I need not dwell at length on these relations because they are recognized today by all geneticists. It is important, nevertheless, to take cognizance of them, because the whole problem of the physiology of development is involved.

The coming together of the chromosomes at the maturation devision, and their subsequent movement apart to opposite poles of the meiotic figure, insures the regular distribution of one set of chromosomes to each daughter cell and the fulfilment of Mendel's second law. These movements have the appearance of physical events. Cytologists speak of these two phenomena as *attraction* and *repulsion* of the members of individual chromosomes, but we have no knowledge of the kind of physical processes involved. The terms attraction and repulsion are purely descriptive, and mean no more at present than that like chromosomes come together and later separate.

In earlier times, when the constitution of the chromosomes was not known, it was supposed that the chromosomes come together at random in pairs. There was the implication that any two chromosomes may mate. The comparison with conjugation of male and female protozoa, or egg and sperm cell, was obvious, and since in all diploid cells one member of each pair of chromosomes has come from the father and one from the mother, it must have seemed that somehow maleness and femaleness are involved in the conjugation of the chromosomes also. But today we have abundant evidence to prove that this idea is entirely erroneous, since there are cases where both chromosomes that conjugate have come from the female, and even where both have been sister strands of the same chromosome.

Recent genetic analysis shows not only that the conjugating chromosomes

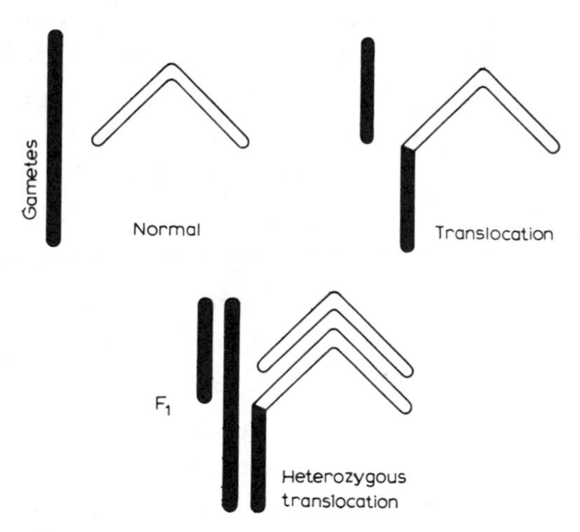

Fig. 1. Diagram to illustrate the case when a piece of one chromosome (*black*) has been translocated to another chromosome (*white*). In the lower part of the figure the method of conjugation of these chromosomes is shown.

are like chromosomes, i.e., chains of the same genes, but also that a very exact process is involved. The genes come together, point for point, unless some physical obstacle prevents. The last few years have furnished some beautiful illustrations showing that it is genes rather than whole chromosomes that come to lie side by side when the chromosomes come together. For example: occasionally a chromosome may have a piece broken off (Fig. 1 above) which becomes attached to another chromosome. A new linkage group is thus established. When conjugation takes place, this piece has no corresponding piece in the sister chromosome. It has been shown (Fig. 1 below) that it then conjugates with that part of the parental chromosomes from which it came.

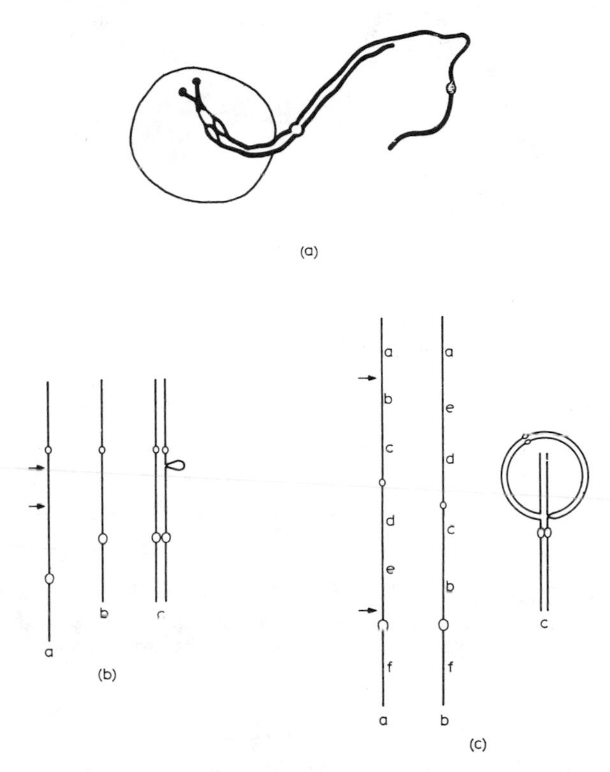

Fig. 2. (*a*) Two conjugating chromosomes of Indian corn (after McClintock). One chromosome has a terminal deficiency. (*b*) Two chromosomes of Indian corn, one having a deficiency near its middle. When these two chromosomes conjugate there is a loop in the longer chromosome opposite the deficiency in the other one. (*c*) Two chromosomes of Indian corn, one having a long inverted region. When they conjugate they come together as shown in the figure to the right (after McClintock), like genes coming together.

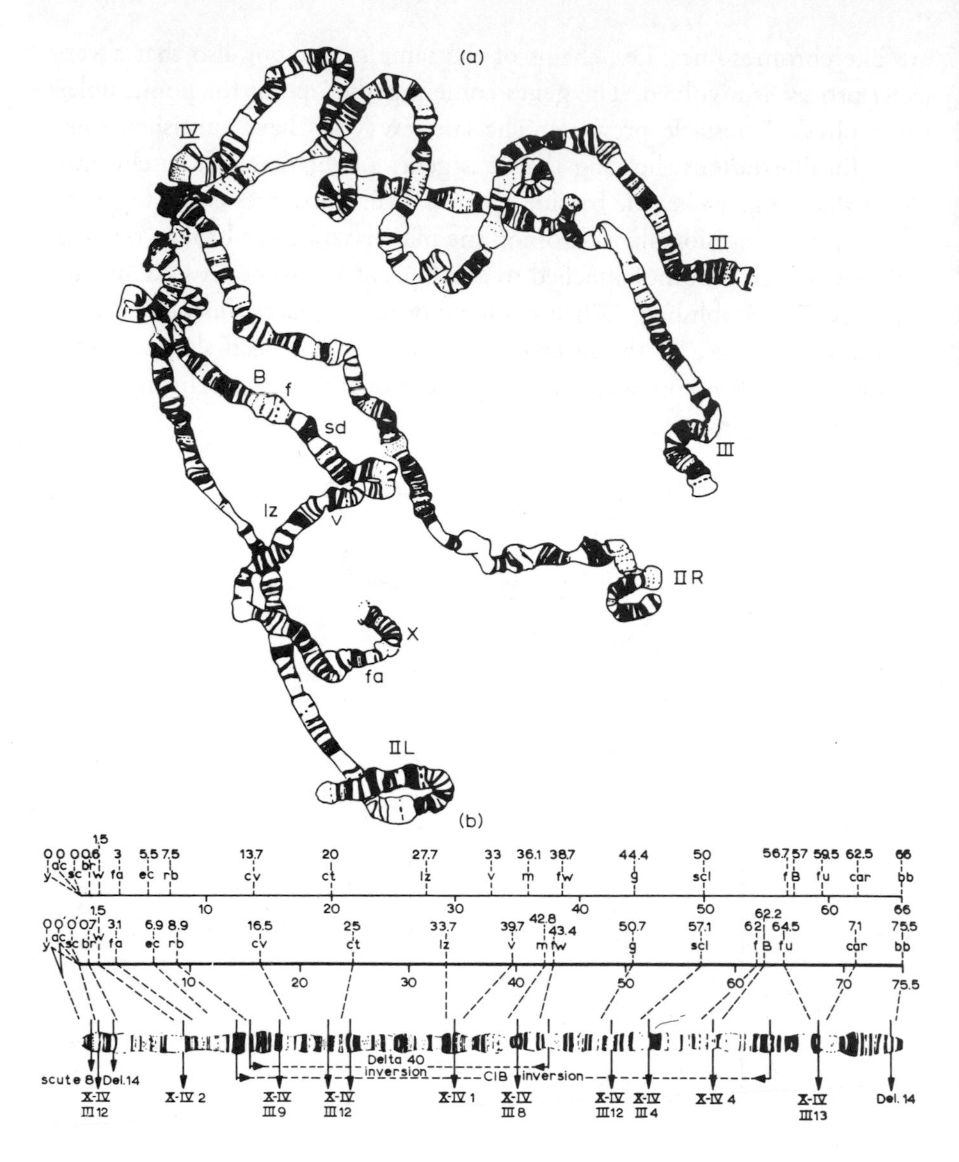

Fig. 3. (*a*) The chromosomes of the salivary gland of the female larva of *Drosophila melanogaster* (after Painter). The two X-chromosomes are fused into a single body. This chromosome is attached at one end to the common chromidial mass at its «attachment end». The 2nd and 3rd chromosomes have the attached point near the middle and are fused with the common chromidial mass at this point, leaving two free ends of each chromosome. Like limbs of each of these free ends are fused, giving four free ends in all. (*b*) The banded salivary X-chromosome of *Drosophila melanogaster* is below, with the genetic map above (after Painter). Oblique broken lines connect the loci of the genetic map with corresponding or homologous loci of the salivary chromosome.

When a chromosome has lost one end, it conjugates with its mate only in part (Fig. 2 a), i.e., where like genes are present. When a chromosome has lost a small region, somewhere along its length, so that it is shorter than the original chromosome, the larger chromosome shows a loop which is opposite the region of deficiency in the shorter chromosome as shown in Fig. 2 b. Thus like genes, or corresponding loci, are enabled to come together through the rest of the chromosome. More remarkable still is the case where the middle region of a chromosome has become turned around (inversion). When such a chromosome is brought together with its normal homologue, as shown in Fig. 2 c, like regions come together by the inverted piece reversing itself, so to speak, so that like genes come together as shown to the right in Fig. 2 c. In this same connection the conjugation of the chromosomes in species of *Oenothera* furnish beautiful examples of the way in which like series of genes find each other, even when halves of different chromosomes have been interchanged.

The very recent work of Heitz, Painter, and Bridges has brought to light some astonishing evidence relating to the constitution of the chromosomes in the salivary glands of *Drosophila melanogaster*.

The nuclei of the cells of the salivary glands of the old larvae are very large and their contained chromosomes (Fig. 3) may be 70 to 150 times as large as those of the ordinary chromosomes in process of division. Heitz has shown that there are regions of some of the chromosomes of the ganglion cells – more especially of the X- and the Y-chromosomes – that stain deeply, and other regions faintly, and that these regions correspond to regions of the genetic map that do not and do contain genes. Painter has made the further important contribution that the series of bands of the salivary chromosomes can be homologized with the genetically known series of genes of the linkage maps (Fig. 3 a, b), and that the empty regions of the X and Y do not have the banded structure. He has further shown that when a part of the linkage map is reversed, the sequence of the bands is also reversed; that when pieces are translocated they can be identified by characteristic bands: and that when pieces of linked genes are lost there is a corresponding loss of bands. Bridges has carried the analysis further by an intensive study of regions of particular chromosomes, and has shown a close agreement between bands and gene location. With improved methods he has identified twice as many bands, thus making a more complete analysis of the relation of bands and gene location. Thus, whether or not the bands are the actual genes, the evidence is clear in showing a remarkable agreement between the location of genes and the location of corresponding bands. The analysis of the banded structure has confirmed the genetic evidence, showing that when certain alterations of the order of the genes takes place, there is a corresponding change in the sequence of the bands which holds for the finest details of the bands.

The number of chromosomes in the salivary nuclei is half that of the full number (as reported by Heitz) which Painter interprets as due to homologous chromosomes conjugating (Fig. 3 a). Moreover, the bands in each of the component halves show an identical sequence which is strikingly evi-

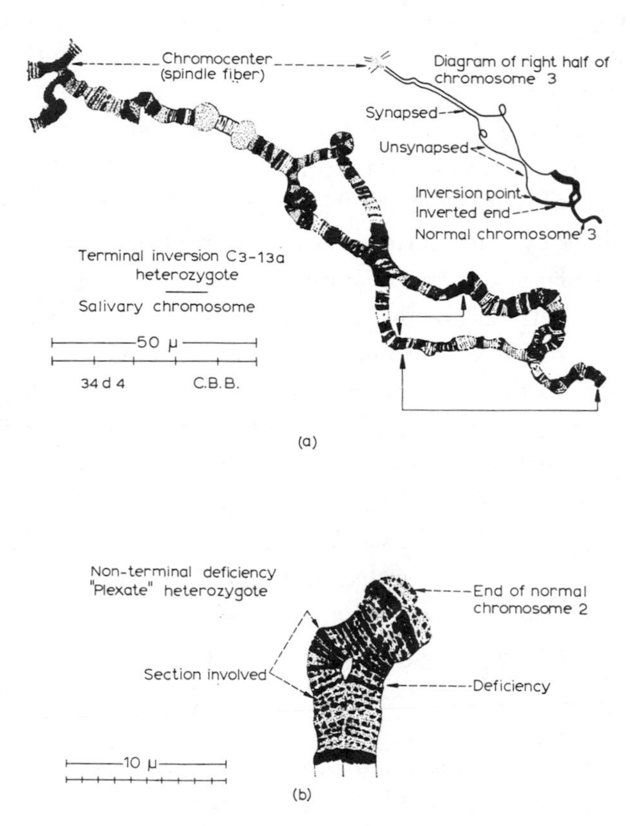

Fig. 4. (*a*) Salivary gland preparation of the right half of the third chromosome. The two components are united through a part of their length (*above left*). One component had a terminal inversion. This part conjugated with the corresponding normal chromosome by turning back on itself, as shown in the small diagram above (*to the right*). (*b*) Salivary gland preparation showing a part of chromosome 2; one component is «deficient». At the level of the deficiency the other component is bent outward so that above and below these level like bands meet. (After Bridges.)

dent when the halves are not closely apposed. It has been suggested by Bridges and by Koltzoff that homologous chromosomes have not only united, but that they have each divided two or three times, giving in some cases as many as 16 or 32 strands (Fig. 4 a, b). The bands may then be said to be composed each of 16 or 32 genes; or, if this identification of the bands as genes is questioned in so far as the genes are concerned, the bands are multiples of some kind of unit of which the chromosomes are composed.

A few examples may serve to illustrate the way in which the banded chromosomes confirm the genetic conclusions as to occasional changes that have taken place in the serial order of the genes. In Fig. 4a the right half of chromosome 3 from the salivary gland is represented. In part the two components are fused, in part are separate. In the lower part of the figure a reversed piece of one component is present (terminal inversion). Like bands conjugate with like and, as shown in the smaller diagram above, in Fig. 4a, this is made possible by the end of one component turning back on itself. In Fig. 4b is drawn a short region of chromosome 2. One component has a deficiency for certain genes; the opposite normal chromosome forms a bulge in the region of the deficiency, allowing like bands to come together above and below the deficiency level.

The physiological properties of the genes

If, as is generally implied in genetic work (although not often explicitly stated), all of the genes are active all the time; and if the characters of the individual are determined by the genes, then why are not all the cells of the body exactly alike?

The same paradox appears when we turn to the development of the egg into an embryo. The egg appears to be an unspecialized cell, destined to undergo a prescribed and known series of changes leading to the differentiation of organs and tissues. At every division of the egg, the chromosomes split lengthwise into exactly equivalent halves. Every cell comes to contain the same kind of genes. Why then, is it that some cells become muscle cells, some nerve cells, and others remain reproductive cells?

The answer to these questions seemed relatively simple at the end of the last century. The protoplasm of the egg is visibly different at different levels. The fate of the cells in each region is determined, it was said, by the differences in different protoplasmic regions of the egg.

Such a view is consistent with the idea that the genes are all acting; the initial stages of development being the outcome of a reaction between the identical output of the genes and the different regions of the egg. This seemed to give a satisfactory *picture* of development, even if it did not give us a *scientific explanation* of the kind of reactions taking place.

But there is an alternative view that can not be ignored. It is conceivable that different batteries of genes come into action one after the other, as the embryo passes through its stages of development. This sequence might be assumed to be an automatic property of the chain of genes. Such an assumption would, without proof, beg the whole question of embryonic development, and could not be regarded as a satisfactory solution. But it might be that in different regions of the egg there is a reaction between the kind of protoplasm present in those regions and specific genes in the nuclei; certain genes being more affected in one region of the egg, other genes in other regions. Such a view might give also a purely formal hypothesis to account for the differentiation of the cells of the embryo. The initial steps would be given in the regional constitution of the egg.

The first responsive output of the genes would then be supposed to affect the protoplasm of the cells in which they lie. The changed protoplasm would now act reciprocally on the genes, bringing into activity additional or other batteries of genes. If true this would give a pleasing picture of the developmental process. A variation of this view would be to assume that the product of one set of genes is gradually in time overtaken and nullified or changed by the slower development of the output of other genes, as Goldschmidt, for example, has postulated for the sex genes. In the last case the theory is dealing with the development of hybrid embryos whose sex genes are assumed to have different rates of activity.

A third view may also be permissible. Instead of all the genes acting in the same way all the time, or instead of certain kinds of genes coming successively into action, we might postulate that the kind of activity of all the genes is changed in response to the kind of protoplasm in which they lie. This interpretation may seem less forced than the others, and in better accord with the functional activity of the adult organ systems.

We must wait until experiments can be devised that will help us to discriminate between these several possibilities. In fact, geneticists all over the world are today trying to find methods that will help to determine the relation of genes to embryonic and adult characters. The problem (or problems) is being approached both from a study of chemical changes that take place near the final steps in organ formation, especially in the development of pigments, and from a study of the early differentiation of the cell groups of the embryo.

We have come to realize that the problem of development is not as simple as I have so far assumed to be the case, for it depends, not only on independent cell differentiation of individual cells, but also on interactions between cells, both in the early stages of development and on the action of hormones on the adult organ systems. At the end of the last century, when experimental embryology greatly flourished, some of the most thoughtful students of embryology laid emphasis on the importance of the interaction of the parts on each other, in contrast to the theories of Roux and Weismann that attempted to explain development as a progressive series of events that are the outcome of self-differentiating processes, or as we would say today, by the sorting out of genes during the cleavage of the egg. At that time there was almost no experimental evidence as to the nature of the postulated interaction of the cells. The idea was a generalization rather than an experimentally determined conclusion, and, unfortunately, took a metaphysical turn.

Today this has changed, and owing mainly to the extensive experiments of the Spemann school of Germany, and to the brilliant results of Hörstadius of Stockholm, we have positive evidence of the far-reaching importance of interactions between the cells of different regions of the developing egg. This implies that original differences are already present, either in the undivided egg, or in the early formed cells of different regions. From the point of view under consideration, results of this kind are of interest because they bring up once more, in a slightly different form, the problem as to whether the organizer acts first on the protoplasm of the neighboring region with which it comes in contact, and through the protoplasm of the cells on the genes; or whether the influence is more directly on the genes. In either case the problem under discussion remains exactly where it was before. The evidence from the organizer has not as yet helped to solve the more fundamental relation between genes and differentiation, although it certainly marks an important step forward in our understanding of embryonic development.

The physiological action of the genes on the protoplasm, and reciprocally that of the protoplasm on the genes, is a problem of functional physiology in a very profound sense. For it is a problem that involves not only the irreversible changes of embryonic development, but also the recurrent changes in the organ systems of the adult body.

That man inherits his characters in the same way as do other animals there can be no doubt. The medical literature contains hundreds of family pedigrees, in which certain characters, usually malformations, appear more frequently than in the general population. Most of these are structural defects; a few are physiological traits (such as haemophilia); others are psychopathic. Enough is already known to show that they follow genetic principles.

Man is a poor breeder – hence many of these family pedigrees are too meagre to furnish good material for genetic analysis. When an attempt is made to combine pedigrees from different sources in order to insure sufficient data, the question of correct diagnosis sometimes presents serious difficulties, especially in the older materials; but with the very great advances that have been made in medical diagnosis in recent years this difficulty will certainly be less serious in the future.

The most important contribution to medicine that genetics has made is, in my opinion, intellectual. I do not mean to imply that the practical applications are unimportant, and I shall in a moment point out some of the more obvious connections, but the whole subject of human heredity in the past (and even at the present time in uninformed quarters) has been so vague and tainted by myths and superstitions that a scientific understanding of the subject is an achievement of the first order. Owing to genetic knowledge, medicine is today emancipated from the superstition of the inheritance of maternal impressions: it is free from the myth of the transmission of acquired characters, and in time the medical man will absorb the genetic meaning of the role of internal environment in the coming to expression of genetic characters.

The importance of this relation will be seen when it is recalled that the germ plasm or, as we say, the genic composition of man is a very complex mixture – much more so than that of most other animals, because in very recent times there has been a great amalgamation of many different races owing to the extensive migration of the human animal, and also because man's social institutions help to keep alive defective types of many kinds that would be eliminated in wild species through competition. Medicine has been, in fact, largely instrumental in devising means for the preservation of weak types of individuals, and in the near future medical men will, I suggest, often be asked for advice as to how to get rid of this increasing load of defectives. Possibly the doctor may then want to call in his genetic friends for consultation! The point I want to make clear is that the complexity of the genic composition of man makes it somewhat hazardous to apply only the simpler rules of Mendelian inheritance; for, the development of many inherited characters depends both on the presence of modifying factors and on the external environment for their expression.

I have already pointed out that the gene generally produces more than one visible effect on the individual, and that there may be also many invisible effects of the same gene. In cases where a condition of susceptibility to certain diseases is present, it may be that a careful scrutiny will detect some minor visible effects produced by the same gene. As yet our knowledge on this score is inadequate, but it is a promising field for further medical investigation. Even the phenomenon of linkage may some day be helpful in diagnosis. It is true there are known as yet in man no certain cases of linkage, but there can be little doubt that there will in time be discovered hundreds of linkages and some of these, we may anticipate, will tie together visible and invisible hereditary characteristics. I am aware, of course, of the ancient attempts to identify certain gross physical human types – the bilious, the lymphatic, the nervous and the sanguine dispositions, and of more modern attempts to classify human beings into the cerebral, respiratory, digestive and muscular, or, more briefly, into asthenics and pycnics. Some of these are supposed to be more susceptible to certain ailments or diseases than are other types, which in turn have their own constitutional characteristics. These well-intended efforts are, however, so far in advance of our genetic information that the geneticist may be excused if he refuses to discuss them seriously.

In medial practice the physician is often called upon for advice as to the suitability of certain marriages where a hereditary taint is present in the ancestry. He is often called upon to decide as to the risk of transmitting certain abnormalities that have appeared in the first-born child. Here genetics will, I think, be increasingly helpful in making known the risk incurred, and in distinguishing between environmental and hereditary traits.

Again, a knowledge of the laws of transmission of hereditary characters may sometimes give information that may be helpful in the diagnosis of certain diseases in their incipient stages. If, for example, certain stigmata appear, whose diagnosis is uncertain, an examination of the family pedigree of the individual may help materially in judging as to the probability of the diagnosis.

I need scarcely point out those legal questions concerning the paternity of an illegitimate child. In such cases a knowledge of the inheritance of blood groups, about which we now have very exact genetic information, may often furnish the needed information.

Geneticists can now produce by suitable breeding, strains of populations of animals and plants that are free from certain hereditary defects; and they can also produce, by breeding, plant populations that are resistant or immune to certain diseases. In man it is not desirable, in practice, to attempt to do this, except in so far as here and there a hereditary defective may be discouraged from breeding. The same end is accomplished by the discovery and removal of the external causes of the disease (as in the case of yellow fever and malaria) rather than by attempting to breed an immune race. Also, in another way the same purpose is attained in producing immunity by inoculation and by various serum treatments. The claims of a few enthusiasts that the human race can be entirely purified or renovated, at this later date, by proper breeding, have I think been greatly exaggerated. Rather must we look to medical research to discover remedial measures to insure better health and more happiness for mankind.

While it is true, as I have said, some little amelioration can be brought about by discouraging or preventing from propagating well-recognized hereditary defects (as has been done for a long time by confinement of the insane), nevertheless it is, I think, through public hygiene and protective measures of various kinds that we can more successfully cope with some of the evils that human flesh is heir to. Medical science will here take the lead – but I hope that genetics can at times offer a helping hand.

Biography

Thomas Hunt Morgan was born on September 25, 1866, at Lexington, Kentucky, U.S.A. He was the eldest son of Charlton Hunt Morgan.

He was educated at the University of Kentucky, where he took his B.S. degree in 1886, subsequently doing postgraduate work at Johns Hopkins University, where he studied morphology with W. K. Brooks, and physiology with H. Newell Martin.

As a child he had shown an immense interest in natural history and even at the age of ten, he collected birds, birds' eggs, and fossils during his life in the country; and in 1887, the year after his graduation, he spent some time at the seashore laboratory of Alphaeus Hyatt at Annisquam, Mass. During the years 1888–1889, he was engaged in research for the United States Fish Commission at Woods Hole, a laboratory with which he was continuously associated from 1902 onwards, making expeditions to Jamaica and the Bahamas. In 1890 he obtained his Ph.D. degree at Johns Hopkins University. In that same year he was awarded the Adam Bruce Fellowship and visited Europe, working especially at the Marine Zoological Laboratory at Naples which he visited again in 1895 and 1900. At Naples he met Hans Driesch and Curt Herbst. The influence of Driesch with whom he later collaborated, no doubt turned his mind in the direction of experimental embryology.

In 1891 he became Associate Professor of Biology at Bryn Mawr College for Women, where he stayed until 1904, when he became Professor of Experimental Zoology at Columbia University, New York. He remained there until 1928, when he was appointed Professor of Biology and Director of the G. Kerckhoff Laboratories at the California Institute of Technology, at Pasadena. Here he remained until 1945. During his later years he had his private laboratory at Corona del Mar, California.

During Morgan's 24-years period at Columbia University his attention was drawn toward the bearing of cytology on the broader aspects of biological interpretation. His close contact with E. B. Wilson offered exceptional opportunities to come into more direct contact with the kind of work which was being actively carried out in the zoological department, at that time.

Morgan was a many-sided character who was, as a student, critical and independent. His early published work showed him to be critical of Mendelian conceptions of heredity, and in 1905 he challenged the assumption then current that the germ cells are pure and uncrossed and, like Bateson was sceptical of the view that species arise by natural selection. « Nature », he said, « makes new species outright. » In 1909 he began the work on the fruit-fly *Drosophila melanogaster* with which his name will always be associated.

It appears that Drosophila was first bred in quantity by C. W. Woodworth, who was working from 1900–1901, at Harvard University, and Woodworth there suggested to W. E. Castle that Drosophila might be used for genetical work. Castle and his associates used it for their work on the effects of inbreeding, and through them F. E. Lutz became interested in it and the latter introduced it to Morgan, who was looking for less expensive material that could be bred in the very limited space at his command. Shortly after he commenced work with this new material (1909), a number of striking mutants turned up. His subsequent studies on this phenomenon ultimately enabled him to determine the precise behaviour and exact localization of genes.

The importance of Morgan's earlier work with Drosophila was that it demonstrated that the associations known as *coupling* and *repulsion*, discovered by English workers in 1909 and 1910 using the Sweet Pea, are in reality the obverse and reverse of the same phenomenon, which was later called *linkage*. Morgan's first papers dealt with the demonstration of sex linkage of the gene for white eyes in the fly, the male fly being heterogametic. His work also showed that very large progenies of Drosophila could be bred. The flies were, in fact, bred by the million, and all the material thus obtained was carefully analysed. His work also demonstrated the important fact that spontaneous mutations frequently appeared in the cultures of the flies. On the basis of the analysis of the large body of facts thus obtained, Morgan put forward a theory of the *linear arrangement* of the genes in the chromosomes, expanding this theory in his book, *Mechanism of Mendelian Heredity* (1915).

In addition to this genetical work, however, Morgan made contributions of great importance to experimental embryology and to regeneration. So far as embryology is concerned, he refuted by a simple experiment the theory of Roux and Weismann that, when the embryo of the frog is in the two-cell stage, the blastomeres receive unequal contributions from the parent blastoderm, so that a «mosaic» results. Among his other embryological discoveries was the demonstration that gravity is not, as Roux's work had suggested, important in the early development of the egg.

Although so much of his time and effort was given to genetical work, Morgan never lost his interest in experimental embryology and he gave it, during his last years increasing attention.

To the study of regeneration he made several important contributions, an outstanding one being his demonstration that parts of the organism which are not subject to injury, such as the abdominal appendages of the hermit crab, will nevertheless regenerate, so that regeneration is not an adaptation evolved to meet the risks of loss of parts of the body. On this part of his work he wrote his book *Regeneration*.

Apart from the books previously mentioned Morgan wrote: *Heredity and Sex* (1913), *The Physical Basis of Heredity* (1919), *Embryology and Genetics* (1924), *Evolution and Genetics* (1925), *The Theory of the Gene* (1926), *Experimental Embryology* (1927), *The Scientific Basis of Evolution* (2nd. ed., 1935), all of them classics in the literature of genetics.

Morgan was made a Foreign Member of the Royal Society of London in 1919, where he delivered the Croonian Lecture in 1922. In 1924, he was awarded the Darwin Medal, and in 1939 the Copley Medal of the Society.

For his discoveries concerning the role played by the chromosome in heredity, he was awarded the Nobel Prize in 1933.

Among his collaborators at Columbia may be mentioned H. J. Muller, who was awarded the Nobel Prize in 1946 for his production of mutations by means of X-rays.

Morgan married Lilian Vaughan Sampson, in 1904, who had been a student at Bryn Mawr College, and who often assisted him in his research. They had one son and three daughters.

Professor Morgan died in 1945.

1946

Hermann J. Muller

HERMANN J. MULLER

The Production of Mutations

December 12, 1946

If as Darwin maintained the adaptiveness of living things results from natural selection, rather than from a teleological tendency in the process of variation itself, then heritable variations must, under most conditions, occur in numerous directions, so as to give a wide range of choice for the selective process. Such a state of affairs seems, however, in more or less contradiction to the commonly held idea, to which Darwin also gave some credence, that heritable variations of given kinds tend to be produced, in a fairly regular way, by given kinds of external conditions. For then we are again confronted with the difficulty, how is it that the « right kinds » of variations (i.e. the adaptive ones) manage to arise in response to the « right kinds » of conditions (i.e. those they are adapted to)? Moreover, the de Vriesian notion of mutations does not help us in this connection. On that view, there are sudden jumps, going all the way from one « elementary species » to another, and involving radical changes in numerous characters at once, and there are relatively few different jumps to choose between. This obviously would fail to explain how, through such coarse steps, the body could have come to be so remarkably streamlined in its internal and external organization, or, in other words, so thoroughly adaptive.

The older selectionists, thinking in terms of chemical reactions on a molar scale when they thought in terms of chemistry at all, did not realize sufficiently the ultramicroscopic randomness of the processes causing inherited variations. The earliest mutationists failed, in addition, to appreciate the qualitative and quantitative multiplicity of mutations. It was not long, however, before the results of Baur on *Antirrhinum* and of Morgan on *Drosophila*, supplemented by scattered observations on other forms, gave evidence of the occurence of numerous Mendelizing mutations, many of them small ones, in varied directions, and they showed no discoverable relation between the type of mutation and the type of environment or condition of living under which it arose. These observations, then, came closer to the statistical requirements for a process of evolution which has its basis in accidents. In what sense, however, could the events be regarded as accidental? Were they perhaps expressions of veiled forces working in a more regular manner? It was more than ever evident that further investigation of the manner of occurrence of mutations was called for.

If the mutations were really non-teleological, with no relation between type of environment and type of change, and above all no adaptive relation, and if they were of as numerous types as the theory of natural selection would demand, then the great majority of the changes should be harmful in their effects, just as any alterations made blindly in a complicated apparatus are usually detrimental to its proper functioning, and many of the larger changes should even be totally incompatible with the functioning of the whole, or, as we say, lethal. That is, strange as it may seem at first sight, we should expect most mutations to be disadvantageous if the theory of natural selection is correct. We should also expect these mainly disadvantageous changes to be highly diversified in their genetic basis.

To get exact evidence on these points required the elaboration of special genetic methods, adapted to the recognition of mutations that ordinarily escape detection: (1) lethals, (2) changes with but small visible effects, and (3) changes without any externally visible effects but influencing the viability more or less unfavorably. It would take us too far afield to explain these techniques here. Suffice it to say that they made use of the principle according to which a chromosome is, as we say, « marked », by having had inserted into it to begin with one or more known mutant genes with conspicuous visible effects, to differentiate it from the homologous chromosome. An individual with two such differentiated chromosomes, when appropriately bred, will then be expected to give two groups of visibly different offspring, holding certain expected ratios to one another. If, however, a lethal mutation has occurred in one of the two chromosomes, its existence will be made evident by the absence of the corresponding expected group of offspring. Similarly, a mutated gene with invisible but somewhat detrimental action, though not fully lethal, will be recognized by the fact that the corresponding group of offspring are found in smaller numbers than expected. And a gene with a very small visible effect, that might be overlooked in a single individual, will have a greatly increased chance of being seen because the given group of offspring as a whole will tend to be distinguished in this regard from the corresponding group derived from a non-mutant.

In this way, it was possible in the first tests of this kind, which Altenburg and the writer conducted, partly in collaboration, in 1918–19, to get definite evidence in *Drosophila* that the lethal mutations greatly outnumbered those with visible effects, and that among the latter the types having an obscure manifestation were more numerous than the definite conspicuous ones used in ordinary genetic work. Visible or not, the great majority had lowered viability. Tests of their genetic basis, using the newly found facts of linkage, showed them to be most varied in their locus in the chromosomes, and it could be calculated by a simple extrapolative process that there must be at least hundreds, and probably thousands, of different kinds arising in the course of spontaneous mutation. In work done much later, employing induced mutations, it was also shown (in independent experiments both of the present writer and Kerkis, and of Timoféeff and his co-workers, done in 1934) that «invisible» mutations which by reason of one or another physiological change lower viability without being fully lethal form the most abundant group of any detected thus far, being at least 2 to 3 times as numerous as the complete lethals. No doubt there are in addition very many, perhaps even more, with effects too small to have been detected at all by our rather crude methods. It is among these that we should be most apt to find those rare accidents which, under given conditions or in given combinations with others, may happen to have some adaptive value. Tests of Timoféeff, however, have shown that even a few of the more conspicuous visible mutations do in certain combinations give an advantage in laboratory breeding.

Because of the nature of the test whereby it is detected – the absence of an entire group of offspring bearing certain conspicuous expected characters – a lethal is surer of being detected, and detected by any observer, than is the inconspicuous or invisible, merely detrimental, mutation. Fortunately, there are relatively few borderline cases, of nearly but not quite completely lethal genes. It was this objectivity of recognition, combined with the fact that they were so much more numerous than conspicuous visible mutations, that made it feasible for lethals to be used as an index of mutation frequency, even though they suffer from the disadvantage of requiring the breeding of an individual, rather than its mere inspection, for the recognition that it carries a lethal. In the earliest published work, we (Altenburg and the author) attempted not only to find a quantitative value for the «normal» mutation frequency, but also to determine whether a certain condition, which we considered of special interest, affected the mutation frequency. The plan was ultimately to use the method as a general one for studying the effects of various conditions. The condition chosen for the first experiment was temperature, and the results, verified by later work of the writer's, indicated that a rise of temperature, within limits normal to the organism, produced an increase of mutation frequency of about the amount to be expected if mutations were, in essentials, orthodox chemical reactions.

On this view, however, single mutations correspond with individual molecular changes, and an extended series of mutations, in a great number of identical genes in a population, spread out over thousands of years, is what corresponds with the course of an ordinary chemical reaction that takes place in a whole collection of molecules in a test tube in the course of a fraction of a second or a few seconds. For the individual gene, in its biological setting, is far more stable than the ordinary chemical molecule is, when the latter is exposed to a reagent in the laboratory. Thus, mutations, when taken collectively, should be subject to the statistical laws applying to mass reactions, but the individual mutation, corresponding to a change in one molecule, should be subject to the vicissitudes of ultramicroscopic or atomic events, and the apparition of a mutant individual represents an enormous amplification of such a phenomenon. This is a principle which gives the clue to the fact, which otherwise seems opposed to a rational, scientific and molarly deterministic point of view, that differences in external conditions or conditions of living do not appear to affect the occurrence of mutations, while on the other hand, even in a normal and sensibly constant environment, mutations of varied kinds do occur. It is also in harmony with our finding, of about the same time, that when a mutation takes place in a given gene, the other gene of identical type present nearby in the same cell usually remains unaffected, though it must of course have been subjected to the same macroscopic physico-chemical conditions. On this conception, then, the mutations ordinarily result from submicroscopic accidents, that is, from caprices of thermal agitation, that occur on a molecular and submolecular scale. More recently Delbrück and Timoféeff, in more extended work on temperature, have shown that the amount of increase in mutation frequency with rising temperature is not merely that of an ordinary test-tube chemical reaction, but in fact corresponds closely with that larger rise to be expected of a reaction as slow in absolute time rate (i.e. with as small a proportion of molecular changes per unit of time) as the observed mutation frequency shows this reaction to be, and this quantitative correspondence helps to confirm the entire conception.

Now this inference concerning the non-molar nature of the individual mutation process, which sets it in so different a class from most other grossly observable chemical changes in nature, led naturally to the expectation that some of the « point effects » brought about by high-energy radiation like X-rays would also work to produce alterations in the hereditary material. For if even the relatively mild events of thermal agitation can, some of them, have such consequences, surely the energetically far more potent point changes caused by powerful radiation should succeed. And, as a matter of fact, our trials of X-rays, carried out with the same kind of genetic methods as previously used for temperature, proved that such radiation is extremely effective, and inordinately more so than a mere temperature rise, since by this method it was possible to obtain, by a half-hour's treatment, over a hundred times as many mutations in a group of treated germ cells as would have occurred in them spontaneously in the course of a whole generation. These mutations too were found ordinarily to occur pointwise and randomly, in one gene at a time, without affecting an identical gene that might be present nearby in a homologous chromosome.

In addition to the individual gene changes, radiation also produced rearrangements of parts of chromosomes. As our later work (including that with co-workers, especially Raychaudhuri and Pontecorvo) has shown, these latter were caused in the first place by breakages of the chromosomes, followed afterwards by attachments occurring between the adhesive broken ends, that joined them in a different order than before. The two or more breaks involved in such a rearrangement may be far apart, caused by independent hits, and thus result in what we call a *gross* structural change. Such changes are of various kinds, depending upon just where the breaks are and just which broken ends become attached to which. But, though the effects of the individual « hits » are rather narrowly localized, it is not uncommon for two breaks to be produced at nearby points by what amounts to one local change (or at any rate one localized group of changes) whose influence becomes somewhat spread out. By the rejoining, in a new order, of broken ends resulting from two such nearby breaks, a *minute* change of sequence of the genes is brought about. More usually, the small piece between the two breaks becomes lost (a « deficiency »), but sometimes it becomes inverted, or even becomes transferred into a totally different position, made available by a separate hit.

Both earlier and later work by collaborators (Oliver, Hanson, etc.) showed definitely that the frequency of the gene mutations is directly and simply proportional to the dose of irradiation applied, and this despite the wavelength used, whether X- or gamma- or even beta-rays, and despite the timing of the irradiation. These facts have since been established with great exactitude and detail, more especially by Timoféeff and his co-workers. In our more recent work with Raychaudhuri (1939, 1940) these principles have been extended to total doses as low as 400 *r*, and rates as low as 0.01 *r* per minute, with gamma rays. They leave, we believe, no escape from the conclusion that there is no threshold dose, and that the individual mutations result from individual « hits », producing genetic effects in their immediate neighborhood. Whether these so-called « hits » are the individual ionizations, or may even be the activations that occur at lower energy levels, or whether, at the other end of the scale, they require the clustering of ionizations that occurs at the termini of electron tracks and of their side branches (as Lea and Fano point out might be the case), is as yet undecided. But in any case they are, even when microscopically considered, what we have termed « point mutations », as they involve only disturbances on an ultramicroscopically localized scale. And whether or not they are to occur at any particular point is entirely a matter of accident, using this term in the sense in which it is employed in the mathematics of statistics.

Naturally, other agents than photons which produce effects of this kind must also produce mutations, as has been shown by students and collaborators working under Altenburg in Houston for alpha rays (Ward, 1935) and for neutrons (Nagai and Locher, 1937), and extended in regard to the quantitative relations concerned by Zimmer and others working with Timoféeff (1936, 1937, 1938), and by others. Moreover, as Altenburg (1930, 1935) showed, even the smaller quantum changes induced by ultraviolet exert this effect on the genes. They cause, however, only a relatively small amount of rearrangement of chromosome parts (Muller and Mackenzie, 1939) and, in fact, they also tend to inhibit such rearrangement, as Swanson (1944), followed by Kaufmann and Hollaender (1944 *et seq.*), has found. Since the effective ultraviolet hits are in the form of randomly scattered single-atom changes in the purines and pyrimidines of the chromosomes, rather than in groups of atom changes, it seems likely that clusters of ionizations are not necessary for the gene mutation effects, at any rate, although we cannot be sure of this until the relation of mutation frequency to dosage is better known for this agent.

Inasmuch as the changes brought about in the genes by radiation must certainly be of an accidental nature, unpremeditated, ateleological, without reference to the value of the end result for the organism or its descendants, it is of interest to compare them with the so-called spontaneous or natural mutations. For in the radiation mutations we have a yardstick of what really random changes should be. Now it is found in *Drosophila* that the radiation-induced mutations of the genes (we exclude here the demonstrable chromosome rearrangements) are in every respect which has been investigated of the same essential nature as those arising naturally in the laboratory or field. They usually occur in one gene without affecting an identical one nearby. They are distributed similarly in the chromosomes. The effects, similarly, may be large or small, and there is a similar ratio of fully lethal to so-called visible gene mutations. That is, the radiation mutations of the genes do not give evidence of being more deleterious. And when one concentrates attention upon given genes one finds that a whole series of different forms, or alleles, may be produced, of a similar and in many cases sensibly identical nature in the two cases. In fact, every natural mutation, when searched for long enough, is found to be producible also by radiation. Moreover, under any given condition of living tried, without radiation, the effects appear as scattered as when radiation is applied, even though of much lower frequency. All this surely means then, does it not, that the natural mutations have in truth no innate tendency to be adaptive, nor even to be different, as a whole group, under some natural conditions than under others? In other words, they cannot be determinate in a molar sense, but must themselves be caused by the ultramicroscopic accidents of the molecular and submolecular motions, i.e. by the individual quantum exchanges of thermal agitation, taking this word in a broad sense. The only escape from this would be to suppose that they are caused by the radiation present in nature, resulting from natural radioactive substances and cosmic rays, but a little calculation (by Mott-Smith and the writer, 1930, corroborated by others) has shown that this radiation is quite inadequate in amount to account for the majority of mutations occurring in most organisms.

But to say that most natural mutations are the results of the quantum exchanges of thermal agitation, and, further, that a given energy level must be reached to produce them, does not, as some authors have seemed to imply, mean that the physico-chemical conditions in and around the organism, other than temperature, have no influence upon their chance of occurrence. That such circumstances may play a decided role was early evident from the studies of spontaneous mutation frequency, when it was found (1921, reported 1928) that the frequency in one experiment, with one genetic stock, might be ten times as high as in another, with another stock. And more recently (1945) we have found that, in different portions of the natural life cycle of the same individual, the mutation frequency may be very different. Finally, in the work of Auerbach and Robson (1941–46), with mustard gas and related substances, it has been proved that these chemicals may induce mutations at as high a frequency as a heavy dose of X-rays. In all these cases, however, the effects are similarly scattered at random, individually uncontrolled, and similarly non-adaptive.

It should also be noted in this connection that the genes are not under all conditions equally vulnerable to the mutating effects of X-rays themselves. Genes in the condensed chromosomes of spermatozoa, for example, appear to be changed more easily than those in the more usual « resting » stages. We have mentioned that, as Swanson has shown, ultraviolet exerts besides its own mutating effect an inhibition on the process of chromosome breakage, or at any rate on that of reunion of the broken parts in a new viable order, while infrared, in Hollaender's and Kaufmann's recent experiments, has a contrary action. And Stadler, in his great work on the production of mutations in cereals, started independently of our own, has obtained evidence that in this material X-radiation in the doses used is unable to produce a sensible rise in the gene mutation frequency, though numerous chromosome breakages do arise, leading to both gross and minute rearrangements of chromosome parts. Either the genes are more resistant in this material to permanent changes by X-rays, as compared with their responsiveness to thermal agitation, or a break or loss must usually be produced by X-rays along with the gene change. The milder ultraviolet quanta, on the other hand, do produce gene mutations like the natural ones in these plants.

Such variations in effectiveness are, I believe, to have been expected. They do not shake our conclusion as to the accidental, quantum character of the event which usually initiates a gene mutation. But they give rise to the hope that, through further study of them, more may be learned concerning the nature of the mutation process, as well as of the genetic material that undergoes the changes.

No one can answer the question whether some special means may not be found whereby, through the application of molar influences, such as specific antibodies, individual genes could be changed to order. Certainly the search for such influences, and for increasing control of things on a microscopic and submicroscopic scale as well, must be carried further. But there is as yet no good evidence that anything of the sort has been done artificially, or that it occurs naturally. Even if possible, there could be no generalized method of control of gene composition without far greater knowledge than we now have of the intimate chemical structure and the mode of working of the most complicated and diverse substances that exist, namely, nucleoproteins, proteins in general, and enzymes. The works of Sumner, Northrop, and Stanley, together with those of other protein chemists, point the way in this direction, but everyone will agree that it is a long and devious system of roads which is beginning here.

It is true that some cases are known of mutable genes which change selectively in response to special conditions. Such cases may be very informative in shedding light on gene structure, but we have as yet no indication that the alterations of these genes, which in the great majority of instances known are abnormal genes, have anything in common with ordinary natural mutations. It is also true that cases are known among bacteria and viruses of the induction of particular kinds of hereditary changes by application of particular substances, but here the substances applied are in each case the same as those whose presence is later found to have been induced, and so there is every reason to infer that they have in fact become implanted in some way, that is, that we do not really have a specifically induced mutation.

So far, then, we have no means, or prospect of means, of inducing given mutations at will in normal material, though the production of mutations in abundance at random may be regarded as a first step along such a path, if there is to be such a path. So long as we cannot direct mutations, then, selection is indispensable, and progress in the hereditary constitution of a living thing can be made only with the aid of a most thoroughgoing selection of the mutations that occur since, being non-adaptive except by accident, an overwhelming majority is always harmful. For a sensible advance, usually a considerable number of rare steps must be accumulated in the painful selective process. By far the most of these are individually small steps, but, as species and race crossings have shown, there may be a few large distinctive steps that have been, as Huxley terms it, «buffered», by small changes that readjust the organism to them. Not only is this accumulation of many rare, mainly tiny changes the chief means of artificial animal and plant improvement, but it is, even more, the way in which natural evolution has occurred, under the guidance of natural selection. Thus the Darwinian theory becomes implemented, and freed from the accretions of directed variation and of Lamarckism that once encumbered it.

It is probable that, in a state of nature, most species have a not very much (though somewhat) lower frequency of gene mutation than would be most advantageous for them, in consideration of the degree of rigor of the natural selection that occurs in the given species. A much higher frequency would probably lead to faster genetic degenerative processes than the existing selection could well cope with. But, under conditions of artificial breeding, where selection can be made more effective, a higher mutation frequency can for a time at least be tolerated in some cases, and larger mutations also can be nursed through to the point where they become suitably buffered. Here it may become of practical use to apply X-rays, ultraviolet, or other means of inducing mutations, as Gustafsson especially has demonstrated for X-rays. This will be especially true in species which naturally undergo much inbreeding, or in which there is a well-expressed haploid phase, or a considerable haploid portion of the genotype, for under these circumstances many of the spontaneous mutations that might otherwise have accumulated in the population and that could be brought to light by inbreeding, will have become eliminated before being found, and the natural mutation rate itself will be lower.

We have above largely confined ourselves to considering the relation of the production of gene mutations to the problems of the general method of evolution, including that of the nature of hereditary variation, because this has been, historically, the main line of approach to the subject of artificial mutations. It was from the first evident, however, that the production of mutations would, as we once stated, provide us with tools of the greatest nicety, wherewith to dissect piece by piece the physiological, embryological, and biochemical structure of the organism and to analyze its workings. Already with natural mutations, such works as those of Bonnevie, Grueneberg, Scott-Moncrief, Ephrussi, and Beadle, etc., have shown how the intensive tracing of the effects, and interrelations of effects, of just one or a few mutations, can lead to a deeper understanding of the complex processes whereby the genes operate to produce the organism. But there are thousands of genes, and it is desirable to be able to choose them for study in an orderly fashion as we proceed with our dissection process. For this purpose we have thought that it would often be advantageous to produce mutations artificially in abundance, so as then to take our pick of those more suited for successive steps in our analysis. The work of Beadle and his co-workers on *Neurospora* in recent years, followed by similar work of Malin and Fries and of others, has brilliantly shown the applicability of this method for studies of the paths of biochemical synthesis of amino acids, vitamins, purines, and pyrimidines. And yet, in a sense, the surface of the subject as a whole has barely been scratched, and we may look forward with confidence to the combination of this technique with that of tracer substances and with all the other techniques of biochemistry, physiology and experimental embryology, for the increasing unravelling of that surpassingly intricate tangle of processes of which the living organism is constituted. There is not time, however, to go further into this subject here.

For we cannot neglect here a brief outline of another phase of the artificial mutation work, more specifically of interest to geneticists: that is, the further analysis of the properties of the chromosomes and their parts, gained chiefly from studies in which parts have been removed, added, or rearranged. We have already spoken, in passing, of the studies of the mechanism of such structural change, in which a relatively simple general scheme lying at the basis of all such alterations has emerged: namely, breakage first, followed by adhesion of broken ends. It was early evident that by the use of such rearranged chromosomes additional proof of the physical validity of the linkage maps could be obtained, and this was done (Muller and Painter, 1929 *et seq.*). Furthermore, it has been possible to throw light on problems of crossing-over, as in the demonstration (Muller, Stone, and Offermann, 1930 *et seq.*) that to whatever position the centromere is moved, it causes a strong inhibition of crossing-over, the strength of which gradually diminishes with distance. Moreover, the same proves to be true of any point of discontinuity in pairing, caused by heterozygosity in regard to a structural change. Such studies on crossing-over, and on the pairing forces that affect segregation, are still capable of considerable extension.

We must remember, in speaking of the centromere and other apparently distinctive chromosome parts, that we have no right to infer them to be autonomous, locally determined structures, dependent on the genes of the regions in which they are seen to lie, before observations have been made that show the effects of removing or displacing those regions. Therefore, it has in the main been necessary to wait for the study of induced inversions, deletions and translocations of chromosomes, before the inference could be secure that the centromere is, in most instances, such an autonomous organelle, dependent upon a gene or genes in its immediate neighborhood (but not in all instances in its neighborhood, as Rhoades has recently shown in a special strain of maize). Similarly, it has been possible to show (despite some contrary claims, the validity or invalidity of which cannot be discussed here) that the free end of the chromosome, or telomere, constitutes in much material a locally determined distinctive structure.

By a combined genetic and cytological analysis of various cases of breakage and rearrangement of parts, in work done in collaboration with Painter, Stone, Prokofyeva, Gershenson, and others, it was found that there are distinctive, largely locally determined regions of the chromosomes, usually most markedly developed near the centromeres, which we at first called «inactive» but which are now usually referred to as «heterochromatic». These were also found independently in purely cytological studies by Heitz. It would be fascinating to enter here into a discussion of the remarkable peculiarities which the cytogenetic studies have shown these regions to have – the evidence of repetition of more or less similar parts, of a tendency to conjugation between the differently placed parts, of distinctive cytological appearance correlated with whether or not such conjugation occurs, of inordinately high tendency to structural change, of strong influence of certain of their genes upon segregation, etc. – and then to go on to discuss hypotheses of their evolutionary origin and their functions. This would unfortunately take us too far afield. We must however insist upon one point – as it is not yet generally enough recognized – namely, that the evidence is very strong that what in the *Drosophila* chromosome, as seen at mitosis, is called «the heterochromatic region», is simply a large temporary body of accessory, non-genic nucleoprotein, produced under the influence of one or two particular genes from among the dozen or more that constitute the whole heterochromatic region, as detected by genetic analysis and by the chromosome as seen at the resting stage (as in the salivary gland). And it is not these conspicuous non-genic blocks which are responsible for the other known peculiarities of the heterochromatin, above mentioned – the function of the blocks is still undetermined. In other words, the so-called «heterochromatin» with which the cytologist deals in studying mitotic chromosomes is a quite different thing from, although in the neighborhood of, the heterochromatin proper having the above described complex of properties. Moreover, it has been possible to show (Sutton-Gersh in collaboration with the author, unpublished) that the conspicuous nucleoli often associated with the heterochromatin are produced under the influence of still other autonomous genes in it, that are separate from those for the mitotically visible blocks.

One of the most interesting findings which has come out of the study of *Drosophila* chromosomes that underwent rearrangement of parts as a result of irradiation has been the generalization of the existence of the phenomenon known as «position effect». This effect was first found by Sturtevant (1925) in the case of the spontaneous mutant known as Bar eye, but it was not known to what extent the effect might be a special one until numerous rearrangements could be studied. The term position effect implies that the functioning of a gene is to a certain extent at least dependent upon what other genes lie in its neighborhood. There is now adequate evidence that this is a general principle applying to very many if not all of the genes in *Drosophila*, and that their functioning can be qualitatively as well as quantitatively conditioned by the character of the genes in their vicinity, some of the genes having much more effect than others and different genes working in different ways and to different extents.

It is possible that, as Sturtevant (*ibid.*) suggested, the position effect is caused by the interaction between gene products in the vicinity of the genes producing them, assuming that such products are more concentrated there and under such circumstances tend to react more with one another than when dispersed. However, the interpretation which we favor is that the functioning of the gene is affected by its shape and that this, in turn, varies with the strength and nature of synaptic forces acting on the region of the chromosome in which it lies. These might consist of forces directly exerted on the gene by other genes, whether allelic or not (Muller, 1935), or they might be resultants of the state of spiralization, etc., of the chromosome region, circumstances which in their turn are in part dependent on synaptic forces (Ephrussi and Sutton, 1944 *et seq.*). This interpretation, in either of its variants, would explain why position effects are so much more general in *Drosophila*, an organism in which the synaptic forces are known to operate strongly even in somatic cells, than in other organisms tested, in which such forces are much weaker or absent in somatic cells. It would also fit in with the findings (Muller, 1935) that the heterochromatic regions tend to have especially strong, extensive, and distinctive kinds of position effects, effects varying in degree with the total amount of heterochromatin present in a cell, as well as with vacillating embryological factors. For these genetic findings are in conformity with the cytological effects of heterochromatin, observed first by Prokofyeva, on the degree of extension, synaptic properties, etc., of euchromatin in its neighborhood, effects which she showed to be subject to similar vacillations, that are correlated with the variations in the phænotypically observed position effects. Recent observations, both by Ephrussi and Sutton (*ibid.*), undertaken jointly with the author, and by Stern (1944, 1946), also seem to point in this direction, for they show an influence, on the position effects exhibited by given parts, of the arrangement of *homologous* chromosome parts. If this interpretation based on gene shape should hold, it would open up a new angle of attack on the structure and method of functioning of the gene, perhaps ultimately relating it to nucleoprotein composition and properties.

Another use to which the process of breakage and rearrangement of chromosome parts by irradiation has been put is for the study of the effects of adding and of subtracting small pieces of chromosomes, in order to determine the relation of gene dosage to gene expression. In this way, it has been found out (1) that most normal genes are, even in single dose, near the upper limit of their effectiveness, and (2) that most mutant genes have a final effect qualitatively similar to but quantitatively less than that of their allelic normal genes. The dominance of normal genes over their mutant alleles, then, turns out in most instances to be a special case of the principle that one dose of a normal gene usually produces nearly though not quite as much effect as two doses. This in turn is best understood as resulting from a long course of selection of the normal gene and its modifiers for stability of expression, when under the influence of environmental and genetic conditions which would affect the gene's operation quantitatively, i.e. in a manner similar to that of dosage changes. This does not mean that selection has specifically worked to produce dominance of the normal gene over its alleles, however, because (3) not all mutant genes behave merely like weaker normal genes, and (4) those which the dosage tests show to produce qualitatively different effects from the normal genes seem oftener to escape from the principle of being dominated over by the normals, just as would be expected on our hypothesis.

Among the further results of gene dosage studies carried out by the use of chromosome fragments produced by irradiation, attention should be especially called to the findings coming under the head of « dosage compensation ». These have shown: (1) that, when the dosage of virtually all genes in the X-chromosome except a given one is held constant, the expression of that one is usually so very nearly the same when present in one dose as in two that no difference in the character can ordinarily be seen, and (2) that nevertheless this invisible difference has been so important for the organism that, in the course of the past natural selection, a system of modifying genes, called compensators, has been established, having the function of making the effects of the one and two doses normally present in the two respective sexes much more nearly equal still, when these dosage differences in the given genes are present simultaneously with those in all the other X-chromosomal genes. Each gene seems to have acquired a different system of compensators, the interrelations of all together being extremely complicated. This then gives evidence from a new angle of the meticulousness of natural selection, of the very precise adaptiveness of the characters existing in a species, and of the final grade of a character having ordinarily become established through the accumulation of numerous small mutations having very complex functional relations with one another. It is in line with our previous thesis of evolution through the selection of multitudinous tiny accidental changes.

When attention is concentrated on a given very circumscribed region of a chromosome, by a comparison of various induced rearrangements all of which have a point of breakage within that region, other facts come to light, bearing on the problems of chromosome and gene divisibility. By means of special genetic methods, which cannot be detailed here, evidence has been obtained that the breaks in any such limited region tend to occur at specific points, giving indication that discrete units or segments lie between these points, and thus arguing against the idea of the chromosome being a continuum and in favor of its genes corresponding to physical entities rather than merely to concepts arbitrarily set up for the convenience of geneticists. We are also enabled in this way to make estimates of the probable number of genes in the chromosome, as well as to get maximally limiting figures for their size. These estimates agree as closely as could have been expected with those based on previous genetic work, using entirely different methods, although not with the estimates based on the « sensitive volume » hypothesis.

Another finding made in studies of cases having a small fragment of chromosome moved, as a result of irradiation, to another position, was that individuals are frequently able to survive and reproduce even when they have the given chromosome part present in its original position as well as in the new position. In fact, it was in work of this kind that the effect of extra doses of genes was determined. Now, in some of these cases stocks could even be obtained which were homozygous for the duplicated piece as well as for the original piece. This led to the idea that duplications of chromosome material might in this manner have become established in the previous course of evolution. When in the analysis of a limited region of the X-chromosome, including the locus of the so-called « scute » effect, it was found that there are in fact, within the normal X-chromosome, two genes of closely related effect (« achaete » and « scute ») very close or adjacent to one another, it became evident that this was in all probability an example of the above postulated occurrence. This then showed the way, and apparently the main if not the only way (aside from the far rarer phenomena of polyploidy and « tetrasomy »), by which the number of genes has become increased during the course of evolution. By a curious coincidence, Bridges was at the same time making his studies of salivary chromosomes and finding direct cytological evidence for the existence of such « repeats », as he called them, in the normal chromosomes, and he interpreted these in the same manner. In the twelve years since that time, various other clear cases of the same kind have been demonstrated. Thus, increase in gene number, brought about by the duplication of small parts of chromosomes, more usually in positions near their original ones, must be set down as one of the major processes in evolution, in addition to the mutations in the individual genes. By itself, this process would not be of great importance, but it becomes important because, by allowing gene mutations to come afterwards that differentiate the genes in one position from the originally identical ones in the other position, the number of different kinds of genes is increased and so the germ plasm, and with it the processes of development and the organism as a whole, are eventually enabled to grow more complex.

Rearrangements of chromosome parts which do not lead to an increase in gene number can of course also occur in evolution, although it is unlikely that their role is so fundamental. By producing such changes in the laboratory it has been possible to find out a good deal more about what types can arise, and what their properties are. Various inferences can then be drawn concerning the viability and fertility that the different types would have, under varied genetic circumstances, and whether they would tend to become eliminated or to accumulate in a population of a given type. Some of them can be shown to have, under given conditions, an evolutionary survival value, both by aiding in the process of genetic isolation and in other ways, as by affecting heterosis. In this manner, evolutionary inferences have been drawn which have later been confirmed by comparison of the chromosome differences actually existing between related races, sub-species, and species.

Probably of greater ultimate interest will be the results of studies of gene mutations occurring at individual loci. Radiation mutations are frequent enough to lend themselves to comparisons of the potentialities of different loci, although not nearly enough has yet been done along these lines. Similarly, a comparison of the different mutations which can occur at the same locus can lead to very important results, especially since it has been shown that the different alleles may have very complex relationships to one another, so as even, in some cases, to reconstitute the normal type when they are crossed together. The way in which genes change as a result of successive mutations remains to be gone into at much greater length. So too does the question of changes in gene mutability, brought about by gene mutation itself.

The further the analysis of the genetic effects of irradiation, particularly of the breakage and rearrangement of chromosome parts, has gone, the more does our conviction grow that a large proportion if not the great majority of the somatic effects of irradiation that have been observed by medical men and by students of embryology, regeneration, and general biology, arise secondarily as consequences of genetic effects produced in the somatic cells. The usefulness of this interpretation has been shown in recent studies of Koller, dealing with improved methods of irradiation of mammalian carcinoma. This is too large a subject to digress upon here, but it is to be noted that it has been the analyses based in the first place on genetic and cytogenetic studies of the reproductive cells, as shown by subsequent generations, which are thus helping to clear the way for an understanding of the mechanism by which radiation acts in inhibiting growth, in causing sterilization, in producing necrosis and burns, in causing recession of malignant tissue, and perhaps also, on occasion at least, in inducing the initiation of such tissue.

During the war years, a curious confirmation of the correctness of the above inference regarding the nature of the somatic effects of irradiation has come to light. While working with mustard gas in Edinburgh, J. M. Robson was struck with the remarkable similarity between the somatic effects of this agent and those produced by X-ray and radium irradiation. This led him to wonder whether perhaps mustard gas might produce genetic changes of essentially the same kind as those known to be brought about by radiation. Comprehensive experiments were thereupon undertaken by C. Auerbach, working in collaboration with Robson, and (as mentioned on p. 160) she succeeded in showing that in fact this substance does produce mutations, both in the individual genes and by breakage and rearrangement of chromosome parts, such as X-rays and radium do, and in similar abundance. Other substances of the same general group were then found to have a similar effect. This constitutes the first decided break in the chemical attack on mutation. The fact that these findings were made as a direct result of the above inference, when so many previous attempts to produce mutations by chemical means had failed, appears to provide strong evidence that these peculiar somatic effects are in truth consequences of the more underlying ones which, when occurring in the germ cells, are analyzed by the geneticist in his breeding tests. There are, however, some very interesting differences between the nature of the genetic effects of irradiation and of these chemicals, which we cannot go into here, but which give promise of allowing an extension of the genetic and somatic analyses.

We see then that production of mutations by radiation is a method, capable of being turned in various directions, both for the analysis of the germ plasm itself, and of the organism which is in a sense an outgrowth of that germ plasm. It is to be hoped that it may also, in certain fields, prove of increasing practical use in plant and animal improvement, in the service of man. So far as direct practical application in man himself is concerned, however, we are as yet a long way from practicing any intentional selection over our own germ plasm, although like most species we are already encumbered by countless undesirable mutations, from which no individual is immune. In this situation we can, however, draw the practical lesson, from the fact of the great majority of mutations being undesirable, that their further random production in ourselves should so far as possible be rigorously avoided. As we can infer with certainty from experiments on lower organisms that all high-energy radiation must produce such mutations in man, it becomes an obligation for radiologists – though one far too little observed as yet in most countries – to insist that the simple precautions are taken which are necessary for shielding the gonads, whenever people are exposed to such radiation, either in industry or in medical practice. And, with the coming increasing use of atomic energy, even for peace-time purposes, the problem will become very important of insuring that the human germ plasm – the all-important material of which we are the temporary custodians – is effectively protected from this additional and potent source of permanent contamination.

Biography

Hermann Joseph Muller was born in New York City on December 21, 1890. His grandparents on his father's side were of artisan and professional background and, though at first Catholics, had emigrated from the Rhineland during the wave of reaction of 1848 to seek the greater freedom of America. His father, born in New York, had continued the grandfather's art metal works (the first in the U.S.A.), but was not by inclination a business man, and, although he died in 1900, he early awoke in the boy a lively interest in the nature of the universe and in the process of evolution, as well as in the welfare of men in general. The boy's mother, Frances Lyons Muller, had also been born in New York City. Her parents had come from Britain, but were in the main descended from Spanish and Portuguese Jews who, as an after-effect of the Inquisition, had settled generations earlier in England and Ireland. She, as well as the father, encouraged in the boy a broad sympathy, an interest in living things, and a love of nature.

He was brought up in Harlem, first attending public school there and later Morris High School (also public) in the Bronx. There he and his classmates Lester Thompson and Edgar Altenburg founded what was perhaps the first high-school science club. Though his family (mother, sister Ada, and himself) had very limited means, they were fortunate in usually being able to spend their summers in the country while he was of school age. But he was enabled to attend a first class college – Columbia – only through the unexpected award of a scholarship (the Cooper Hewitt), automatically granted to him in 1907 on the basis of entrance examination grades. He spent his summers, during his college years, at such jobs as bank runners and hotel clerk (the latter at $25 a month, plus board, for a 14-hour work-day).

At Columbia College he was before the end of his first year fascinated by the subject of biology. Reading by himself in the summer of 1908 R.H. Lock's (1906) book on genetics, his interests became centered in that field. Courses soon afterwards taken under E.B. Wilson influenced him profoundly, as did also his reading, independently of courses, of works by Jacques Loeb and by other writers on experimental biology and physiology. In 1909 he founded a students' biology club, which was participated in, among others, by Altenburg, and by two students, Bridges and Sturtevant, who had entered Columbia a year later.

For his first two years of graduate work, since there was no opening offered to him in zoology, he managed to obtain a scholarship (1910–1911) and then a teaching fellowship (1911–1912) in physiology, the latter at Cornell Medical College, while keeping up with genetics on the side and doing various extra jobs, such as teaching English to foreigners in night school. Finally, however, he obtained a teaching assistantship in zoology at Columbia (1912–1915). The first summer (1911) of graduate work was spent in studies at Woods Hole, the rest in laboratory teaching at Columbia. During these five years he was seriously overworked. In all this period he was chiefly interested in the *Drosophila* work which Morgan had opened up, and from 1910 on he closely followed this research and was an intimate member of the group, although he did not have opportunity for much experimental work of his own on this material until 1912. Then he was able to begin his investigation of the simultaneous inter-relationships of many linked genes, which supported the theory of crossing-over and constituted his thesis. At the same time he undertook his analysis of variable, multiple-factor, characters by means of the device of « marker genes ». This extended the validity both of chromosomal inheritance and of gene stability, and led later (1916) to his theory of balanced lethals.

Called to the Rice Institute, Houston, as Instructor, by Julian Huxley, he taught varied biological courses (1915–1918), and began studies on mutation. During this time and the two years following, when he was again at Columbia (1918–1920), now as instructor, he elaborated methods for quantitative mutation study. Altenburg, who had meanwhile moved to the Rice Institute, and he, partly in collaboration, obtained the first results in this field (1918–1919), including evidence that made probable an effect of temperature. He then (1920) returned to Texas, this time to the University, at Austin, as Associate Professor, and from 1925 on as Professor, teaching mainly genetics and evolution, and doing research mainly on mutation. He formulated in 1918, 1920, 1921, and 1926 the chief principles of spontaneous gene mutation as now recognized, including those of most mutations being detrimental and recessive, and being point effects of ultramicroscopic physico-chemical accidents arising in the course of random molecular motions (thermal agitation). At the same time he put forward the conception of the gene as constituting the basis of life, as well as of evolution, by virtue of its possessing the property of reproducing its own changes, and he represented this phenomenon as the cardinal problem of living matter.

In late 1926 he obtained critical evidence of the abundant production of gene mutations and chromosome changes by X-rays (published 1927). This opened the door to numerous researches, many of them carried on with the aid of students and co-workers, both at his own and other institutions, in the twenty years that followed. These have been briefly outlined in his Nobel Lecture, since they, together with the first discovery of the effect, constitute the work for which the Nobel Award was granted. They include studies on the mechanisms of the gene mutation effects and of the structural changes, on the roles which each kind of changes, when spontaneously occurring, play in evolution, and on the properties of genes and of chromosome parts (e.g. eu- versus hetero-chromatin), as disclosed by studies in which the chromosomes were broken and rearranged.

This later work was carried on at a succession of places. In 1932 he was awarded a Guggenheim Fellowship and for a year worked at Oscar Vogt's institute in Berlin, in Timoféeff's department of genetics. At the request of N. I. Vavilov, he then spent 3½ years as Senior Geneticist at the Institute of Genetics of the Academy of Sciences of the U.S.S.R., first in Leningrad, later (1934–1937) in Moscow, with a considerable staff of co-workers. With the rise of the Lysenko anti-genetics movement, he moved to the Institute of Animal Genetics, University of Edinburgh (1937–1940); here numerous graduate students, largely from India, took part. Then, from 1940 to 1945, he did both teaching and research at Amherst College, being professor ad interim there from 1942 to 1945. At Amherst he completed a large-scale experiment showing the relationship of ageing to spontaneous mutations. Finally, in 1945, he accepted a professorship in the Zoology Department at Indiana University, Bloomington, Indiana. Here he is again devoting his time chiefly to work on radiation-induced mutations, using them on the one hand for purposes of genetic analysis and on the other hand in the study of how radiation produces its biological effects.

One group of studies, participated in by J. I. and Ruby M. Valencia, I. H. Herskowitz, I. I. Oster, S. Zimmering, S. Abrahamson, A. Schalet, J. D. Telfer, Helen U. Meyer, Sara Frye, Helen Byers, and others, has been concerned with the influence on mutation frequency in the fruit fly *Drosophila* of diverse conditions and agents, when these were used before, after, or with radiation, or without radiation, on the influence of dose-rate and total dose of the radiation, and on the relative sensitivities of different cell stages to induced or natural mutagenesis. The types of mutations studied included «point» changes and both minute and gross structural changes of chromosomes. In another group of studies, since carried much further by E. A. Carlson, the interrelations among independently arisen mutations of the same gene were studied intensively, their intra-genic arrangement determined, and principles governing their functional interactions worked out.

The incidence of radiation damage to the bodies of the individuals that have themselves been exposed, as manifested in a long-term mortality or, in other words, life-span shortening or accelerated «ageing», was also investigated, first by I. I. Oster and then by W. Ostertag and Helen U. Meyer in collaboration with Muller. Evidence was obtained that these effects are for the most part consequences of losses of chromosomes from dividing somatic cells, after these chromosomes have been broken by the radiation. Natural ageing, however, gave evidence of not being caused in this way.

Another group of researches, also carried out with cooperation from students, more especially Margaret Lieb and S. L. Campbell, had concerned itself with problems of dominance and related subjects. It was shown that most mutant genes are incompletely recessive (not «overdominant») and are acted upon by selection while heterozygous. Studies of dominance in relation to «dosage compensation» disclosed that selection usually acts with high precision, tending to establish homozygous «normal» types. Most genetic variation within populations was deduced to depend on the recurrence of detrimental mutations which, balanced by selective elimination, constitute a «load». Estimates of this load were formed for both *Drosophila* and man (in the latter case in cooperation with Drs. Newton E. Morton and James F. Crow of the University of Wisconsin).

Included in the studies were calculations concerned with both the «spontaneous» and the radiation-induced mutation frequencies, and of the consequences of selection. Estimates were made of the effects of changes in mutation frequency, on the one hand, and of selection pressure, on the other hand, on the load. It was shown that eugenic policies are needed to avoid genetic degeneration in man as well as to bring about the genetic enhancement called for by his advances in technology and in other aspects of his culture. It was pointed out that modern reproductive technologies, such as germ-cell banks, and liberalized mores now make possible the exercise of voluntary germinal choice in human reproduction, and that this procedure affords the practical solution necessary to enable cultural evolution to promote the biological evolution of man instead of perverting it.

Prof. Muller will retire from Indiana University, in June, 1964, to take an appointment at the Institute for Advanced Learning in the Medical Sciences, The City of Hope, Duarte, California – for one year.

Muller has contributed over 300 articles on biological subjects to the scientific publications of learned societies. His principal books are *The Mechanism of Mendelian Heredity* with T. H. Morgan and others, 1915 and 1922, *Out of the Night – a Biologist's View of the Future*, 1935, 1936, and 1938, and *Genetics, Medicine and Man* with C. C. Little and L. H. Snyder, 1947.

He was President of the 8th International Congress of Genetics in 1948 and of the American Humanist Association during 1956–1958. He has received Doctor of Science degrees from the Universities of Edinburgh (1940), Columbia (1949) and Chicago (1959), the honorary Doctor of Medicine from Jefferson Medical College (1963), the Annual Award of the American Association for Advancement of Science (1927), the Kimber Genetics Award (1955) and the Darwin-Wallace Commemoration Medal (1958). He was Pilgrim Trust Lecturer (Royal Society) and Messenger Lecturer (Cornell University) in 1945, and was designated Humanist of the Year by the American Humanist Association in 1963. He has also received honorary memberships and fellowships of many learned societies in the United States, England, Scotland, Sweden, Denmark, India, Japan, Italy, etc.

Muller married his first wife, formerly Jessie M. Jacobs, in 1923 – they had one son, David Eugene. In 1939 he married Dorothea Kantorowicz – they have one daugther, Helen Juliette.

1958

George W. Beadle
Edward L. Tatum
Joshua Lederberg

George W. Beadle

Genes and Chemical Reactions in Neurospora

December 11, 1958

On this occasion of sharing the high honor of a Nobel Award with Edward L. Tatum for our «... discovery that genes act by regulating chemical events», and with Joshua Lederberg for his related «... discoveries concerning the organization of the genetic material of bacteria», it seems appropriate that I sketch briefly the background events that led to the work on *Neurospora* that Tatum and I initiated in 1940. I shall leave to my co-recipients of the award the task of describing in detail the developments in *Neurospora* that followed our first success, and the relation of this to the rise of bacterial genetics, which has depended largely on studies of genetic recombination following conjugation and transduction.

I shall make no attempt to review the entire history of biochemical genetics, for this has been done elsewhere[2,13,22,23].

Anthocyanins and Alcaptonuria

Soon after De Vries, Correns, and Tschermak «rediscovered» Mendel's 1865 paper and appreciated its full significance, investigators in the exciting new field, which was to be called genetics, naturally speculated about the physical nature of the «elements» of Mendel and the manner of their action. Renamed genes, these units of inheritance were soon found to be carried in the chromosomes.

One line of investigation that was destined to reveal much about what genes do was started by Wheldale (later Onslow) in 1903. It began with a genetic study of flower pigmentation in snapdragons. But soon the genetic observations began to be correlated with the chemistry of the anthocyanin and related pigments that were responsible. The material was favorable for both genetic and chemical studies and the work has continued to yield new information ever since and almost without interruption. Many workers and many species of plants have been involved[2,4,13,22,23].

It became clear very soon that a number of genes were involved and that they acted by somehow controlling the onset of various identifiable and specific chemical reactions. Since an understanding of the genetics helped in interpreting the chemistry and *vice versa*, the anthocyanin work was well-known to both geneticists and biochemists. It significantly influenced the thinking in both fields, and thus had great importance in further developments.

A second important line of investigation was begun even earlier by the Oxford physician-biochemist Sir Archibald E. Garrod. At the turn of the century he was interested in a group of congenital metabolic diseases in man, which he later named, « inborn errors of metabolism ». There are now many diseases described as such; in fact, they have come to be recognized as a category of diseases of major medical importance.

One of the first inborn errors to be studied by Garrod was alcaptonuria. Its most striking symptom is blackening of urine on exposure to air. It had been recorded medically long before Garrod became interested in it and important aspects of its biochemistry were understood. The substance responsible for blackening of the urine is alcapton or homogentisic acid (2,5-dihydroxyphenylacetic acid). Garrod suggested early that alcaptonuria behaved in inheritance as though it were differentiated by a single recessive gene.

By 1908 a considerable body of knowledge about alcaptonuria had accumulated. This was brought together and interpreted by Garrod in his Croonian lectures and in the two editions of his book, *Inborn Errors of Metabolism*, which were based on them[11]. It was his belief that alcaptonuria was the result of inability of affected individuals to cleave the ring of homogentisic acid as do normal individuals. He believed this to be due to absence or inactivity of the enzyme that normally catalyzes this reaction. This in turn was dependent on the absence of the normal form of a specific gene.

Thus Garrod had clearly in mind the concept of a gene–enzyme–chemical reaction system in which all three entities were interrelated in a very specific way. In the 1923 edition of « Inborn Errors »[11] he wrote:

« We may further conceive that the splitting of the benzene ring of homogentisic acid in normal metabolism is the work of a special enzyme, that in congenital alcaptonuria this enzyme is wanting... »

Failure to metabolize an intermediate compound when its normal pathway is thus blocked by a gene–enzyme defect was a part of the interpretation and accounted for the accumulation and excretion of homogentisic acid. Garrod recognized this as a means of identifying an intermediate compound that might otherwise not appear in sufficient amounts to be detected.

He also clearly appreciated that alcaptonurics would be used experimentally to explore the metabolic pathways by which homogentisic acid was formed. He summarized a large body of evidence indicating that when normal precursors of homogentisic acid are fed to alcaptonurics there is an almost quantitative increase in homogentisic acid excretion. In this way evidence was accumulated that phenylalanine, tyrosine, and the keto acid analogue of the latter were almost certainly the direct precursors of homogentisic acid.

Despite the simplicity and elegance of Garrod's interpretation of alcaptonuria and other inborn errors of metabolism as gene defects which resulted in inactivity of specific enzymes and thus in blocked reactions, his work had relatively little influence on the thinking of the geneticists of his time. Bateson's *Mendel's Principles of Heredity* and a few other books of its time discuss the concept briefly. But up to the 1940's, no widely used later textbook of genetics that I have examined even so much as refers to alcaptonuria. It is true that a number of other workers had seriously considered that genes might act in regulating chemical reactions by way of enzymes[2,13,17,21,23]. But there was no other known instance as simple as alcaptonuria. It is interesting – and significant, I think – that it was approximately 50 years after Garrod proposed his hypothesis before it was anything like fully verified through the resolution into six enzymatically catalyzed steps of phenylalanine–tyrosine metabolism via the homogentisic acid pathway, and by the clear demonstration that homogentisate oxidase is indeed lacking in the liver of an alcaptonuric[17]. Perhaps it is also well to recall that it was not until 1926 that the first enzyme was isolated in crystalline form and shown in a convincing way to consist solely of protein.

Eye pigments of Drosophila

I shall now shift to a consideration of an independent line of investigation that ended up with conclusions very much like those of Garrod and which led directly to the work with *Neurospora* that Tatum and I subsequently began.

In 1933, Boris Ephrussi came to the California Institute of Technology to work on developmental aspects of genetics. During his stay he and I had many long discussions in which we deplored the lack of information about the manner in which genes act on development. This we ascribed to the fact that the classical organisms of experimental embryology did not lend themselves readily to genetic investigation. Contrariwise, those plants and animals about which most was known genetically had been little used in studies of development.

It would be worth-while, we believed, to attempt to remedy this situation by finding new ways experimentally to study *Drosophila melanogaster* – which, genetically, was the best understood organism of the time. Tissue culture technics seemed to offer hope. In the spring of 1935 we joined forces in Euphrussi's section of *l'Institut de Biologie physico-chimique* in Paris, resolved to find ways of culturing tissues of the larvae of *Drosophila*.

After some discouraging preliminary attempts, we followed Ephrussi's suggestion and shifted to a transplantation technic. It was our hope that in this way we could make use of non-autonomous genetic characters as a means of investigating gene action in development.

Drosophila larvae are small. And we were told by a noted Sorbonne authority on the development of Diptera that the prospects were not good. In fact, he said, they were terrible.

But we were determined to try, so returned to the laboratory, made micropipettes, dissected larvae, and attempted to transfer embryonic buds from one larva to the body cavity of another. The results were discouraging. But we persisted, and finally one day discovered we had produced a fly with three eyes. Although our joy was great with this small success, we immediately began to worry about three points: First, could we do it again? Second, if we could, would we be able to characterize the diffusible substances responsible for interactions between tissues of different genetic types? And, third, how many non-autonomous characters could we find?

We first investigated the sex-linked eye-color mutant vermilion because of the earlier finding of Sturtevant that in gynandromorphs genetically vermilion eye tissue often fails to follow the general rule of autonomy[20].

Gynandromorphs may result if in an embryo that begins development as a female from an egg with two X-chromosomes, one X-chromosome is lost during an early cleavage, giving rise to a sector that has one X-chromosome and is male. If the original egg is heterozygous for a sex-linked gene, say vermilion, and the lost chromosome carries the normal allele, the male sector will be genetically vermilion, whereas the female parts are normal or wild type. (Other sex-linked characters like yellow body or forked bristles can be used as markers to independently reveal genetic constitution in most parts of the body.)

Yet in Sturtevant's gynandromorphs in which only a small part of the body including eye tissue was vermilion, the appearance of that tissue was usually not vermilion but wild type – as though some substance had diffused from wild-type tissue to the eye and caused it to become normally pigmented.

It was on the basis of this observation that Ephrussi and I transplanted vermilion eyes into wild-type larvae. The result was as expected – the transplanted eyes were indeed wild type.

At that time there were some 26 separate eye-color genes known in *Drosophila*. We obtained stocks of all of them and made a series of transplants of mutant eyes into wild-type hosts. We found only one other clear-cut non-autonomous eye character. This was cinnabar, a bright-red eye color, like vermilion but differentiated by a second chromosome recessive gene. We had a third less clear case, claret, but this was never entirely satisfactory from an experimental point of view because it was difficult to distinguish claret from wild-type eyes in transplants.

The vermilion and cinnabar characters are alike in appearance; both lack the brown pigment of the wild-type fly but retain the bright-red component. Were the diffusible substances that caused them to develop brown pigment when grown in wild-type hosts the same or different? If the same, reciprocal transplants between the two mutants should give mutant transplanted eyes in both cases. If two separate and independent substances were involved, such reciprocal transplants should give wild-type transplanted eyes in both instances.

We made the experiment and were much puzzled that neither of these results was obtained. A cinnabar eye in a vermilion host remained cinnabar, but a vermilion eye in a cinnabar host became wild type.

To explain this result we formulated the hypothesis that there must be two diffusible substances involved, one formed from the other according to the scheme: → Precursor → v^+ substance → cn^+ substance → Pigment... where v^+ substance is a diffusible material capable of making a vermilion eye become wild type, and cn^+ substance is capable of doing the same to a cinnabar eye[9].

The vermilion (v) mutant gene blocks the first reaction and the cinnabar (cn) mutant gene interrupts the second. A vermilion eye in a cinnabar host makes pigment because it can, in its own tissues, convert the v^+ substance into cn^+ substance and pigment. In it, the second reaction is not blocked.

This scheme involves the following concepts:

(*a*) A sequence of two gene-regulated chemical reactions, one gene identified with each.

(*b*) The accumulation of intermediates prior to blocked reactions.

(*c*) The ability of the mutant blocked in the first reaction to make use of an intermediate accumulated as a result of a genetic interruption of the second reaction. The principle involved is the same as that employed in the cross-feeding technic later so much used in detecting biosynthetic intermediates in micro-organisms.

What was later called the « one gene–one enzyme » concept was clearly in our minds at this time although as I remember, we did not so designate it.

Ours was a scheme closely similar to that proposed by Garrod for alcaptonuria, except that he did not have genes that blocked an adjacent reaction in the sequence. But at the time we were oblivious of Garrod's work, partly because geneticists were not in the habit of referring to it, and partly through failure of ourselves to explore the literature. Garrod's book was available in many libraries.

We continued the eye-color investigations at the California Institute of Technology, Ephrussi having returned there to spend part of 1936. Late in the year, Ephrussi returned to Paris and I went for a year to Harvard, both continuing to work along similar lines. We identified the source of diffusible substances – fat bodies and malpighian tubercules – and began to devise ways of determining their chemical nature. In this I collaborated to some extent with Professor Kenneth Thimann.

In the fall of 1937 I moved to Stanford, where Tatum shortly joined me to take charge of the chemical aspects identifying the eye-color substances. Dr. Yvonne Khouvine worked in a similar role with Ephrussi. We made progress slowly. Ephrussi and Khouvine discovered that under certain conditions feeding tryptophan had an effect on vermilion eye color. Following this lead, Tatum found – through accidental contamination of an aseptic culture containing tryptophan and test flies – an aerobic *Bacillus* that converted tryptophan into a substance highly active in inducing formation of brown pigment in vermilion flies. He soon isolated and crystallized this, but its final identification was slowed down by what later proved to be a sucrose molecule esterified with the active compound.

Professor Butenandt and co-workers[6] in Germany who had been collaborating with Professor Kühn on an analogous eye-color mutant in the meal moth *Ephestia*, and Amano *et al.*[1], working at Osaka University, showed that v^+ substance was kynurenine. Later, Butenandt and Hallmann[5], and Butenandt *et al.*[7] showed that our original cn^+ substance was 3-hydroxykynurenine.

Thus was established a reaction series of the kind we had originally conceived. Substituting the known chemical, it is as follows:

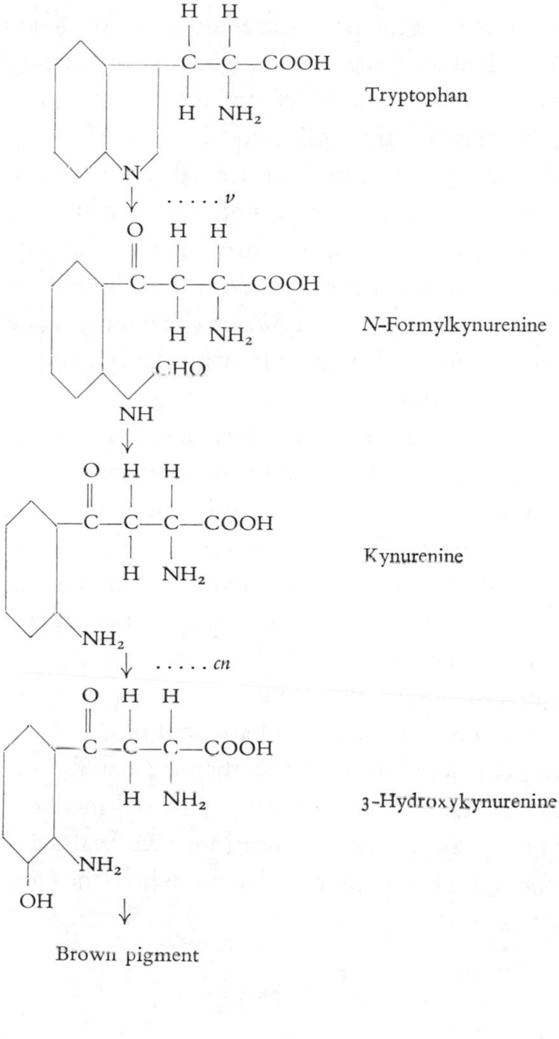

Tryptophan

↓ ..... *v*

N-Formylkynurenine

↓

Kynurenine

↓ ..... *cn*

3-Hydroxykynurenine

↓

Brown pigment

A new approach

Isolating the eye-pigment precursors of *Drosophila* was a slow and discouraging job. Tatum and I realized this was likely to be so in most cases of attempting to identify the chemical disturbances underlying inherited abnormalities; it would be no more than good fortune if any particular example chosen for investigation should prove to be simple chemically. Alcaptonuria was such a happy choice for Garrod, for the chemistry had been largely worked out and the homogentisic acid isolated and identified many years before.

Our idea – to reverse the procedure and look for gene mutations that influence known chemical reactions – was an obvious one. It followed logically from the concept that, in general, enzymatically catalyzed reactions are gene-dependent, presumably through genic control of enzyme specificity. Although we were without doubt influenced in arriving at this approach by the anthocyanin investigations, by Lwoff's demonstrations that parasites tend to become specialized nutritionally through loss of ability to synthesize substances that they can obtain readily from their hosts[18], and by the speculations of others as to how genes might act, the concepts on which it was based developed in our minds fairly directly from the eye-color work Ephrussi and I had started five years earlier.

The idea was simple: Select an organism like a fungus that has simple nutritional requirements. This will mean it can carry out many reactions by which amino acids and vitamins are made. Induce mutations by radiation or other mutagenic agents. Allow meiosis to take place so as to produce spores that are genetically homogeneous. Grow these on a medium supplemented with an array of vitamins and amino acids. Test them by vegetative transfer to a medium with no supplement. Those that have lost the ability to grow on the minimal medium will have lost the ability to synthesize one or more of the substances present in the supplemented medium. The growth requirements of the deficient strain would then be readily ascertained by a systematic series of tests on partially supplemented media.

In addition to the above specifications, we wanted an organism well-suited to genetic studies, preferably one on which the basic genetic work had already been done.

Neurospora

As a graduate student at Cornell, I had heard Dr. B. O. Dodge of the New York Botanical Garden give a seminar on inheritance in the bread mold *Neurospora*. So-called second-division segregation of mating types and of albinism were a puzzle to him. Several of us who had just been reviewing the evidence for 4-strand crossing-over in *Drosophila* suggested that crossing-over between the centromere and the segregating gene could well explain the result.

Dodge was an enthusiastic supporter of *Neurospora* as an organism for genetic work. « It's even better than *Drosophila* », he insisted to Thomas Hunt Morgan, whose laboratory he often visited. He finally persuaded Morgan to take a collection of *Neurospora* cultures with him from Columbia to the new Biology Division of the California Institute of Technology, which he established in 1928.

Shortly thereafter when Carl C. Lindegren came to Morgan's laboratory to become a graduate student, it was suggested that he should work on the genetics of *Neurospora* as a basis for his thesis. This was a fortunate choice, for Lindegren had an abundance of imagination, enthusiasm and energy and at the same time had the advice of E. G. Anderson, C. B. Bridges, S. Emerson, A. H. Sturtevant and others at the Institute who at that time were actively interested in problems of crossing-over as a part of the mechanism of meiosis. In this favorable setting, Lindegren soon worked out much of the basic genetics of *Neurospora*. New characters were found and a good start was made toward mapping the chromosomes.

Thus, Tatum and I realized that *Neurospora* was genetically an almost ideal organism for use in our new approach.

There was one important unanswered question. We did not know the mold's nutritional requirements. But we had the monograph of Dr. Nils Fries, which told us that the nutritional requirements of a number of related filamentous fungi were simple. Thus encouraged, we obtained strains of *Neurospora crassa* from Lindegren and from Dodge. Tatum soon discovered that the only growth factor required, other than the usual inorganic salts and sugar, was the recently discovered vitamin, biotin. We could not have used *Neurospora* for our purposes as much as a year earlier, for biotin would not then have been available in the quantities we required.

It remained only to irradiate asexual spores, cross them with a strain of the opposite mating type, allow sexual spores to be produced, isolate them, grow them on a suitably supplemented medium and test them on the unsupplemented medium. We believed so thoroughly that the gene–enzyme reaction relation was a general one that there was no doubt in our minds that we would find the mutants we wanted. The only worry we had was that their frequency might be so low that we would get discouraged and give up before finding one.

We were so concerned about the possible discouragement of a long series of negative results that we prepared more than thousand single-spore cultures on supplemented medium before we tested them. The 299th spore isolated gave a mutant strain requiring vitamin B_6 and the 1,085th one required B_1. We made a vow to keep going until we had 10 mutants. We soon had dozens.

Because of the ease of recovery of all the products of a single meiotic process in *Neurospora*, it was a simple matter to determine whether our newly induced nutritional deficiencies were the result of mutations in single genes. If they were, crosses with the original should yield four mutant and four non-mutant spores in each spore sac. They did[3,21].

In this long, roundabout way, first in *Drosophila* and then in *Neurospora*, we had rediscovered what Garrod had seen so clearly so many years before. By now we knew of his work and were aware that we had added little if anything new in principle. We were working with a more favorable organism and were able to produce, almost at will, inborn errors of metabolism for almost any chemical reaction whose product we could supply through the medium. Thus we were able to demonstrate that what Garrod had shown for a few genes and a few chemical reactions in man was true for many genes and many reactions in *Neurospora*.

In the fall of 1941 Francis J. Ryan came to Stanford as a National Research Council Fellow and was soon deeply involved in the Neurospora work. A year later David M. Bonner and Norman H. Horowitz joined the group. Shortly thereafter Herschel K. Mitchell did likewise. With the collaboration of a number of capable graduate students and a group of enthusiastic and able research assistants the work moved along at a gratifying pace.

A substantial part of the financial support that enabled us thus to expand our efforts was generously made available by the Rockefeller Foundation and the Nutrition Foundation.

The directions of our subsequent investigations and their accomplishments I shall leave to Professor Tatum to summarize.

One gene–one enzyme

It is sometimes thought that the *Neurospora* work was responsible for the « one gene–one enzyme » hypothesis – the concept that genes in general have single primary functions, aside from serving an essential role in their own replication, and that in many cases this function is to direct specificities of enzymatically active proteins. The fact is that it was the other way around – the hypothesis was clearly responsible for the new approach.

Although it may not have been stated explicitly, Ephrussi and I had some such concept in mind. A more specific form of the hypothesis was suggested by the fact that of all the 26 known eye-color mutants in *Drosophila*, there was only one that blocked the first of our postulated reactions and one that similarly interrupted the second. Thus it seemed reasonable to assume that the *total* specificity of a particular enzyme might somehow be derived from a single gene. The finding in *Neurospora* that many nutritionally deficient mutant strains can be repaired by supplying single chemical compounds was a verification of our prediction and as such reinforced our belief in the hypothesis, at least in its more general form.

As I hope Professor Tatum will point out in detail, there are now known a number of instances in which mutations of independent origin, all abolishing or reducing the activity of a specific enzyme, have been shown to involve one small segment of genetic material[8,12,24]. To me these lend strong support to the more restricted form of the hypothesis.

Regardless of when it was first written down on paper, or in what form, I myself am convinced that the one gene–one enzyme concept was the product of gradual evolution beginning with Garrod and contributed to by many including Moore, Goldschmidt, Troland, Haldane, Wright, Grüneberg and many others[2,13,19,22,23]. Horowitz and his co-workers[15,16] have given it, in both forms referred to above, its clearest and most explicit formulation. They have summarized and critically evaluated the evidence for and against it, with the result that they remain convinced of its continued value.

In addition Horowitz has himself made an important application of the concept in arriving at a plausible hypothesis as to how sequences of biosynthetic reactions might originally have evolved[14]. He points out that many biologically important compounds are known to be synthesized in a stepwise manner in which the intermediate compounds as such seem not to serve useful purposes. How could such a synthetic pathway have evolved if it serves no purpose unless complete? Simultaneous appearance of several independent enzymes would of course be exceedingly improbable.

Horowitz proposes that the end product of such a series of reactions was at first obtained directly from the environment, it having been produced there in the first place by non-biological reactions such as have been postulated by a number of persons, including Darwin, Haldane, Oparin and Urey and demonstrated by Miller, Fox and others[10]. It is then possible reasonably to assume that the ability to synthesize such a compound biologically could arise by a series of separate single mutations, each adding successive enzymatically catalyzed steps in the synthetic sequence, starting with the one immediately responsible for the end product. In this way each mutational step could confer a selective advantage by making the organism dependent on one less exogenous precursor of a needed end product. Without some such mechanism, by which no more than a single gene mutation is required for the origin of a new enzyme, it is difficult to see how complex synthetic pathways could have evolved. I know of no alternative hypothesis that is equally simple and plausible.

The place of genetics in modern biology

In a sense, genetics grew up as an orphan. In the beginning botanists and zoologists were often indifferent and sometimes hostile toward it. « Genetics deals only with superficial characters », it was often said. Biochemists likewise paid it little heed in its early days. They, especially medical biochemists, knew of Garrod's inborn errors of metabolism and no doubt appreciated them in the biochemical sense and as diseases; but the biological world was inadequately prepared to appreciate fully the significance of his investigations and his thinking. Geneticists, it should be said, tended to be preoccupied mainly with the mechanisms by which genetic material is transmitted from one generation to the next.

Today, happily, the situation is much changed. Genetics has an established place in modern biology. Biochemists recognize the genetic material as an integral part of the systems with which they work. Our rapidly growing knowledge of the architecture of proteins and nucleic acids is making it possible – for the first time in the history of science – for geneticists, biochemists, and biophysicists to discuss basic problems of biology in the common language of molecular structure. To me, this is most encouraging and significant.

1. T. Amano, M. Torii, and H. Iritani, *Med. J. Osaka Univ.*, 2 (1950) 45.
2. G. W. Beadle, *Chem. Rev.*, 37 (1945) 15.
3. G. W. Beadle and E. L. Tatum, *Proc. Natl. Acad. Sci. U.S.*, 27 (1941) 499.
4. G. H. J. Beale, *Genetics*, 42 (1941) 196.
5. A. Butenandt and G. Hallmann, *Z. Naturforsch.*, 5b (1950) 444.
6. A. Butenandt, W. Weidel, and E. Becker, *Naturwiss.*, 28 (1940) 63.
7. A. Butenandt, W. Weidel, and H. Schlossberger, *Z. Naturforsch.*, 4b (1949) 242.
8. M. Demerec, Z. Hartman, P. E. Hartman, T. Yura, J. S. Gots, H. Ozeki, and S. W. Glover, *Carnegie Inst. Wash. Publ.*, No. 612 (1956).
9. B. Ephrussi, *Quart. Rev. Biol.*, 17 (1942) 327.
10. S. W. Fox, *Am. Scientist*, 44 (1956) 347.
11. A. E. Garrod, *Inborn Errors of Metabolism*, Oxford Univ. Press, 1923.
12. N. H. Giles, *Proc. 10th Intern. Congr. Genet., 1958*, Univ. Toronto Press, 1959.
13. J. B. S. Haldane, *The Biochemistry of Genetics*, Allen & Unwin, London, 1954.
14. N. H. Horowitz, *Proc. Natl. Acad. Sci. U.S.*, 31 (1945) 153.
15. N. H. Horowitz and M. Fling, in *Enzymes* (Ed. by O. H. Gaebler), Academic Press, New York, 1956, p. 139.
16. N. H. Horowitz and U. Leopold, *Cold Spring Harbor Symp. Quant. Biol.*, 16 (1951) 65.
17. W. E. Knox, *Am. J. Human Genet.*, 10 (1958) 95.
18. A. Lwoff, *L'évolution physiologique*, Hermann et Cie, Paris, 1944.
19. H. J. Muller, *Proc. Royal Soc. London*, B 134 (1947) 1.
20. A. H. Sturtevant, *Proc. 6th Intern. Congr. Genet.*, 1 (1932) 304.
21. E. L. Tatum and G. W. Beadle, *Proc. Natl. Acad. Sci. U.S.*, 28 (1942) 234.
22. R. P. Wagner and H. K. Mitchell, *Genetics and Metabolism*, John Wiley, New York, 1955.
23. S. Wright, *Physiol. Rev.*, 21 (1941) 487.
24. C. Yanofsky, in *Enzymes* (Ed. by O. H. Gaebler), Academic Press, New York, 1956, p. 147.

Biography

George Wells Beadle was born at Wahoo, Nebraska, U.S.A., October 22, 1903, the son of Chauncey Elmer Beadle, a farmer, and his wife Hattie Albro. George was educated at the Wahoo High School and might himself have become a farmer if one of his teachers at school had not directed his mind towards science and persuaded him to go to the College of Agriculture at Lincoln, Nebraska. In 1926 he took his B.Sc. degree at the University of Nebraska and subsequently worked for a year with Professor F. D. Keim, who was studying hybrid wheat. In 1927 he took his M.Sc. degree, and Professor Keim secured for him a post as Teaching Assistant at Cornell University, where he worked, until 1931, with Professors R. A. Emerson and L. W. Sharp on Mendelian asynopsis in *Zea mays*. For this work he obtained, in 1931, his Ph.D. degree. In 1931 he was awarded a National Research Council Fellowship at the California Institute of Technology at Pasadena, where he remained from 1931 until 1936. During this period he continued his work on Indian corn and began, in collaboration with Professors Th. Dobzhansky, S. Emerson, and A. H. Sturtevant, work on crossing-over in the fruit fly, *Drosophila melanogaster*.

In 1935 Beadle visited Paris for six months to work with Professor Boris Ephrussi at the *Institut de Biologie physico-chimique*. Together they began the study of the development of eye pigment in *Drosophila* which later led to the work on the biochemistry of the genetics of the fungus *Neurospora* for which Beadle and Edward Lawrie Tatum were together awarded the 1958 Nobel Prize for Physiology or Medicine.

In 1936 Beadle left the California Institute of Technology to become Assistant Professor of Genetics at Harvard University. A year later he was appointed Professor of Biology (Genetics) at Stanford University and there he remained for nine years, working for most of this period in collaboration with Tatum. In 1946 he returned to the California Institute of Technology as Professor of Biology and Chairman of the Division of Biology. Here he remained until January 1961, when he was elected Chancellor of the University of Chicago and, in the autumn of the same year, President of this University.

During his career, Beadle has received many honours. These include the Hon. D.Sc. of the following Universities: Yale (1947), Nebraska (1949), Northwestern University (1952), Rutgers University (1954), Kenyon College (1955), Wesleyan University (1956), Birmingham University and Oxford University, England (1959), Pomona College (1961), and Lake Forest College (1962). In 1962 he was also given the honorary degree of LL.D. by the University of California, Los Angeles. He also received the Lasker Award of the American Public Health Association (1950), the Dyer Award (1951), the Emil Christian Hansen Prize of Denmark (1953), the Albert Einstein Commemorative Award in Science (1958), the Nobel Prize in Physiology or Medicine 1958 with E. L. Tatum and J. Lederberg, the National Award of the American Cancer Society (1959), and the Kimber Genetics Award of the National Academy of Sciences (1960).

He is a member of several learned societies, among which the National Academy of Sciences (Chairman of Committee on Genetic Effects of Atomic Radiation), the Genetics Society of America (President in 1946), the American Association for the Advancement of Science (President in 1955), the American Cancer Society (Chairman of Scientific Advisory Council), the Royal Society of London, and the Danish Royal Academy of Science.

Beadle has married twice. By his first wife he had a son, David, who now lives at The Hague, the Netherlands. His second wife, Muriel McClure, a well-known writer, was born in California. Beadle's chief hobbies are rock-climbing, skiing, and gardening.

EDWARD L. TATUM

A Case History in Biological Research

December 11, 1958

In casting around in search of a new approach, an important consideration was that much of biochemical genetics has been and will be covered by Professor Beadle and Professor Lederberg, and in many symposia and reviews, in which many aspects have been and will be considered in greater detail and with greater competence than I can hope to do here. It occurred to me that perhaps it might be instructive, valuable, and interesting to use the approach which I have attempted to define by the title « A Case History in Biological Research ». In the development of this case history I hope to point out some of the factors involved in all research, specifically the dependence of scientific progress: on knowledge and concepts provided by investigators of the past and present all over the world; on the free interchange of ideas within the international scientific community; on the hybrid vigor resulting from cross-fertilization between disciplines; and last but not least, also dependent on chance, geographical proximity, and opportunity. I would like finally to complete this case history with a brief discussion of the present status of the field, and a prognosis of its possible development.

Under the circumstances, I hope I will be forgiven if this presentation is given from a personal viewpoint. After graduating from the University of Wisconsin in chemistry, I was fortunate in having the opportunity of doing graduate work in biochemistry and microbiology at this University under the direction and leadership of W. H. Peterson and E. B. Fred. At that time, in the early 30's, one of the exciting areas being opened concerned the so-called « growth-factors » for micro-organisms, for the most part as yet mysterious and unidentified. I became deeply involved in this field, and was fortunate to have been able, in collaboration with H. G. Wood, then visiting at Wisconsin, to identify one of the required growth-factors for propionic acid bacteria, as the recently synthesized vitamin B_1 or thiamine[1]. This was before the universality of need for the B vitamins, and the enzymatic basis of this requirement, had been clearly defined. The vision of Lwoff and Knight had already indicated a correlation of the need of micro-organisms for « growth-factors » with failure of synthesis, and correlated this failure with evolution, particularly in relation to the complex environment of « fastidious » pathogenic micro-organisms. However, the tendency at this time was to consider « growth-factors » as highly individual requirements, peculiar to particular strains or species of micro-organisms as isolated from nature, and their variation in these respects was not generally considered as related to gene mutation and variation in higher organisms. Actually my ignorance of and naïveté in genetics was probably typical of that of most biochemists and microbiologists of the time, with my only contact with genetic concepts being a course primarily on vertebrate evolution.

After completing graduate work at Wisconsin I was fortunate in being able to spend a year studying at the University of Utrecht with F. Kögl, the discoverer of the growth-factor biotin, and to work in the same laboratory with Nils Fries, who already had contributed significantly in the field of nutrition and growth of fungi.

At this time, Professor Beadle was just moving to Stanford University, and invited me as a biochemist to join him in the further study of the eye-color hormones of *Drosophila*, which he and Ephrussi in their work at the California Institute of Technology and at Paris had so brilliantly established as diffusible products of gene-controlled reactions. During this, my first contacts with modern genetic concepts, as a consequence of a number of factors – the observation of Khouvine, Ephrussi, and Chevais[2] in Paris that dietary tryptophan was concerned with *Drosophila* eye-color hormone production; our studies on the nutrition of *Drosophila* in aseptic culture[3]; and the chance contamination of one of our cultures of *Drosophila* with a particular bacterium – we were able to isolate the v^{+} hormone in crystalline state from a bacterial culture supplied with tryptophan[4], and with A. J. Haagen-Smit to identify it as kynurenine[5], originally isolated by Kotake, and later structurally identified correctly by Butenandt. It might be pointed out here that kynurenine has since been recognized to occupy a central position in tryptophan metabolism in many organisms aside from insects, including mammals and fungi.

At about this time, as the result of many discussions and considerations of the general biological applicability of chemical genetic concepts, stimulated by the wealth of potentialities among the micro-organisms and their variation in nature with respect to their nutritional requirements, we began our work with the mold *Neurospora crassa*.

I shall not renumerate the factors involved in our selection of this organism for the production of chemical or nutritionally deficient mutants, but must take this opportunity of reiterating our indebtedness to the previous basic findings of a number of investigators. Foremost among these, to B. O. Dodge for his establishment of this *Ascomycete* as a most suitable organism for genetic studies[6]; and to C. C. Lindegren[7], who became interested in *Neurospora* through T. H. Morgan, a close friend of Dodge.

Our use of *Neurospora* for chemical genetic studies would also have been much more difficult, if not impossible, without the availability of synthetic biotin as the result of the work of Kögl[8] and of du Vigneaud[9]. In addition, the investigations of Nils Fries on the nutrition of *Ascomycetes*[10] were most helpful, as shown by the fact that the synthetic minimal medium used with *Neurospora* for many years was that described by him and supplemented only with biotin, and has ordinarily since been referred to as « Fries medium ». It should also be pointed out that the experimental feasibility of producing the desired nutritionally deficient mutant strains depended on the early pioneering work of Roentgen, with X-rays, and on that of H. J. Muller, on the mutagenic activity of X-rays and ultraviolet light on *Drosophila*. All that was needed was to put these various facts and findings together to produce in the laboratory with irradiation, nutritionally deficient (auxotrophic) mutant strains of *Neurospora*, and to show that each single deficiency produced was associated with the mutation of a single gene[11].

Having thus successfully tested with *Neurospora* the basic premise that the biochemical processes concerned with the synthesis of essential cell constituents are gene controlled, and alterable as a consequence of gene mutation, it then seemed a desirable and natural step to carry this approach to the bacteria, in which so many and various naturally occurring growth-factor requirements were known, to see if analogous nutritional deficiencies followed their exposure to radiation. As is known to all of you, the first mutants of this type were successfully produced in *Acetobacter* and in *Escherichia coli*[12], and the first step had been taken in bringing the bacteria into the fold of organisms suitable for genetic study.

Now to point out some of the curious coincidences or twists of fate as involved in science: One of the first series of mutants in *Neurospora* which was studied intensily from the biochemical viewpoint was that concerned with the biosynthesis of tryptophan. In connection with the role of indole as a precursor of tryptophan, we wanted also to study the reverse process, the breakdown of tryptophan to indole, a reaction typical of the bacterium *E. coli*. For this purpose we obtained, from the Bacteriology Department at Stanford, a typical *E. coli* culture, designated K-12. Naturally, this strain was later used for the mutation experiments just described so that a variety of biochemically marked mutant strains of *E. coli* K-12 were soon available. It is also of interest that Miss Esther Zimmer, who later became Esther Lederberg, assisted in the production and isolation of these mutant strains.

Another interesting coincidence is that F. J. Ryan spent some time on leave from Columbia University at Stanford, working with *Neurospora*. Shortly after I moved to Yale University in 1945, Ryan encouraged Lederberg, then a medical student at Columbia who had worked some time with Ryan on *Neurospora*, to spend some time with me at Yale University. As all of you know, Lederberg was successful in showing genetic recombination between mutant strains of *E. coli* K-12[13] and never returned to medical school, but continued his brilliant work on bacterial recombination at Wisconsin. In any case, the first demonstration of a process analogous to a sexual process in bacteria was successful only because of the clear-cut nature of the genetic markers available which permitted detection of this very rare event, and because of the combination of circumstances which had provided those selective markers in one of the rare strains of *E. coli* capable of recombination. In summing up this portion of this case history, then, I wish only to emphasize again the role of coincidence and chance played in the sequence of developments, but yet more strongly to acknowledge the even greater contributions of my close friends and associates, Professor Beadle and Professor Lederberg, with whom it is a rare privilege and honor to share this award.

Now for a brief and necessarily somewhat superficial mention of some of the problems and areas of biology to which these relatively simple experiments with *Neurospora* have led and contributed. First, however, let us review the basic concepts involved in this work. Essentially these are: (*1*) that all biochemical processes in all organisms are under genic control; (*2*) that these overall biochemical processes are resolvable into a series of individual stepwise reactions; (*3*) that each single reaction is controlled in a primary fashion by a single gene, or in other terms, in every case a 1:1 correspondence of gene and biochemical reaction exists, such that (*4*) mutation of a single gene results only in an alteration in the ability of the cell to carry out a single primary chemical reaction. As has repeatedly been stated, the underlying hypothesis, which in a number of cases has been supported by direct experimental evidence, is that each gene controls the production, function, and specificity of a particular enzyme. Important experimental implications of these relations are that each and every biochemical reaction in a cell of any organism, from a bacterium to man, is theoretically alterable by gene mutation, and that each such mutant cell strain differs in only one primary way from the non-mutant parental strain. It is probably unnecessary to point out that these experimental expectations have been amply supported by the production and isolation, by many investigators during the last 15 or more years, of biochemical mutant strains of micro-organisms in almost every species tried: bacteria, yeasts, algae, and fungi.

It is certainly unnecessary for me to do more than point out that mutant strains such as those produced and isolated first in *Neurospora* and *E. coli* have been of primary utility as genetic markers in detecting and elucidating the details of the often exotic mechanisms of genetic recombination of micro-organisms.

Similarly, it seems superfluous even to mention the proven usefulness of mutant strains of micro-organisms in unraveling the detailed steps involved in the biosynthesis of vital cellular constituents. I would like to list, however, a few of the biosynthetic sequences and biochemical interrelationships which owe their discovery and elucidation largely to the use of biochemical mutants. These include: the synthesis of the aromatic amino acids via dehydroshikimic and shikimic acids[14,15], by way of prephenic acid to phenylalanine[16], and by way of anthranilic acid, indole glycerol phosphate[17], and condensation of indole with serine to give tryptophan[18]; the conversion of tryptophan via kynurenine and 3-OH anthranilic acid to niacin[19,20]; the biosynthesis of histidine[21]; of isoleucine and valine via the analogous di-OH and keto acids[22]; the biosynthesis of proline and ornithine from glutamic acid[23]; and the synthesis of pyrimidines via orotic acid[24].

If the postulated relationship of gene to enzyme is correct, several consequences can be predicted. First, mutation should result in the production of a changed protein, which might either be enzymatically inactive, of intermediate activity, or have otherwise detectably altered physical properties. The production of such proteins changed in respect to heat stability, enzymatic activity, or other properties such as activation energy, by mutant strains has indeed been demonstrated in a number of instances[25–31]. Recognition of the molecular bases of these changes must await detailed comparison of their structures with those of the normal enzyme, using techniques similar to the elegant methods of Professor Sanger. That the primary effect of gene mutation may be as simple as the substitution of a single amino acid by another and may lead to profound secondary changes in protein structure and properties has recently been strongly indicated by the work of Ingram on hemoglobin[32]. It seems inevitable that induced mutant strains of micro-organisms will play a most important part in providing material for the further examination of these problems.

A second consequence of the postulated relationship stems from the concept that the genetic constitution defines the potentialities of the cell, the time and degree of expression of which are to a certain extent modifiable by the cellular environment. The analysis of this type of secondary control at the biochemical level is one of the important and exciting new areas of biochemistry. This deals with the regulation and integration of biochemical reactions by means of feed-back mechanisms restricting the synthesis or activities of enzymes[33-36] and through substrate induced biosynthesis of enzymes[37]. It seems probable that some gene mutations may affect biochemical activities at this level, (modifiers, and suppressors), and that chemical mutants will prove of great value in the analysis of the details of such control mechanisms.

An equally fascinating newer area of genetics, opened by Benzer[38] with bacteriophage, is that of the detailed correlation of fine structure of the gene in terms of mutation and recombination, with its fine structure in terms of activity. Biochemical mutants of micro-organisms have recently opened this area to investigation at two levels of organization of genetic material. The higher level relates to the genetic linkage of non-allelic genes concerned with sequential biosynthetic reactions. This has been shown by Demerec and by Hartman in the biosynthesis of tryptophan and histidine by *Salmonella*[39].

At a finer level of organization of genetic material, the biological versatility of *Neurospora* in forming heterocaryotic cells has permitted the demonstration[40-42] that genes damaged by mutation in different areas, within the same locus and controlling the same enzyme, complement each other in a heterocaryon in such a way that synthesis of enzymatically active protein is restored, perhaps, in a manner analogous to the reconstitution of ribonuclease from its a and b constituents, by the production in the cytoplasm of an active protein from two gene products defective in different areas. This phenomenon of complementation, which appears also to take place in *Aspergillus*[43], permits the mapping of genetic fine structure in terms of function, and should lead to further information on the mechanism of enzyme production and clarification of the role of the gene in enzyme synthesis.

The concepts of biochemical genetics have already been, and will undoubtedly continue to be, significant in broader areas of biology. Let me cite a few examples in microbiology and medicine.

In microbiology the roles of mutation and selection in evolution are coming to be better understood through the use of bacterial cultures of mutant strains. In more immediately practical ways, mutation has proven of primary importance in the improvement of yields of important antibiotics – such as in the classic example of penicillin, the yield of which has gone up from around 40 units per ml of culture shortly after its discovery by Fleming to approximately 4,000, as the result of a long series of successive experimentally produced mutational steps. On the other side of the coin, the mutational origin of antibiotic-resistant micro-organisms is of definite medical significance. The therapeutic use of massive doses of antibiotics to reduce the numbers of bacteria which by mutation could develop resistance, is a direct consequence of the application of genetic concepts. Similarly, so is the increasing use of combined antibiotic therapy, resistance to both of which would require the simultaneous mutation of two independent characters.

As an important example of the application of these same concepts of microbial genetics to mammalian cells, we may cite the probable mutational origin of resistance to chemotherapeutic agents in leukemic cells[44], and the increasing and effective simultaneous use of two or more chemotherapeutic agents in the treatment of this disease. In this connection it should be pointed out that the most effective cancer chemotherapeutic agents so far found are those which interfere with DNA synthesis, and that more detailed information on the biochemical steps involved in this synthesis is making possible a more rational design of such agents. Parenthetically, I want to emphasize the analogy between the situation in a bacterial culture consisting of two or more cell types and that involved in the competition and survival of a malignant cell, regardless of its origin, in a population of normal cells. Changes in the cellular environment, such as involved in chemotherapy, would be expected to affect the metabolic efficiency of an altered cell, and hence its growth characteristics. However, as in the operation of selection pressures in bacterial populations, based on the interaction between cell types, it would seem that the effects of chemotherapeutic agents on the efficiency of selective pressures among mammalian cell populations can be examined most effectively only in controlled mixed populations of the cell types concerned.

In other areas in cancer, the concepts of genetics are becoming increasingly important, both theoretically and practically. It seems probable that neoplastic changes are directly correlated with changes in the biochemistry of the cell. The relationships between DNA, RNA, and enzymes which have evolved during the last few decades, lead one to look for the basic neoplastic change in one of these intimately interrelated hierarchies of cellular materials.

In relation to DNA, hereditary changes are now known to take place as a consequence of mutation, or of the introduction of new genetic material through virus infection (as in transduction) or directly (as in transformation). Although each of these related hereditary changes may theoretically be involved in cancer, definite evidence is available only for the role of viruses, stemming from the classic investigations of Rous on fowl sarcoma[45]. At the RNA level of genetic determination, any one of these classes of change might take place, as in the RNA containing viruses, and result in an heritable change, perhaps of the cytoplasmic type, semi-autonomous with respect to the gene. At the protein level, regulatory mechanisms determining gene activity and enzyme synthesis as mentioned earlier, likewise provide promising areas for exploration.

Among the many exciting applications of microbial-genetic concepts and techniques to the problems of cancer, may I mention in addition the exploration by Klein[46] of the genetic basis of the immunological changes which distinguish the cancer cell from the normal, and the studies on the culture, nutrition, morphology, and mutation of isolated normal and malignant mammalian cells of Puck[47] and of Eagle[48]. Such studies are basic to our exploration and to our eventual understanding of the origin and nature of the change to malignancy.

Regardless of the origin of a cancer cell, however, and of the precise genetic level at which the primary change takes place, it is not too much to hope and expect eventually to be able to correct or alleviate the consequences of the metabolic defect, just as a closer understanding of a heritable metabolic defect in man permits its correction or alleviation. In terms of biochemical genetics, the consequences of a metabolic block may be rectified by dietary limitation of the precursor of an injurious accumulation product, aromatic amino acids in phenylketonuria; or by supplying the essential end product from without the cell, the specific blood protein in hemophilia, or a specific essential nutrient molecule such as a vitamin.

Time does not permit the continuation of these examples. Perhaps, however, I will be pardoned if I venture briefly on a few more predictions and hopes for the future.

It does not seem unrealistic to expect that as more is learned about control of cell machinery and heredity, we will see the complete conquering of many of man's ills, including hereditary defects in metabolism, and the momentarily more obscure conditions such as cancer and the degenerative diseases, just as disease of bacterial and viral etiology are now being conquered.

With a more complete understanding of the functioning and regulation of gene activity in development and differentiation, these processes may be more efficiently controlled and regulated, not only to avoid structural or metabolic errors in the developing organism, but also to produce better organisms.

Perhaps within the lifetime of some of us here, the code of life processes tied up in the molecular structure of proteins and nucleic acids will be broken. This may permit the improvement of all living organisms by processes which we might call biological engineering.

This might proceed in stages from the *in vitro* biosynthesis of better and more efficient enzymes, to the biosynthesis of the corresponding nucleic acid molecules, and to the introduction of these molecules into the genome of organisms, whether via injection, viral introduction into germ cells, or via a process analogous to transformation. Alternatively, it may be possible to reach the same goal by a process involving directed mutation.

As a biologist, and more particularly as a geneticist, I have great faith in the versatility of the gene and of living organisms in providing the material with which to meet the challenges of life at any level. Selection, survival, and evolution take place in response to environmental pressures of all kinds, including sociological and intellectual. In the larger view, the dangerous and often poorly understood and controlled forces of modern civilization, including atomic energy and its attendant hazards, are but more complex and sophisticated environmental challenges of life. If man cannot meet those challenges, in a biological sense he is not fit to survive.

However, it may confidently be hoped that with real understanding of the roles of heredity and environment, together with the consequent improvement in man's physical capacities and greater freedom from physical disease, will come an improvement in his approach to, and understanding of, sociological and economic problems. As in any scientific research, a problem clearly seen is already half solved. Hence, a renaissance may be foreseen, in which the major sociological problems will be solved, and mankind will take a big stride towards the state of world brotherhood and mutual trust and well-being envisaged by that great humanitarian and philanthropist Alfred Nobel.

1. E. L. Tatum, H. G. Wood, and W. H. Peterson, *Biochem. J.*, 30 (1936) 1898.
2. Y. Khouvine, B. Ephrussi, and S. Chevais, *Biol. Bull.*, 75 (1938) 425.
3. E. L. Tatum, *Proc. Natl. Acad. Sci. U.S.*, 27 (1941) 193.
4. E. L. Tatum and G. W. Beadle, *Science*, 91 (1940) 458.
5. E. L. Tatum and A. J. Haagen-Smit, *J. Biol. Chem.*, 140 (1941) 575.
6. B. O. Dodge, *J. Agr. Res.*, 35 (1927) 289.
7. C. C. Lindegren, *Bull. Torrey Botan. Club.*, 59 (1932) 85.
8. F. Kögl, *Ber.*, 68 (1935) 16.
9. V. du Vigneaud, *Science*, 96 (1942) 455.
10. N. Fries, *Symbolae Botan. Upsalienses*, 3 (1938) 1–188.
11. G. W. Beadle and E. L. Tatum, *Proc. Natl. Acad. Sci. U.S.*, 27 (1941) 499.
12. E. L. Tatum, *Cold Spring Harbor Symp. Quant. Biol.*, 11 (1946) 278.
13. J. Lederberg and E. L. Tatum, *Nature*, 158 (1946) 558.
14. B. D. Davis, in *Amino Acid Metabolism*, Baltimore, 1955, p. 799.
15. E. L. Tatum, S. R. Gross, G. Ehrensvärd, and L. Garnjobst, *Proc. Natl. Acad. Sci. U.S.*, 40 (1954) 271.
16. R. L. Metzenberg and H. K. Mitchell, *Biochem. J.*, 68 (1958) 168.
17. C. Yanofsky, *J. Biol. Chem.*, 224 (1957) 783.
18. E. L. Tatum and D. M. Bonner, *Proc. Natl. Acad. Sci. U.S.*, 30 (1944) 30.
19. D. Bonner, *Proc. Natl. Acad. Sci. U.S.*, 34 (1948) 5.
20. H. K. Mitchell and J. F. Nye, *Proc. Natl. Acad. Sci. U.S.*, 34 (1948) 1.
21. B. N. Ames, in *Amino Acid Metabolism*, Baltimore, 1955.
22. E. A. Adelberg, *J. Bacteriol.*, 61 (1951) 365.
23. H. J. Vogel, in *Amino Acid Metabolism*, Baltimore, 1955.
24. H. K. Mitchell, M. B. Honlahan, and J. F. Nye, *J. Biol. Chem.*, 172 (1948) 525.
25. W. K. Maas and B. D. Davis, *Proc. Natl. Acad. Sci. U.S.*, 38 (1952) 785.
26. N. H. Horowitz and M. Fling, *Genetics*, 38 (1953) 360.
27. T. Yura and H. J. Vogel, *Biochim. Biophys. Acta*, 17 (1955) 582.
28. J. R. S. Fincham, *Biochem. J.*, 65 (1957) 721.
29. D. R. Suskind and L. I. Kurek, *Science*, 126 (1957) 1068.
30. N. H. Giles, C. W. H. Partridge, and N. J. Nelson, *Proc. Natl. Acad. Sci. U.S.*, 43 (1957) 305.
31. T. Yura, *Proc. Natl. Acad. Sci. U.S.*, 45 (1959) 197.
32. V. M. Ingram, *Nature*, 180 (1957) 326.
33. H. J. Vogel, in *Symposium on the Genetic Basis of Heredity*, Baltimore, 1957.
34. L. Gorini and W. K. Maas, *Biochim. Biophys. Acta*, 25 (1957) 208.
35. R. A. Yates and A. B. Pardee, *J. Biol. Chem.*, 221 (1956) 757.
36. H. E. Umbarger and B. Brown, *J. Biol. Chem.*, 233 (1958) 415.
37. M. Cohn and J. Monod, *Symp. Soc. Gen. Microbiol.*, 3 (1953) 132.
38. S. Benzer, in *Symposium on the Chemical Basis of Heredity*, Baltimore, 1957.
39. P. E. Hartman, in *Symposium on the Chemical Basis of Heredity*, Baltimore, 1957.
40. N. H. Giles, C. W. H. Partridge, and N. J. Nelson, *Proc. Natl. Acad. Sci. U.S.*, 43 (1957) 305.
41. M. E. Case and N. H. Giles, *Proc. Natl. Acad. Sci. U.S.*, 44 (1958) 378.
42. J. A. Pateman and J. R. S. Fincham, *Heredity*, 12 (1958) 317.

43. E. Calef, *Heredity*, 10 (1956) 83.
44. L. W. Law, *Nature*, 169 (1952) 628.
45. P. Rous, *J. Exp. Med.*, 12 (1910) 696.
46. G. Klein, E. Klein, and L. Révész, *Nature*, 178 (1956) 1389.
47. T. T. Puck, in *Symposium on Growth and Development*, Princeton, 1957.
48. H. Eagle, V. I. Oyama, M. Levy, and A. E. Freiman, *Science*, 123 (1956) 845.

Biography

Edward Lawrie Tatum was born on December 14th, 1909, at Boulder, Colorado, U.S.A. He was the eldest son of Arthur Lawrie Tatum, Professor of Pharmacology at the University of Wisconsin Medical School, and Mabel Webb Tatum. After the death of his mother, his father married the former Celia Harriman.

Tatum was educated at the University of Chicago and Wisconsin, taking his A.B. degree in Chemistry in 1931, his M.S. degree in Microbiology in 1932 and his Ph.D. degree in Biochemistry in 1934. For the Ph.D. degree his thesis was on work on the nutrition and metabolism of bacteria which he had done under the direction of Edwin Broun Fred and William Harold Peterson. This work no doubt laid the foundations of his later work with George Wells Beadle, which was to earn for their book, in 1958, the Nobel Prize in Physiology or Medicine.

After taking his doctor's degree, Tatum studied for a year at the University of Wisconsin and then was awarded a General Education Fellowship at the University of Utrecht, Holland. He then joined the Department of Biological Sciences at Stanford University, California, where he was Research Associate from 1937 until 1941, and Assistant Professor from 1941 until 1945. From 1945 until 1948 he was successively Assistant Professor of Botany and Professor of Microbiology at Yale University. In 1948 he returned to Stanford University as Professor of Biology and later became Professor of Biochemistry there. It was during this period of his life and work at Stanford University that he collaborated with George Wells Beadle, who was Professor of Biology (Genetics) at that University until 1946.

Tatum's research has been concerned primarily with the biochemistry, nutrition and genetics of microorganisms and of the fruit fly, *Drosophila melanogaster*. During his fruitful collaboration with George Wells Beadle he took charge of the chemical aspects of their joint work on the genetics of eye-colour in *Drosophila* and, when he and Beadle decided to give up their work on *Drosophila* and to work instead with the fungus *Neurospora crassa*, it was Tatum who discovered that biotin was necessary for the successful cultivation of this fungus on simple inorganic media and thus provided these two workers with the genetic material that they needed for the work which gained them, together with Joshua Lederberg, the Nobel Prize.

In 1953, he received the Remsen Award of the American Chemical Society. He is a member of the Advisory Committee of the National Foundation and has served on research advisory panels of the American Committee of the National Research Council on Growth. He also served for 10 years on the Editorial Board of the *Journal of Biological Chemistry*. He is now a member of the Editorial Board of *Science* and of *Biochimica et Biophysica Acta*.

Tatum is married to Viola Kantor. He has two daughters, Margaret and Barbara, born to him and his first wife, June Alton.

JOSHUA LEDERBERG

A View of Genetics

May 29, 1959

The Nobel Statutes of 1900 charge each prize-winner to give a public lecture in Stockholm within six months of Commemoration Day. That I have fully used this margin is not altogether ingenuous, since it furnishes a pleasant occasion to revisit my many friends and colleagues in your beautiful city during its best season.

The charge might call for a historical account of past « studies on genetic recombination and organization of the genetic material in bacteria », studies in which I have enjoyed the companionship of many colleagues, above all my wife. However, this subject has been reviewed regularly[36,37,38,41,42,45,49,54,55,58] and I hope you will share my own inclination to assume a more speculative task, to look at the context of contemporary science in which bacterial genetics can be better understood, and to scrutinize the future prospects of experimental genetics.

The dispersion of a Nobel Award in the field of genetics symbolizes the convergent efforts of a world-wide community of investigators. That genetics should now be recognized is also timely – for its axial role in the conceptual structure of biology, and for its ripening yield for the theory and practice of medicine. However, experimental genetics is reaching its full powers in coalescence with biochemistry: in principle, each phenotype should eventually be denoted as an exact sequence of amino acids in protein[79], and the genotype as a corresponding sequence of nucleotides in DNA*[63]. The precise demarcation of genetics from biochemistry is already futile: but when genetics has been fully reduced to its molecular foundations, it may continue to serve in the same relation as thermodynamics to mechanics[69]. The coordination of so many adjacent sciences will be a cogent challenge to the intellectual powers of our successors.

That bacteria and their genetics should now be so relevant to general biology is already a fresh cycle in our scientific outlook. When thought of at all, they have often been relegated to some obscure by-way of evolution,

* No reader who recognizes *deoxyribonucleic acid* will need to be reminded what DNA stands for.

their complexity and their homology with other organisms grossly underrated. « Since Pasteur's startling discoveries of the important role played by microbes in human affairs, microbiology as a science has always suffered from its eminent practical applications. By far the majority of the microbiological studies were undertaken to answer questions connected with the well-being of mankind »[30]. The pedagogic cleavage of academic biology from medical education has helped sustain this distortion. Happily, the repatriation of bacteria and viruses is only the first measure of the repayment of medicine's debt to biology[6,7,8].

Comparative biochemistry has consummated the unification of biology revitalized by Darwin one hundred years ago. Throughout the living world we see a common set of structural units – amino acids, coenzymes, nucleins, carbohydrates, and so forth – from which every organism builds itself. The same holds for the fundamental process of biosynthesis and of energy metabolism. The exceptions to this rule thus command special interest as meaningful tokens of biological individuality, e.g., the replacement of cytosine by hydroxymethyl cytosine in the DNA of T2 phage[12].

Nutrition has been a special triumph. Bacteria which required no vitamins had seemed simpler than man. But deeper insights[32,61] interpret nutritional simplicity as a greater power of synthesis. The requirements of more exacting organisms comprise just those metabolites they can not synthesize with their own enzymatic machinery.

Species differ in their nutrition: if species are delimited by their genes, then genes must control the biosynthetic steps which are reflected in nutritional patterns. This syllogism, so evident once told, has been amplified by Beadle and Tatum from this podium. Its implications for experimental biology and medicine are well-known: among these, the methodology of bacterial genetics. Tatum has related how his early experience with bacterial nutrition reinforced the foundations of the biochemical genetics of *Neurospora*. Then, disregarding the common knowledge that bacteria were too simple to have genes, Tatum took courage to look for the genes that would indeed control bacterial nutrition. This conjunction marked the start of my own happy association with him, and with the fascinating challenges of bacterial genetics.

Contemporary genetic research is predicated on the role of DNA as the genetic material, of enzyme proteins as the cell's working tools, and of RNA as the communication channel between them[63]. Three lines of evidence substantiate the genetic function of DNA. Two are related to bacterial genetics; the third and most general is the cytochemical observation of DNA in the chromosomes, which are undeniably strings of genes. But chromosomes also contain other constituents besides DNA: we want a technique to isolate a chromosome or a fragment of one, to analyse it and to retransplant it to verify its functional capacity. The impressive achievements of nuclear transplantation[29] should encourage the audacity needed to try such experiments. The constructive equivalent to chromosome transplantation was discovered by a bacteriologist thirty years ago[20] but the genetic implications of the « pneumococcus transformation » in the minds of some of Griffith's successors were clouded by its involvement with the gummy outer capsule of the bacteria. However, by 1943, Avery and his colleagues had shown that this inherited trait was transmitted from one pneumococcal strain to another by DNA. The general transmission of other traits by the same mechanism[25] can only mean that DNA comprises the genes*.

To reinforce this conclusion, Hershey and Chase[23] proved that the genetic element of a bacterial virus is also DNA. Infection of a host cell requires the injection of just the DNA content of the adsorbed particle. This DNA controls not only its own replication in the production of new phage but also the specificity of the protein coat, which governs the serological and host range specificity of the intact phage.

At least in some small viruses, RNA also displays genetic functions. However, the hereditary autonomy of gene-initiated RNA of the cytoplasm is now very doubtful – at least some of the plasmagenes that have been proposed as fulfilling this function are now better understood as feedback-regulated systems of substrate-transport[65,72,81].

The work of the past decade thus strongly supports the simple doctrine that genetic information is nucleic, i.e., is coded in a linear sequence of nucleotides. This simplification of life may appear too facile, and has furnished a tempting target for agnostic criticism[37,41,44,74]. But, while no scientific theory would decry continual refinement and amplification, such criticism has little value if it detracts from the evident fruitfulness of the doctrine in experimental design.

The cell may, of course, carry information other than nucleic either in the cytoplasm or, accessory to the polynucleotide sequence, in the chromosomes. Epinucleic information has been invoked, without being more precisely de-

* One might be tempted to write: «One DNA molecule = one gene». However, the quanta of factorial genetics, based on mutation, recombination and enzymatic function are all smaller[4] than the DNA unit of molecular weight $\sim 6 \times 10^6$. There is increasing evidence that such a molecule is a natural unit rather than an artefact of fragmentation[64].

fined, in many recent speculations on cytodifferentiation and on such models of this as antigenic phase variation in *Salmonella*[71,52,56,47]. Alternative schemes have so much less information capacity than the nucleic cycle that they are more likely to concern the regulation of genic functions than to mimic their specificities.

DNA as a substance

The chemistry of DNA deserves to be exposed by apter craftsmen[86,31,13] and I shall merely recapitulate before addressing its biological implications. A

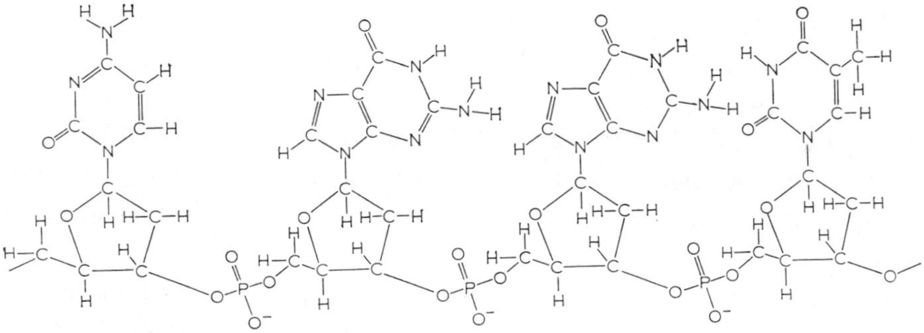

Fig. 1. Primary structure of DNA – a segment of a polynuclotide sequence CGGT. (From Ref. 13.)

segment of DNA is illustrated in Fig. 1. This shows a linear polymer whose backbone contains the repeating unit:

$$[\text{—O—P(OO}^-\text{)—O—CH}_2\text{—}\overset{|}{\text{C}}\text{H—}\overset{|}{\text{C}}\text{H—]O—P(OO}^-\text{)—O—CH}_2\text{—}$$

[diester phosphate C_5' C_4' C_3'] diester phosphate ....

The carbon atoms are conventionally numbered according to their position in the furanose ring of deoxyribose, which is coupled as an N-glycoside to one of the nucleins: adenine, guanine, cytosine, or thymine, symbolized A, G, C, or T, the now well-known alphabet in which genetic instructions are composed. With a chain length of about 10,000 residues, one molecule of DNA contains 20,000 « bits of information », comparable to the text of this article, or in a page of newsprint.

Pyrophosphate-activated monomer units (e.g. thymidine triphosphate) have been identified as the metabolic precursors of DNA[31]. For genetic replication, the monomer units must be assembled in a sequence that reflects that of the parent molecule. A plausible mechanism has been forwarded by Watson and Crick[87] as a corollary to their structural model whereby DNA occurs as a two-stranded helix with the bases being centrally oriented. When their relative positions are fixed by the deoxyribose-phosphate backbones, just two pairs of bases are able to form hydrogen bonds between their respective NH and CO groups; these are A:T and G:C. This pairing of bases would tie the two strands together for the length of the helix. In conformity with this model, extensive analytical evidence shows a remarkable equality of A with T and of G with C in DNA from various sources. The two strands of any DNA are then mutually complementary, the A, T, G, and C of one strand being represented by T, A, C, and G, respectively, of the other. The information of one strand is therefore equivalent to, because fully determined by, the other. The determination occurs at the replication of one parent strand by the controlled stepwise accretion of monomers to form a complementary strand. At each step only the monomer which is complementary to the template would fit for a chain-lengthening esterification with the adjacent nucleotide. The model requires the unravelling of the intertwined helices to allow each of them to serve as a template. This might, however, occur gradually, with the growth of the daughter chain – a concept embedded in Fig. 2 which symbolizes the new Cabala. The discovery of a single-stranded configuration of DNA[85] makes complete unravelling more tenable as an alternative model.

For the vehicle of life's continuity, DNA may seem a remarkably undistinguished molecule. Its overall shape is controlled by the uniform deoxyribose phosphate backbone whose monotony then gives X-ray diffraction patterns of high crystallinity. The nucleins themselves are relatively unreactive, hardly different from one to the other, and in DNA introverted and mutually saturated. Nor are any of the hydroxyls of deoxyribose left unsubstituted in the polymer. The structure of DNA befits the solipsism of its function.

The most plausible function of DNA is ultimately to specify the amino acid sequence in proteins. However, as there are twenty amino acids to choose among, there cannot be a 1:1 correspondence of nucleotide to amino acid. Taking account of the code-duplication in complementary structures and the need to indicate spacing of the words in the code sequence, from three to four nucleins may be needed to spell one amino acid[19].

While a protein is also defined by the sequence of its monomeric units, the amino acids, the protein molecule lacks the «aperiodic crystallinity»[80] of DNA. The *differentiæ* of the amino acids vary widely in size, shape, and ionic charge (e.g., H· ; CH_3· ; H_2N·CH_2·CH_2·CH_2·CH_2· ; COOH·CH_2·CH_2· ; HO·C_6H_4·CH_2·) and in the case of proline, bond angles. The biological action of a protein is, therefore, attributable to the shape of the critical surface into which the polypeptide chain folds[73]. The one-dimensional specificity of the DNA must therefore be translated into the three-dimensional specificity of an enzyme or antibody surface. The simplest assumption would be that the amino acid sequence of the extended polypeptide, as it is released from the protein-building template in the cytoplasm, fully determines the folding pattern of the complete protein, which may, of course, be stabilized by non-peptide linkages. If not, we should have to interpose some accessory mechanism to govern the folding of the protein. This issue has reached a climax in speculations about the mechanism of antibody formation. If antibody globulins have a common sequence on which specificity is superimposed by directed folding, an antigen could directly mold the corresponding antibody. However, if sequence determines folding,

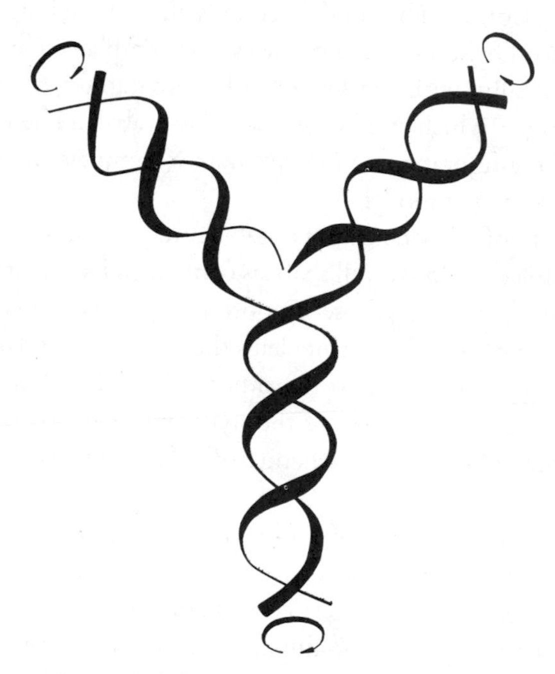

Fig. 2. The scheme of Watson and Crick for DNA replication. «Unwinding and replication proceed *pari passu*. All three arms of the Y rotate as indicated.»[14]

it should in turn obey nucleic information. As this should be independent of antigenic instruction, we may look instead to a purely selective role of antigens to choose among nucleic alternatives which arise by spontaneous mutation[8,50].

The correspondence between amino acids and clusters of nucleotides has no evident basis in their inherent chemical make-up and it now appears more probable that this *code* has evolved secondarily and arbitrarily to be translated by some biological intermediary. The coding relationship would then be analogous to, say, Morse-English (binary linear) to Chinese (pictographic). Encouragingly, several workers have reported the enzymatic reaction of amino acids with RNA fragments[22,75]. Apparently each amino acid has a different RNA receptor and an enzyme whose twofold specificity thus obviates any direct recognition of amino acid by polynucleotides. The alignment of amino-acyl residues for protein synthesis could then follow controlled assembly of their nucleotidates on an RNA template, by analogy with the model for DNA replication. We then visualize the following modes of information transfer:

(1) DNA replication – assembly of complementary deoxyribonucleotides on a DNA template.
(2) Transfer to RNA by some comparable mechanism of assembling ribonucleotides. Our understanding of this is limited by uncertainties of the structure of RNA[16].
(3) Protein synthesis:
    (a) aminoacylation of polynucleotide fragments;
    (b) assembly of the nucleotidates on an RNA template by analogy with step (1);
    (c) peptide condensation of the amino acid residues.

Some workers have suggested that RNA is replicated in step (3) concurrently with protein synthesis, in addition to its initiation from DNA.

The chief difference in primary structure between DNA and RNA is the hydroxylation of C_2' in the ribose, so that a reactive sugar hydroxyl is available in RNA. This may prove to be important in the less ordered secondary structure of RNA, and in its function as an intermediary to protein. It remains to be determined whether the aminoacyl nucleotidates are esterified at C_2' or at C_3' which is also available in the terminal residue. From this resume we may observe that the DNA backbone constitutes an inert but rigid framework on which the differential nucleins are strung. Their spatial constraint lends specificity to the pattern of hydrogen bonding exposed at each level. This extended pattern is a plausible basis for replication; it is difficult to visualize any reagents besides other nucleotides to which this pattern would be relevant. These conditions are quite apt for a memory device – rubber and guncotton are poor choices for a computing tape.

DNA and bacterial mutation

The *ignis fatuus* of genetics has been the specific mutagen, the reagent that would penetrate to a given gene, recognize and modify it in a specific way. Directed mutation has long been discredited for higher organisms and the « molar indeterminacy » of mutation established both for its spontaneous occurrence and its enhancement by X-rays[68]. However, the development of resistance apparently induced by drugs revived illusions that bacterial genes might be alterable, an inference that would inevitably undermine the conception of « gene » for these organisms. No wonder that the mechanism of drug resistance has excited so much controversy[89]!

What sort of molecule could function as a specific mutagen, a reagent for a particular one of the bacterium's complement of genes, which can hardly number less than a thousand targets? On the nucleic hypothesis, the smallest segment capable of this variety would be a *hexa*nucleotide, all possible configurations of which must be discriminated by the specific mutagen. How could this be generally accomplished except by another molecule of conforming length and periodicity, that is an analogous polynucleotide? Certainly there is nothing in the chemistry of penicillin or streptomycin to support their direct intervention in nucleic instructions.

In addition, we recognize no chemical reagent capable of substituting one nuclein for another in the structure of existent DNA. However, as the modification of a nuclein, even to give an unnatural base, could have mutagenic effect, the chief limitation for specific mutagenesis is the recognition of the appropriate target.

Of course the origin of drug resistance, for all its theoretical implications, poses an experimental challenge of its own. Concededly, experiments cannot decide untried situations. Nevertheless, the mechanism whereby resistant mutants arise spontaneously and are then selected by the drug can account for every well-studied case of inherited resistance[5,10]. Furthermore, in favorable instances the spontaneous origin of drug-resistant mutants can be verified unambiguously by contriving to isolate them without their ever being exposed to the drug. One method entails indirect selection. To illustrate its application, consider a culture of *Escherichia coli* containing 10^9 bacteria per ml. By plating samples on agar containing streptomycin, we infer that one per million bacteria or 10^3 per ml produce resistant clones. But to count these clones they were selected in the presence of streptomycin which hypothetically might have induced the resistance. We may however dilute the original bacteria in plain broth to give samples containing 10^5 per ml. Since 10^{-6} of the bacteria are resistant, each sample has a mathematical expectation of 0.1 of including a resistant bacterium. The individual bacterium being indivisible by dilution, nine samples in ten will include no resistants; the tenth will have one, but now augmented to $1:10^5$. Which one this is can be readily determined by retrospective assay on the incubated samples. The procedure can be reiterated to enrich for the resistant organisms until they are obtained in pure culture[11]. The same result is reached more conveniently if we spread the original culture out on a nutrient agar plate rather than distribute samples into separate test tubes. Replica plating, transposing a pattern of surface growth from plate to plate with a sheet of velvet, takes the place of assaying inocula distributed in tubes[53]. Dilution sampling and replica plating are then alternative methods of indirect selection whereby the test line is spared direct contact with the drug. Selection is accomplished by saving sublines whose *sibling* clones show the resistant reaction. This proof merely reinforces the incisive arguments that had already been forwarded by many other authors.

If mutations are not specific responses to the cellular environment, how do they arise? We still have very little information on the proximate causes of spontaneous, even of radiation and chemically induced, mutation. Most mutagenic chemicals are potent alkylating agents, e.g. formaldehyde or nitrogen mustard, which attack a variety of reactive groups in the cell. Similar compounds may occur in normal metabolism and account for part of the spontaneous mutation rate; they may also play a role as chemical intermediates in radiation effects. For the most part, then, studies on mutagenesis, especially by the more vigorous reagents, have told us little about the chemistry of the gene. Probably any agent which can penetrate to the chromosomes and have a localized chemical effect is capable of introducing random errors into the genetic information. If the cell were not first killed by other mechanisms most toxic agents would then probably be mutagenic.

Another class of mutagenic chemicals promises more information: analogues of the natural nucleins which are incorporated into DNA. For example, bromouracil specifically replaces thymine in phage DNA when furnished as bromodeoxyuridine to infected bacteria. Freese has shown, by genetic analyses of the utmost refinement, that the loci of resulting mutations in T4 phage are distributed differently from the mutants of spontaneous origin or those induced by other chemicals[18]. This method presumably maps the locations of thymine in the original DNA. In order to account for wide variations in mutation rate for different loci, further interactions among the nucleotides must be supposed. So far, these studies represent the closest approach to a rational basis for chemical mutagenesis. However, every gene must present many targets to any nuclein analogue and the specificity of their mutagenesis can be detected only in systems where the resolution of genetic loci approximates the spacing of single nucleotides[4]. At present this is feasible only in micro-organisms; similar studies with bacteria and fungi would be of the greatest interest.

More specific effects might result from the insertion of oligo- and polynucleotides, a program which, however, faces a number of technical difficulties: even if the requisite polymers were to be synthesized, there are obstacles to their penetration into cells. The use of DNA extracted from mutant bacteria to transfer the corresponding genetic qualities is discussed as « genetic transduction ».

RNA is the one other reagent that may be expected to recognize particular genes. As yet we have no direct evidence that the transfer of information from DNA to RNA is reversible. However, the anti-mutagenic effect of nuclein ribosides[21,71] may implicate RNA in mutation. The reversibility of DNA $\rightleftharpoons$ RNA information is also implicit in Stent's closely reasoned scheme for DNA replication[82]. The needed experiment is the transfer of DNA information by some isolated RNA. Although not reported, this has probably not been fairly tried.

One motivation for this approach is the difficult problem of finding sources of homogeneous nucleic acids. DNA occurs biologically as sets of different molecules presumably in equimolar proportions. (A useful exception may be a remarkably small phage which seems to be unimolecular[85].) The species of RNA, however, may vary with the predominant metabolic activity of the cells. If so, some molecular species may be sufficiently exaggerated in specialized cells to facilitate their isolation. A purified RNA would have many potential applications, among others as a vehicle for the recognition of the corresponding DNA implied by our theory of information transfer. Pending such advances, *specific* mutagenesis is an implausible expectation.

Adaptive mutations, of which drug resistance is a familiar example, are crucial to the methodology of microbial genetics. Once having connected adaptive variation with gene mutation[78], we could proceed to exploit these systems for the detection of specific genotypes in very large test populations. The genotypes of interest may arise, as in the previous examples, by mutation: the most extensive studies of the physiology of mutation now use these methods for precise assay. For, in order to count the number of mutants of a given kind, it suffices to plate large numbers of bacteria into selective media and count the surviving colonies which appear after incubation. In this way, mutation rates as low as one per 10^9 divisions can be treated in routine fashion.

Genetic recombination in bacteria

The selective isolation of designed genotypes is also the most efficient way to detect genetic recombination. For example, the sexual mechanism of *Escherichia coli* was first exposed when prototrophic (nutritionally self-sufficient) recombinants arose in mixed cultures of two auxotrophic (nutritionally dependent) mutants[35,84,57]. At first only one recombinant appeared per million parental bacteria and the selective procedure was quite obligatory. Later, more fertile strains were discovered which have been most helpful to further analysis[45,51]. This has shown that typical multinucleate vegetative bacteria unite by a conjugation bridge through which part or all of a male genome migrates into the female cell[43]. The gametic cells then separate. The exconjugant male forms an unaltered clone, surviving by virtue of its remaining nuclei. The exconjugant female generates a mixed clone including recombinants[1,46]. Wollman, Jacob, and Hayes[88] have since demonstrated that the paternal chromosome migrates during fertilization in an orderly, progressive way. When fertilization is prematurely interrupted, the chromosome may be broken so that only anterior markers appear among the recombinants. All of the genetic markers are arranged in a single linkage group and their order can be established either by timing their passage during fertilization or by their statistical association with one another among the recombinants. Finally, the transfer of genetic markers can be correlated[27] with the transfer of DNA as inferred from the lethal effect of the radioactive decay of incorporated ^{32}P.

Sexual recombination is one of the methods for analysing the gene–enzyme relationship. The studies so far are fragmentary but they support the conception that the gene is a string of nucleotides which must function as a coherent unit in order to produce an active enzyme[4,15,33,67,90]. However, metabolic blocks may originate through interference with accessory regulatory mechanisms instead of the fundamental capacity to produce the enzyme. For example, many « lactase-negative » mutants have an altered pattern of enzyme induction or a defective permease system for substrate-transport[55,65]. Several laboratories are now working to correlate the relative sequence of genetic defects with the sequence of corresponding alterations in enzyme proteins; this may be the next best approach to the coding problem short of a system where a pure DNA can be matched with its protein phenotype.

At first these recombination experiments were confined to a single strain of *E. coli*, K-12. For many purposes this is a favorable choice of material – perhaps the main advantage is the accumulation of a library of many thousands of substrains carrying the various markers called for by the design of genetic tests. However, strain K-12 is rather unsuitable for serological studies, having lost the characteristic surface antigens which are the basis of serological typing. In any event it would be important to know the breeding structure of the group of enteric bacteria. Systematic studies have therefore been made of the interfertility of different strains of bacteria, principally with a convenient tester of the K-12 strain[39,93]. About one-fourth of the serotype strains of *E. coli* are fertile with strain K-12, and in at least some instances with one another. Whether the remaining three-fourths of strains are completely sterile, or whether they include different, closed, breeding groups (i.e., different genetic species) has not been systematically tested, partly because of the preliminary work needed to establish suitable strains.

E. coli K-12 is also interfertile with a number of strains of *Shigella* spp.[59] Finally, although attempted crosses of *E. coli* with many *Salmonella* types and of *Salmonellas* with one another have usually failed, Baron has demonstrated crosses of *E. coli* with a unique strain of *Salmonella typhimurium*[3]. This may be especially useful as a means of developing hybrids which can be used to bridge the studies of sexuality in *E. coli* and transduction in *Salmonella*.

Bacteria furnish a unique opportunity to study the genetic relationships with their host cells. Another treasure of strain K-12 was for a time hidden: it carries a temperate phage, λ, which is technically quite favorable for genetic work. In accord with Burnet's early predictions we had anticipated that the provirus for λ would behave as a genetic unit but Dr. Esther Lederberg's first crosses were quite startling in their implication that the prophage segregated as a typical chromosomal marker[34]. This was shown quite unambiguously by the segregation of lysogenicity *vs.* sensitivity from persistent heterozygous cells, a test which bypassed the then controversial details of fertilization. The viability of such heterozygous cells supports the hypothesis that lysogenicity depends in part on the development of a cytoplasmic immunity to the cytopathic effects of infecting phage as a result of the establishment of the prophage in a bacterial chromosome. This picture is also brought out by «zygotic induction»[26], whereby the fertilization of a sensitive cell by a prophage-bearing chromosome may provoke the maturation and aggressive growth of the phage and lysis of the complex. On the other hand, the introduction of a sensitive chromosome into a lysogenic bacterium does not result in this induction. The mode of attachment of prophage to its chromosomal site is as unsettled as the general picture of the higher organization of DNA, but most students favor a lateral rather than an axial relationship for the prophage. The isolation of intact chromosomes of bacteria would give a new approach to this question but has so far been inconclusive.

Another infectious particle that has jumped out of our Pandora's box determines the very capacity of *E. coli* to function as a male partner in fertilization[51]. For lack of a better inspiration, we call this particle «F». Two kinds of male strains are now recognized according to whether the F particle has a chromosomal or a cytoplasmic location. F^+ strains, like the original K-12, are highly contagious for F and will rapidly convert populations of female, F^- strains in which they are introduced. Hfr males, on the other hand, have a chromosomal localization of the F factor resulting from occasional transpositions in F^+ strains. The different localization of the F particle in the two cases is diagnosed primarily by the behavior of the particle in crosses. In addition, Hirota and Iijima[24] found that the F particle could be eliminated from F^+ strains by treatment with acridine dyes. Hfr clones are unaffected by acridine orange, but when they revert to the F^+ state, as occasionally happens, the F particle again becomes vulnerable to the dye. The accessibility of extrachromosomal F is paralleled by several other examples of plasmid disinfection (reviewed in Ref. 40); perhaps the most notable is the bleaching of green plant cells by streptomycin[17,76]. No reagent is known to inactivate F or prophage while bound to the chromosome.

The virus λ and the plasmagene F are analogous in many features[28,48]. Their main differences are:

(1) Cytopathogenicity – A bacterium cannot long tolerate λ in its cytoplasmic state and remain viable. The vegetative λ must promptly reduce itself to a chromosomal state or multiply aggressively and lyse the host bacterium. F has no known cytopathic effect.

(2) Maturation – Vegetative λ organizes a protein coat and matures into an infective phage particle. F is known only as an intracellular vegetative element; however, the coat of the F^+ cell may be analogous to that of the phage.

(3) Transmission – λ is infective, i.e., forms a free particle which can penetrate susceptible cells. F is transmitted only by cell-to-cell conjugation.

(4) Fixation – λ has a fore-ordained site of fixation on the bacterial chromosome; F has been identified at a variety of sites. However, this difference may be illusory. In special situations, F does have preferential sites of fixation[77], and generally, translocations of F to different sites are more readily discovered than those of λ would be.

(5) Induction – Exposure of lysogenic bacteria to small doses of ultraviolet light causes the prophage to initiate a lytic cycle with the appearance first of vegetative, then of mature phage[62]. Hfr bacteria make no analogous response. However, the kinetics of the reversion, Hfr $\rightarrow F^+$, has not been carefully studied.

The genetic function of bacteriophages is further exemplified by *transduction* whereby genes are transferred from cell to cell by the intervention of phage particles[42,91]. In our first studies we concluded that the bacterial genes were adventitiously carried in normal phage particles[66,83,92]. Further studies favor the view that the transducing particle has a normal phage coat but a *defective* phage nucleus. This correlation has suggested that a gene becomes transducible when a prophage segment is translocated to its vicinity[2,9,60].

Transduction focuses special attention on the phenomenon of specific pairing of homologous chromosome segments. Howsoever a transduced gene is finally integrated into the bacterial genome, at some stage it must locate the homologous gene in the recipient chromosome. For in transduction, as in sexual recombination, new information is not merely added to the complement; it must also replace the old. This must involve the confrontation of the two homologues prior to the decision which one is to be retained. Synapsis is even more puzzling as between chromosomes whose DNA is in the stabilized double helix and then further contracted by supercoiling. Conceivably gene products rather than DNA are the agency of synaptic pairing.

The integration of a transduced fragment raises further issues[41]. The competing hypotheses are the physical incorporation of the fragment in the recipient chromosome, or the use of its information when new DNA is replicated. The same issues still confound models of crossing-over at meiosis in higher forms; once again the fundamentals of chromosome structure are needed for a resolution.

Virus versus gene

The homology of gene and virus in their fundamental aspects makes their overt differences even more puzzling. According to the simplest nucleic doctrine, DNA plays no active role in its own replication other than furnishing a useful pattern. Various nucleotide sequences should then be equally replicable. What then distinguishes virus DNA, which replicates itself at the expense of the other pathways of cellular anabolism? For the T-even phages, the presence of the unique glucosylated hydroxymethylcytosine furnishes a partial answer[12]. However, other viruses such as λ display no unique constituents; furthermore, as prophage they replicate coordinately with bacterial DNA. Does the virus have a unique element of structure, either chemical or physical, so far undetected? Or does it instruct its own preferential synthesis by a code for supporting enzymes?

The creation of Life

The mutualism of DNA, RNA, and proteins as just reviewed is fundamental to all contemporary life. Viruses are simpler as infective particles but must, of course, parasitize the metabolic machinery of the host cell. What would be the least requirements of a primeval organism, the simplest starting-point for progressive replication of DNA in terms of presently known or conjectured mechanisms? They include at least:

(1) DNA.
(2) The four deoxyribotide pyrophosphates in abundance.
(3) One molecule of the protein, DNA polymerase.
(4) Ribotide phosphates as precursors for RNA.
(5) One molecule of the protein RNA polymerase.
(6) A supply of the twenty amino acyl nucleotidates.
    (Failing these, each of the twenty enzymes which catalyze the condensation of an amino acid and corresponding RNA fragments together with sources of these components.)
(7) One molecule of the protein aminoacyl-RNA polymerase.

In principle, this formidable list might be reduced to a single polynucleotide polymerized by a single enzyme. However, any scheme for the enzymatic synthesis of nucleic acid calls for the coincidence of a particular nucleic acid and of a particular protein. This is a far more stringent improbability than the sudden emergence of an isolated DNA such as many authors have suggested, so much more so that we must look for alternative solutions to the problem of the origin of life. These are of two kinds. The primeval organism could still be a nucleic cycle if nucleic replication occurs, however imperfectly, without the intervention of protein. The polymerase enzyme, and the transfer of information from nucleic acid to protein, would then be evolved refinements. Alternatively, DNA has evolved from a simpler, spontaneously condensing polymer. The exquisite perfection of DNA makes the second suggestion all the more plausible.

The nucleoprotein cycle is the climax of biochemical evolution. Its antiquity is shown by its adoption by all phyla. Having persisted for $\sim 10^9$ years, nucleoprotein may be the most durable feature of the geochemistry of this planet.

At the present time, no other self-replicating polymers are known or understood. Nevertheless, the nucleic system illustrates the basic requirements for such a polymer. It must have a rigid periodic structure in which two or more alternative units can be readily substituted. It must allow for the reversible sorption of specific monomers to the units in its own sequence. Adjacent, sorbed monomers must then condense to form the replica polymer, which must be able to desorb from the template. Primitively, the condensation must be spontaneous but reliable. In DNA, the sorption depends on the hydrogen bonding of nuclein molecules constrained on a rigid helical backbone. This highly specific but subtle design would be difficult to imitate. For the more primitive stages, both of biological evolution and of our own experimental insight, we may prefer to invoke somewhat cruder techniques of complementary attachment. The simplest of these is perhaps the attraction between ionic groups of opposite charge, for example, NH_3^+ and COO^- which are so prevalent in simple organic compounds. If the ingenuity and craftmanship so successfully directed at the fabrication of organic polymers for the practical needs of mankind were to be concentrated on the problem of constructing a self-replicating assembly along these lines I predict that the construction of an artificial molecule having the essential function of primitive life would fall within the grasp of our current knowledge of organic chemistry.

Conclusions

The experimental control of cellular genotype is one of the measures of the scope of genetic science. However, nucleic genes will not be readily approached for experimental manipulation except by reagents that mimic them in periodic structure. Specifically induced mutation, if ever accomplished, will then consist of an act of genetic recombination between the target DNA and the controlled information specified by the reagent. Methods for the step-wise analysis and re-assembly of nucleic acids are likely to be perfected in the near future in pace with the accessibility of nucleic acid preparations which are homogeneous enough to make their use worth-while. For the immediate future, it is likely that the greatest success will attend the use of biological reagents to furnish the selectivity needed to discriminate one among innumerable classes of polynucleotides. Synthetic chemistry is, however, challenged to produce model polymers that can emulate the essential features of genetic systems.

1. T. F. Anderson, Recombination and segregation in Escherichia coli, *Cold Spring Harbor Symp. Quant. Biol.*, 23 (1958) 47–58.
2. W. Arber, Transduction des caractères Gal par le bactériophage Lambda, *Arch. Sci. Geneva*, 11 (1958) 259–338.

3. L. S. Baron, W. F. Carey, and W. M. Spilman, Hybridization of Salmonella species by mating with Escherichia coli, *Abstr. 7th Intern. Congr. Microbiol., Stockholm*, 1958, pp. 50–51.
4. S. Benzer, *The elementary units of heredity*, in *The Chemical Basis of Heredity*, W. D. McElroy and B. Glass (Eds.), Johns Hopkins Press, Baltimore, Md., 1957, pp. 70–93.
5. V. Bryson and W. Szybalski, Microbial drug resistance, *Advan. Genet.*, 7 (1955) 1–46.
6. F. M. Burnet, *Biological Aspects of Infectious Disease*, Cambridge University Press, 1940.
7. F. M. Burnet, *Virus as Organism*, Harvard University Press, Cambridge, Mass., 1945.
8. Sir MacFarlane Burnet, *The Clonal Selection Theory of Immunity* (Abraham Flexner Lectures 1958), Vanderbilt University Press, Nashville, 1959.
9. A. Campbell, Transduction and segregation in Escherichia coli K-12, *Virology*, 4 (1957) 366–384.
10. L. L. Cavalli-Sforza and J. Lederberg, Genetics of resistance to bacterial inhibitors, in Symposium on Growth Inhibition and Chemotherapy, *Intern. Congr. Microbiol., Rome*, 1953, pp. 108–142.
11. L. L. Cavalli-Sforza, Isolation of preadaptive mutants in bacteria by sib selection, *Genetics*, 41 (1956) 367–381.
12. S. S. Cohen, Molecular bases of parasitism of some bacterial viruses, *Science*, 123 (1956) 653–656.
13. F. H. C. Crick, The structure of the hereditary material, *Sci. Am.*, 151 (1954) 54ff.
14. M. Delbrück and G. S. Stent, *On the mechanism of DNA replication*, in *The Chemical Basis of Heredity*, W. D. McElroy and B. Glass (Eds.)., Johns Hopkins Press, Baltimore, Md., 1957, pp. 699–736.
15. M. Demerec, Z. Hartman, P. E. Hartman, T. Yura, J. S. Gots, H. Ozeki, and S. W. Glover, Genetic studies with bacteria, *Carnegie Inst. Wash. Publ.*, 612 (1956).
16. P. Doty, H. Boedtker, J. R. Fresco, R. Haselkorn, and M. Litt, Secondary structure in ribonucleic acids, *Proc. Natl. Acad. Sci. U.S.*, 45 (1959) 482–499.
17. H. von Euler, Einfluss des Streptomycins auf die Chlorophyllbildung, *Kem. Arb.*, 9 (1947) 1–3.
18. E. Freese, The difference between spontaneous and base-analogue induced mutations of phage T_4, *Proc. Natl. Acad. Sci. U.S.*, 45 (1959) 622–633.
19. S. W. Golomb, L. R. Welch, and M. Delbrück, Construction and properties of comma-free codes, *Biol. Medd. Can. Vid. Selsk.*, 23 (1958) 1–34.
20. F. Griffith, The significance of pneumococcal types, *J. Hyg.*, 27 (1928) 113–159.
21. F. L. Haas and C. O. Doudney, A relation of nucleic acid synthesis to radiation-induced mutation frequency in bacteria, *Proc. Natl. Acad. Sci. U.S.*, 43 (1957) 871–883.
22. L. I. Hecht, M. L. Stephenson, and P. C. Zamecnik, Binding of amino acids to the end group of a soluble ribonucleic acid, *Proc. Natl. Acad. Sci. U.S.*, 45 (1959) 505–518.
23. A. D. Hershey and M. Chase, Independent function of viral protein and nucleic acid in growth of bacteriophage, *J. Gen. Physiol.*, 36 (1951) 39–56.

24. Y. Hirota and T. Iijima, Acriflavine as an effective agent for eliminating F-factor in Escherichia coli K-12, *Nature*, 180 (1957) 655–656.
25. R. D. Hotchkiss, The genetic chemistry of the pneumococcal transformations, *Harvey Lectures*, 49 (1955) 124–144.
26. F. Jacob and E. L. Wollman, Sur les processus de conjugaison et de recombinaison chez Escherichia coli. I. L'induction par conjugaison ou induction zygotique, *Ann. Inst. Pasteur*, 91 (1956) 486–510.
27. F. Jacob, Genetic and physical determinations of chromosomal segments in Escherichia coli, *Symp. Soc. Exptl. Biol.*, 7 (1958) 75–92.
28. F. Jacob, Les épisomes, éléments génétiques ajoutés, *Compt. rend.*, 247 (1958) 154–156.
29. T. J. King and R. Briggs, Serial transplantation of embryonic nuclei, *Cold Spring Harbor Symp. Quant. Biol.*, 21 (1956) 271–290.
30. A. J. Kluyver and C. B. van Niel, *The Microbe's Contribution to Biology*, Harvard University Press, Cambridge, Mass. 1956.
31. A. Kornberg, Enzymatic synthesis of deoxyribonucleic acid, *Harvey Lectures*, 53 (1959) 83–112.
32. B. C. J. G. Knight, Bacterial Nutrition, *Med. Res. Council, Spec. Rept. Ser.*, No. 210 (1936).
33. E. M. Lederberg, Allelic relationships and reverse mutation in Escherichia coli, *Genetics*, 37 (1952) 469–483.
34. E. M. Lederberg and J. Lederberg, Genetic studies of lysogenicity in Escherichia coli, *Genetics*, 38 (1953) 51–64.
35. J. Lederberg, Gene recombination and linked segregations in Escherichia coli, *Genetics*, 32 (1947) 505–525.
36. J. Lederberg, Problems in microbial genetics, *Heredity*, 2 (1948) 145–198.
37. J. Lederberg, Bacterial variation, *Ann. Rev. Microbiol.*, (1949) 1–22.
38. J. Lederberg, *Genetic studies with bacteria*, in *Genetics in the 20th Century*, L. C. Dunn (Ed.), MacMillan, New York, 1951, pp. 263–289.
39. J. Lederberg, Prevalence of Escherichia coli strains exhibiting genetic recombination, *Science*, 114 (1951) 68–69.
40. J. Lederberg, Cell genetics and hereditary symbiosis, *Physiol. Rev.*, 32 (1952) 403–430.
41. J. Lederberg, Recombination mechanisms in bacteria, *J. Cellular Comp. Physiol.*, 45, Suppl. 2 (1955) 75–107.
42. J. Lederberg, Genetic transduction, *Am. Scientist*, 44 (1956) 264–280.
43. J. Lederberg, Conjugal pairing in Escherichia coli, *J. Bacteriol.*, 71 (1956) 497–498.
44. J. Lederberg, Comments on gene–enzyme relationship, in *Enzymes: Units of Biological Structure and Function*, O. H. Gaebler (Ed.), Academic Press, New York, 1956, pp. 161–169.
45. J. Lederberg, Viruses, genes, and cells, *Bacteriol. Rev.*, 21 (1957) 133–139.
46. J. Lederberg, Sibling recombinants in zygote pedigrees of Escherichia coli, *Proc. Natl. Acad. Sci. U.S.*, 43 (1957) 1060–1065.
47. J. Lederberg, Genetic approaches to somatic cell variation: Summary comment, *J. Cellular Comp. Physiol.*, 52, Suppl. 1 (1958) 383–402.

48. J. Lederberg, Extranuclear transmission of the F compatibility factor in Escherichia coli, *Abstr. 7th Intern. Congr. Microbiol., Stockholm*, 1958, pp. 58–60.
49. J. Lederberg, Bacterial reproduction, *Harvey Lectures*, 53 (1959) 69–82.
50. J. Lederberg, Genes and antibodies, *Science*, 129 (1959) 1649–1653.
51. J. Lederberg, L. L. Cavalli, and E. M. Lederberg, Sex compatibility in Escherichia coli, *Genetics*, 37 (1952) 720–730.
52. J. Lederberg and P. R. Edwards, Serotypic recombination in Salmonella, *J. Immunol.*, 71 (1953) 232–240.
53. J. Lederberg and E. M. Lederberg, Replica plating and indirect selection of bacterial mutants, *J. Bacteriol.*, 63 (1952) 399–406.
54. J. Lederberg, Infection and heredity, *Symp. Soc. Growth Develop.*, 14 (1956) 101–124.
55. J. Lederberg, E. M. Lederberg, N. D. Zinder, and E. R. Lively, Recombination analysis of bacterial heredity, *Cold Spring Harbor Symp. Quant. Biol.*, 16 (1951) 413–443.
56. J. Lederberg and T. Iino, Phase variation in Salmonella, *Genetics*, 41 (1956) 743–757.
57. J. Lederberg and E. L. Tatum, Gene recombination in Escherichia coli, *Nature*, 158 (1946) 558.
58. J. Lederberg, Sex in bacteria: Genetic studies 1945–1952, *Science*, 118 (1954) 169–175.
59. S. E. Luria and J. W. Burrous, Hybridization between Escherichia coli and Shigella, *J. Bacteriol.*, 74 (1957) 461–476.
60. S. E. Luria, D. K. Fraser, J. N. Adams, and J. W. Burrous, Lysogenization, transduction, and genetic recombination in bacteria, *Cold Spring Harbor Symp. Quant. Biol.*, 23 (1958) 71–82.
61. A. Lwoff, Les facteurs de croissance pour les microorganismes, *Ann. Inst. Pasteur*, 61 (1938) 580–634.
62. A. Lwoff, L. Siminovitch, and N. Kjeldgaard, Induction de la production de bactériophages chez une bactérie lysogène, *Ann. Inst. Pasteur*, 79 (1950) 815–859.
63. W. D. McElroy and B. Glass (Eds.), *The Chemical Basis of Heredity*, Johns Hopkins Press, Baltimore, Md., 1957.
64. M. Meselson and F. W. Stahl, The replication of DNA in Escherichia coli, *Proc. Natl. Acad. Sci. U.S.*, 44 (1958) 671–682.
65. J. Monod, *Remarks on the mechanism of enzyme induction*, in *Enzymes: Units of Biological Structure and Function*, O. H. Gaebler (Ed.), Academic Press Inc., New York, 1956, pp. 7–28.
66. M. L. Morse, E. M. Lederberg, and J. Lederberg, Transduction in Escherichia coli K-12, *Genetics*, 41 (1956) 142–156.
67. M. L. Morse, Transductional heterogenotes in Escherichia coli, *Genetics*, 41 (1956) 758–779.
68. H. J. Muller, The production of mutations, *This Volume*, pp. 154–171.
69. E. Nagel, *The meaning of reduction in the natural sciences*, in *Science and Civilization*, R. C. Stauffer (Ed.), University of Wisconsin Press, Madison, 1949, pp. 99–138.

70. D. L. Nanney, Epigenetic control systems, *Proc. Natl. Acad. Sci. U.S.*, 44 (1958) 712–717.
71. A. Novick, Mutagens and antimutagens, *Brookhaven Symp. Biol.*, *8 (Mutation)*, Office of Tech. Serv., U.S. Dept. Commerce, Washington, D.C., 1956, pp. 201–215.
72. A. Novick and A. McCoy, *Quasi-genetic regulation of enzyme level. Physiological Adaptation*, Am. Physiol. Soc., Washington, D.C., 1958, pp. 140–150.
73. L. Pauling, *Molecular structure and intermolecular forces*, in *The Specificity of Serological Reactions*, K. Landsteiner (Ed.), Harvard University Press, Cambridge, Mass., 1945, pp. 275–293.
74. N. W. Pirie, Some aspects of the origins of life considered in the light of the Moscow International Symposium, *ICSU Rev.*, 1 (1959) 40–48.
75. J. P. Preiss, P. Berg, E. J. Ofengand, F. H. Bergmann, and M. Dieckmann, The Chemical nature of the RNA-amino compound formed by amino acid activating enzymes, *Proc. Natl. Acad. Sci. U.S.*, 45 (1959) 319–328.
76. L. Provasoli, S. H. Hutner, and I. J. Pintner, Destruction of chloroplasts by streptomycin, *Cold Spring Harbor Symp. Quant. Biol.*, 16 (1951) 113–120.
77. A. A. Richter, Determinants of mating type in Escherichia coli, *Ph.D. dissertation*, University of Wisconsin (University Microfilms, Ann Arbor, Mich.), 1959.
78. F. J. Ryan and J. Lederberg, Reverse-mutation in leucineless Neurospora, *Proc. Natl. Acad. Sci. U.S.*, 32 (1946) 163–173.
79. F. Sanger, *Les Prix Nobel en 1958*, Stockholm, 1959, pp. 134–146; *Nobel Lectures Chemistry 1942–1962*, Elsevier, Amsterdam, 1964, pp. 544 ff.
80. E. Schrödinger, *What is Life?* Cambridge University Press, 1944.
81. S. Spiegelman, C. C. Lindegren, and G. Lindegren, Maintenance and increase of a genetic character by a substrate-cytoplasmic interaction in the absence of the specific gene, *Proc. Natl. Acad. Sci. U.S.*, 31 (1945) 95–102.
82. G. S. Stent, Mating in the reproduction of bacterial viruses, *Advan. Virus Res.*, 5 (1958) 95–149.
83. B. A. D. S. Stocker, N. D. Zinder, and J. Lederberg. Transduction of flagellar characters in Salmonella, *J. Gen. Microbiol.*, 9 (1953) 410–433.
84. E. L. Tatum and J. Lederberg, Gene recombination in the bacterium Escherichia coli, *J. Bacteriol.*, 53 (1947) 673–684.
85. I. Tessman, Some unusual properties of the nucleic acid in bacteriophages S13 and ØX174, *Virology*, 7 (1959) 263–275.
86. Sir Alexander Todd, Synthesis in the study of nucleotides, *Les Prix Nobel en 1957*, Stockholm, 1958, pp. 119–133; *Nobel Lectures Chemistry 1942–1962*, Elsevier, Amsterdam, 1964, pp. 522 ff.
87. J. D. Watson and F. H. C. Crick, The structure of DNA, *Cold Spring Harbor Symp. Quant. Biol.*, 23 (1958) 123–131.
88. E. L. Wollman, F. Jacob, and W. Hayes, Conjugation and genetic recombination in Escherichia coli K-12, *Cold Spring Harbor Symp. Quant. Biol.*, 21 (1956) 141–162.
89. G. E. W. Wolstenholme and C. M. O'Connor (Eds.), *Ciba Foundation Symposium on Drug Resistance in Microorganisms*, J. and A. Churchill, Ltd., London, 1957.
90. C. Yanofsky and I. P. Crawford, Effects of deletions, point mutations, suppressor

mutations and reversions on the two components of tryptophan synthetase of Escherichia coli, *Proc. Natl. Acad. Sci. U.S.*, 45 (1959) 1016–1026.

91. N. D. Zinder, Bacterial transduction, *J. Cellular Comp. Physiol.*, 45, Suppl. 2 (1955) 23–49.
92. N. D. Zinder and J. Lederberg, Genetic exchange in Salmonella, *J. Bacteriol.*, 64 (1952) 679–699.
93. F. Ørskov and I. Ørskov, unpublished observations.

The experimental work from my laboratory summarized in this paper has been generously supported by research grants from the National Institutes of Health, U.S. Public Health Service, the National Science Foundation, the Rockefeller Foundation, the Wisconsin Alumni Research Foundation, the University of Wisconsin and, most recently, Stanford University. It is also a pleasure to record my thanks to the Jane Coffin Childs Fund for Medical Research for a research fellowship which supported my first association with Professor E. L. Tatum.

Biography

Joshua Lederberg was born in Montclair, N. J. on May 23, 1925. He was brought up in the Washington Heights District of Upper Manhattan, New York City, where he received his education in Public School 46, Junior High School 164 and Stuyvesant High School. From 1941 to 1944 he studied at Columbia College, where he obtained his B.A. with honours in Zoology (premedical course), and from 1944 to 1946 at the College of Physicians and Surgeons of Columbia University Medical School. Here he carried out part-time research with Professor F. J. Ryan in the Department of Zoology. Subsequently he went to the Department of Microbiology and Botany at Yale University, New Haven, Conn., as Research Fellow of the Jane Coffin Childs Fund for Medical Research and, during 1946–1947, as a graduate student with Professor E. L. Tatum. He was awarded his Ph.D. degree in 1948.

In 1947, he was appointed Assistant Professor of Genetics at the University of Wisconsin, where he was promoted to Associate Professor in 1950 and Professor in 1954. He organized the Department of Medical Genetics in 1957, of which he was Chairman during 1957–1958.

Stanford University Medical School entrusted to him the organization of its Department of Genetics and appointed him Professor and Executive Head in 1959. Since 1962, he has been Director of the Kennedy Laboratories for Molecular Medicine.

Lederberg was Visiting Professor of Bacteriology at the University of California, Berkeley, in 1950; and Fulbright Visiting Professor of Bacteriology at Melbourne University, Australia, in 1957. In the latter year, he was also elected to the National Academy of Sciences (USA).

While at Yale, Lederberg married Esther M. Zimmer in 1946. They have no children. Mrs. Lederberg had obtained her M.A. at Stanford with Professor G. W. Beadle during 1944–1946, and her Ph.D. degree at the University of Wisconsin in 1950. She is working full time as research associate.

Biography

1959

Severo Ochoa

Arthur Kornberg

Severo Ochoa

Enzymatic Synthesis of Ribonucleic Acid

December 11, 1959

I am deeply conscious of the great distinction with which I have been honored and deem it a special privilege to review the recent studies of the biosynthesis of ribonucleic acid on this occasion.

The nucleic acids have considerable biological importance because of their role in cell growth and in the transmission of hereditary characters. As first suggested by the pioneer work of Caspersson and Brachet the former function is performed by ribonucleic acid (RNA) through its participation in the biosynthesis of proteins. The second is carried out by deoxyribonucleic acid (DNA), the main component of the nuclear chromosomes. It is of interest, however, that in certain viruses such as tobacco mosaic, influenza, and poliomyelitis virus, which consist of RNA and protein, RNA is the carrier of genetic information.

Most of the cell's RNA is present in the cytoplasm. There are two kinds of cytoplasmic RNA. One, of relatively small molecular size, is found in the cytoplasmic fluid and is, therefore, referred to as soluble RNA; the other, of much higher molecular weight, is a component of the microsomal ribonucleo-protein particles. Both play an essential role in protein synthesis. There is, in addition, a small amount of RNA in the cell nucleus, most of it located in the nucleolus. There are indications that most, if not all, of the cytoplasmic RNA is synthesized in the nucleus and subsequently transported to the cytoplasm. In transmitting genetic information, nuclear DNA is supposed to determine the nature of the nuclear RNA which, on entering the cytoplasm, determines in turn the nature of the proteins synthesized.

Although notable advances had been made in our knowledge of the way in which the nucleotides, the nucleic acid building stones, are synthesized, little was known until recently of the mechanism of synthesis of the giant molecules of the nucleic acids themselves. We owe our present information to the discovery of enzymes capable of catalyzing the synthesis of RNA and DNA in the test tube from simple, naturally occurring precursors. These precursors are the nucleoside di- and triphosphates, the nucleotide moieties of which undergo polymerization with release of orthophosphate in the first case or of pyrophosphate in the second.

Polynucleotide phosphorylase

In 1955 we isolated a bacterial enzyme capable of catalyzing the synthesis of high molecular weight polyribonucleotides from nucleoside diphosphates with release of orthophosphate[1,2]. The reaction, which requires magnesium ions and is reversible, can be formulated by the equation

$$n\,X - R - P - P \overset{Mg^{++}}{\rightleftharpoons} (X - R - P)_n + n\,P$$

where R stands for ribose, P — P for pyrophosphate, P for orthophosphate, and X for one or more bases including, among others, adenine, hypoxanthine, guanine, uracil, or cytosine. In the reverse direction the enzyme brings about a cleavage of polyribonucleotides by phosphate, i.e. a phosphorolysis, to yield ribonucleoside diphosphates. The reaction is similar to the reversible synthesis and cleavage of polysaccharides, catalyzed by phosphorylase; for this reason the new enzyme was named polynucleotide phosphorylase. Because of its reversibility, the reaction leads to an incorporation or « exchange » of orthophosphate into the terminal phosphate group of nucleoside diphosphates. It was through this exchange that polynucleotide phosphorylase was discovered by the use of radioactive phosphate. In our early work with Grunberg-Manago, polynucleotide phosphorylase was partially purified, by use of the radiophosphate exchange reaction, from the microorganism *Azotobacter vinelandii*. The enzyme has the unique feature of catalyzing not only the synthesis of RNA from mixtures of the four naturally occurring ribonucleoside diphosphates, but also that of non-naturally occurring polyribonucleotides containing only one, two, or three different kinds of nucleotides in their chains. The nature of the product depends on the kind and variety of the nucleoside diphosphate substrates utilized for the synthesis[3,4]. Table 1 lists the main types of polyribonucleotides which have been prepared with polynucleotide phosphorylase. The preparation of polyribothymidylic acid from synthetic ribothymidine diphosphate has recently been reported[5].

Structure of polynucleotides – In joint experiments with L. A. Heppel[6-8] it has been established that the synthetic polyribonucleotides conform in all respects to the structural pattern of natural RNA. Thus, it was found by degradation with alkali, or such enzymes as snake venom and spleen phosphodiesterase or pancreatic ribonuclease, that they consist of linear chains in which the component nucleoside units are linked to one another through 3′,5′-phosphodiester bridges. The structural identity with natural RNA can briefly be illustrated by the effect of pancreatic ribonuclease on the synthetic polymer poly AU. From what is known of the action of ribonuclease on RNA, poly AU would be degraded at the points indicated by the arrows in Fig. 1. Uridylic acid (uridine 3′-monophosphate) would be released as the

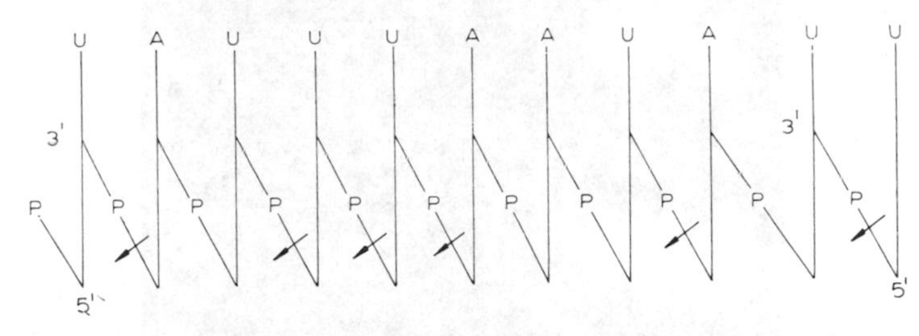

Fig. 1. Scheme of cleavage of poly AU by pancreatic ribonuclease. The vertical lines represent the ribose residues; A and U represent the bases adenine and uracil, respectively. The cleavage points are indicated by arrows[3].

Table 1. Synthetic polyribonucleotides.

Substrate	*Polymer*
ADP	Poly A
GDP	Poly G
UDP	Poly U
CDP	Poly C
IDP	Poly I
Ribothymidine diphosphate	Polyribothymidylic acid
ADP + UDP	Poly AU
GDP + CDP	Poly GC
ADP + GDP + CDP + UDP	Poly AGUC (synthetic RNA)

only mononucleotide, along with a series of small oligonucleotides, each consisting of one uridylic acid residue and one or more adenylic acid residues. Fig. 2 is an ultraviolet print of a chromatogram illustrating the separation of mono-, di,- tri-, tetra-, and pentanucleotides each with decreasing *Rf* values from a ribonuclease digest of poly AU. The individual spots were eluted and the respective oligonucleotides identified following hydrolysis with alkali. Digestion of synthetic RNA with ribonuclease yields, along with uridylic and cytidylic acids, mixtures of oligonucleotides which, as far as

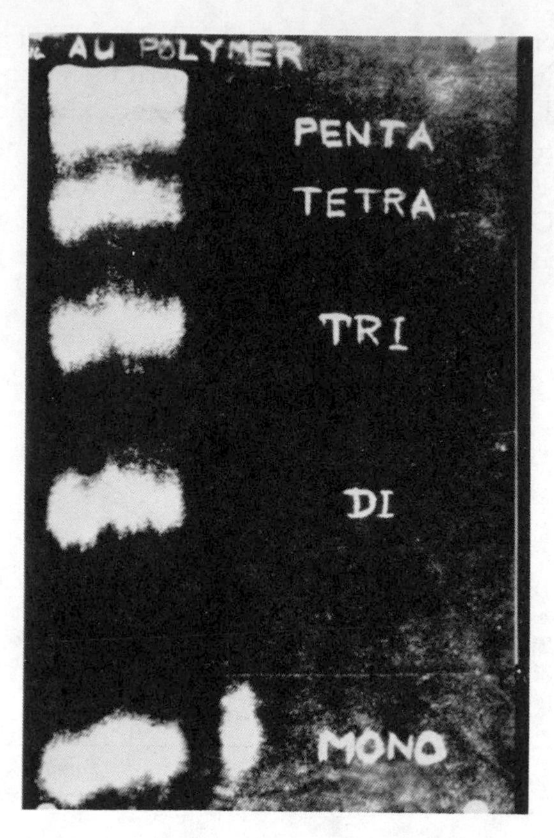

Fig. 2. Products of hydrolysis of poly AU by ribonuclease (L. A. Heppel). Origin is at top of chromatogram. Spot at bottom center is a marker of uridine 3′-phosphate[3].

they have been identified, are identical to those obtained from natural RNA under the same conditions.

The question whether a given nucleotide species is linked to the various other nucleotides in the polynucleotide chains, as is the case with natural RNA, can be easily answered through degradation of synthetic RNA labeled with radioactive phosphate[9]. Fig. 3 presents the structure of a polynucleotide (poly A*GUC), prepared from a mixture of adenosine diphosphate labeled with ^{32}P in the first phosphate group (adenosine-^{32}P-P) and non-labeled guanosine-, uridine-, and cytidine diphosphates. If the labeled adenylic acid is randomly distributed as shown, hydrolysis of such a polymer with snake venom phosphodiesterase (Fig. 3 A) will yield nucleoside 5′-monophosphates of which only adenosine 5′-monophosphate will be labeled. Hydrolysis with spleen phosphodiesterase, on the other hand (Fig. 3 B) will release

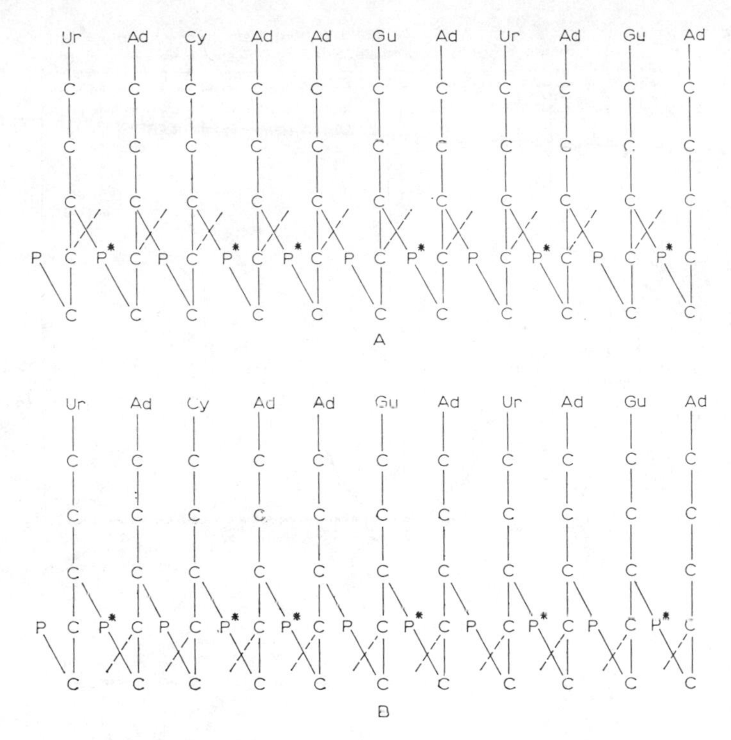

Fig. 3. Scheme of hydrolysis of ^{32}P-labeled RNA (poly A*GUC) by snake venom (*A*) or spleen (*B*) phosphodiesterase. The bonds hydrolyzed are indicated by dashed lines. The asterisk denotes ^{32}P labeling. Ad, Gu, Ur and Cy represent the bases adenine, guanine, uracil and cytosine, respectively[9].

nucleoside 3′-monophosphates all of which will be labeled. That such is indeed the case is shown in Fig. 4. The separation of the hydrolysis products in this experiment was effected by ion-exchange chromatography.

Although large variations in the relative proportions of the different nucleoside diphosphate substrates has a rather marked influence on the nucleotide composition of the resulting polymer, when synthetic RNA is prepared

Table 2. Base ratios of natural and synthetic RNA[9].

Base	*Azotobacter RNA*	*Poly AGUC (sample 1)*	*Poly AGUC (sample 2)*
Adenine	1.00	1.00	1.00
Guanine	1.30	1.16	1.25
Uracil	0.73	0.66	0.69
Cytosine	0.90	0.72	0.73

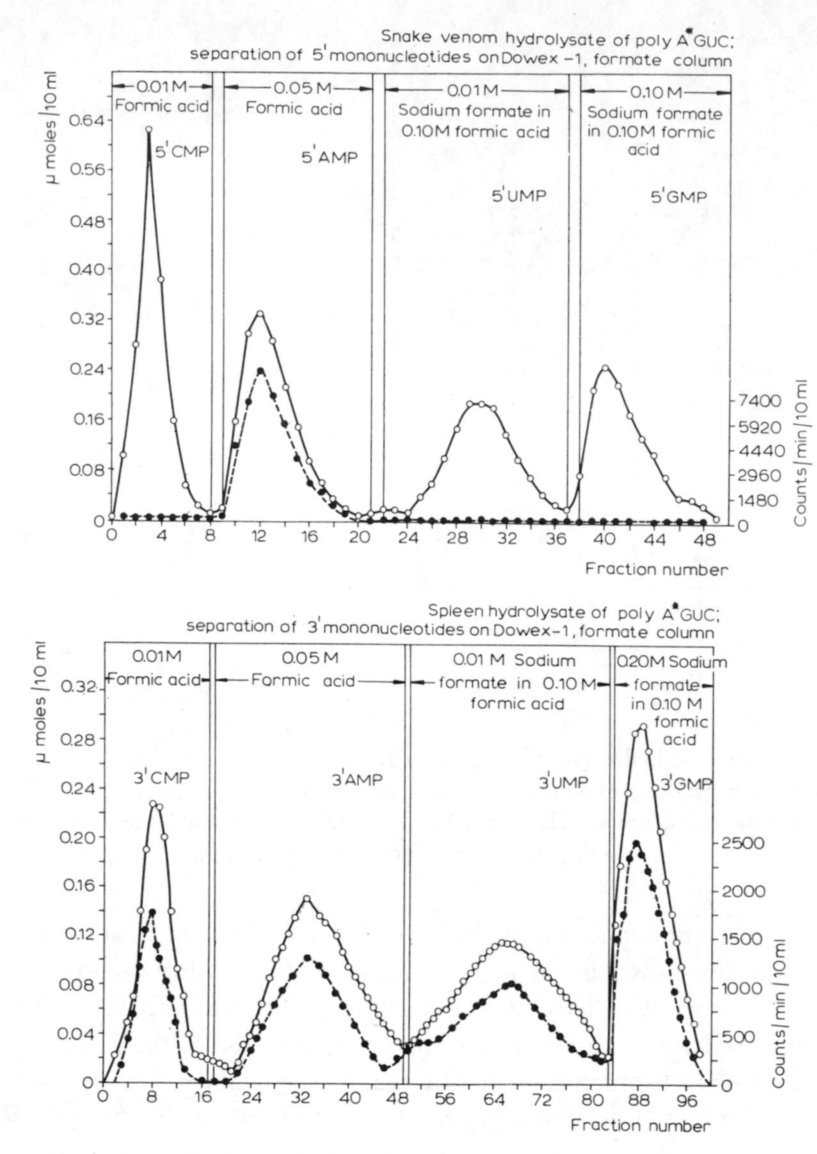

Fig. 4. Hydrolysis of poly A*GUC with snake venom (*top*) or spleen (*bottom*) phosphodiesterase with separation of mononucleotides by ion-exchange chromatography. Plot of nucleotide concentration (*solid lines* -○-○-○-) and radioactivity (*dashed lines* -●-●-●-) against effluent fraction number[12].

from equimolar mixtures of adenosine, guanosine, uridine, and cytidine diphosphate, the nucleotide composition of the product is very similar to that of natural *Azotobacter* RNA. This is shown in Table 2 which gives the base ratios of *Azotobacter* RNA and of two different samples of the synthetic product. It is noteworthy that the base ratios differ widely from unity in spite of the fact that equimolar concentrations of the nucleoside diphosphate precursors were used.

The synthetic polyribonucleotides also resemble natural RNA in size. Their molecular weight varies between about 30,000 and one to two millions. The sedimentation constant of samples of synthetic RNA was similar to that of RNA isolated from whole *Azotobacter* cells. Polynucleotides containing only one kind of nucleotide unit, such as polyadenylic or polycytidylic acid, are often of very large size and confer high viscosity to their solutions. It is possible to follow the course of synthesis visually by the marked increase in viscosity that takes place on incubation of nucleoside diphosphates with a few micrograms of enzyme. Natural RNA has a nonspecific biological activity which is also exhibited by synthetic RNA. The latter is as effective as the former in stimulating the formation of streptolysin S (a lecithinase) by hemolytic streptococci[10].

Reaction mechanism – To study the mechanism of action of polynucleotide phosphorylase it was essential to obtain highly purified preparations of the enzyme. The last step (Fig. 5) was chromatography on a hydroxyl apatite column, a method developed by Tiselius and collaborators[11]. Only through chromatography was it possible to separate the enzyme from a contaminating yellow protein of unknown nature which, as seen in the figure, is eluted at higher buffer concentrations than is phosphorylase. The purification of the enzyme after chromatography is some six hundred fold over the initial extract of *Azotobacter* cells. The highly purified enzyme contains a firmly bound oligonucleotide[12] which cannot be removed by such methods as treatment with charcoal or ribonuclease. Since it has so far not been possible to remove the oligonucleotide without destroying the enzyme protein it remains undecided whether this compound, which represents about 3.5 per cent of the enzyme, is a prosthetic group or a contaminant. The oligonucleotide consists of about twelve nucleotide residues of adenylic, guanylic, uridylic and cytidylic acid in roughly the same molar ratios as in *Azotobacter* RNA. It can be isolated after denaturation of the protein with perchloric acid or phenol.

Since the *Azotobacter* enzyme can synthesize RNA as well as polynucleotides with only one nucleotide species, it is important to decide whether one is dealing with a mixture of enzymes, each reacting with a different nucleoside diphosphate, or with a single enzyme. Although this question cannot be answered unequivocally, if the activity toward each of several nucleoside diphosphates increases to the same extent on purification, it is very likely that a single enzyme is involved. That this is so is shown in Table 3 in which the

Table 3. ^{32}P-exchange assay with different nucleoside diphosphates* (S. Mii and S. Ochoa, unpublished).

Enzyme fraction	*ADP*	*GDP*	*UDP*	*CDP*	*IDP*
$Ca_3(PO_4)_2$ gel eluate	39	33	46	35	41
After chromatography	316	300	370	280	370
Purification ratio	8.1	9.1	8.0	8.0	9.0

* Results, micromoles of ^{32}P-exchange per mg of enzyme protein, 15 minutes at 30°.

activity toward each of five nucleoside diphosphates, as assayed by the radioactive phosphate exchange method, is seen to increase approximately to the same extent between two advanced purification steps. The same was true for earlier stages of purification.

With partially purified preparations of polynucleotide phosphorylase po-

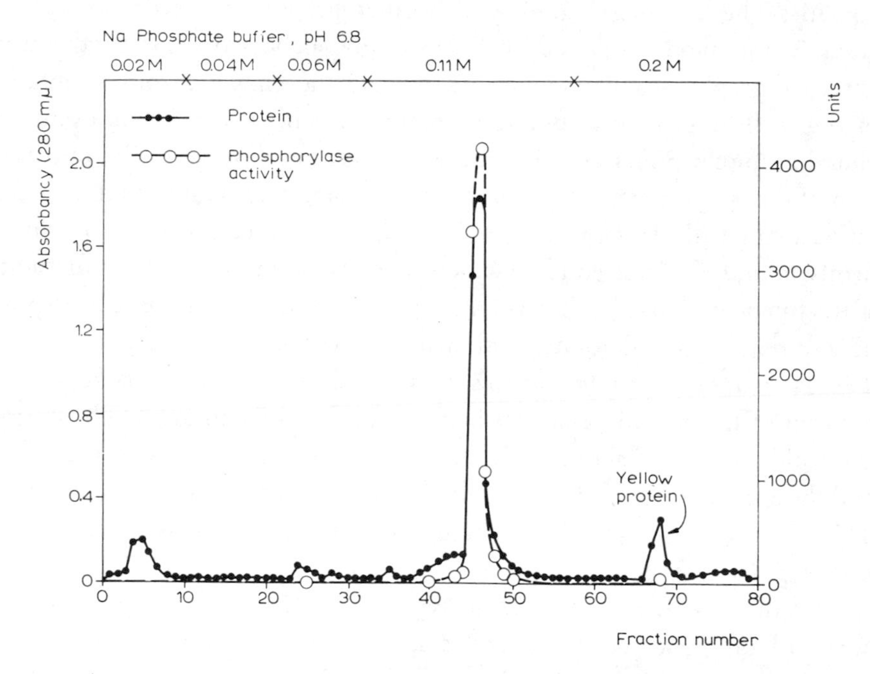

Fig. 5. Chromatography of *Azotobacter* polynucleotide phosphorylase on hydroxyl apatite (S. Ochoa and S. Mii, unpublished).

lynucleotide synthesis starts immediately after adding the enzyme to an otherwise complete system. This is not the case, however, with highly purified preparations. In this case, there is mostly a more or less pronounced lag period although eventually the reaction starts and gradually increases in rate. Equilibrium is not reached even after many hours of incubation. The reaction rate is markedly stimulated by addition of small amounts of oligo- or polynucleotides which, as is the case with glycogen in polysaccharide synthesis from glucose-1-phosphate by (polysaccharide) phosphorylase, serve as primers of the reaction. The priming effect of oligoribonucleotides was discovered by Heppel and collaborators[13], the priming by polynucleotides was disclosed in our laboratory[14]. The main oligonucleotide primers used have been di-, tri-, or tetraadenylic acids isolated as reaction products of hydrolysis of polyadenylic acid by a nuclease from liver nuclei[15].

The priming by oligonucleotides is not specific. The oligoadenylic acids can prime the synthesis of polyadenylic and polyuridylic acid as well as that of RNA or any of the other polynucleotides. Priming by polynucleotides, on the other hand, shows a certain degree of specificity. Thus, polyadenylic acid primes only its own synthesis and the same is true of polyuridylic acid. On the other hand RNA, whether natural or synthetic, primes the synthesis of RNA as well as that of polyadenylic acid or polyuridylic acid. Very curious and completely unexplained is the effect of polycytidylic acid which primes the synthesis of all the polynucleotides os far tested (Table 4). Fig. 6 shows the priming of RNA synthesis with highly purified polynucleotide phosphorylase by triadenylic acid (TAA), liver RNA, and polycytidylic

Table 4. Specificity of priming by polynucleotides (Ref. 14, and unpublished data).

Polymer synthesized	*Effect of*					
	Poly A	*Poly U*	*Poly C*	*Poly I*	*Poly AU*	*RNA (natural or synthetic)*
Poly A	+	–	+	o	+	+
Poly U	–	+	+	o	+	+
Poly C	–	–	+			–
Poly G	o	o	+			
Poly I	–	o	+	+		
Poly AU					+	
Poly AGUC	o	o	+			+

+ Denotes priming, – denotes inhibition, o denotes no effect. Blank spaces, no information.

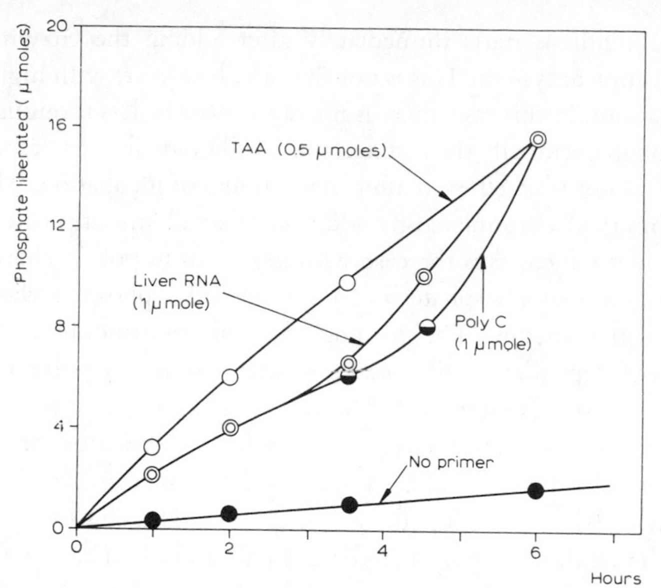

Fig. 6. Priming of RNA synthesis (S. Ochoa and S. Mii, unpublished). Diphosphates (A+G+U+C), 5 μ moles each; Mg^{++}, 2.0 μ moles; enzyme, 140 μ g; TRIS, pH 8.1, 150 μ moles; volume, 1.0 ml.

acid (poly C). The effect of different polynucleotides on the synthetic reaction is illustrated in Table 4. It is to be noted that polyadenylic acid is not only without effect but actually inhibits the slow synthesis of polyuridylic acid which occurs in the absence of added primer; the converse is also true. It should further be noted that the synthesis of polycytidylic acid is primed only by polycytidylic acid itself.

The mechanism of priming by polynucleotides and the cause and significance of the specificity just described are as yet unexplained. The possibility that polyribonucleotides might function as templates for their own replication has not been substantiated experimentally. On the other hand, the mode of action of the oligonucleotide primers has been elucidated in elegant experiments by Heppel and his collaborators[13,16]. They proved that the oligonucleotides serve as nuclei for growth of the polynucleotide chains by successive addition of mononucleotide units. It may be recalled that polysaccharide phosphorylase acts in a similar way by catalyzing the successive addition of glucosyl residues to the terminal units of a polysaccharide primer. When the synthesis of polyuridylic acid is primed by di- or triadenylic acid, the new polynucleotide chains should consist of a number of uridylic acid residues preceded by two or three adenylic acid residues. This is shown schemat-

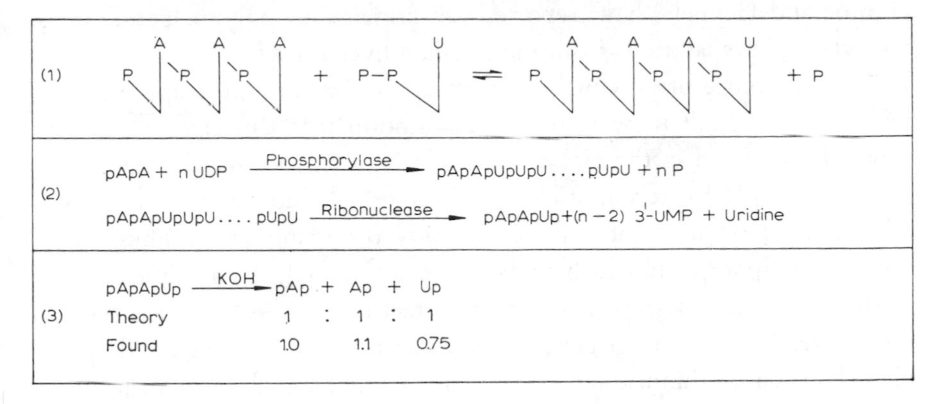

Fig. 7. Mechanism of priming by oligonucleotides (From data of M. F. Singer, L. A. Heppel, and R. J. Hilmoe[13]).

ically in the upper and middle portions of Fig. 7. That such is in fact the case was proved following ribonuclease digestion of polyuridylic acid synthesized in the presence of diadenylic acid as primer[13,16]. As shown in the middle portion of Fig. 7, ribonuclease would release a trinucleotide (pApApUp) from the origin, one molecule of uridine from the end, and a number of uridine 3′-monophosphate residues from the remainder of the chain. The trinucleotide was isolated by chromatography and digested with potassium hydroxide. As shown in the lower section of Fig. 7, this should yield equimolar amounts of adenosine 5′,3′-diphosphate, adenosine 3′-monophosphate and uridine 3′-monophosphate. The figure also shows that the amounts actually recovered were in good agreement with the theory.

It appears justified to conclude that polynucleotide phosphorylase may be unable to start the synthesis of a polynucleotide chain from nucleoside diphosphates as the only reactants and that the presence of an oligonucleotide to serve as nucleus for growth of new polynucleotide chains is probably indispensable. If the enzyme could be obtained completely free of oligonucleotide primer material, it might prove to be completely inactive in the absence of added oligonucleotides.

Biological significance – Polynucleotide phosphorylase is widely distributed in bacteria. The enzyme has been partially purified from microorganisms other than *Azotobacter vinelandii*[17–19] and polyribonucleotides have been synthesized with these enzyme preparations. Indications have also been obtained for the presence of the enzyme in green leaves[17]. On the other hand, it has been difficult to detect the enzyme in animal tissues. Recently, however,

Hilmoe and Heppel[20] have reported the presence of polynucleotide phosphorylase in preparations from mammalian liver nuclei.

The occurrence of polynucleotide phosphorylase in nature appears to be widespread enough to warrant the assumption that this enzyme may be generally involved in the biosynthesis of RNA. This possibility appears to be strengthened by recent studies with ribonucleoside diphosphates containing different analogues of the naturally occurring bases. Thus 5-bromouridine diphosphate which contains 5-bromouracil, an analogue of uracil or thymine which in experiments with intact bacterial cells is incorporated into DNA but not into RNA, is not a substrate of polynucleotide phosphorylase, while thiouridine diphosphate containing the uracil analogue thiouracil, which in similar experiments is incorporated into RNA but not into DNA, is a substrate for phosphorylase. In line with these observations is the fact that azauridine diphosphate containing the uracil analogue azauracil which is not incorporated *in vivo* into RNA, does not react with polynucleotide phosphorylase[21]. However, in spite of the fact that polynucleotide phosphorylase can bring about the synthesis of an RNA of the same nucleotide composition and molecular weight as that isolated from *Azotobacter*, and despite the intriguing specificity of priming by polyribonucleotides, there is so far no evidence that the enzyme is able to replicate the primer molecules as is the case with Kornberg's DNA polymerase. Since there must be mechanisms in the cell capable of synthesizing individual ribonucleic acid molecules with a determined nucleotide sequence, it is likely that enzymes capable of performing this function are still to be discovered.

Enzymes catalyzing the addition of a few nucleotide units to a preexisting RNA chain have recently been described from various laboratories[22–24]. These enzymes catalyze the transfer of cytidylic and adenylic acid residues from the corresponding nucleoside triphosphates to the end of polynucleotide chains with release of pyrophosphate. There are no indications that they can bring about a net synthesis of RNA. Other investigators[25,26] have described the incorporation of ribonucleotides in the interior of RNA chains by particulate cell fractions from animal tissues. In a recent report[26] a fraction from rat liver nuclei brought about this incorporation optimally from a mixture of all four ribonucleoside triphosphates of adenosine, guanosine, uridine and cytidine, and the reaction was markedly decreased after treatment with ribonuclease suggesting requirement for an RNA primer. These experiments suggest that an enzyme similar to Kornberg's DNA polymerase might be involved.

ELECTROPHORETIC PATTERNS
Glycine buffer, pH 9.6

SEDIMENTATION PATTERNS
Phosphate buffer, pH 7, $\Gamma/2 = 0.2$
59780 rpm

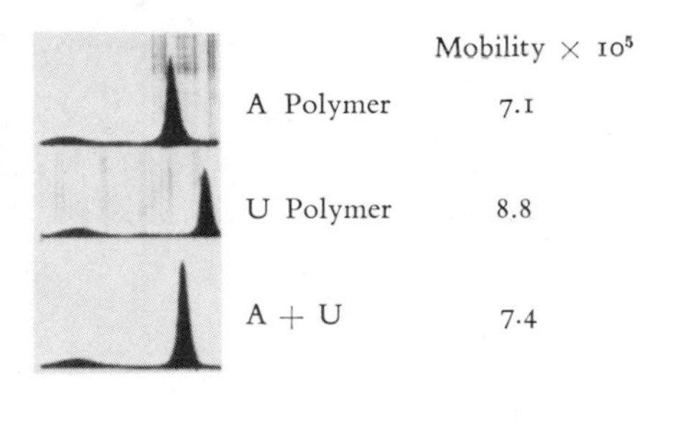

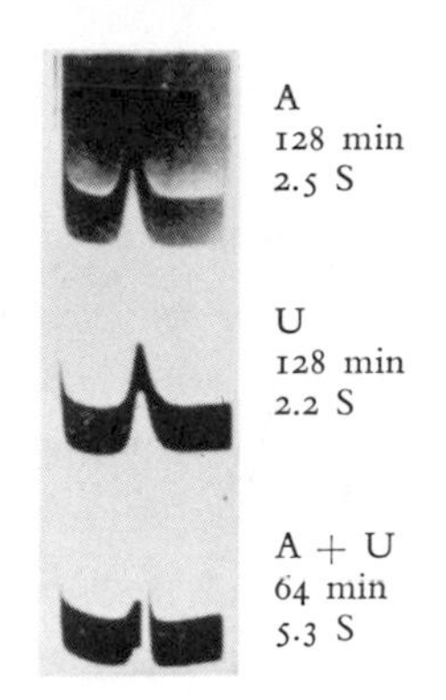

Fig. 8. Electrophoresis and sedimentation patterns of poly A, poly U and poly A + U[27].

Polynucleotide interactions

Physical-chemical studies on a variety of synthetic polyribonucleotides have thrown much light on their macromolecular structure and may greatly further our understanding of the biological properties of RNA and DNA. Warner, in our laboratory, found that polyadenylic and polyuridylic acids interact in solution to form a stable complex[27]. At suitable pH values this complex migrates on electrophoresis with a sharp single boundary of mobility intermediate between that of poly A and poly U. On ultracentrifugation, it has a higher sedimentation constant than that of the parent polynucleotides (Fig. 8). Warner further observed that formation of the complex is accompanied by a marked decrease of the absorption of ultraviolet light (Fig. 9). Formation of the complex can in fact be observed visually through the marked increase in viscosity that takes place on mixing solutions of poly A and poly U.

X-ray diffraction studies of Rich and collaborators[28] showed that fibers made from the poly A + U complex give rise to a crystalline pattern not unlike that of DNA[29], indicating that this complex has a double-stranded helical structure. Further studies[30] demonstrated that the strands are held together by hydrogen bonds between the complementary pairs of bases adenine and uracil. These observations provided the first experimental dem-

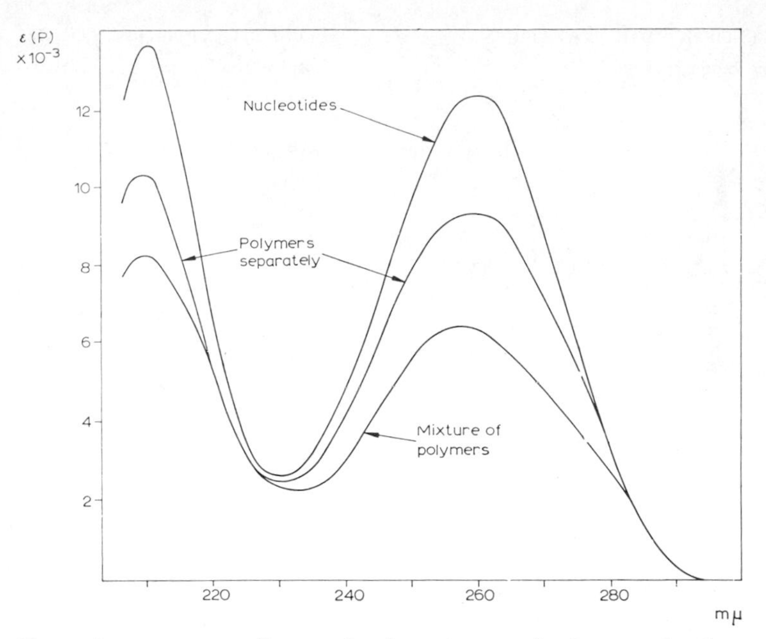

Fig. 9. Absorption spectrum of an equimolar mixture of poly A and poly U. The *upper curve* refers to the mononucleotides obtained by alkaline hydrolysis of the mixture. The *middle curve* is that calculated for the separately measured spectra of the individual polymers. The *lower curve* is the measured curve for the mixture of polymers[27].

onstration that polynucleotides can interact to form double-stranded helical structures similar to that proposed by Watson and Crick for DNA and, moreover, that the same configuration may be assumed by RNA.

The decrease of ultraviolet light absorption accompanying polynucleotide complex formation has facilitated an extensive study of interactions between different polynucleotides by Rich and collaborators[31–33]. These investigators made the further important observation that triple-stranded helical polyribonucleotide structures can also be formed. As illustrated in Fig. 10, a complex consisting of one poly A and two poly U molecules is formed in the presence of Mg^{++}. The figure shows the optical density at 259 mμ on adding increasing amounts of a solution of poly U to one of poly A. In the absence of magnesium the absorbancy goes through a minimum when equimolecular amounts of the two polynucleotides are present; in the presence of magnesium, minimum absorbancy is reached when the solution contains 2 moles of poly U per mole of poly A. Fig. 11 illustrates the type of hydrogen bonding postulated by Rich for the poly A + U and the poly A + U + U

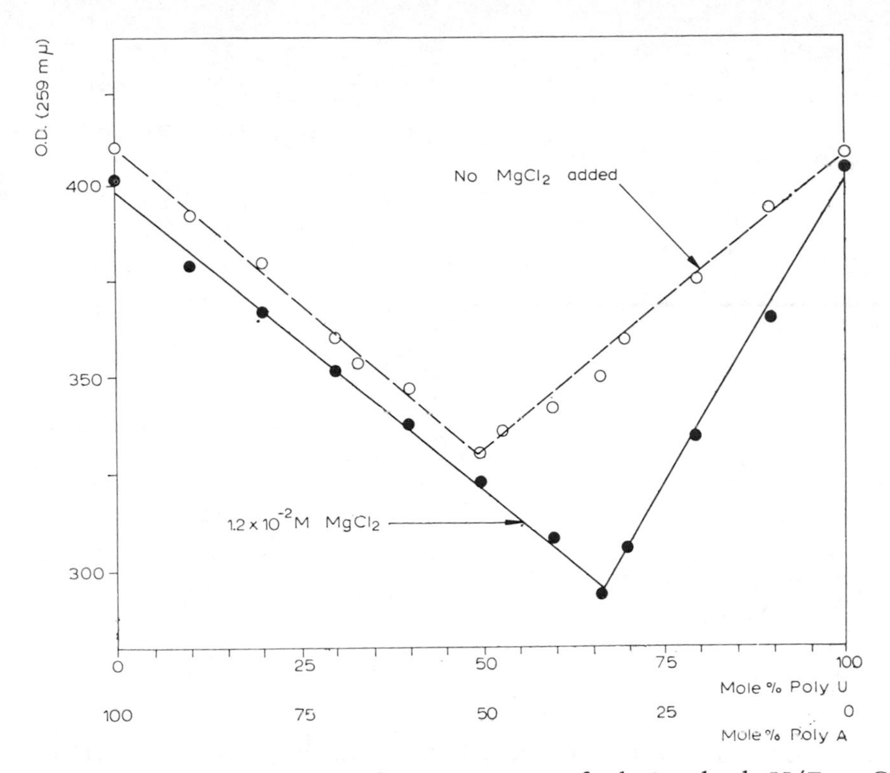

Fig. 10. Optical density (× 10³) of various mixtures of poly A and poly U (From G. Felsenfeld, D. R. Davies, and A. Rich, *J. Am. Chem. Soc.*, 79 (1957) 2023.

complexes. A double-stranded helical complex between poly A and polyribo-thymidylic acid has more recently been obtained by Rich and collaborators. Its formation is illustrated diagrammatically in Fig. 12. I am greatly indebted to Dr. Rich for permission to use these illustrations. Double-stranded helical structures can be formed by poly-adenylic acid in solution. The elegant experiments of Doty and collaborators[34] have shown that, above

Table 5. Double- and triple-stranded polynucleotide complexes.

Double-stranded	Poly A + Y
	Poly A + polyribo-thymidylic acid
	Poly I + C
	Poly A + A
Triple-stranded	Poly A + U + U
	Poly A + I + I
	Poly I + I + I

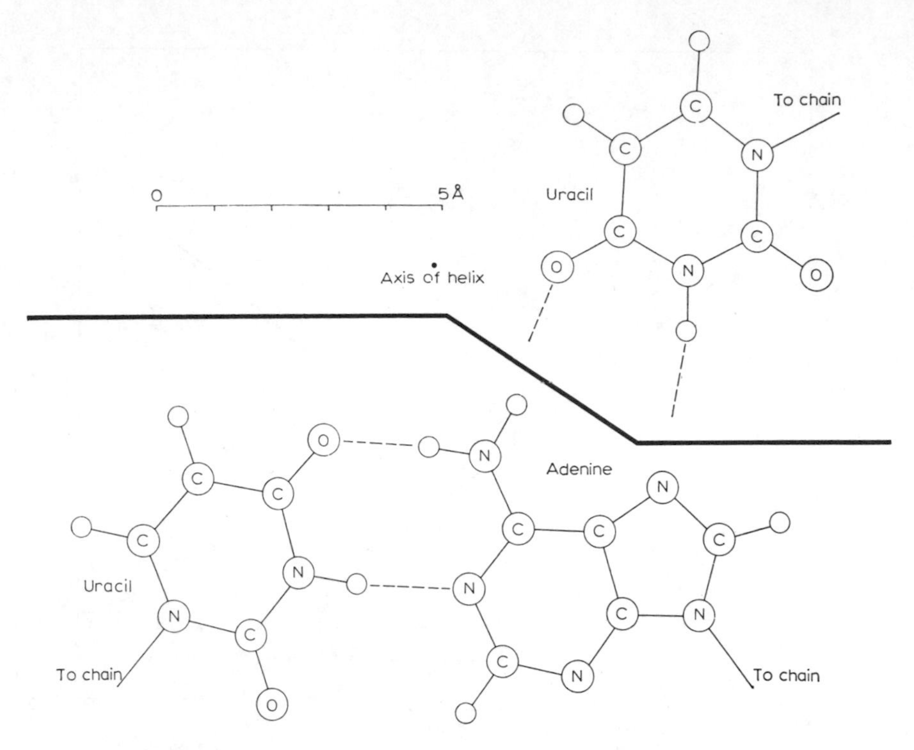

Fig. 11. Hydrogen-bonding system in poly A + U and poly A + U + U. The hydrogen bonds are represented by dashed lines (*Courtesy of A. Rich*).

neutral pH, poly A exists in solution as a random coil but the chains are joined to form a helical complex at pH values below neutrality. This transformation is reversible and has a sharp transition point. The various types of polyribonucleotide complexes thus far obtained are listed in Table 5.

The above studies may be of importance for a better understanding of the physical-chemical interactions underlying the role of DNA in cell division. These interactions may also play a role in the biological behavior of RNA. Since there are good indications that the genetic information stored in DNA is first transmitted to RNA, it is believed that DNA may function as a template for RNA replication. Oncoming ribonucleotide residues could be linked into an organized polynucleotide chain growing in the helical groove of the native, double-stranded DNA molecule to form a triple-stranded helix. Alternatively, a single-stranded DNA template might be used to give a double-stranded DNA-RNA helix[35–37].

The work of Kornberg and his collaborators[38] has given us deep insight

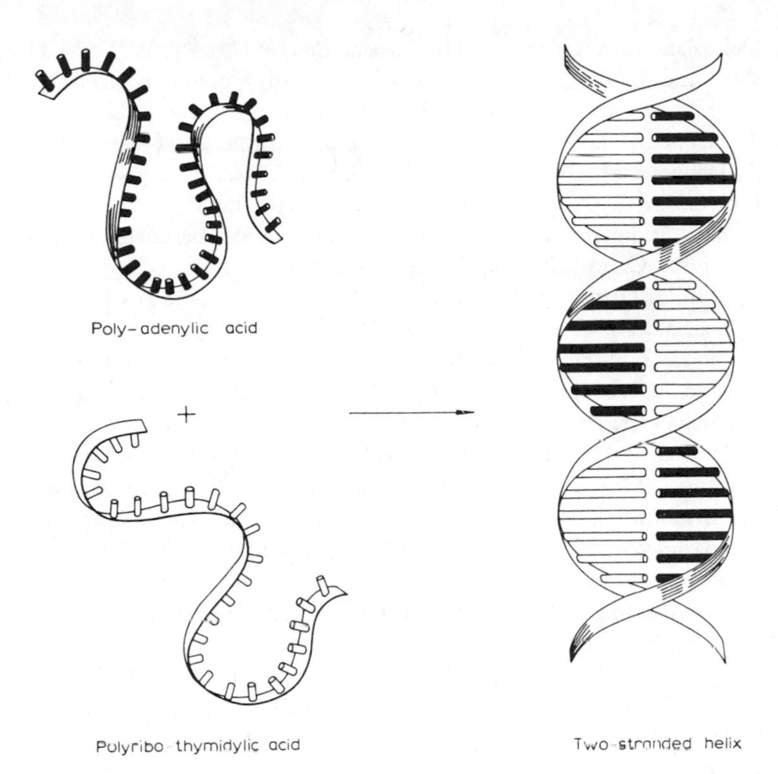

Fig. 12. Diagram showing the combination of two random coil molecules of poly-adenylic and polyribo-thymidylic acid to form a double-stranded helix. The bases are represented by short rods (*Courtesy of A. Rich*).

into the mode of replication of DNA and may lead in the not too distant future to the synthesis of genetic material in the test tube. Since RNA is the genetic material of some viruses, the work reviewed in this lecture may help to pave the way for the artificial synthesis of biologically active viral RNA and the synthesis of viruses. These particles are at the threshold of life and appear to hold the clue to a better understanding of some of its most fundamental principles.

1. M. Grunberg-Manago and S. Ochoa, *J. Am. Chem. Soc.*, 77 (1955) 3165.
2. M. Grunberg-Manago, P. J. Ortiz, and S. Ochoa, *Science*, 122 (1955) 907.
3. S. Ochoa, *Federation Proc.*, 15 (1956) 832.
4. M. Grunberg-Manago, P. J. Ortiz, and S. Ochoa, *Biochim. Biophys. Acta*, 20 (1956) 269.
5. B. E. Griffin, A. Todd, and A. Rich, *Proc. Natl. Acad. Sci. U.S.*, 44 (1958) 1123.

6. S. Ochoa and L. A. Heppel, in *The Chemical Basis of Heredity*, W. D. McElroy and B. Glass (Eds.), Johns Hopkins Press, Baltimore, 1957, p. 615; S. Ochoa, in *Cellular Biology, Nucleic Acids, and Viruses*, N.Y. Acad. of Sci., Spec. Publ., 5 (1957) 191.
7. L. A. Heppel, P. J. Ortiz, and S. Ochoa, *J. Biol. Chem.*, 229 (1957) 679.
8. L. A. Heppel, P. J. Ortiz, and S. Ochoa, *J. Biol. Chem.*, 229 (1957) 695.
9. P. J. Ortiz and S. Ochoa, *J. Biol. Chem.*, 234 (1959) 1208.
10. K. Tanaka, F. Egami, T. Hayashi, J. E. Winter, A. W. Bernheimer, S. Mii, P. J. Ortiz, and S. Ochoa, *Biochim. Biophys. Acta*, 25 (1957) 663.
11. A. Tiselius, S. Hjertén, and Ö. Levin, *Arch. Biochem. Biophys.*, 65 (1956) 132.
12. S. Ochoa, *XI Conseil de Chimie Solvay, Bruxelles*, June, 1959.
13. M. F. Singer, L. A. Heppel, and R. J. Hilmoe, *Biochim. Biophys. Acta*, 26 (1957) 447.
14. S. Mii and S. Ochoa, *Biochim. Biophys. Acta*, 26 (1957) 445; S. Ochoa, S. Mii, and M. C. Schneider, *Proc. Intern. Symp. Enzyme Chem., Tokyo Kyoto*, 2 (1957) 44.
15. L. A. Heppel, P. J. Ortiz, and S. Ochoa, *Science*, 123 (1956) 415.
16. M. F. Singer, L. A. Heppel, and R. J. Hilmoe, *J. Biol. Chem.*, 235 (1960) 738.
17. D. O. Brummond, M. Staehelin, and S. Ochoa, *J. Biol. Chem.*, 225 (1957) 835; S. Ochoa, in *Recent Progress in Microbiology, Symp. 7th Intern. Congr. Microbiol., Stockholm*, 2 (1958) 122.
18. U. Z. Littauer and A. Kornberg, *J. Biol. Chem.*, 226 (1957) 1077.
19. R. F. Beers, Jr., *Nature*, 177 (1956) 790.
20. R. J. Hilmoe and L. A. Heppel, *J. Am. Chem. Soc.*, 79 (1957) 4810.
21. J. Škoda, J. Kára, A. Šormova, and F. Šorm, *Biochim. Biophys. Acta*, 33 (1959) 579.
22. L. I. Hecht, P. C. Zamecnik, M. L. Stephenson, and J. F. Scott, *J. Biol. Chem.*, 233 (1958) 954.
23. E. S. Canellakis, *Biochim. Biophys. Acta*, 25 (1957) 217.
24. J. Hurwitz, A. Bresler, and A. Kaye, *Biochem. Biophys. Res. Commun.*, 1 (1959) 3.
25. E. Goldwasser, *J. Am. Chem. Soc.*, 77 (1955) 6083.
26. S. B. Weiss and L. Gladstone, *J. Am. Chem. Soc.*, 81 (1959) 4118.
27. R. C. Warner, *Federation Proc.*, 15 (1956) 379; *J. Biol. Chem.*, 229 (1957) 711.
28. A. Rich and D. R. Davies, *J. Am. Chem. Soc.*, 78 (1956) 3548; A. Rich, in *The Chemical Basis of Heredity*, W. D. McElroy and B. Glass (Eds.), Johns Hopkins Press, Baltimore, 1957, p. 557.
29. J. D. Watson and F. H. C. Crick, *Nature*, 171 (1953) 737.
30. R. C. Warner and E. Breslow, *Symp. 4th Intern. Congr. Biochem., Vienna*, 9 (1958) 157.
31. G. Felsenfeld and A. Rich, *Biochim. Biophys. Acta*, 26 (1957) 457; A. Rich, in *Cellular Biology, Nucleic Acids, and Viruses*, N.Y. Academy of Sciences, Spec. Publ., 5 (1957) 186.
32. A. Rich, *Biochim. Biophys. Acta*, 29 (1958) 502.
33. A. Rich, *Nature*, 181 (1958) 521.
34. J. R. Fresco and P. Doty, *J. Am. Chem. Soc.*, 79 (1957) 3928; J. R. Fresco and E. Klemperer, *Ann. N.Y. Acad. Sci.*, 81 (1959) 730.
35. G. Stent, *Advan. Virus Res.*, 5 (1958) 138.
36. G. Zubay, *Nature*, 182 (1958) 1290.
37. A. Rich, *Ann. N.Y. Acad. Sci.*, 81 (1959) 709.
38. A. Kornberg, *This Volume*, pp. 665–680.

Biography

Severo Ochoa was born at Luarca, Spain, on September 24th, 1905. He is the son of Severo Ochoa, a lawyer and business man, and Carmen de Albornoz.

Ochoa was educated at Malaga College, where he took his B.A. degree in 1921. His interest in biology was greatly stimulated by the publications of the great Spanish neurologist, Ramón y Cajal, and he went to the Medical School of the University of Madrid, where he obtained his M.D. degree (with honours) in 1929. While he was at the University he was Assistant to Professor Juan Negrin and he paid, during the summer of 1927, a visit to the University of Glasgow to work under Professor D. Noel Paton.

After graduating in 1929 Ochoa went, with the aid of the Spanish Council of Scientific Research, to work under Otto Meyerhof at the Kaiser Wilhelm Institut für Medizinische Forschung at Heidelberg. During this period he worked on the biochemistry and physiology of muscle, and his outlook and training were decisively influenced by Meyerhof.

In 1931, Ochoa was appointed Lecturer in Physiology at the University of Madrid, a post he held until 1935. In 1932 he went to the National Institute for Medical Research, London, where he worked with Dr. H. W. Dudley on his first problem in enzymology.

Returning to Madrid in 1934, he was appointed Lecturer in Physiology and Biochemistry there and later became Head of the Physiology Division of the Institute for Medical Research, Madrid. In 1936 he was appointed Guest Research Assistant in Meyerhof's Laboratory at Heidelberg, where he worked on some of the enzymatic steps of glycolysis and fermentation. In 1937 he held a Ray Lankester Investigatorship at the Plymouth Marine Biological Laboratory and from 1938 until 1941 he worked on the biological function of vitamin B_1 with Professor R. A. Peters at Oxford University, where he was appointed Demonstrator and Nuffield Research Assistant.

While he was at Oxford he became interested in the enzymatic mechanisms of oxidative metabolism and in 1941 he went to America and worked, until 1942, at the Washington University School of Medicine, St. Louis, where he was appointed Instructor and Research Associate in Pharmacology and worked with Carl and Gerty Cori on problems of enzymology. In 1942 he was appointed Research Associate in Medicine at the New York University School of Medicine and there subsequently became Assistant Professor of Biochemistry (1945), Professor of Pharmacology (1946), Professor of Biochemistry (1954), and Chairman of the Department of Biochemistry. In 1956 he became an American citizen.

Ochoa's research has dealt mainly with enzymatic processes in biological oxidation and synthesis and the transfer of energy. It has contributed much to the knowledge of the basic steps in the metabolism of carbohydrates and fatty acids, the utilization of carbon dioxide, and the biosynthesis of nucleic acids. It has included the biological functions of vitamin B_1, oxidative phosphorylation, the reductive carboxylation of ketoglutaric and pyruvic acids, the photochemical reduction of pyridine nucleotides in photosynthesis, condensing enzyme – which is the key enzyme of the Krebs citric acid cycle, polynucleotide phosphorylase and the genetic code.

Ochoa holds honorary degrees of the Universities of St. Louis (Washington University), Glasgow, Oxford, Salamanca, Brazil, and the Wesleyan University. He is Honorary Professor of the University of San Marcos, Lima, Peru. He was awarded the Neuberg Medal in Biochemistry in 1951, the Medal of the Société de Chimie Biologique in 1959, and the Medal of New York University in the same year. He is a member of several learned societies in the U.S.A., Germany, Japan, Argentina, Uruguay, and Chile, and President of the International Union of Biochemistry.

In 1931 Ochoa married Carmen Garcia Cobian.

ARTHUR KORNBERG

The Biologic Synthesis of Deoxyribonucleic Acid

December 11, 1959

The knowledge drawn in recent years from studies of bacterial transformation[1] and viral infection of bacterial cells[2,3] combined with other evidence[3], has just about convinced most of us that deoxyribonucleic acid (DNA) is the genetic substance. We shall assume then that it is DNA which not only directs the synthesis of the proteins and the development of the cell but that it must also be the substance which is copied so as to provide for a similar development of the progeny of that cell for many generations. DNA, like a tape recording, carries a message in which there are specific instructions for a job to be done. Also like a tape recording, exact copies can be made from it so that this information can be used again and elsewhere in time and space.

Are these two functions, the expression of the code (protein synthesis) and the copying of the code (preservation of the race) closely integrated or are they separable? What we have learned from our studies over the past five years and what I shall present is that the replication of DNA can be examined and at least partially understood at the enzymatic level even though the secret of how DNA directs protein synthesis is still locked in the cell.

DNA structure

First I should like to review very briefly some aspects of DNA structure which are essential for this discussion. Analysis of the composition of samples of DNA from a great variety of sources and by many investigators[4] revealed the remarkable fact that the purine content always equals the pyrimidine content. Among the purines, the adenine content may differ considerably from the guanine, and among the pyrimidines, the thymine from the cytosine. However, there is an equivalence of the bases with an amino group in the 6-position of the ring, to the bases with a keto group in the 6-position. These facts were interpreted by Watson and Crick[5] in their masterful hypothesis on the structure of DNA. As shown in Fig. 1, they proposed in

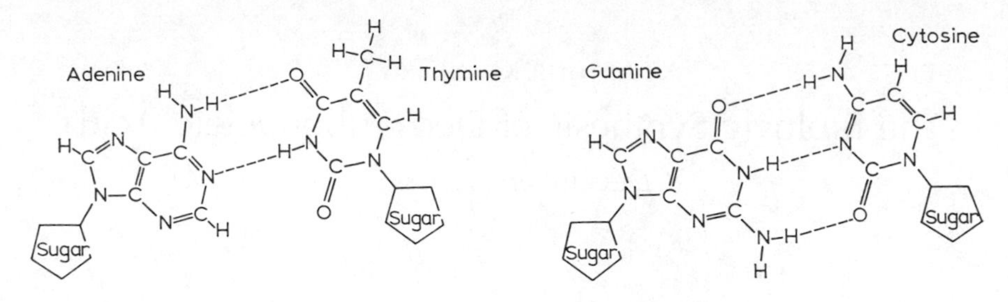

Fig. 1. Hydrogen bonding of bases.

connection with their double-stranded model for DNA, to be discussed presently, that the 6-amino group of adenine is linked by hydrogen bonds to the 6-keto group of thymine and in a like manner guanine is hydrogen-bonded to cytosine, thus accounting for the equivalence of the purines to the pyri-

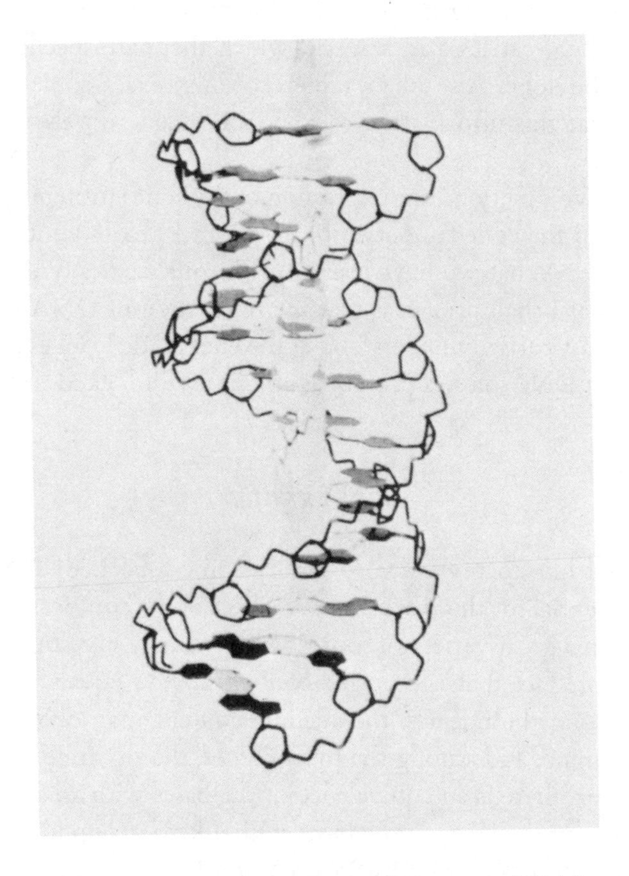

Fig. 2. Double-helical structure of DNA (Watson and Crick model).

midines. On the basis of these considerations and the results of X-ray crystallographic measurements by Wilkins and associates[6], Watson and Crick proposed a structure for DNA in which two long strands are wound about each other in a helical manner. Fig. 2 is diagrammatic representation of a fragment of a DNA chain about ten nucleotide units long. According to physical measurements, DNA chains are on the average 10,000 units long. We see here the deoxypentose rings linked by phosphate residues to form the backbone of the chain; the purine and pyrimidine rings are the planar structures emerging at right angles from the main axis of the chain. Fig. 3 is a more detailed molecular model[7] and gives a better idea of the packing of the atoms in the structure. The purine and pyrimidine bases of one chain are bonded to the pyrimidine and purine bases of the complementary chain by the hydrogen bonds described in Fig. 1. The X-ray measurements have indicated that the space between the opposing chains in the model agrees with

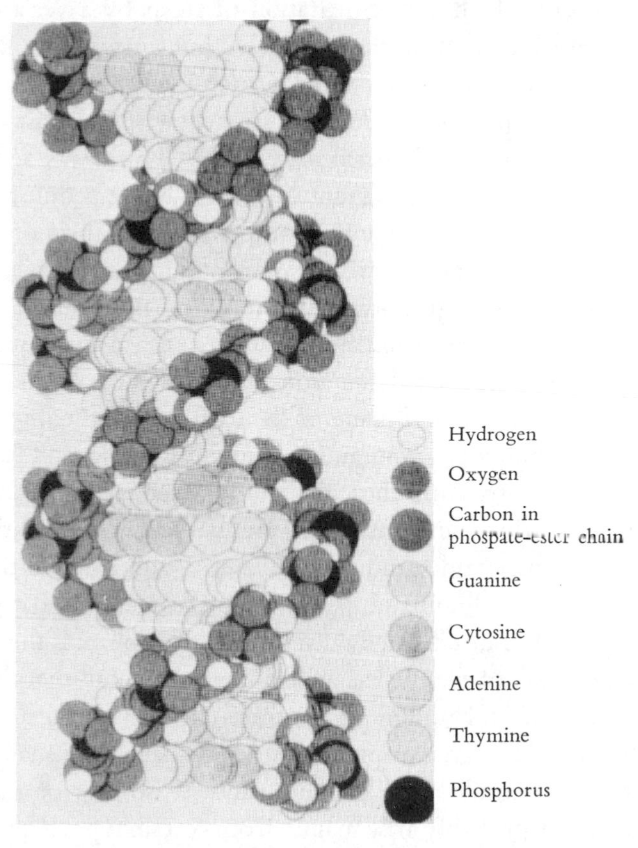

Fig. 3. Molecular model of DNA (After M. Feughelman, *et al.*[7]).

the calculated value for the hydrogen-bond linkage of a purine to a pyrimidine; it is too small for two purines and too large for two pyrimidines. Most rewarding from the biological point of view, the structure provides a useful model to explain how cellular replication of DNA may come about. For, if you imagine that these two chains separate and that a new chain is formed complementary to each of them, the result will be two pairs of strands, each pair identical to the original parent duplex and identical to each other.

Enzymatic approach to the problem of DNA replication

Although we have in the Watson and Crick proposal a mechanical model of replication, we may at this point pose the question: « What is the chemical mechanism by which this super molecule is built up in the cell? » Some sixty years ago the alcoholic fermentation of sugar by a yeast cell was a « vital » process inseparable from the living cell, but through the Buchner discovery of fermentation in extracts and the march of enzymology during the first half of this century, we understand fermentation by yeast as a, now familiar, sequence of integrated chemical reactions. Five years ago the synthesis of DNA was also regarded as a « vital » process. Some people considered it useful for biochemists to examine the combustion chambers of the cell, but tampering with the very genetic apparatus itself would surely produce nothing but disorder. These gloomy predictions were not justified then, nor are similar pessimistic attitudes justified now with regard to the problems of cellular structure and specialized function which face us. High adventures in enzymology lie ahead and many of the explorers will come from the training fields of carbohydrate, fat, amino acid and nucleic acid enzymology.

I feel now, as we did then, that for an effective approach to the problem of nucleic acid biosynthesis it was essential to understand the biosynthesis of the simple nucleotides and the coenzymes and to have these concepts and methodology well in hand. It was from these studies that we developed the conviction that an activated nucleoside 5′-phosphate is the basic biosynthetic building block of the nucleic acids[8]. You will recall that the main pathways of purine and pyrimidine biosynthesis all lead to the nucleoside 5′-phosphate[8]; they do not, except as salvage mechanisms, usually include the free bases or nucleosides. While the 2′ and 3′ isomers of the nucleotides are known, they probably arise mainly from certain types of enzymatic degradation of the nucleic acids. You will also recall from the biosynthesis of coen-

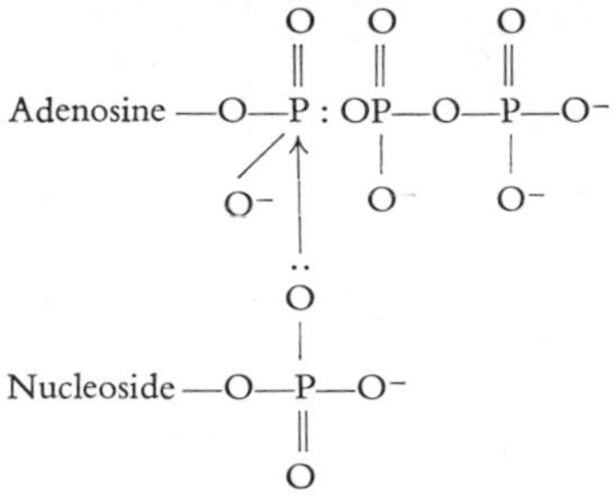

```
                 O    O      O
                 ‖    ‖      ‖
Adenosine —O—P : OP—O—P—O⁻
               /↑    |      |
            O⁻ |    O⁻     O⁻
               |
               :
               O
               |
Nucleoside —O—P—O⁻
               ‖
               O
```

Fig. 4. Nucleophilic attack of a nucleoside monophosphate on ATP.

zymes[9], the simplest of the nucleotide condensation products, that it is ATP which condenses with nicotinamide mononucleotide to form diphosphopyridine nucleotide, with riboflavin phosphate to form FAD, with pantetheine phosphate to form the precursor of coenzyme A and so forth. This pattern has been amplified by the discovery of identical mechanisms for the activation of fatty acids and amino acids and it has been demonstrated further that uridine, cytidine and guanosine coenzymes are likewise formed from the respective triphosphates of these nucleosides.

This mechanism (Fig. 4), in which a nucleophilic attack[10] on the pyrophosphate-activated adenyl group by a nucleoside monophosphate leads to the formation of a coenzyme, was adopted as a working hypothesis for studying the synthesis of a DNA chain. As illustrated in Fig. 5, it was postulated that the basic building block is a deoxynucleoside 5′-triphosphate which is attacked by the 3′-hydroxyl group at the growing end of a polydeoxynucleotide chain; inorganic pyrophosphate is eliminated and the chain is lengthened by one unit. The results of our studies on DNA synthesis, as will be mentioned presently, are in keeping with this type of reaction.

Properties of the DNA-synthesizing enzyme

First let us consider the enzyme and comment on its discovery[8,11,12]. Mixing the triphosphates of the four deoxynucleosides which commonly occur in DNA with an extract of thymus or bone-marrow or of *Escherichia coli* would not be expected to lead to the net synthesis of DNA. Instead, as might be expected, the destruction of DNA by the extracts of such cells and tissues was by far the predominant process and one had to resort to the use of more subtle devices for detection of such a biosynthetic reaction. We used a ^{14}C-

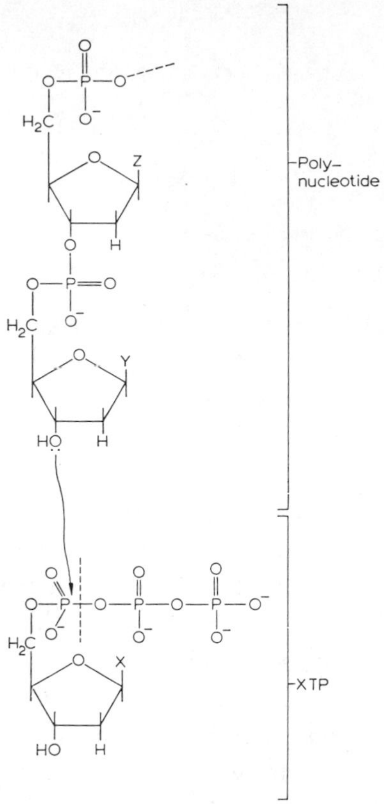

Fig. 5. Postulated mechanism for extending a DNA chain.

labeled substrate of high specific radioactivity and incubated it with ATP and extracts of *Escherichia coli*, an organism which reproduces itself every 20 minutes. The first positive results represented the conversion of only a very small fraction of the acid-soluble substrate into an acid-insoluble fraction (50 or so counts out of a million added). While this represented only a few $\mu\mu$moles of reaction, it was something. Through this tiny crack we tried to drive a wedge, and the hammer was enzyme purification[13]. This has been and still is a major preoccupation. Our best preparations are several thousand-fold enriched with respect to protein over the crude extracts, but there are still contaminating quantities of one or more of the many varieties of nuclease and diesterase present in the *coli* cell. The occurrence of what appears to be a similar DNA-synthesizing system in animal cells as well as in other bacterial species has been observed[14]. We must wait for purification of the enzymes from these sources in order to make valid comparisons with the *coli* system.

The requirements for net synthesis of DNA with the purified *coli* enzyme[15] are shown in the equation in Fig. 6. All four of the deoxynucleotides which form the adenine-thymine and guanine-cytosine couples must be present. The substrates must be the tri- and not the diphosphates and only the deoxy sugar compounds are active. DNA which must be present may be obtained

$$\begin{matrix} n & TPPP \\ n & dGPPP \\ n & dAPPP \\ n & dCPPP \end{matrix} + DNA \rightleftharpoons DNA - \begin{bmatrix} TP \\ dGP \\ dAP \\ dCP \end{bmatrix}_n + 4(n)\,PP$$

Fig. 6. Equation for enzymatic synthesis of DNA.

from animal, plant, bacterial or viral sources and the best indications are that all these DNA samples serve equally well in DNA synthesis provided their molecular weight is high. The product, which we will discuss in further detail, accumulates until one of the substrates is exhausted and may be 20 or more times greater in amount than the DNA added and thus is composed to the extent of 95% or more of the substrates added to the reaction mixture. Inorganic pyrophosphate is released in quantities equimolar to the deoxynucleotides converted to DNA.

Should one of these substrates be omitted, the extent of reaction is diminished by a factor of greater than 10^4 and special methods are now required for its detection. It turns out that when one of the deoxynucleotide substrates is lacking, an extremely small but yet significant quantity of nucleotide is linked to the DNA primer. We have described this so-called «limited reaction»[16], and have shown that under these circumstances a few deoxynucleotides are added to the nucleoside ends of some of the DNA chains but that further synthesis is blocked for lack of the missing nucleotide. Current studies suggest to us that this limited reaction represents the repair of the shorter strand of a double helix in which the strands are of unequal length, and that the reaction is governed by the hydrogen bonding of adenine to thymine and of guanine to cytosine.

When all four triphosphates are present, but when DNA is omitted, no reaction at all takes place. What is the basis for this requirement? Does the DNA function as a primer in the manner of glycogen or does it function as a template in directing the synthesis of exact copies of itself? We have good reason to believe that it is the latter and as the central and restricted theme of this lecture I would like to emphasize that it is the capacity for base-pairing by hydrogen bonding between the preexisting DNA and the nucleotides added as substrates that accounts for the requirement for DNA.

The enzyme we are studying is thus unique in present experience in taking directions from a template – it adds the particular purine or pyrimidine sub-

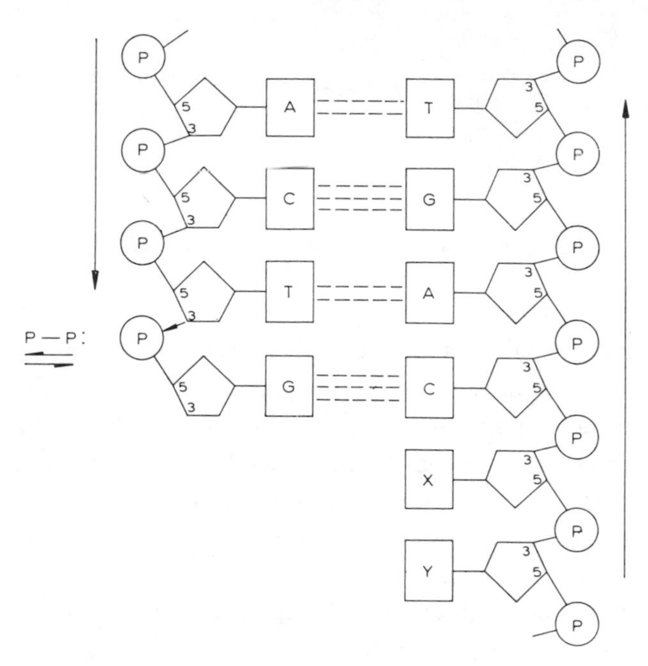

Fig. 7. Mechanism for enzymatic DNA replication.

strate which will form a hydrogen-bonded pair with a base on the template (Fig. 7). There are five major lines of evidence that I would like to present to support this thesis.

Physical properties of enzymatically synthesized DNA

The first line of evidence is derived from studies of the physical nature of the DNA produced by the enzyme. It may be mentioned again that in these descriptions as in those of the chemical nature of DNA, to be discussed shortly, 90–95% of the DNA sample comes from the substrates used in the reaction. From collaborative studies with Dr. Howard K. Schachman, to whom we are greatly indebted, it can be said that the enzymatic product is indistinguishable from high-molecular weight, double-stranded DNA isolated from nature[17]. It has sedimentation coefficients in the neighbourhood of 25, reduced viscosities of 40 deciliters per gram and, on the basis of these measurements, it is believed to be a long, stiff rod with a molecular weight of about 6 million. Upon heating the DNA, the rod collapses and the molecule becomes a compact, randomly coiled structure; it may be inferred that the hydrogen bonds holding the strands together have melted and this is borne out by characteristic changes in the viscometric and optical properties of the molecule. Similar results are found upon cleavage of the molecule by pancreatic deoxyribonuclease. In all these respects the enzymatically synthesized DNA is indistinguishable from the material isolated from nature, and may thus be presumed to have a hydrogen-bonded structure similar to that possessed by natural DNA.

Would one imagine that the collapsed jumbled strands of heated DNA would serve as a primer for DNA synthesis? Very likely one would think not. Guided by intuition derived from everyday experience with a jumbled strand of twine one might regard this as a hopeless template for replication. It turns out that the collapsed DNA is an excellent primer and the non-viscous, randomly coiled, single-stranded DNA leads to the synthesis of highly viscous, double-stranded DNA[18]. Sinsheimer has isolated from the tiny ØX 174 virus a DNA which appears to be single-stranded[19]. Like heated DNA it has proved to be an excellent primer[18] and a favorable material in current studies[20] for demonstrating in density gradient sedimentations that it is progressively converted to a double-stranded condition during the course of enzymatic synthesis.

While a detailed discussion of the physical aspects of replication is not feasible in this lecture, it should be mentioned that the DNA in the single-stranded condition is not only a suitable primer but is the only active form when the most purified enzyme preparations are used. With such *coli* preparations, the native, double-stranded DNA is inert unless it is heated or pretreated very slightly with deoxyribonuclease. Bollum has made similar observations with the enzyme that he has purified from calf thymus[21].

Substitution of analogues in DNA synthesis

The second line of evidence is derived from studies of the activity of the substrates when substitutions are made in the purine and pyrimidine bases

From the many interesting reports on the incorporation of bromouracil[22], azaguanine[23] and other analogues into bacterial and viral DNA, it might be surmised that some latitude in the structure of the bases can be tolerated provided there is no interference with their hydrogen bondings. When experiments were carried out with deoxyuridine triphosphate or 5-bromodeoxyuridine triphosphate, it was found that they supported DNA synthesis when used in place of thymidine triphosphate but not when substituted for the triphosphates of deoxyadenosine, deoxyguanosine or deoxycytidine. As already described[24], 5-methyl- and 5-bromocytosine specifically replaced cytosine; hypoxanthine substituted only for guanine; and, as just mentioned, uracil and 5-bromouracil specifically replaced thymine. These findings are best interpreted on the basis of hydrogen bonding of the adenine-thymine and guanine-cytosine type.

Along these lines it is relevant to mention the existence of a naturally occurring « analogue » of cytosine, hydroxymethyl cytosine (HMC), which is found in place of cytosine in the DNA of the *coli* bacteriophages of the T-even series[25]. In this case the DNA contains equivalent amounts of HMC and guanine and, as usual, equivalent amounts of adenine and thymine. Of additional interest is the fact that the DNA's of T2, T4 and T6 contain glucose linked to the hydroxymethyl groups of the HMC in characteristic ratios[26,27,28] although it is clear that in T2 and T6 some of the HMC groups contain no glucose[27]. These characteristics have posed two problems regarding the synthesis of these DNA's which might appear to be incompatible with the simple base-pairing hypothesis. First, what mechanism is there for preventing the inclusion of cytosine in a cell which under normal conditions has deoxycytidine triphosphate and incorporates it into its DNA? Secondly, how does one conceive of the origin of the constant ratios of glucose to HMC in DNA if the incorporation were to occur via glucosylated and non-glucosylated HMC nucleotides? Our recent experiments have shown that the polymerase reaction in the virus-infected cell is governed by the usual hydrogen-bonding restrictions but with the auxiliary action of several new enzymes developed specifically in response to infection with a given virus[29,30]. Among the new enzymes is one which splits deoxycytidine triphosphate and thus removes it from the sites of polymerase action[30]. Another is a type of glucosylating enzyme which transfers glucose from uridine diphosphate glucose directly and specifically to certain HMC residues in the DNA[30].

Chemical composition of enzymatically synthesized DNA

The third line of evidence is supplied by an analysis of the purine and pyrimidine base composition of the enzymatically synthesized DNA. We may ask two questions. First, will the product have the equivalence of adenine to thymine and of guanine to cytosine that characterize natural DNA? Secondly, will the composition of the natural DNA used as primer influence and determine the composition of the product? In Fig. 8 are the results which answer these two questions[31]. The experiments are identical except that in each case a different DNA primer was used: *Mycobacterium phlei*, *Escherichia coli*, calf thymus and phage T2 DNA. In answer to the first question it is clear that in the enzymatically synthesized DNA, adenine equals thymine and guanine equals cytosine so that the purine content is in every case identical to the pyrimidine. In answer to the second question it is again apparent that the characteristic ratio of adenine-thymine pairs to guanine-cytosine pairs of a given DNA primer is imposed rather faithfully on the product that is synthesized. Whether these measurements are made with isotopic tracers when the net DNA increase is only 1% or if it is 1,000% the results are the same. It can be said further that it has not been possible to distort these base ratios by using widely differing molar concentrations of substrates or by any other means. In the last line of Fig. 8 is a rather novel «DNA» which is synthesized under conditions that I will not describe here[18,32]. Suffice it to say that after very long lag periodes a copolymer of

DNA		*A*	*T*	*G*	*C*	$\frac{A+G}{T+C}$	$\frac{A+T}{G+C}$
Mycobacterium phlei	primer	0.65	0.66	1.35	1.34	1.01	0.49
	product	0.66	0.65	1.34	1.37	0.99	0.48
Escherichia coli	primer	1.00	0.97	0.98	1.05	0.98	0.97
	product	1.04	1.00	0.97	0.98	1.01	1.02
Calf thymus	primer	1.14	1.05	0.90	0.85	1.05	1.25
	product	1.12	1.08	0.85	0.85	1.02	1.29
Bacteriophage T2	primer	1.31	1.32	0.67	0.70	0.98	1.92
	product	1.33	1.29	0.69	0.70	1.02	1.90
A-T Copolymer		1.99	1.93	< 0.05	< 0.05	1.03	> 40

Fig. 8. Chemical composition of enzymatically synthesized DNA with different primers.

deoxyadenylate and thymidylate (A-T) develops which has the physical size and properties of natural DNA and in which the adenine and thymine are in a perfectly alternating sequence. When this rare form of DNA-like polymer is used as a primer, new A-T polymer synthesis starts immediately and even though all four triphosphates be present, no trace of guanine or cytosine can be detected in the product. The conclusion from these several experiments thus seems inescapable that the base composition is replicated in the enzymatic synthesis and that hydrogen-bonding of adenine to thymine and guanine to cytosine is the guiding mechanism.

Enzymatic replication of nucleotide sequences

The fourth line of evidence which I would like to cite is drawn from current studies of base sequences in DNA and their replication. As I have suggested already, we believe that DNA is the genetic code; the four kinds of nucleotides make up a four-letter alphabet and their sequence spells out the message. At present we do not know the sequence; what Sanger has done for peptide sequence in protein remains to be done for nucleic acids. The problem is more difficult, but not insoluble.

Our present attempts at determining the nucleotide sequences[33] will be described in detail elsewhere and I will only summarize them here. DNA is enzymatically synthesized using ^{32}P as label in one of the deoxynucleoside triphosphates; the other three substrates are unlabeled. This radioactive phosphate, attached to the 5-carbon of the deoxyribose, now becomes the bridge between that substrate molecule and the nucleotide at the growing end of the chain with which it has reacted (Fig. 9). At the end of the synthetic reaction (after some 10^{16} diester bonds have been formed), the DNA is isolated and digested enzymatically to yield the 3′ deoxynucleotides quantitatively. It is apparent (Fig. 9) that the P atom formerly attached to the 5-carbon of the deoxynucleoside triphosphate substrate is now attached to the 3-carbon of the nucleotide with which it reacted during the course of synthesis of the DNA chains. The ^{32}P content of each of the 3′ deoxynucleotides, isolated by paper electrophoresis, is a measure of the relative frequency with which a particular substrate reacted with each of the four available nucleotides in the course of synthesis of the DNA chains. This procedure carried out four times, using in turn a different labeled substrate, yields the relative frequencies of all the sixteen possible kinds of dinucleotide (nearest neighbor) sequences.

Such studies have to date been carried out using DNA primer samples from six different natural sources. The conclusions are:

1. All 16 possible dinucleotide sequences are found in each case.
2. The pattern of relative frequencies of the sequences is unique and reproducible in each case and is not readily predicted from the base composition of the DNA.
3. Enzymatic replication involves base-pairing of adenine to thymine and guanine to cytosine and, most significantly:
4. The frequencies also indicate clearly that the enzymatic replication produces two strands of opposite direction, as predicted by the Watson and Crick model.

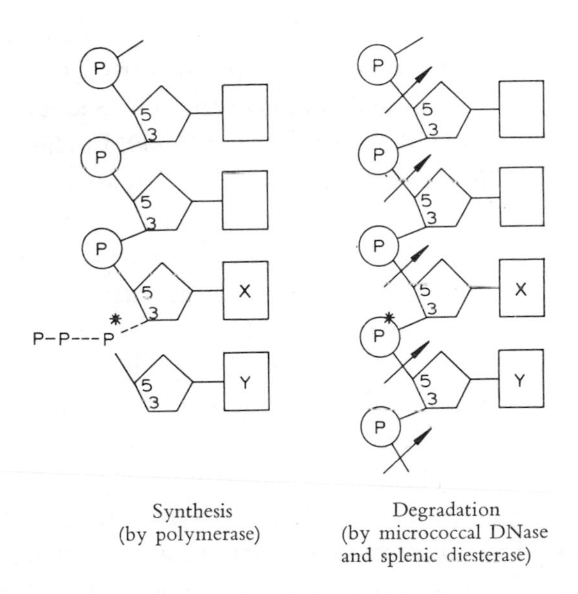

Fig. 9. Method for determining sequences in DNA.

These studies and anticipated extensions of them should yield the dinucleotide frequencies of any DNA sample which can serve as an effective primer for enzymatic replication and thus provide some clues for deciphering the DNA code. Unfortunately this method does not provide information about trinucleotide frequencies but we are hopeful that with the improvement of enzymatic tools for analysis and chromatographic techniques for isolation some start can be made in this direction.

Requirement for four triphosphates and DNA for DNA synthesis

Returning to the earlier-stated requirement for all four deoxynucleoside triphosphates and DNA in order to obtain DNA synthesis, we can now regard and understand these requirements as another and final line of evidence for hydrogen bonding. Without added DNA there is no template for hydrogen bonding and without all four triphosphates synthesis stops early and abruptly for lack of a hydrogen bonding mate for one of the bases in the template.

Summary

The enzymatic approaches to the problem of DNA replication and the properties of the DNA-synthesizing enzyme purified from *Escherichia coli* have been sketched. The unifying and basic generalization about the action of this enzyme is that it catalyzes the synthesis of a new DNA chain in response to directions from a DNA template; these directions are dictated by the hydrogen-bonding relationship of adenine to thymine and guanine to cytosine. The experimental basis for this conclusion is derived from the observations of: (1) The double-stranded character of the enzymatically synthesized DNA and its origin from a single-stranded molecule, (2) the pattern of substitution of analogues for the naturally occurring bases, (3) the replication of the chemical composition, (4) the replication of the nucleotide (nearest neighbor) sequences and the antiparallel direction of the strands, and (5) the requirement for all four deoxynucleoside triphosphates (adenine, thymine, guanine, and cytosine) and DNA for DNA synthesis.

In closing may I repeat what was said at the banquet last night: Any credit for the work cited here is shared by my colleagues in New York, Bethesda, Saint Louis and Stanford, and by the whole international community of chemists, geneticists and physiologists, which is truly responsible for the progress in nucleic acid biochemistry.

1. O. T. Avery, C. M. MacLeod, and M. McCarty, *J. Exptl. Med.*, 79 (1944) 137; R. D. Hotchkiss, in *The Chemical Basis of Heredity*, W. D. McElroy and B. Glass (Eds.), Johns Hopkins Press, Baltimore, 1957, p. 321.
2. A. D. Hershey, *Cold Spring Harbor Symp. Quant. Biol.*, 18 (1953) 135.

3. G. W. Beadle, in *Chemical Basis of Heredity*, W. D. McElroy and B. Glass (Eds.), Johns Hopkins Press, Baltimore, 1957, p. 3.
4. E. Chargaff, in *Nucleic Acids*, E. Chargaff and J. N. Davidson (Eds.), Vol. I, Academic Press, New York, 1955, pp. 307–71.
5. J. D. Watson and F. H. C. Crick, *Nature*, 171 (1953) 737; *Cold Spring Harbor Symp. Quant. Biol.*, 18 (1953) 123.
6. M. H. F. Wilkins, *Biochem. Soc. Symp., Cambridge, Engl.*, 14 (1957) 13.
7. M. Feughelman, R. Langridge, W. E. Seeds, A. R. Stokes, H. R. Wilson, C. W. Hooper, M. H. F. Wilkins, R. K. Barclay, and L. D. Hamilton, *Nature*, 175 (1955) 834.
8. A. Kornberg, in *The Chemical Basis of Heredity*, W. D. McElroy and B. Glass (Eds.), Johns Hopkins Press, Baltimore, 1957, p. 579. *Rev. Mod. Phys.*, 31 (1959) 200.
9. A. Kornberg, in *Phosphorus Metabolism*, W. D. McElroy and B. Glass (Eds.), Johns Hopkins Press, Baltimore, 1951, p. 392. *Advan. Enzymol.*, 18 (1957) 191.
10. D. E. Koshland, Jr., in *The Mechanism of Enzyme Action*, W. D. McElroy and B. Glass (Eds.), Johns Hopkins Press, Baltimore, 1954, p. 608.
11. A. Kornberg, I. R. Lehman, and E. S. Simms, *Federation Proc.*, 15 (1956) 291.
12. A. Kornberg, *Harvey Lectures*, 53 (1957–1958) 83.
13. I. R. Lehman, M. J. Bessman, E. S. Simms, and A. Kornberg, *J. Biol. Chem.*, 233 (1958) 163.
14. F. J. Bollum and V. R. Potter, *J. Am. Chem. Soc.*, 79 (1957) 3603; C. G. Harford and A. Kornberg, *Federation Proc.*, 17 (1958) 515; F. J. Bollum, *Federation Proc.*, 17 (1958) 193; 18 (1959) 194.
15. M. J. Bessman, I. R. Lehman, E. S. Simms, and A. Kornberg, *J. Biol. Chem.*, 233 (1958) 171.
16. J. Adler, I. R. Lehman, M. J. Bessman, E. S. Simms, and A. Kornberg, *Proc. Natl. Acad. Sci. U.S.*, 44 (1958) 641.
17. H. K. Schachman, I. R. Lehman, M. J. Bessman, J. Adler, E. S. Simms, and A. Kornberg, *Federation Proc.*, 17 (1958) 304.
18. I. R. Lehman, *Ann. N.Y. Acad. Sci.*, 81 (1959) 745.
19. R. L. Sinsheimer, *J. Mol. Biol.*, 1 (1959) 43.
20. I. R. Lehman, R. L. Sinsheimer, and A. Kornberg (unpublished observations).
21. F. J. Bollum, *J. Biol. Chem.*, 234 (1959) 2733.
22. F. Weygand, A. Wacker, and Z. Dellweg, *Z. Naturforsch.*, 7 b (1952) 19; D. B. Dunn and J. D. Smith, *Nature*, 174 (1954) 305; S. Zamenhof and G. Griboff, *Nature*, 174 (1954) 306.
23. M. R. Heinrich, V. C. Dewey, R. E. Parks, Jr., and G. W. Kidder, *J. Biol. Chem.*, 197 (1952) 199.
24. M. J. Bessman, I. R. Lehman, J. Adler, S. B. Zimmerman, E. S. Simms, and A. Kornberg, *Proc. Natl. Acad. Sci. U.S.*, 44 (1958) 633.
25. G. R. Wyatt and S. S. Cohen, *Biochem. J.*, 55 (1953) 774.
26. R. L. Sinsheimer, *Science*, 120 (1954) 551; E. Volkin, *J. Am. Chem. Soc.*, 76 (1954) 5892.
27. R. L. Sinsheimer, *Proc. Natl. Acad. Sci. U.S.*, 42 (1956) 502; M. A. Jesaitis, *J. Exptl. Med.*, 106 (1957) 233; *Federation Proc.*, 17 (1958) 250.

28. G. Streisinger and J. Weigle, *Proc. Natl. Acad. Sci. U.S.*, 42 (1956) 504.
29. J. G. Flaks and S. S. Cohen, *J. Biol. Chem.*, 234 (1959) 1501; J. G. Flaks, J. Lichtenstein, and S. S. Cohen, *J. Biol. Chem.*, 234 (1959) 1507.
30. A. Kornberg, S. B. Zimmerman, S. R. Kornberg, and J. Josse, *Proc. Natl. Acad. Sci. U.S.*, 45 (1959) 772.
31. I. R. Lehman, S. B. Zimmerman, J. Adler, M. J. Bessman, E. S. Simms, and A. Kornberg, *Proc. Natl. Acad. Sci. U.S.*, 44 (1958) 1191.
32. C. M. Radding, J. Adler, and H. K. Schachman, *Federation Proc.*, 19 (1960) 307
33. J. Josse and A. Kornberg, *Federation Proc.*, 19 (1960) 305.

Biography

Arthur Kornberg was born in Brooklyn, New York, on 3rd March, 1918, the son of Joseph and Lena Kornberg. He was educated at City College, New York, where he took his B.Sc. degree in 1937, and the University of Rochester, obtaining the M.D. in 1941. He returned to City College in 1960, for LL.D., and Rochester University in 1962, to take D.Sc.

Between 1941–1942, Kornberg served as Intern in the Strong Memorial Hospital of the University of Rochester, leaving in 1942 to serve from that year till 1953 as a Commissioned Officer in the U.S. Public Health Service, being gazetted Lieutenant in the United States Coast Guards. During this time he worked in the Nutrition Section of the Division of Physiology (1942–1945), and during 1947–1953 was also Chief of the Enzyme and Metabolism Section of the National Institutes of Health of Bethesda, Maryland.

Kornberg served as research investigator in the Departments of Chemistry and Pharmacology of New York College of Medicine, with Professor Severo Ochoa, in 1946; in the Department of Biological Chemistry, Washington University School of Medicine, with Professors Carl and Gerty Cori, in 1947; and in the Department of Plant Biology, University of California, Berkeley, with Professor H. A. Barker, in 1951.

From 1953–1959, Kornberg was Professor and Head of the Department of Microbiology in the Washington University School of Medicine, St. Louis, Missouri, and from 1959 he has been Professor and Executive Head of the Department of Biochemistry at Stanford University School of Medicine.

The special fields of interest to Professor Kornberg are biochemistry, especially enzyme chemistry, and synthesis of deoxyribonucleic acid (DNA) – studying the nucleic acids which control heredity in animals, plants, bacteria and viruses. He is a member of the American Academy of Arts and Sciences, the National Academy of Sciences and the American Philosophical Society.

Besides by the Nobel Prize for Physiology or Medicine 1959, which he shared with his former associate Dr. Severo Ochoa of New York University,

Professor Kornberg's work has also been acknowledged by the presentation of the Paul-Lewis Laboratories Award in Enzyme Chemistry from the American Chemical Society in 1951, and an L.H.D. degree from Yeshiva University in 1962.

During his Public Health Service, in 1943, on 21st November, he married Sylvy Ruth Levy, of Rochester, New York, who is now a research associate in the Department of Biochemistry. They have three children – Roger David, born 1947; Thomas Bill, born 1948; and Kenneth Andrew, born 1950.

1962

Maurice H.F. Wilkins
James D. Watson
Francis H.C. Crick

MAURICE H. F. WILKINS

The Molecular Configuration of Nucleic Acids

December 11, 1962

Nucleic acids are basically simple. They are at the root of very fundamental biological processes, growth and inheritance. The simplicity of nucleic acid molecular structure and of its relation to function expresses the underlying simplicity of the biological phenomena, clarifies their nature, and has given rise to the first extensive interpretation of living processes in terms of macromolecular structure. These matters have only become clear by an unprecedented combination of biological, chemical and physical studies, ranging from genetics to hydrogen-bond stereochemistry. I shall not discuss all this here but concentrate on the field in which I have worked, and show how X-ray diffraction analysis has made its contribution. I shall describe some of the background of my own researches, for I suspect I am not alone in finding such accounts often more interesting than general reviews.

Early Background

I took a physics degree at Cambridge in 1938, with some training in X-ray crystallography. This X-ray background was influenced by J. D. Bernal, then at the Cavendish. I began research at Birmingham, under J. T. Randall, studying luminescence and how electrons move in crystals. My contemporaries at Cambridge had mainly been interested in elementary particles, but the organization of the solid state and the special properties which depended on this organization interested me more. This may have been a forerunner of my interest in biological macromolecules and how their structure related to their highly specific properties which so largely determine the processes of life.

During the war I took part in making the atomic bomb. When the war was ending, I, like many others, cast around for a new field of research. Partly on account of the bomb, I had lost some interest in physics. I was therefore very interested when I read Schrödinger's book « What is Life? » and was struck by the concept of a highly complex molecular structure which controlled living processes. Research on such matters seemed more ambitious than solid-state physics. At that time many leading physicists such as Massey, Oliphant, and Randall (and later I learned that Bohr shared their view) believed that physics would contribute significantly to biology; their advice encouraged me to move into biology.

I went to work in the Physics Department at St. Andrews, Scotland, where Randall had invited me to join a biophysics project he had begun. Stimulated by Muller's experimental modification, by means of X-radiation, of genetic substance, I thought it might be interesting to investigate the effects of ultrasonics; but the results were not very encouraging.

The biophysics work then moved to King's College, London, where Randall took the Wheatstone Chair of Physics and built up, with the help of the Medical Research Council, an unusual laboratory for a Physics Department, where biologists, biochemists and others worked with the physicists. He suggested I might take over some ultraviolet microscope studies of the quantities of nucleic acids in cells. This work followed that of Caspersson, but made use of the achromatism of reflecting microscopes. By this time, the work of Caspersson[1] and Brachet[2] had made the scientific world generally aware that nucleic acids had important biological roles which were connected with protein synthesis. The idea that DNA might itself be the genetic substance was, however, barely hinted at. Its function in chromosomes was supposed to be associated with replication of the protein chromosome thread. The work of Avery, MacLeod, and McCarty[3], showing that bacteria could be genetically transformed by DNA, was published in 1944, but even in 1946 seemed almost unknown, or if known, its significance was often belittled.

It was fascinating to look through microscopes at chromosomes in cells, but I began to feel that as a physicist I might contribute more to biology by studying macromolecules isolated from cells. I was encouraged in this by Gerald Oster who came from Stanley's virus laboratory and interested me in particles of tobacco mosaic virus. As Caspersson had shown, ultraviolet microscopes could be used to find the orientation of ultraviolet absorbing groups in molecules as well as to measure quantities of nucleic acids in cells. Bill Seeds and I studied DNA, proteins, tobacco mosaic virus, vitamin B_{12}, etc. While examining oriented films of DNA prepared for ultraviolet dichroism studies, I saw in the polarizing microscope extremely uniform fibres giving clear extinction between crossed nicols. I found the fibres had been produced unwittingly while I was manipulating DNA gel. Each time

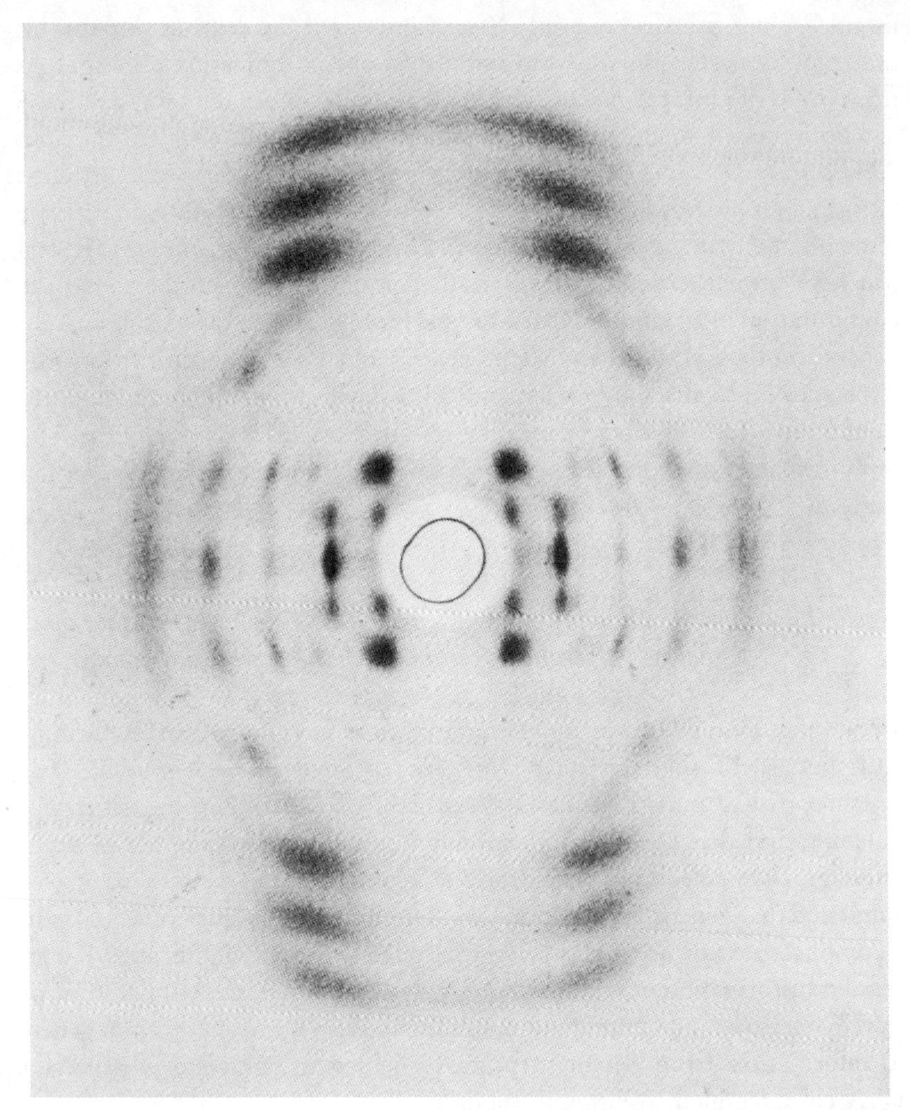

Fig. 1. One of the first X-ray diffraction photographs of DNA taken in our laboratory. This may be compared with the later photograph in Fig. 10. (Photograph with R. Gosling; DNA by R. Signer).

that I touched the gel with a glass rod and removed the rod, a thin and almost invisible fibre of DNA was drawn out like a filament of spider's web. The perfection and uniformity of the fibres suggested that the molecules in them were regularly arranged. I immediately thought the fibres might be excellent objects to study by X-ray diffraction analysis. I took them to Raymond Gosling, who had our only X-ray equipment (made from war-surplus radiography parts) and who was using it to obtain diffraction photographs from heads of ram spermatozoa. This research was directed by Randall, who had been trained under W. L. Bragg and had worked with X-ray diffraction. Almost immediately, Gosling obtained very encouraging diffraction patterns (see Fig. 1). One reason for this success was that we kept the fibres moist. We remembered that, to obtain detailed X-ray patterns from proteins, Bernal had kept protein crystals in their mother liquor. It seemed likely that the configuration of all kinds of water-soluble biological macromolecules would depend on their aqueous environment. We obtained good diffraction patterns with DNA made by Signer and Schwander[4] which Singer brought to London to a Faraday Society meeting on nucleic acids and which he generously distributed so that all workers, using their various techniques, could study it.

Realization that the Genetic Material was a Pure Chemical Substance and Signs that its Molecular Structure was Singularly Simple

Between 1946 and 1950 many lines of evidence were uncovered indicating that the genetic substance was DNA, not protein or nucleoprotein. For instance, it was found that the DNA content of a set of chromosomes was constant, and that DNA from a given species had a constant composition although the nucleotide sequence in DNA molecules was complex. It was suggested that genetic information was carried in the polynucleotide chain in a complicated sequence of the four nucleotides. The great significance of bacterial transformation now became generally recognized, and the demonstration by Hershey and Chase[5] that bacteriophage DNA carried the viral genetic information from parent to progeny helped to complete what was a fairly considerable revolution in thought.

The prospects of elucidating genetic function in terms of molecular structure were greatly improved when it was known that the genetic substance was DNA, which had a well-defined chemical structure, rather than an ill-defined nucleoprotein. There were many indications of simplicity and regularity in DNA structure. The chemists had shown that DNA was a polymer in which the phosphate and deoxyribose parts of the molecule were regularly repeated in a polynucleotide chain with 3′–5′ linkages. Chargaff[6] discovered an important regularity: although the sequence of bases along the polynucleotide chains was complex and the base composition of different DNA's varied considerably, the numbers of adenine and thymine groups were always equal, and so were the numbers of guanine and cytosine. In the electron microscope, DNA was seen as a uniform unbranched thread of diameter about 20 Å. Signer, Caspersson, and Hammarsten[7] showed by flow-birefringence measurements that the bases in DNA lay with their planes roughly perpendicular to the length of the thread-like molecule. Their ultraviolet dichroism measurements gave the same results and showed marked parallelism of the bases in the DNA in heads of spermatozoa. Earlier, Schmidt[8] and Pattri[9] had studied optically the remarkable ordering of the genetic material in sperm heads. Astbury[10] made pioneer X-ray diffraction studies of DNA fibres and found evidence of considerable regularity in DNA; he correctly interpreted the strong 3.4 Å reflection as being due to planar bases stacked on each other. The electro-titrometric study by Gulland and Jordan[11] showed that the bases were hydrogen-bonded together, and indeed Gulland[12] suggested that the polynucleotide chains might be linked by these hydrogen bonds to form multi-chain micelles.

Thus the remarkable conclusion that a pure chemical substance was invested with a deeply significant biological activity coincided with a considerable growth of many-sided knowledge of the nature of the substance. Meanwhile we began to obtain detailed X-ray diffraction data from DNA. This was the only type of data that could provide an adequate description of the 3-dimensional configuration of the molecule.

The Need for Combining X-ray Diffraction Studies of DNA with Molecular Model-Building

As soon as good diffraction patterns were obtained from fibres of DNA, great interest was aroused. In our laboratory, Alex Stokes provided a theory of diffraction from helical DNA. Rosalind Franklin (who died some years later at the peak of her career) made very valuable contributions to the X-ray analysis. In Cambridge, at the Medical Research Council laboratory where structures of biological macromolecules were studied, my friends Francis Crick and Jim Watson were deeply interested in DNA structure. Watson was a biologist who had gone to Cambridge to study molecular structure. He had worked on bacteriophage reproduction and was keenly aware of the great possibilities that might be opened up by finding the molecular structure of DNA. Crick was working on helical protein structure and was interested in what controlled protein synthesis. Pauling and Corey, by their discovery of the protein α-helix, had shown that precise molecular model-building was a powerful analytical tool in its own right. The X-ray data from DNA were not so complete that a detailed picture of DNA structure could be derived without considerable aid from stereochemistry. It was clear that the X-ray studies of DNA needed to be complemented by precise molecular model-building. In our laboratory we concentrated on amplifying the X-ray data. In Cambridge, Watson and Crick built molecular models.

The paradox of the regularity of the DNA molecule

The sharpness of the X-ray diffraction patterns of DNA showed that DNA molecules were highly regular – so regular that DNA could crystallize. The form of the patterns gave clear indications that the molecule was helical, the polynucleotide chains in the molecular thread being regularly twisted. It was known, however, that the purines and pyrimidines of various dimensions were arranged in irregular sequence along the polynucleotide chains. How could such an irregular arrangement give a highly regular structure? This paradox pointed to the solution of the DNA structure problem and was resolved by the structural hypothesis of Watson and Crick.

The Helical Structure of the DNA Molecule

The key to DNA molecular structure was the discovery by Watson and Crick[13] that, if the bases in DNA were joined in pairs by hydrogen-bonding, the overall dimensions of the pairs of adenine and thymine and of guanine and cytosine were identical. This meant that a DNA molecule containing these pairs could be highly regular in spite of the sequence of bases being irregular. Watson and Crick proposed that the DNA molecule consisted of two polynucleotide chains joined together by base-pairs. These pairs are shown in Fig. 2. The distance between the bonds joining the bases to the deoxyribose groups is exactly (within the uncertainty of 0.1 Å or so) the same for both base-pairs, and all those bonds make exactly (within the uncertainty of 1° or so) the same angle with the line joining the C_1 atoms of the deoxyribose (see Fig. 2). As a result, if two polynucleotide chains are

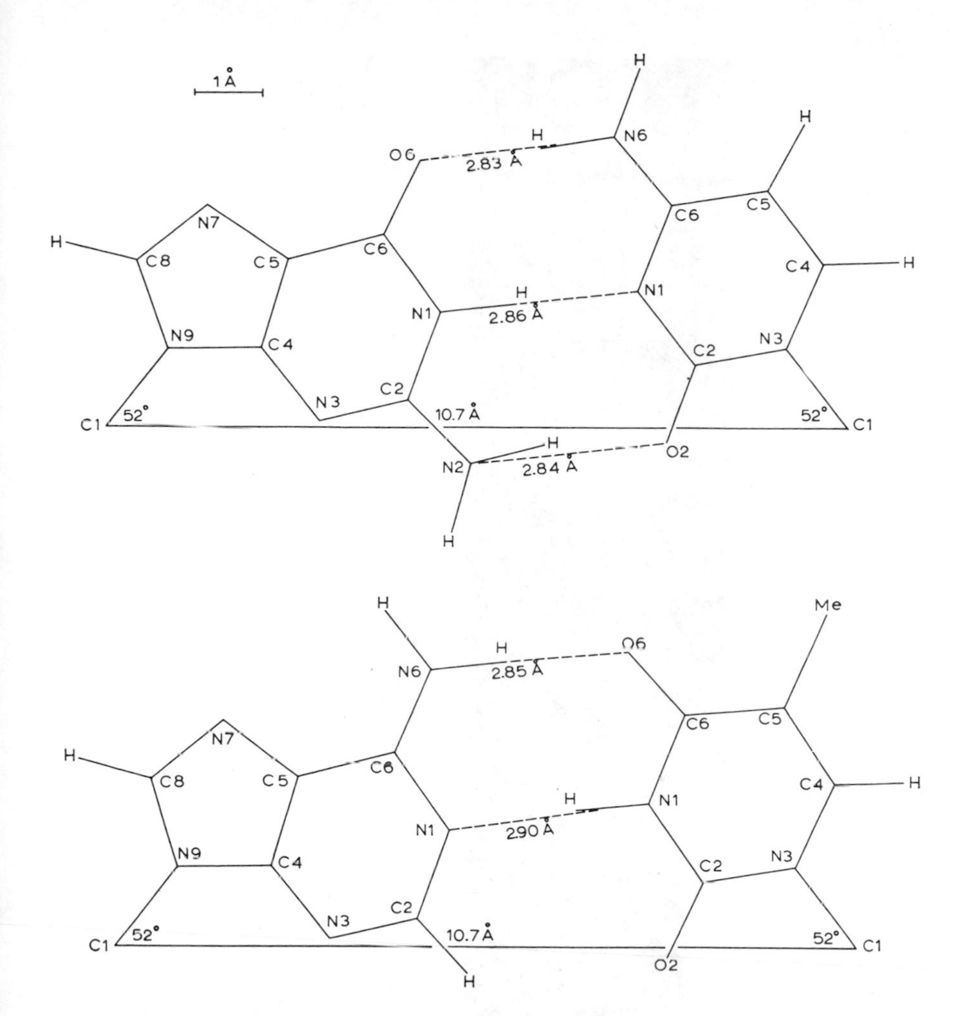

Fig. 2. Watson-Crick base-pairs (revised by S. Arnott). (*Top*): Guanine hydrogen-bonded to cytosine. (*Bottom*): Adenine hydrogen-bonded to thymine. The distances between the ends of the C_I–N_3 and C_I–N_9 bonds are 10.7 Å in both pairs, and all these bonds make an angle of 52° with the C_I–C_I line.

joined by the base-pairs, the distance between the two chains is the same for both base-pairs and, because the angle between the bonds and the C_I–C_I line is the same for all bases, the geometry of the deoxyribose and phosphate parts of the molecule can be exactly regular.

Watson and Crick built a two-chain molecular model of this kind, the chains being helical and the main dimensions being as indicated by the X-ray data. In the model, one polynucleotide chain is twisted round the other and

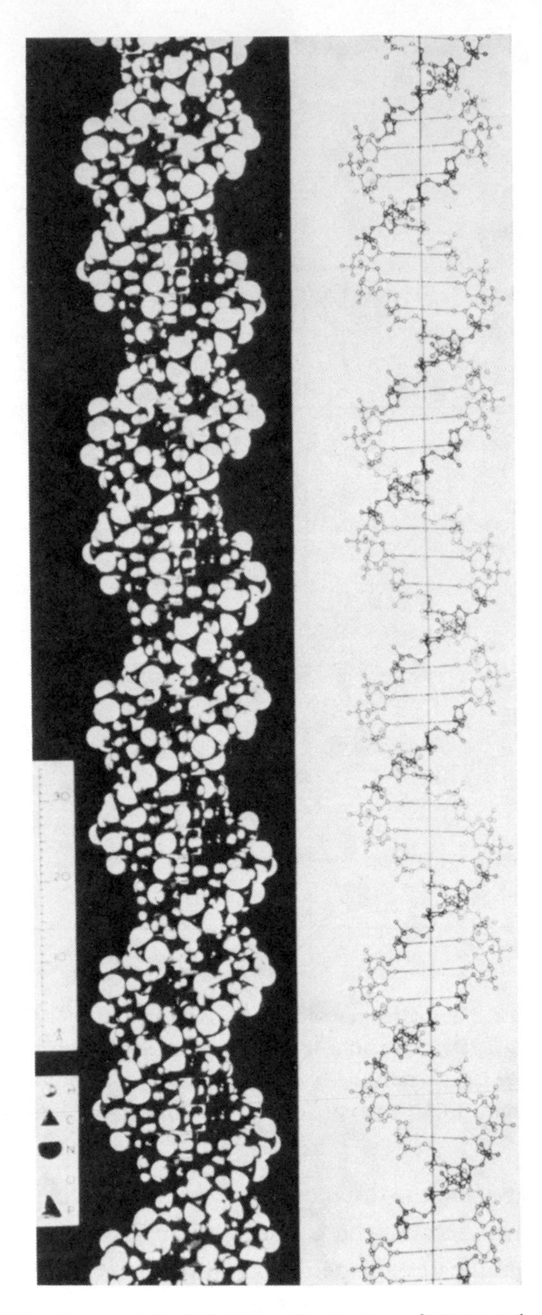

Fig. 3. (*Left*): Molecular model of the *B* configuration of DNA. The sizes of the atoms correspond to Van der Waals diameters. (*Right*): Diagram corresponding to the model. The two polynucleotide chains, joined by hydrogen-bonded bases, may be seen clearly.

the sequence of atoms in one chain runs in opposite direction to that in the other. As a result, one chain is identical with the other if turned upside down, and every nucleotide in the molecule has identical structure and environment. The only irregularities are in the base sequences. The sequence along one chain can vary without restriction, but base-pairing requires that adenine in one chain be linked to thymine in the other, and similarly guanine to cytosine. The sequence in one chain is, therefore, determined by the sequence in the other, and is said to be complementary to it.

The structure of the DNA molecule in the *B* configuration is shown in Fig. 3. The bases are stacked on each other 3.4 Å apart and their planes are almost perpendicular to the helix axis. The flat sides of the bases cannot bind water molecules; as a result there is attraction between the bases when DNA is in an aqueous medium. This hydrophobic bonding, together with the base-pair hydrogen-bonding, stabilizes the structure.

The Watson-Crick Hypothesis of DNA Replication, and Transfer of Information from one Polynucleotide Chain to Another

It is essential for genetic material to be able to make exact copies of itself; otherwise growth would produce disorder, life could not originate, and favourable forms would not be perpetuated by natural selection. Base-pairing provides the means of self-replication (Watson and Crick[14]). It also appears to be the basis of information transfer during various stages in protein synthesis.

Genetic information is written in a four-letter code in the sequence of the four bases along a polynucleotide chain. This information may be transferred from one polynucleotide chain to another. A polynucleotide chain acts as a template on which nucleotides are arranged to build a new chain. Provided that the two-chain molecule so formed is exactly regular, base-pairing ensures that the sequence in the new chain is exactly complementary to that in the parent chain. If the two chains then separate, the new chain can act as a template, and a further chain is formed; this is identical with the original chain. Most DNA molecules consist of two chains; clearly the copying process can be used to replicate such a molecule. It can also be used to transfer information from a DNA chain to an RNA chain (as is believed to be the case in the formation of messenger RNA).

Base-pairing also enables specific attachments to be made between part of one polynucleotide chain and a complementary sequence in another. Such specific interaction may be the means by which amino acids are attached to the requisite portions of a polynucleotide chain that has encoded in it the sequence of amino acids that specifies a protein. In this case the amino acid is attached to a transfer RNA molecule and part of the polynucleotide chain in this RNA pairs with the coding chain.

Since the base-pairs were first described by Watson and Crick in 1953, many new data on purine and pyrimidine dimensions and hydrogen-bond lengths have become available. The most recent refinement of the pairs (due to S. Arnott) is shown in Fig. 2. We now take the distance between C_1 atoms as 10.7 Å instead of the value used recently of 11.0 Å, mainly because new data on N–H...N bonds show that this distance is 0.2 Å shorter between ring nitrogen atoms than between atoms that are not in rings. The linearity of the hydrogen bonds in the base-pairs is excellent and the lengths of the bonds are the same as those found in crystals (these lengths vary by about 0.04 Å).

The remarkable precision of the base-pairs reflects the exactness of DNA replication. One wonders, however, why the precision is so great, for the energy required to distort the base-pairs so that their perfection is appreciably less, is probably no greater than one quantum of thermal energy. The explanation may be that replication is a co-operative phenomenon involving many base-pairs. In any case, it must be emphasized that the specificity of the base-pairing depends on the bonds joining the bases to the deoxyribose groups being correctly placed in relation to each other. This placing is probably determined by the DNA polymerizing enzyme. Whatever the mechanics of the process are, the exact equivalence of geometry and environment of every nucleotide in the double-helix should be conducive to precise replication. Mistakes in the copying process will be produced if there are tautomeric shifts of protons involved in the hydrogen-bonding or chemical alterations of the bases. These mistakes can correspond to mutations.

The Universal Nature and Constancy of the Helical Structure of DNA

After our preliminary X-ray studies had been made, my friend Leonard Hamilton sent me human DNA he and Ralph Barclay had isolated from human leucocytes of a patient with chronic myeloid leukaemia. He was studying nucleic acid metabolism in man in relation to cancer and had pre-

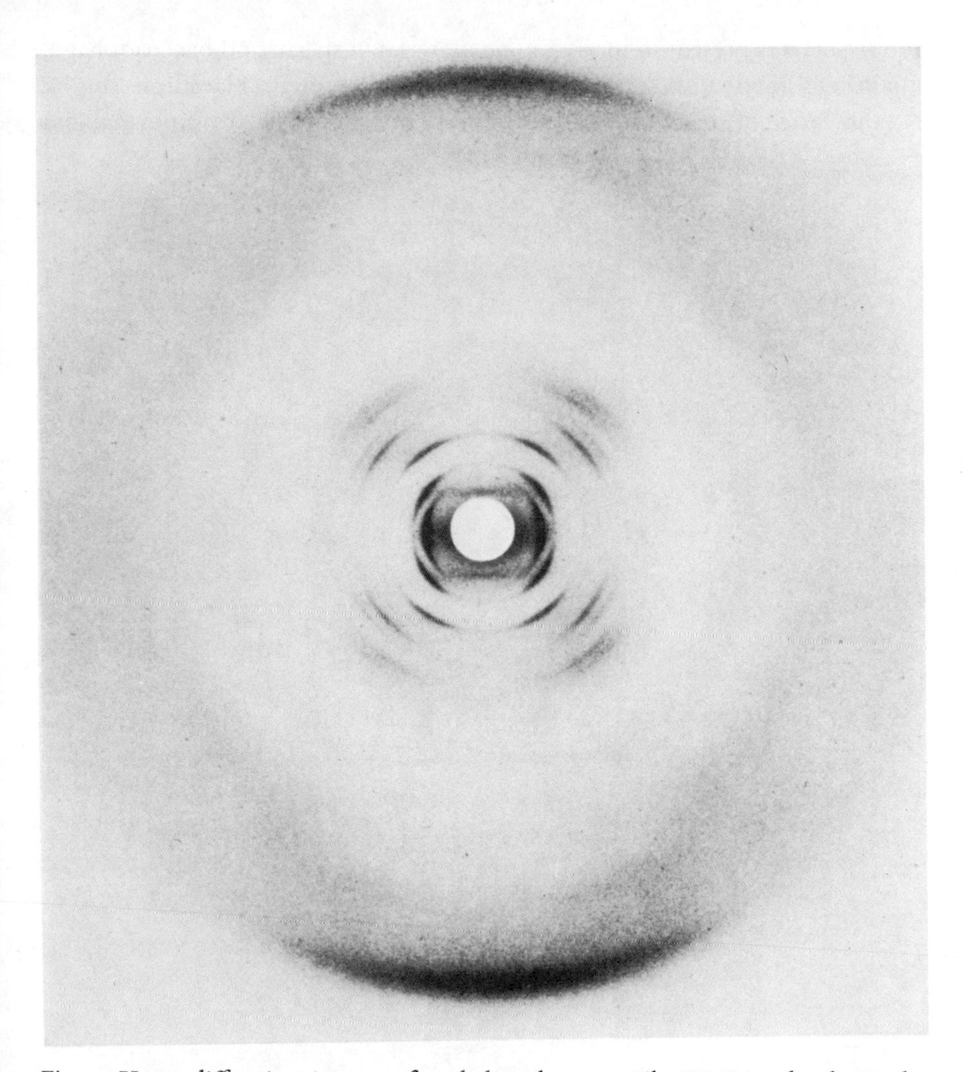

Fig. 4. X-ray diffraction pattern of cephalopod sperm. The DNA molecules in the sperm heads have their axes vertical. The 3.4 Å internucleotide spacing corresponds to the strong diffraction at the top and bottom of the pattern. The sharp reflections in the central part of the pattern show that the molecules are in crystalline array.

pared the DNA in order to compare the DNA of normal and leukaemic leucocytes. The DNA gave a very well-defined X-ray pattern. Thus began a collaboration that has lasted over many years and in which we have used Hamilton's DNA, in the form of many salts, to establish the correctness of the double-helix structure. Hamilton prepared DNA from a very wide range of species and diverse tissues. Thus it has been shown that the DNA double-helix is present in inert genetic material in sperm and bacteriophage, and in cells slowly or rapidly dividing or secreting protein (Hamilton *et al.*[16]). No difference of structure has been found between DNA from normal and

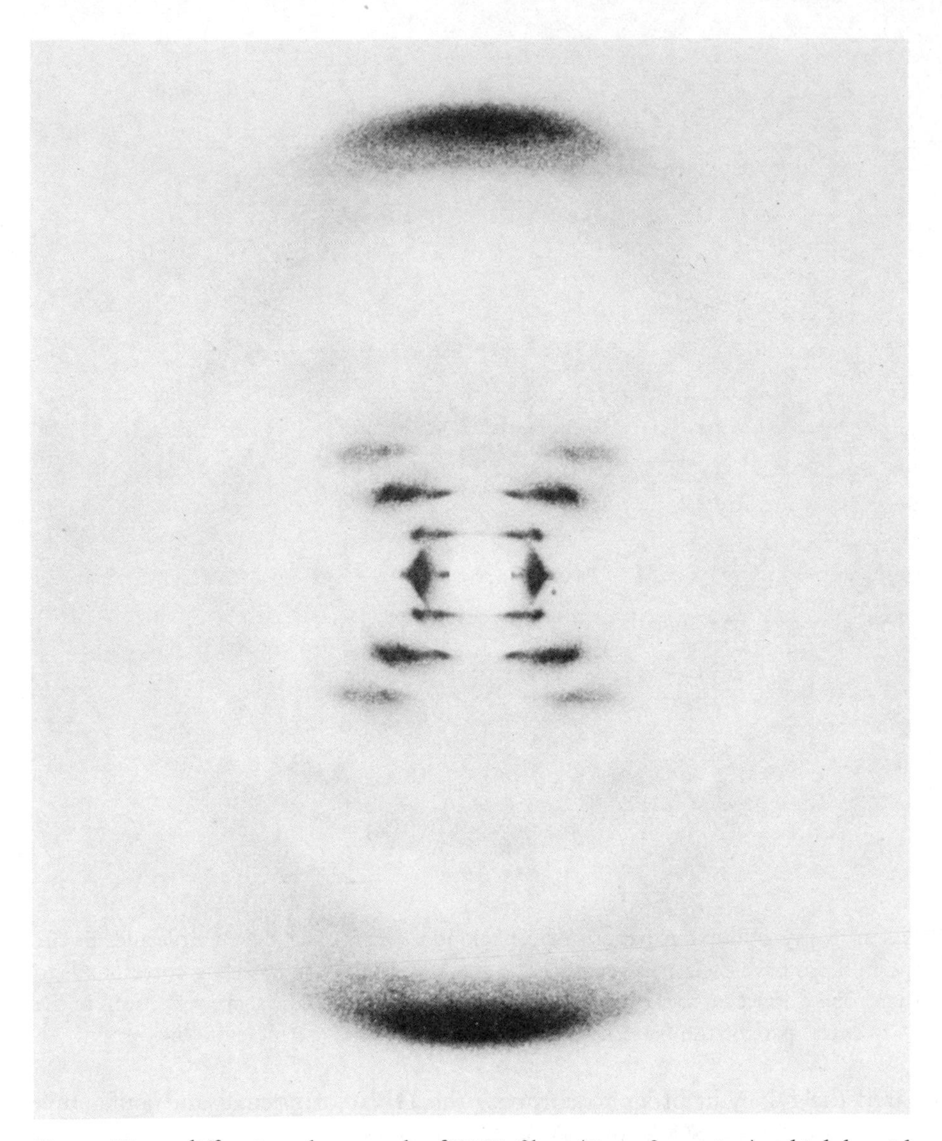

Fig. 5. X-ray diffraction photograph of DNA fibres (*B* configuration) at high humidity. The fibres are vertical. The 3.4 Å reflection is at the top and bottom. The angle in the pronounced X shape, made by the reflections in the central region, corresponds to the constant angle of ascent of the polynucleotide chains in the helical molecule. (Photograph with H. R. Wilson; DNA by L. D. Hamilton.)

from cancerous tissues, or in calf thymus DNA separated into fractions of different base composition by my colleague Geoffrey Brown.

We also made a study, in collaboration with Harriet Ephrussi-Taylor, of active transforming principle from pneumococci, and observed the same DNA structure. The only exception to double-helical DNA so far found is in some very small bacteriophages where the DNA is single-stranded. We have found, however, that DNA, with an unusually high content of adenine, or with glucose attached to hydroxymethylcytosine, crystallized differently.

DNA Structure is Not an Artefact

It did not seem enough to study X-ray diffraction from DNA alone. Obviously one should try to look at genetic material in intact cells. It was possible that the structure of the isolated DNA might be different from that *in vivo*, where DNA was in most cases combined with protein. The optical studies indicated that there was marked molecular order in sperm heads and that they might therefore be good objects for X-ray study, whereas chromosomes in most types of cells were complicated objects with little sign of ordered structure. Randall had been interested in this matter for some years and had started Gosling studying ram sperm. It seemed that the rod-shaped cephalopod sperm, found by Schmidt to be highly anisotropic optically, would be excellent for X-ray investigation. Rinne[17], while making a study of liquid crystals from many branches of Nature, had already taken diffraction photographs of such sperm; but presumably his technique was inadequate, for he came to the mistaken conclusion that the nucleoprotein was liquid-crystalline. Our X-ray photographs (Wilkins and Randall[18]) showed clearly that the material in the sperm heads had 3-dimensional order, i.e. it was crystalline and not liquid-crystalline. The diffraction pattern (Fig. 4) bore a close resemblance to that of DNA (Fig. 5), thus showing that the structure in fibres of purified DNA was basically not an artefact. Working at the Stazione Zoologica in Naples, I found it possible to orient the sperm heads in fibres. Intact wet spermatophore, being bundles of naturally oriented sperm, gave good diffraction patterns. DNA-like patterns were also obtained from T2 bacteriophage given me by Watson.

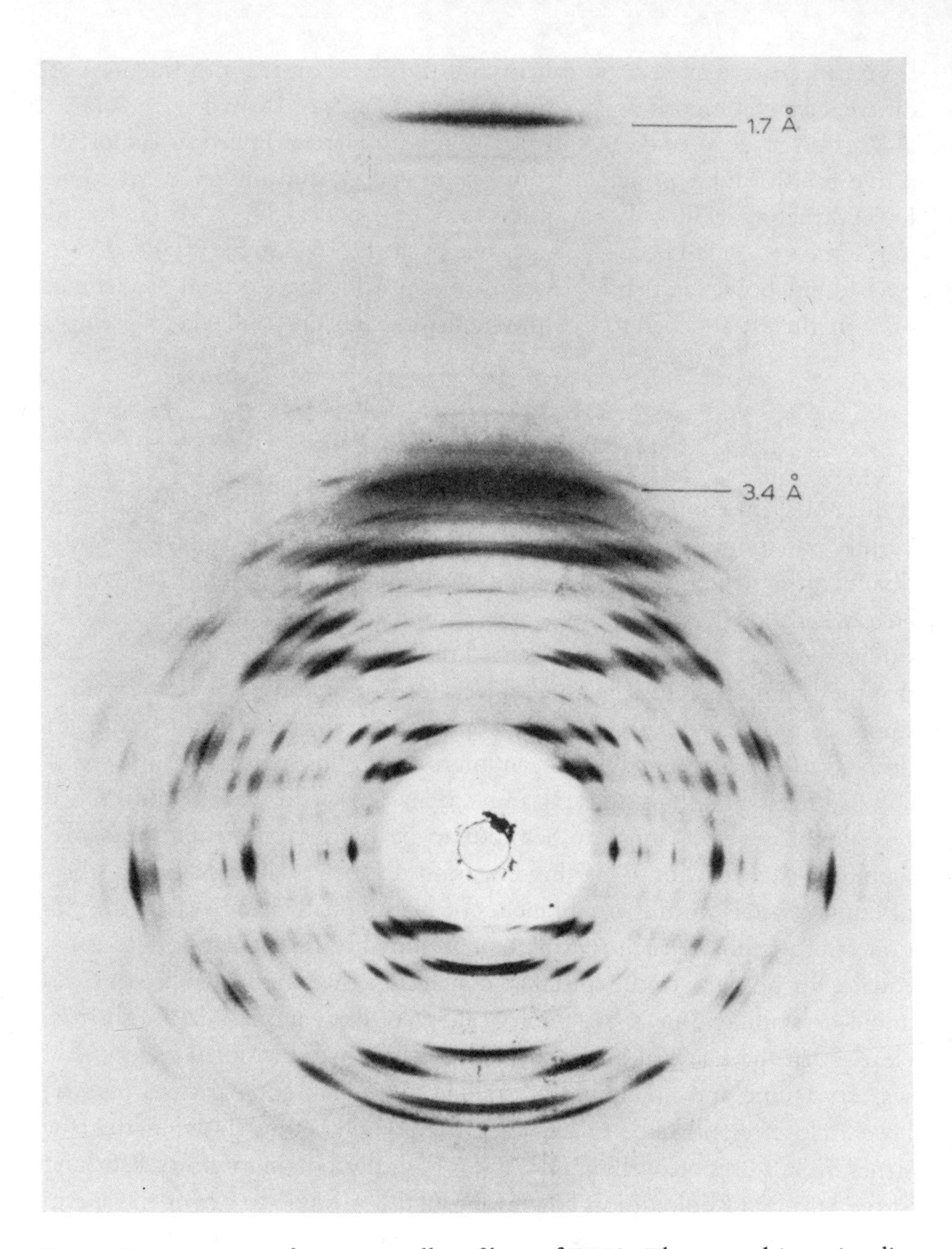

Fig. 6. X-ray pattern of microcrystalline fibres of DNA. The general intensity distribution is similar to that in Fig. 4 but the diffraction is split into sharp reflections owing to the regular arrangement of the molecules in the crystals. Sharp reflections extend to spacings as small as 1.7 Å. (Photograph with N. Chard; DNA by L. D. Hamilton.)

Table 1. Summary of various forms of DNA in fibres

Configuration of molecule	Number of nucleotide pairs per turn of helix	Inclination of base pairs in molecule	Salt	Relative humidity and condition necessary	Crystal class	Crystallinity	Molecular positions	Unit-cell dimensions			
								a (Å)	b (Å)	c (Å)	β
A	11.0	20	Na K Rb	75%	mono-clinic	crystalline	0, 0, 0 $\frac{1}{2}, \frac{1}{2}, 0$	22.24	40.62	28.15	97.0
B	10	~0	Li	66% 3% LiCl in fibre	ortho-rhombic	crystalline	$0, 0, \frac{1}{6}$ $\frac{1}{2}, \frac{1}{2}, -\frac{1}{6}$	22.5	30.9	33.7	—
			Li	75–90%	ortho-rhombic	semi-crystalline	$0, 0, \frac{1}{8}$ $\frac{1}{2}, \frac{1}{2}, -\frac{1}{8}$	24.4	38.5	33.6	—
			Li Na K Rb	92%	hexagonal	semi-crystalline	0, 0, 0 $\frac{1}{3}, \frac{2}{3}, \frac{1}{6}$ $\frac{2}{3}, \frac{1}{3}, -\frac{1}{6}$	46	—	34.6	—
B_2	9.9	0?	Na	75% under tension	tetragonal	semi-crystalline	0, 0, 0 $\frac{1}{2}, \frac{1}{2}, \frac{1}{4}$	27.4	—	33.8	—
C	9.3	−5	Li	44% no LiCl	ortho-rhombic	semi-crystalline	$0, 0, \frac{1}{8}$ $\frac{1}{2}, \frac{1}{2}, -\frac{1}{8}$	20.1	31.9	30.9	—
			Li	44% in some specimens only. No LiCl in fibre	hexagonal	semi-crystalline	$0, 0, 0$, or $\frac{1}{2}$ $\frac{1}{3}, \frac{2}{3}, \frac{1}{6}$, or $-\frac{1}{3}$ $\frac{2}{3}, \frac{1}{3}, -\frac{1}{6}$, or $\frac{1}{3}$	35.0	—	30.9	—

The X-ray Diffraction Patterns of DNA and the Various Configurations of the Molecule

X-ray diffraction analysis is the only technique that can give very detailed information about the configuration of the DNA molecule. Optical techniques, though valuable as being complementary to X-ray analysis, provide much more limited information – mainly about orientation of bonds and groups. X-ray data contributed to the deriving of the structure of DNA at two stages. First, in providing information that helped in building the Watson-Crick model; and second, in showing that the Watson-Crick proposal was correct in its essentials, which involved readjusting and refining the model.

The X-ray studies (e.g. Langridge *et al.*[19], Wilkins[20]) show that DNA molecules are remarkable in that they adopt a large number of different conformations, most of which can exist in several crystal forms. The main factors determining the molecular conformation and crystal form are the water and salt contents of the fibre and the cation used to neutralize the phosphate groups (see Table 1).

I shall describe briefly the three main configurations of DNA. In all cases the diffraction data are satisfactorily accounted for in terms of the same basic Watson-Crick structure. This is a much more convincing demonstration of the correctness of the structure than if one configuration alone were studied. The basic procedure is to adjust the molecular model until the calculated intensities of diffraction from the model correspond to those observed (Langridge *et al.*[19]).

As with most X-ray data, only the intensities, and not the phases, of the diffracted beams from DNA are available. Therefore the structure cannot be derived directly. If the resolution of X-ray data is sufficient to separate most of the atoms in a structure, the structure may be derived with no stereochemical assumption except that the structure is assumed to consist of atoms of known average size. With DNA, however, most of the atoms cannot be separately located by the X-rays alone (see Fig. 7). Therefore, more extensive stereochemical assumptions are made: these take the form of molecular model-building. There are no alternatives to most of these assumptions but where there might be an alternative, e.g. in the arrangement of hydrogen bonds in a base-pair, the X-ray data should be used to establish the correctness of the assumption. In other words, it is necessary to establish that the structure proposed is unique. Most of our work in recent years has

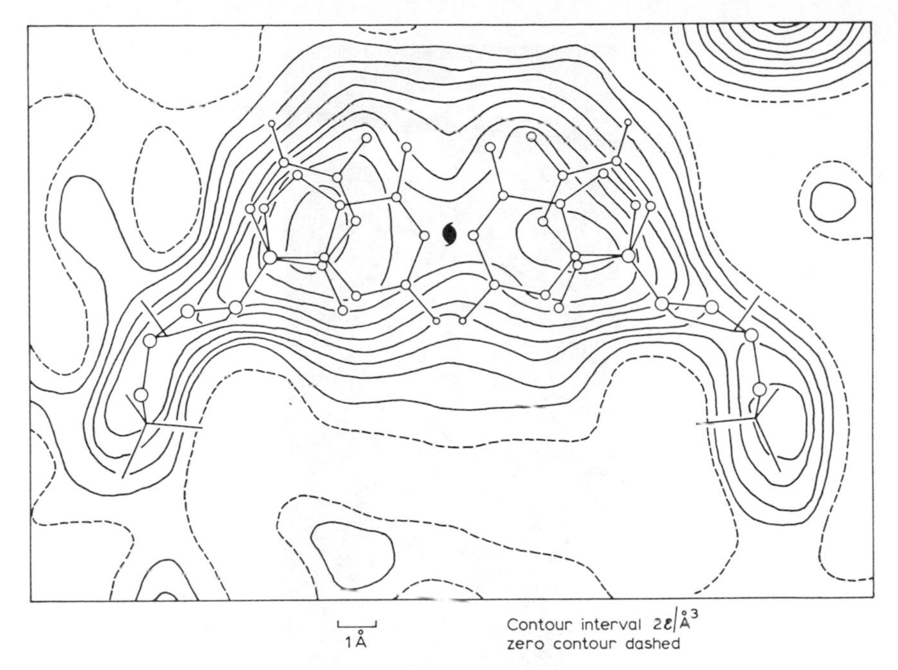

Fig. 7. Fourier synthesis map (by S. Arnott) showing the distribution of electron density in the plane of a base-pair in the *B* configuration of DNA. The distribution corresponds to an average base-pair. The shape of the base-pair appears in the map, but individual atoms in a base-pair are not resolved. (The Fourier synthesis is being revised and the map is subject to improvement.)

been of this nature. To be reasonably certain that the DNA structure was correct, X-ray data, as extensive as possible, had to be collected.

The B configuration

Fig. 5 shows a diffraction pattern of a fibre of DNA at high humidity when the molecules are separated by water and, to a large extent, behave independently of each other. We have not made intensive study of DNA under these conditions. The patterns could be improved, but they are reasonably well-defined, and the sharpness of many of their features shows that the molecules have a regular structure. The configuration is known as *B* (see also Fig. 3); it is observed *in vivo*, and there is evidence that it exists when DNA is in solution in water. There are 10 nucleotide pairs per helix turn. There is no obvious structural reason why this number should be integral; if it is exactly so, the significance of this is not yet apparent.

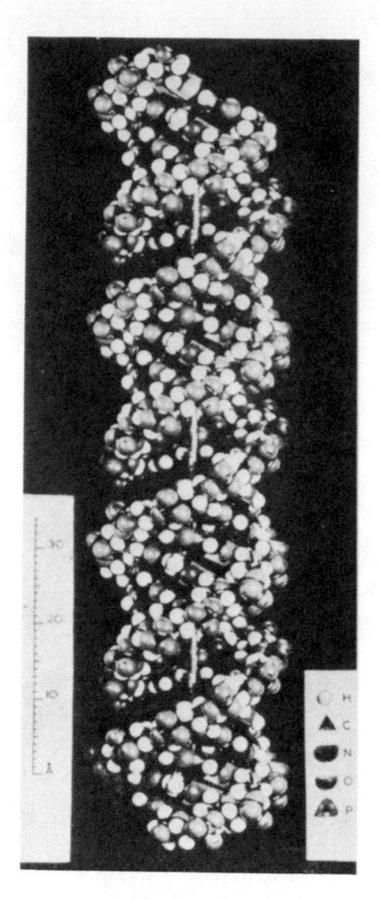

Fig. 8. Molecular model of DNA in the *A* configuration. The base-pairs may be seen inclined 20° to the horizontal.

When DNA crystallizes, the process of crystallization imposes restraints on the molecule and can give it extra regularity. Also, the periodic arrangement of the molecules in the microcrystals in the fibre causes the diffraction pattern to be split into sharp reflections corresponding to the various crystal planes (Fig. 6). Careful measurement of the positions of the reflections and deduction of the crystal lattice enables the directions of the reflections to be identified in three dimensions. Diffraction patterns from most fibrous substances resemble Fig. 5 in that the diffraction data are 2-dimensional. In contrast, the crystalline fibres of DNA give fairly complete 3-dimensional data. These data give information about the appearance of the molecule when viewed from all angles, and are comparable with those from single crystals. Techniques such as 3-dimensional Fourier synthesis (see Fig. 7) can be used and the structure determination made reasonably reliable.

The A configuration

In this conformation, the molecule has 11 nucleotide pairs per helix turn; the helix pitch is 28 Å. The relative positions and orientations of the base, and of the deoxyribose and phosphate parts of the nucleotides differ considerably from those in the *B* form; in particular the base-pairs are tilted 20° from perpendicular to the helix axis (Fig. 8).

The *A* form of DNA (Fig. 1) was the first crystalline form to be observed. Although it has not been observed *in vivo*, it is of special interest because helical RNA adopts a very similar configuration. A full account of *A* DNA will shortly be available. A good photograph of the *A* pattern is shown in Fig. 9.

The C configuration

This form may be regarded as an artefact formed by partial drying. The helix is non-integral, with about $9\frac{1}{3}$ nucleotide pairs per turn. The helices pack together to form a semi-crystalline structure; there is no special relation between the position of one nucleotide in a molecule and that in another. The conformation of an individual nucleotide is very similar to that in the *B* form. The differences between the *B* and *C* diffraction patterns are accounted for by the different position of the nucleotides in the helix. Comparison of the forms provides further confirmation of the correctness of the structures. In a way, the problem is like trying to deduce the structure of a folding chair by observing its shadow: if the conformation of the chair is altered slightly, its structure becomes more evident.

The Helical Structure of RNA Molecules

In contrast to DNA, RNA gave poor diffraction patterns, in spite of much effort by various workers including ourselves. There were many indications that RNA contained helical regions, e.g. optical properties of RNA solutions strongly suggested (e.g. Doty[21]) that parts of RNA molecules resembled DNA in that the bases were stacked on each other and the structure was helical; and X-ray studies of synthetic polyribonucleotides suggested that

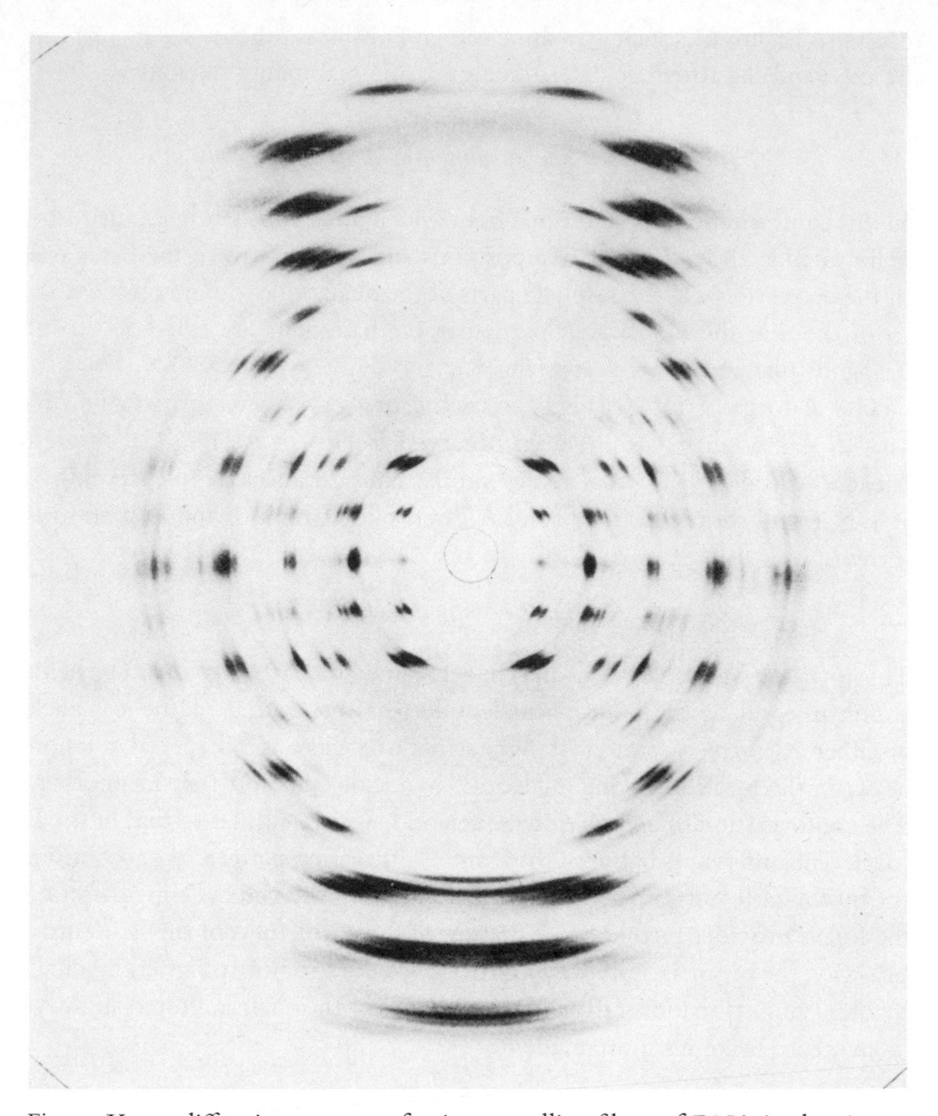

Fig. 9. X-ray diffraction pattern of microcrystalline fibres of DNA in the *A* configuration. (Photograph with H. R. Wilson; DNA by L. D. Hamilton.)

RNA resembled DNA (Rich[22]). The diffraction patterns of RNA (Rich and Watson[23]) bore a general resemblance to those of DNA, but the nature of pattern could not be clearly distinguished because of disorientation and diffuseness. An important difficulty was that there appeared to be strong meridional reflections at 3.3 Å and 4 Å. It was not possible to interpret these in terms of one helical structure.

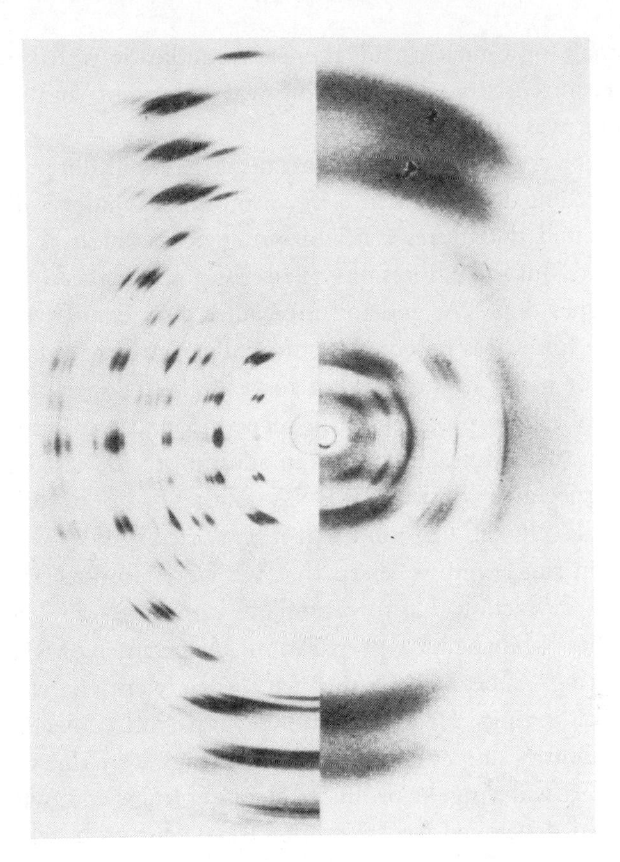

Fig. 10. Comparison of the X-ray diffraction patterns of fibres of DNA in the *A* configuration (*left*) and transfer RNA (*right*). The general distribution of intensity is very similar in both patterns, but the positions of the sharp crystalline reflections differ because the molecular packing in the crystals is different in the two cases. (Photograph with W. Fuller and M. Spencer; RNA by G. L. Brown.)

In early work, many RNA preparations were very heterogeneous. We thought that the much more homogeneous plant virus RNA might give better patterns, but this was not so. However, when preparations of ribosomal RNA and « soluble » RNA became available, we felt the prospects of structure analysis were improved. We decided to concentrate on « soluble » RNA largely because Geoffrey Brown in our laboratory was preparing large quantities of a highly purified transfer RNA component of soluble RNA for his physical and chemical studies, and because he was fractionating it into various transfer RNA's specific for incorporation of particular amino acids into proteins. This RNA was attractive for other reasons: the molecule was unusually small for a nucleic acid, there were indications that it might have a regular structure, its biochemical role was important, and in many ways its functioning was understood.

We found it very difficult to orient transfer RNA in fibres. However, by carefully stretching RNA gels in a dry atmosphere under a dissecting microscope, I found that fibres with birefringence as high as that of DNA could be made. But these fibres gave patterns no better than those obtained with other types of RNA, and the molecules disoriented when the water content of the fibres was raised. Watson, Fuller, Michael Spencer, and myself worked for many months trying to make better specimens for X-ray study. We made little progress until Spencer found a specimen that gave some faint but sharp diffraction rings in addition to the usual diffuse RNA pattern. This specimen consisted of RNA gel that had been sealed for X-ray study in a small cell, and he found that it had dried slowly owing to a leak. The diffraction rings were so sharp that we were almost certain that they were spurious diffraction due to crystalline impurity – this being common in X-ray studies of biochemical preparations. A specimen of RNA had given very similar rings due to DNA impurity. We were therefore not very hopeful about the rings. However, after several weeks Spencer eliminated all other possibilities: it seemed clear that the rings were due to RNA itself. By controlled slow drying, he produced stronger rings; and, with the refined devices we had developed for stretching RNA and with gels slowly concentrated by Brown, Fuller oriented the RNA without destroying its crystallinity. These fibres gave clearly defined diffraction patterns, and the orientation did not disappear when the fibres were hydrated. It appeared that the methods I had been using earlier, of stretching the fibres as much as possible, destroyed the crystallinity. If instead, the material was first allowed to crystallize slowly, stretching oriented the microcrystals and the RNA molecules in them. Single molecules were too small to be oriented well unless aggregated by crystallization. It was rather unexpected that, of all the different types of RNA we had tried, transfer RNA which had the lowest molecular weight, oriented best.

The diffraction patterns of transfer RNA were clearly defined and well-oriented (Spencer, Fuller, Wilkins, and Brown[24]). These improvements revealed a striking resemblance between the patterns of RNA and *A* DNA (Fig. 10). The difficulty of the two reflections at 3.3 Å and 4 Å was resolved (Fig. 11): in the RNA pattern the positions of reflections on three layer-lines differed from those in DNA; as a result, when the patterns were poorly

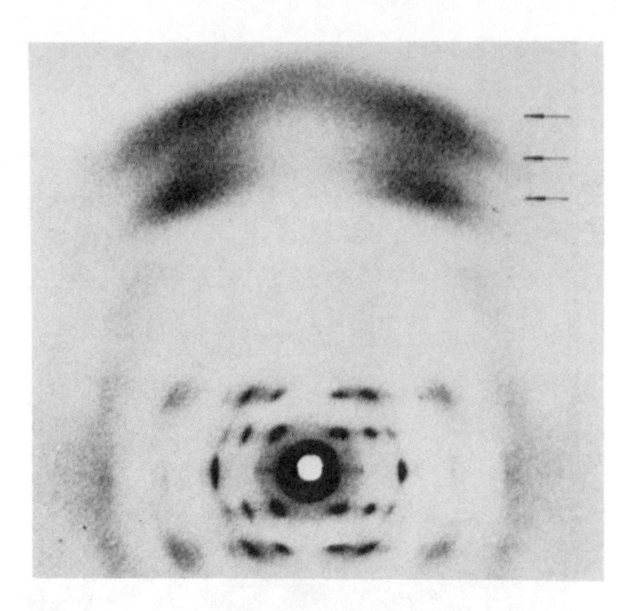

Fig. 11. Diffraction pattern of transfer RNA showing resolution of diffraction, in the regions of 3.3 Å and 4 Å, into three layer-lines indicated by the arrows and corresponding to the *A* DNA pattern. (Photograph with W. Fuller and M. Spencer; RNA by G. L. Brown.)

oriented, the three reflections overlapped and gave the impression of two. There was no doubt that the RNA had a regular helical structure almost identical with that of *A* DNA. The differences between the RNA and DNA patterns could be accounted for in terms of small differences between the two structures.

An important consequence of the close resemblance of the RNA structure to that of DNA is that the RNA must contain base sequences that are largely or entirely complementary. The number of nucleotides in the molecule is about 80. The simplest structure compatible with the X-ray results consists of a single polynucleotide chain folded back on itself, one half of the chain being joined to the other by base-pairing. This structure is shown in Fig. 12. While we are certain the helical structure is correct, it must be emphasized that we do not know whether the two ends of the chain are at the end of the molecule. The chain might be folded at both ends of the molecule with the ends of the chain somewhere along the helix. It is known that the amino acid attaches to the end of the chain terminated by the base sequence cytosine–cytosine–adenine.

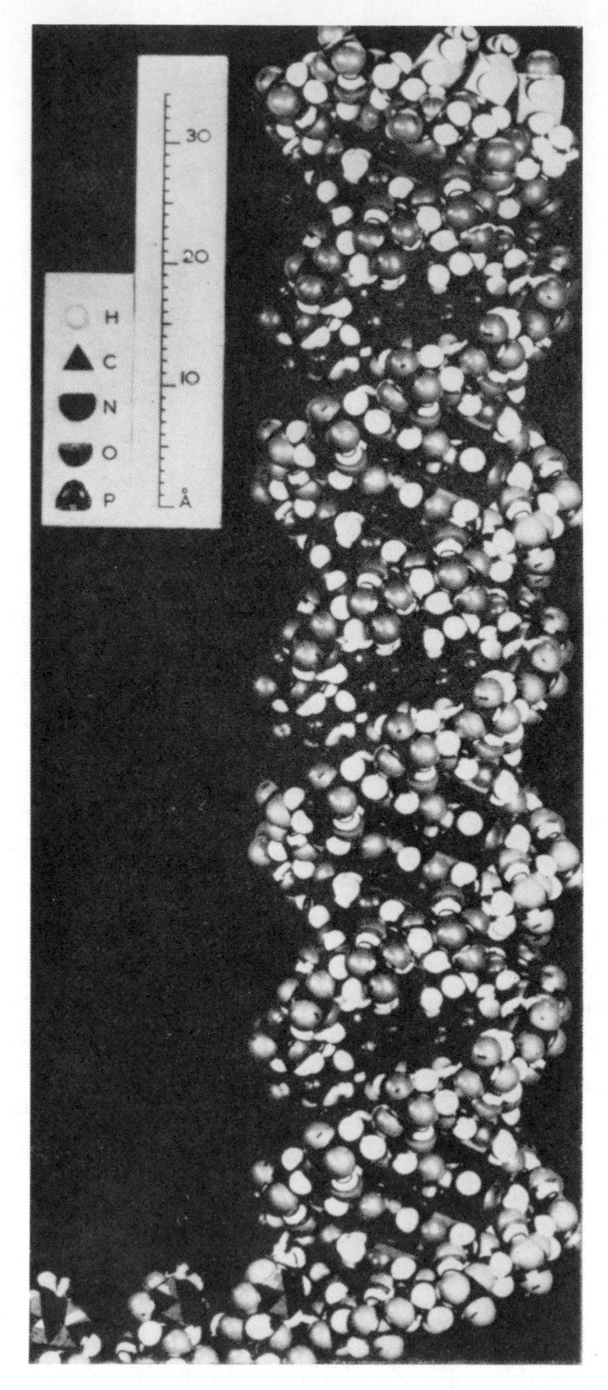

Fig. 12. Molecular model and diagram of a transfer RNA molecule.

Molecular model-building shows that the number of nucleotides forming the fold at the end of a transfer RNA molecule must be three or more. In our model, the fold consists of three nucleotides, each with an unpaired base. It might be that this base-triplet is the part of the molecule that attaches to the requisite part of the coding RNA polynucleotide chain that determines the sequence of amino acids in the polypeptide chain of a protein. It is believed that a base-triplet in the coding RNA corresponds to each amino acid. The triplet in the transfer RNA could attach itself specifically to the coding triplet by hydrogen-bonding and formation of base-pairs. It must be emphasized, however, that these ideas are speculative.

We suppose that part of the transfer RNA molecule interacts specifically with the enzyme that is involved in attaching the amino acid to the RNA; but we do not know how this takes place. Similarly, we know little of the way in which the enzyme involved in DNA replication interacts with DNA, or of other aspects of the mechanics of DNA replication. The presence of complementary base sequences in the transfer RNA molecule, suggests that it might be self-replicating like DNA; but there is at present little evidence to support this idea. The diffraction patterns of virus and ribosome RNA show that these molecules also contain helical regions; the function of these are uncertain too.

In the case of DNA, the discovery of its molecular structure led immediately to the replication hypothesis. This was due to the simplicity of the structure of DNA. It seems that molecular structure and function are in most cases less directly related. Derivation of the helical configuration of RNA molecules is a step towards interpreting RNA function; but more complete structural information, e.g. determination of base sequences, and more knowledge about how the various kinds of RNA interact in the ribosome, will probably be required before an adequate picture of RNA function emerges.

The Possibility of Determining the Base Sequence of Transfer RNA by X-ray Diffraction Analysis

Since the biological specificity of nucleic acids appears to be entirely determined by their base sequences in them, determination of these sequences is probably the most fundamental problem in nucleic acid research today. The

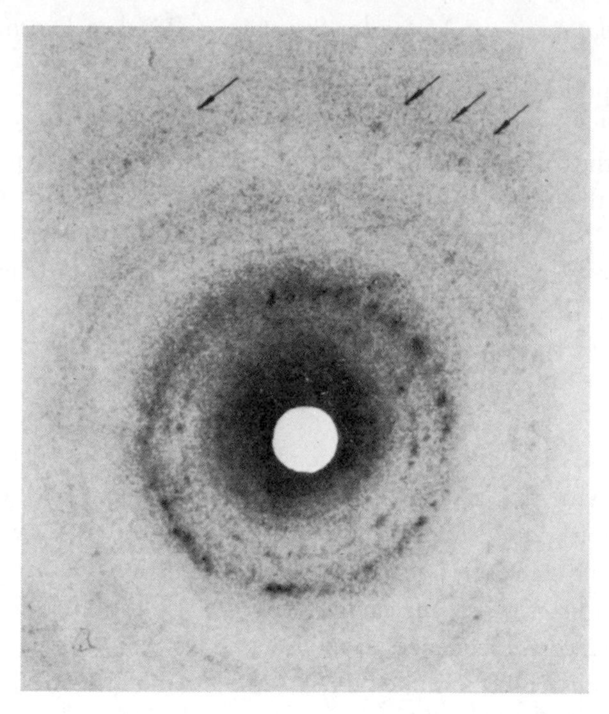

Fig. 13. Diffraction pattern of unoriented transfer RNA, showing diffraction rings with spots corresponding to reflections from single crystals of RNA. The arrows point to reflections from planes ~ 6 Å apart.

number of bases in a DNA molecule is too large for determination of base sequence by X-ray diffraction to be feasible. However, in transfer RNA the number of bases is not too large. The possibility of complete structure analysis of transfer RNA by means of X-rays is indicated by two observations. First, we have observed (Fig. 13), in X-ray patterns of transfer RNA, separate spots each corresponding to a single crystal of RNA. We estimated their size to be about 10μ and have confirmed this estimate by observing, in the polarizing microscope, birefringent regions that probably are the crystals. It should not be too difficult to grow crystals several times larger, which is large enough for single-crystal X-ray analysis.

The second encouraging observation is that the X-ray data from DNA have restricted resolution almost entirely on account of disorientation of the microcrystals in DNA fibres. The DNA intensity data indicate that the temperature factor ($B = 4$ Å) is the same for DNA as for simple compounds. It thus appears that DNA crystals have fairly perfect crystallinity and that, if single crystals of DNA could be obtained, the intensity data would be adequate for precise determination of all atomic positions in DNA (apart from the non-periodic base sequence).

We are investigating the possibility of obtaining single crystals of DNA, but the more exciting problem is to obtain single crystals of transfer RNA with crystalline perfection equal to that of DNA, and thereby analyse base sequence. At present, the RNA crystals are much less perfect than those of DNA. However, most of our experiments have been made with RNA that is a mixture of RNA's specific for different amino acids. We have seldom used RNA that is very largely specific for one amino acid only. We hope that good preparations of such RNA may be obtained consisting of one type of molecule only. We might expect such RNA to form crystals as perfect as those of DNA. If so, there should be no obstacle to the direct analysis of the whole structure of the molecule, including the sequence of the bases and the fold at the end of the helix. We may be over-optimistic, but the recent and somewhat unexpected successes of X-ray diffraction analysis in the nucleic acid and protein fields, are cause for optimism.

Acknowledgements

During the past twelve years, while studying molecular structure of nucleic acids, I have had so much help from so many people that all could not be acknowledged properly here. I must, however, thank the following:

Sir John Randall, for his long-standing help and encouragement, and for his vision and energy in creating and directing a unique laboratory;

all my co-workers at various times over the past twelve years; first, Raymond Gosling, Alex Stokes, Bill Seeds, and Herbert Wilson, then Bob Langridge, Clive Hooper, Max Feughelman, Don Marvin, and Geoffrey Zubay; and at present, Michael Spencer, Watson Fuller, and Struther Arnott, who with much ability, skill and persistence (often through the night) carried out the X-ray, molecular model-building, and computing studies;

my late colleague Rosalind Franklin who, with great ability and experience of X-ray diffraction, so much helped the initial investigations on DNA;

Leonard Hamilton for his constant encouragement and friendly cooperation, and for supplying us with high-quality DNA isolated in many forms and from many sources; Geoffrey Brown for giving me moral and intellectual support throughout the work and for preparing RNA for X-ray study; Harriet Ephrussi-Taylor for the privilege of collaborating with her in studying crystallization of transforming principle; the laboratory technicians, mechanics and photographers, including P. J. Cooper, N. Chard, J. Hayward, Mrs. F. Collier, Z. Gabor, and R. Lerner, for having played a valuable part in the work at various stages.

I also wish to thank:

the Medical Research Council for their far-sighted and consistent support of our work; King's College for being our base; I.B.M. United Kingdom Limited and I.B.M. World Trade Corporation and the London University Computer Unit for help with computing; The Rockefeller Foundation and The British Empire Cancer Campaign for financial support; the Sloan-Kettering Institute, New York, and the Stazione Zoologica, Naples, for use of facilities.

More generally, I thank:

Francis Crick and Jim Watson for stimulating discussion; Norman Simmons for having refined techniques of isolating DNA and thereby helping a great many workers including ourselves; many other workers for supplying us with DNA and RNA; and especially, Erwin Chargaff for laying foundations for nucleic acid structural studies by his analytical work and his discovery of the equality of base contents in DNA, and for generously helping us newcomers in the field of nucleic acids.

1. T. Caspersson, *Naturwiss.*, 29 (1941) 33.
2. J. Brachet, *Arch. Biol. Liege*, 53 (1942) 207.
3. O. T. Avery, C. M. MacLeod, and M. McCarty, *J. Exp. Med.*, 79 (1944) 137.
4. R. Signer and H. Schwander, *Helv. Chim. Acta*, 32 (1949) 853.
5. A. D. Hershey and M. Chase, *J. Gen. Physiol.*, 36 (1952) 39.
6. E. Chargaff, *Experientia*, 6 (1950) 201.
7. R. Signer, T. Caspersson, and E. Hammarsten, *Nature*, 141 (1938) 122.
8. W. J. Schmidt, *Die Doppelbrechung von Karyoplasma, Zytoplasma, und Metaplasma*, Borntraeger, Berlin, 1937.
9. H. O. E. Pattri, *Z. Zellforsch. Mikroskop. Anat.*, 16 (1932) 723.
10. W. T. Astbury, *Symp. Soc. Exptl. Biol., I. Nucleic Acid*, Cambridge Univ. Press, 1947, p. 66.
11. J. M. Gulland and D. O. Jordan, *Symp. Soc. Exptl. Biol., I. Nucleic Acid*, Cambridge Univ. Press, 1947.
12. J. M. Gulland, *Cold. Spring Harbor Symp. Quant. Biol.*, 12 (1947) 95.
13. J. D. Watson and F. H. C. Crick, *Nature*, 171 (1953a) 737.

14. J. D. Watson and F. H. C. Crick, *Nature*, 171 (1953b) 964.
15. K. Hoogsteen, *Acta Cryst.*, 12 (1959) 822.
16. L. D. Hamilton, R. K. Barclay, M. H. F. Wilkins, G. L. Brown, H. R. Wilson, D. A. Marvin, H. Ephrussi-Taylor, and N. S. Simmons, *J. Biophys. Biochem. Cytol.*, 5 (1959) 397.
17. F. Rinne, *Trans. Faraday Soc.*, 29 (1933) 1016.
18. M. H. F. Wilkins and J. T. Randall, *Biochim. Biophys. Act.*, 10 (1953) 192.
19. R. Langridge, H. R. Wilson, C. W. Hooper, M. H. F. Wilkins, and L. D. Hamilton, *J. Mol. Biol.*, 2 (1960) 19.
20. M. H. F. Wilkins, *J. Chim. Phys.*, 58 (1961) 891.
21. P. Doty, *Biochem. Soc. Symp.*, No. 21 (1961) 8.
22. A. Rich, in *A Symposium on Molecular Biology* (Ed. Zirkle), Univ. Chicago Press, 1959, p. 47.
23. A. Rich and J. D. Watson, *Nature*, 173 (1954) 995.
24. M. Spencer, W. Fuller, M. H. F. Wilkins, and G. L. Brown, *Nature*, 194 (1962) 1014.

Biography

Maurice Hugh Frederick Wilkins was born at Pongaroa, New Zealand, on December 15th, 1916. His parents came from Ireland; his father Edgar Henry Wilkins was a doctor in the School Medical Service and was very interested in research but had little opportunity for it.

At the age of 6, Wilkins was brought to England and educated at King Edward's School, Birmingham. He studied physics at St. John's College, Cambridge, taking his degree in 1938. He then went to Birmingham University, where he became research assistant to Dr. J. T. Randall in the Physics Department. They studied the luminescence of solids. He obtained a Ph.D. in 1940, his thesis being mainly on a study of thermal stability of trapped electrons in phosphors, and on the theory of phosphorescence, in terms of electron traps with continuous distribution of trap depths. He then applied these ideas to various war-time problems such as improvement of cathode-ray tube screens for radar. Next he worked under Professor M. L. E. Oliphant on mass spectrograph separation of uranium isotopes for use in bombs and, shortly after, moved with others from Birmingham to the Manhattan Project in Berkeley, California, where these studies continued.

In 1945, when the war was over, he was lecturer in physics at St. Andrews' University, Scotland, where Professor J. T. Randall was organizing biophysical studies. He had spent seven years in physics research and now began in biophysics. The biophysics project moved in 1946 to King's College, London, where he was a member of the staff of the newly formed Medical Research Council Biophysics Research Unit. He was first concerned with genetic effects of ultrasonics; after one or two years, he changed his research to development of reflecting microscopes for ultraviolet microspectrophotometric study of nucleic acids in cells. He also studied the orientation of purines and pyrimidines in tobacco mosaic virus and in nucleic acids, by measuring the ultraviolet dichroism of oriented specimens, and he studied, with the visible-light polarizing microscope, the arrangement of virus particles in crystals of TMV and measured dry mass in cells with interference microscopes. He then began X-ray diffraction studies of DNA and sperm heads. The discovery of the well-defined patterns led to the deriving of the molecular structure of DNA. Further X-ray studies established the correctness of the Watson-Crick proposal for DNA structure. Relevant publications are « The molecular configuration of deoxyribonucleic acid. I. X-ray diffraction study of a crystalline form of the lithium salt », by R. Langridge, H. R. Wilson, C. W. Hooper, M. H. F. Wilkins, and L. D. Hamilton in *J. Mol. Biol.*, 2 (1960) 19, and « Determination of the helical configuration of ribonucleic acid molecules by X-ray diffraction study of crystalline amino-acid-transfer ribonucleic acid », by M. Spencer, W. Fuller, M. H. F. Wilkins, and G. L. Brown in *Nature*, 194 (1962) 1014.

Wilkins became Assistant Director of the Medical Research Council Unit in 1950 and Deputy Director in 1955. A sub-department of Biophysics was formed in King's College, and he was made Honorary Lecturer in it. In 1961 a full Department of Biophysics was established.

He was elected F.R.S. in 1959, given the Albert Lasker Award (jointly with Watson and Crick) by the American Public Health Association in 1960, and made Companion of the British Empire in 1962.

He married Patricia Ann Chidgey in 1959; they have a daughter Sarah and a son George. He finds his recreations in his collection of sculptures and in gardening.

JAMES D. WATSON

The Involvement of RNA in the Synthesis of Proteins

December 11, 1962

Prologue

I arrived in Cambridge in the fall of 1951. Though my previous interests were largely genetic, Luria had arranged for me to work with John Kendrew. I was becoming frustrated with phage experiments and wanted to learn more about the actual structures of the molecules which the geneticists talked about so passionately. At the same time John needed a student and hoped that I should help him with his X-ray studies on myoglobin. I thus became a research student of Clare College with John as my supervisor.

But almost as soon as I set foot in the Cavendish, I inwardly knew I would never be of much help to John. For I had already started talking with Francis. Perhaps even without Francis, I would have quickly bored of myoglobin. But with Francis to talk to, my fate was sealed. For we quickly discovered that we thought the same way about biology. The center of biology was the gene and its control of cellular metabolism. The main challenge in biology was to understand gene replication and the way in which genes control protein synthesis. It was obvious that these problems could be logically attacked only when the structure of the gene became known. This meant solving the structure of DNA. Then this objective seemed out of reach to the interested geneticists. But in our cold, dark Cavendish lab, we thought the job could be done, quite possibly within a few months. Our optimism was partly based on Linus Pauling's feat[1] in deducing the α-helix, largely by following the rules of theoretical chemistry so persuasively explained in his classical *The Nature of the Chemical Bond*. We also knew that Maurice Wilkins had crystalline X-ray diffraction photographs from DNA and so it must have a well-defined structure. There was thus an answer for somebody to get.

During the next eighteen months, until the double-helical structure became elucidated, we frequently discussed the necessity that the correct structure have the capacity for self-replication. And in pessimistic moods, we often worried that the correct structure might be dull. That is, it would suggest absolutely nothing and excite us no more than something inert like collagen.

The finding of the double helix[2] thus brought us not only joy but great relief. It was unbelievably interesting and immediately allowed us to make a serious proposal[3] for the mechanism of gene duplication. Furthermore, this replication scheme involved thoroughly understood conventional chemical forces. Previously, some theoretical physicists, among them Pascual Jordan[4], had proposed that many biological phenomena, particularly gene replication, might be based on still undiscovered long-range forces arising from quantum mechanical resonance interactions. Pauling[5] thoroughly disliked this conjecture and firmly insisted that known short-range forces between complementary surfaces would be the basis of biological replication.

The establishment of the DNA structure reinforced our belief that Pauling's arguments were sound and that long-range forces, or for that matter any form of mysticism, would not be involved in protein synthesis. But for the protein replication problem mere inspection of the DNA structure then gave no immediate bonus. This, however, did not worry us since there was much speculation that RNA, not DNA, was involved in protein synthesis.

Introduction

The notion that RNA is involved in protein synthesis goes back over twenty years to the pioneering experiments of Brachet and Caspersson[6] who showed that cells actively synthesizing protein are rich in RNA. Later when radioactive amino acids became available, this conjecture was strengthened by the observation[7] that the cellular site of protein synthesis is the microsomal component, composed in large part of spherical particles rich in RNA. Still later experiments[8] revealed that these ribonucleoprotein particles (now conveniently called ribosomes), not the lipoprotein membranes to which they are often attached, are the sites where polypeptide bonds are made. Most ribosomes are found in the cytoplasm and correspondingly most cellular protein synthesis occurs without the direct intervention of the nuclear-located DNA. The possibility was thus raised that the genetic specificity present in DNA is first transferred to RNA intermediates which then function as templates controlling assembly of specific amino acids into proteins.

We became able to state this hypothesis in more precise form when the structure of DNA became known in 1953. We then realized that DNA's genetic specificity resides in the complementary base sequences along its two intertwined chains. One or both of these complementary chains must serve as templates for specific RNA molecules whose genetic information again must reside in specific base sequences. These RNA molecules would then assume 3-dimensional configurations containing surfaces complementary to the side groups of the 20 specific amino acids.

X-ray Studies on RNA and RNA-containing Viruses

The direct way to test this hypothesis was to solve the RNA structure. Already in 1952, I had taken some preliminary X-ray diffraction pictures of RNA. These, however, were very diffuse, and it was not until I returned to the United States in the fall of 1953 that serious X-ray studies on RNA began. Alexander Rich and I, then both at the California Institute of Technology, obtained RNA samples from various cellular sources. We[9] were first very encouraged that all the RNA samples, no matter their cellular origin, give similar X-ray diffraction pattern. A general RNA structure thus existed. This gave us hope that the structure, when solved, would be interesting. Our first pictures already showed large systematic absence of reflections on the meridian, suggesting a helical structure. But despite much effort to obtain native undegraded high molecular weight samples, no satisfactory X-ray diffraction pattern was obtained. The reflections were always diffuse, no evidence of crystallinity was seen. Though there were marked similarities to the DNA pattern, we had no solid grounds for believing that these arose from a similar helical molecule. The problem whether RNA was a one- or several-chained structure remained unanswered.

We then considered the possibility that RNA might have a regular structure only when combined with protein. At that time (1955) there was no good evidence for RNA existing free from protein. All RNA was thought to exist either as a viral component or to be combined with protein in ribonucleoprotein particles. It thus seemed logical to turn attention to a study of ribonucleoprotein particles (ribosomes) since upon their surfaces protein was synthesized. Our hope again was that the establishment of their structure would reveal the long-sought-after cavities specific for the amino acids.

Then we were struck by the morphological similarity between ribosomes and small RNA-containing viruses like Turnip Yellow Mosaic Virus or Poliomyelitis Virus. By then (1955–1956) I was back in Cambridge with

Crick to finish formulating some general principles on viral structure[10]. Our main idea was that the finite nucleic acid content of viruses severely restricted the number of amino acids they could code for. As a consequence, the protein coat could not be constructed from a very large number of different protein molecules. Instead it must be constructed from a number of identical small sub-units arranged in a regular manner. These ideas already held for Tobacco Mosaic Virus, a rod-shaped virus, and we were very pleased when D. L. D. Caspar[11], then working with us at the Cavendish, took some elegant diffraction pictures of Bushy Stunt Virus crystals and extended experimental support to the spherical viruses.

Structural Studies on Ribosomes

At that time almost no structural studies had been done with ribosomes. They were chiefly characterized by their sedimentation constants; those from higher organisms[12] in the 70s–80s range, while those from bacteria[13] appeared smaller and to be of two sizes (30s and 50s). Because the bacterial particles seemed smaller, they seemed preferable for structural studies. Thus when Alfred Tissières and I came to Harvard's Biological Laboratories in 1956, we initiated research on the ribosomes of the commonly studied bacteria *Escherichia coli*. We hoped that their structure would show similarities with the small spherical RNA viruses. Then we might have a good chance to crystallize them and to eventually use X-ray diffraction techniques to establish their 3-dimensional structure.

Ribosome sub-units

But from the beginning of our Harvard experiments, it was obvious that ribosome structure would be more complicated than RNA virus structure. Depending upon the concentration of divalent cations (in all our experiments Mg^{++}), 4 classes of *E. coli* ribosomes were found, characterized by sedimentation constants of 30s, 50s, 70s, and 100s. Our first experiments in 10^{-4} *M* Mg^{++} revealed 30s and 50s ribosomes. At the same time Bolton[14], at the Carnegie Institute of Washington employing higher Mg^{++} levels, saw faster sedimenting ribosomes and suggested that they were observing aggregates of the smaller particles. Soon after, our experiments[15] revealed that, as the Mg^{++} concentration is raised, one 30s particle and one 50s particle

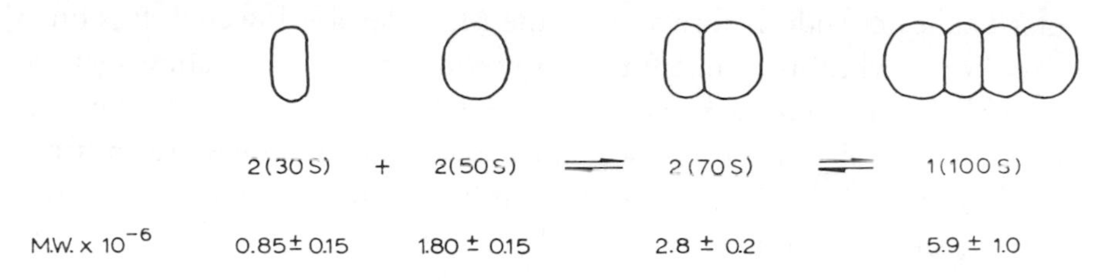

Fig. 1. Diagrammatic representation of *E. coli* ribosome sub-units and their aggregation products. (The molecular weight data are from Tissières *et al.*[15])

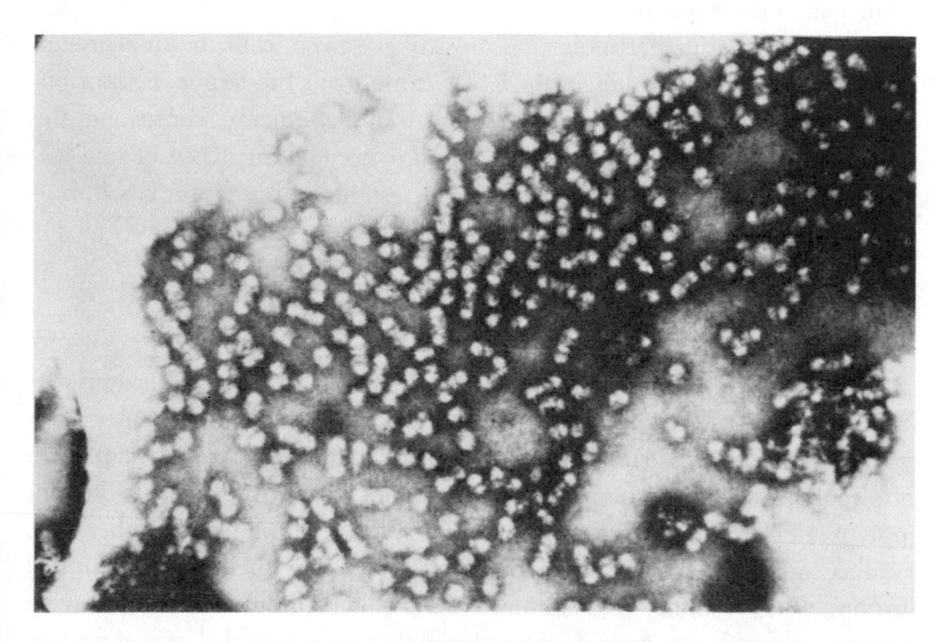

Fig. 2. Electron micrograph of negatively stained *E. coli* ribosomes (Huxley and Zubay[15]). Two particle types are predominant: (*1*) 70s containing two sub-units of unequal size, and (*2*) 100s consisting of two 70s ribosomes joined together at their smaller (30s) sub-units.

combine to form a 70s ribosome. At still higher Mg^{++} concentrations, two 70s ribosomes dimerize to form a 100s ribosome. (Figs. 1 and 2).

Ribosomes from every cellular source have a similar sub-unit construction. As with *E. coli* ribosomes, the level of divalent cations determines which ribosomes exist. Bacterial ribosomes seem to require higher Mg^{++} levels in order to aggregate into the larger sizes. Conversely they break down much faster to the 30s and 50s forms when the Mg^{++} level is lowered. It is often convenient[16] when using mammalian ribosomes to add a chelating agent to rapidly break down the 80s ribosomes (homologous to the 70s ribosomes of bacteria) to their 40s and 60s sub-units. Bacterial ribosomes are thus not significantly smaller than mammalian ribosomes. It is merely easier to observe the smaller sub-units in bacterial systems.

Ribosomal RNA

Already in 1958 there were several reports[17] that ribosomal RNA from higher organisms sedimented as two distinct components (18s and 28s). We thought that the smaller molecules most likely arose from the smaller sub-unit while the faster sedimenting RNA came from the larger of the ribosomal sub-units. Experiments of Mr. Kurland[18] quickly confirmed this hunch. The *E. coli* 30s ribosome was found to contain one RNA chain (16s) with a molecular weight of 5.5×10^5. Correspondingly a larger RNA molecule (23s) of mol. wt. 1.1×10^6 was found in most 50s ribosomes (Fig. 3).

Ribosome proteins

Analysis of the protein component revealed a much more complicated picture. In contrast to the small RNA viruses, where the protein coat is constructed from the regular arrangement of a large number of identical protein molecules, each ribosome most likely contains a large number of different polypeptide chains. At first, our results suggested a simple answer when Drs. Waller and J. I. Harris analysed *E. coli* ribosomes for their amino terminal groups. Only alanine, methionine, with smaller amounts of serine, were present in significant amounts. This hinted that only several classes of protein molecules were used for ribosomal construction. Further experiments of Dr. Waller[19], however, suggested the contrary. When ribosomal protein fractions were analysed in starch-gel electrophoresis, more than 20 distinct bands were seen. Almost all these proteins migrated towards the anode at pH 7 confirming the net basic charge of ribosomal protein[20]. A variety of control experiments suggested that these bands represent distinct polypeptide chains, not merely aggregated states of several fundamental sub-units. Moreover, the band pattern from 30s ribosomes was radically different from that of 50s proteins.

As yet we have no solid proof that each 70s ribosome contains all the various protein components found in the total population. But so far, all attempts by Dr. Waller to separate chromatographically intact ribosomes into fractions with different starch-gel patterns have failed. The total protein component of a 70s ribosome amounts to about 9×10^5 daltons. Since the end group analysis suggests an average mol. wt. of about 30,000, approximately 20 polypeptide chains are used in 50s construction and 10 for the 30s ribosome. It is possible that all the polypeptide chains in a 30s particle are different. Waller already has evidence for 10 distinct components in 30s ribosomes and the present failure to observe more in the 50s protein fraction may merely mean that the same electrophoretic mobility is shared by several polypeptide chains.

We believe that all these proteins have primarily a structural role. That is, they are not enzymes but largely function to hold the ribosomal RNA and necessary intermediates in the correct position for peptide bond formation. In addition a number of enzymes are bound tightly to ribosomes. As yet their function is unclear. One such is a bacterial ribonuclease, found by Elson[21] to be specifically attached to 30s ribosomes in a latent form. No ribonuclease activity is present until ribosome breakdown. Dr. Spahr[22] in our laboratories has purified this enzyme, shown its specificity and from specific activity measurements, concludes that it is present on less than one in twenty

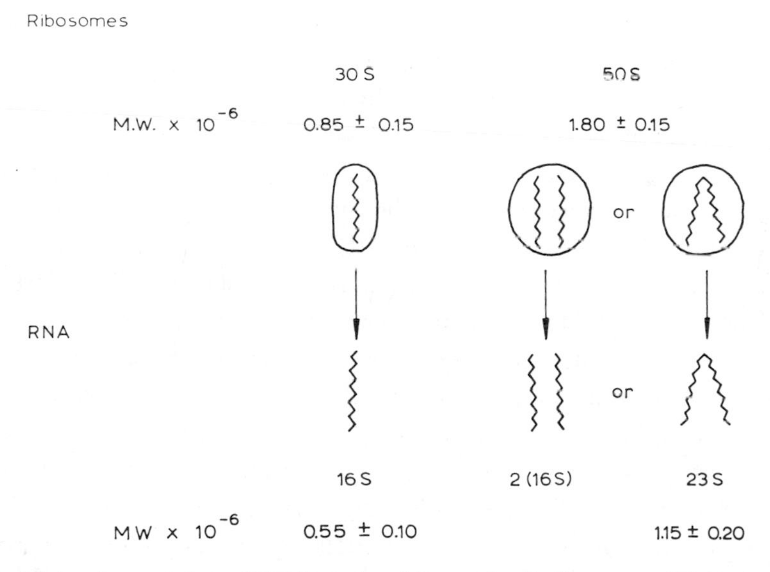

Fig. 3. Molecular weights of RNA isolated from *E. coli* ribosomes. (This picture is diagrammatic and does not represent the true conformation of ribosomal RNA.)

30s particles. It is clear that this enzyme if present in a free active form, would be rapidly lethal to its host cell. Thus its presence in latent form is expected. But why it is stuck to ribosomes is still a complete mystery.

Chemical Intermediates in Protein Synthesis

Our early experiments with ribosomes were almost unrelated to the efforts of biochemists. At that time our research objects seemed very different. The enzymologically oriented biochemists hoped to find the intermediates and enzymes necessary for peptide bond formation. On the contrary, those of us with a genetic orientation wanted to see the template and discover how it picked out the correct amino acid. Very soon, however, these separate paths came together, partly because of a breakthrough in the nature of the amino acid intermediates, and partly from an incisive thought by Crick.

The biochemical advances arose from work in Paul Zamecnik's laboratory at the Massachusetts General Hospital. There was developed a reproducible *in vitro* system[23] containing ribosomes, supernatant factors, and ATP which incorporated amino acids into protein. Using these systems Hoagland made two important discoveries. Firstly, he[24] showed that amino acids are initially activated by ATP to form high-energy AA–AMP complexes. Secondly, he demonstrated[25] that the activated amino acids are then transferred to low molecular weight RNA molecules (now known as soluble or transfer RNA), again in an activated form. These amino–acyl–sRNA compounds then function as the direct intermediate for peptide bond formation (Fig. 4).

It had previously been obvious that amino acid activation would have to occur. However, Hoagland's second discovery (in 1956) of the involvement of a hitherto undiscovered RNA form (sRNA) was unanticipated by almost everybody. Several years previously (in 1954), Leslie Orgel and I spent a quite frustrating fall attempting to construct hypothetical RNA structures which contained cavities complementary in shape to the amino acid side groups. Not only did plausible configurations for the RNA backbone fail to result in good cavities, but even when we disregarded the backbone, we also failed to find convincing holes which might effectively distinguish between such amino acids as valine and isoleucine. Crick, at the same time (early 1955) sensed the same dilemma, and suggested a radical solution to the paradox. He proposed[26] that the amino acids do not combine with the template. Instead each should first combine with a specific adaptor molecule,

$$\text{(a)} \quad AA + ATP \longrightarrow AMP \sim AA + PP$$

$$\text{(b)} \quad AMP \sim AA + SRNA \longrightarrow AA \sim SRNA + AMP$$

$$\text{(c)} \quad (AA \sim SRNA)_n + GTP \xrightarrow{\text{Ribosomes}} AA_1 - AA_2 \cdots AA_n + GDP \ (GMP?)$$

Fig. 4. Enzymatic steps in protein peptide bond formation. Steps (a) and (b) are catalyzed by single enzymes. The number of enzymes required in (c) is unknown.

capable of selectively interacting with the hydrogen bonding surfaces provided by RNA's purine and pyrimidine bases. This scheme requires at least twenty different adaptors, each specific for a given amino acid. These are very neatly provided by the specific sRNA molecules. Soon after Hoagland's discovery of sRNA, many experiments, particularly by Hoagland and Paul Berg[27], established that the sRNA molecules are in fact specific for a given amino acid. It thus became possible to imagine, following Crick's reasoning, that the ribosomal template for protein synthesis combined not with the amino acid side groups, but instead with a specific group of bases on the soluble RNA portion of the amino-acyl-sRNA precursors.

Participation of Active Ribosomes in Protein Synthesis

Very little protein synthesis occurred in the cell-free system developed by the Massachusetts General Hospital Group. Only by using radioactive amino acids could they convincingly demonstrate amino acid incorporation into proteins. This fact, initially seemed trivial and there was much hope that when better experimental conditions were found, significant net synthesis would occur. But despite optimistic claims from several laboratories, no real improvement in the efficiency of cell-free synthesis resulted. Some experiments (1959) of Dr. Tissières and Mr. Schlessinger[28] with *E. coli* extracts illustrate well this point. At 30°C, cell-free synthesis occurs linearly for 5–10 minutes and then gradually stops. During this interval the newly synthesized protein amounts to 1–3 γ of protein per mg of ribosomes. Of this about one third was released from the ribosomes, the remainder being ribosomal bound.

Cell-free synthesis in *E. coli* extracts requires the high ($\sim 10^{-2}\ M$) Mg^{++}

levels which favor the formation of 70s ribosomes from their 30s and 50s sub-units. Following incorporation, those ribosomes possessing nascent polypeptide chains become less susceptible to breakdown to 30s and 50s ribosomes. When cell-free extracts (following synthesis) are briefly dialyzed against 10^{-4} *M* Mg^{++}, about 80–90% of the 30s and 50s ribosomes become free. There remain, however, 10–20% of the original 70s ribosomes and it is upon these «stuck» ribosomes that most ribosomal bound nascent protein is located. This firstly suggests that protein synthesis occurs on 70s ribosomes, not upon free 30s or 50s ribosomes. Secondly, in the commonly studied *E. coli* extract, only a small ribosome fraction is functional. Tissières and Schlessinger named these particles «active ribosomes» and suggested, they contained a functional component lacking in other ribosomes.

Each active ribosome synthesizes on the average between 15,000 and 50,000 daltons of protein. This is in the size range of naturally occurring polypeptide chains. Thus while we remained unsatisfied by the small net synthesis, sufficient synthesis occurs to open the possibility that some complete protein molecules are made. This encouraged us to look for synthesis of β-galactosidase. None, however, was then found[29] despite much effort.

Another important point emerged from these early (1959) incorporation studies with *E. coli* extracts. Addition of small amounts of purified deoxyribonuclease decreased protein synthesis to values 20–40% that found in untreated extracts[28]. This was completely unanticipated, for it suggested that high molecular weight DNA functions in the commonly studied bacterial extracts. But since a basal synthetic level occurs after DNA is destroyed by deoxyribonuclease, the DNA itself must not be directly involved in peptide bond formation. Instead, this suggested synthesis of new template RNA upon DNA in untreated extracts. If true, this would raise the possibility, previously not seriously considered by biochemists that the RNA templates themselves might be unstable, and hence a limiting factor in cell-free protein synthesis.

Metabolic Stability of Ribosomal RNA

All our early ribosome experiments had assumed that the ribosomal RNA was the template. Abundant evidence existed that proteins were synthesized on ribosomes and since the template must be RNA, it was natural to assume that it was ribosomal RNA. Under this hypothesis ribosomal RNA was a collection of molecules of different base sequences, synthesized on the functioning regions of chromosomal DNA. Following their synthesis, they combined with the basic ribosomal proteins to form ribosomes. We thus visualized that the seemingly morphological identical ribosomes were, in fact, a collection of a very large number of genetically distinct particles masked by the similarity of their protein component.

Then there existed much suggestive evidence that ribosomal RNA molecules were stable in growing bacteria. As early as 1949, experiments showed that RNA precursors, once incorporated into RNA, remained in RNA. Then the distinction between ribosomal and soluble RNA was not known, but later experiments by the ribosome group of the Carnegie Institute of Washington and at Harvard indicated similar stabilities of both fractions. These experiments, however, did not follow the fate of single molecules, and the possibility remained that a special trick allowed ribosomal RNA chains to be broken down to fragments that were preferentially re-used to make new ribosomal RNA molecules. Davern and Meselson[30], however, ruled out this possibility by growing ribosomal RNA in heavy (^{13}C,^{15}N) medium, followed by several generations of growth in light (^{12}C, ^{14}N) medium. They then separated light from heavy ribosomal RNA in cesium formate density gradients and showed that the heavy molecules remained completely intact for at least two generations. This result predicts, assuming ribosomes to be genetically specific, that the protein templates should persist indefinitely in growing bacteria.

Experiments Suggesting Unstable Protein Templates

But already by the time of the Davern & Meselson experiment (1959), evidence began to accumulate, chiefly at the Institut Pasteur, that some, if not all, bacterial templates were unstable with lives only several per cent of a generation time. None of these experiments, by themselves, were convincing. Each could be interpreted in other ways which retained the concept of stable templates. But taken together, they argued a strong case.

These experiments were of several types. One studied the effect of suddenly adding or destroying specific DNA molecules. Sudden introduction was achieved by having a male donor introduce a specific chromosomal region absent in the recipient female. Simultaneously the ability of the male gene to function (produce an enzymatically active protein) in the female cell was measured. Riley, Pardee, Jacob, and Monod[31] obtained the striking finding that β-galactosidase, genetically determined by a specific male gene, began to be synthesized at its maximum rate within several minutes after gene transfer. Thus the steady state number of β-galactosidase templates was achieved almost immediately. Conversely when the *E. coli* chromosome was inactivated by decay of ^{32}P atoms incorporated into DNA, they observed that active enzyme formation stops within several minutes. It thus appeared that the ribosomal templates could not function without concomitant DNA function.

At the same time, François Gros discovered[32] that bacteria grown in 5-fluorouracil produced abnormal proteins, most likely altered in amino acid sequences. 5-Fluorouracil is readily incorporated into bacterial RNA and its presence in RNA templates may drastically raise the mistake level. More unexpected was the observation that following 5-fluorouracil addition the production of all normal proteins ceases within several minutes. Again this argues against the persistance of *any* stable templates.

Unstable RNA Molecules in Phage Infected Cells

At first it was thought that no RNA synthesis occurred in T2 infected cells. But in 1952 Hershey[33] observed that new RNA molecules are synthesized at a rapid rate. But no net accumulation occurs since there is a correspondingly fast breakdown. Surprisingly almost everybody ignored this discovery. This oversight was partly due to the tendency, still then prevalent, to suspect that the metabolism of virus infected cells might be qualitatively different from that of uninfected cells.

Volkin and Astrachan[34] were the first (1956) to treat Hershey's unstable fraction seriously. They measured its base composition and found it different from that of uninfected *E. coli* cells. It bore a great resemblance to the infecting viral DNA which suggested that it was synthesized on T2 DNA templates. Moreover, and *most importantly*, this RNA fraction must be the template for phage specific proteins. Unless we assume that RNA is not involved in phage protein synthesis, it necessarily follows that the Volkin-Astrachan DNA-like RNA provides the information for determining amino acid sequences in phage specific proteins.

Not till the late summer of 1959 was its physical form investigated. Then Nomura, Hall, and Spiegelman[35] examined its relationship to the already characterized soluble and ribosomal RNA's. Immediately they observed that none of the T2 RNA was incorporated into stable ribosomes. Instead, in low Mg^{++} (10^{-4} M) it existed free while in 10^{-2} M Mg^{++} they thought it became part of 30s ribosomal like particles. At the same time, Mr. Risebrough in our laboratories began studying T2 RNA, also using sucrose gradient centrifugation. He also found that T2 RNA was not typical ribosomal RNA. In addition, he was the first to notice (in early spring 1960) that in 10^{-2} M Mg^{++}, most T2 RNA sedimented not with 30s particles but with the larger 70s and 100s ribosomes.

His result leads naturally to the hypothesis that phage protein synthesis takes place on genetically non-specific ribosomes to which are attached metabolically unstable template RNA molecules. Independently of our work, Brenner and Jacob motivated by the above-mentioned metabolic and genetic experiments from the Institut Pasteur, were equally convinced that conditions were ripe for the direct demonstration of metabolically unstable RNA templates to which Jacob and Monod[36] gave the name *messenger RNA*. In June of 1960, they travelled to Pasadena for a crucial experiment in Meselson's laboratory. They argued that all the T2 messenger RNA should be attached to old ribosomes synthesized before infection. This they elegantly demonstrated[37] by T2 infecting heavy (^{13}C and ^{15}N) labeled bacteria in light (^{12}C and ^{14}N) medium. Subsequent CsCl equilibrium centrifugation revealed that most of the T2 messenger RNA was indeed attached to « old » ribosomes, as was all the ribosomal bound nascent protein, labeled by pulse exposure to radioactive amino acids.

Demonstration of Messenger RNA Molecules in Uninfected Bacteria

We were equally convinced that similar messenger RNA would be found in uninfected bacteria. Its demonstration then presented greater problems, because of the simultaneous synthesis of ribosomal and soluble RNA. François Gros had then (May 1960) just arrived for a visit to our laboratory. Together with Mr. Kurland and Dr. Gilbert, we decided to look for labeled messenger molecules in cells briefly exposed to a radioactive RNA precursor. Experiments with T2 infected cells suggested that the T2 messenger comprised about 2–4% of the total RNA and that most of its molecules had lives less than several minutes. If a similar situation, held for uninfected cells, then during any short interval, most RNA synthesis would be messenger. There would be no significant accumulation since it would be broken down almost as fast as it was made.

Again the messenger hypothesis was confirmed[38]. The RNA labeled during pulse exposures was largely attached to 70s and 100s ribosomes in 10^{-2} M Mg^{++}. In low Mg^{++} (10^{-4} M), it came off the ribosomes and sedimented free with an average sedimentation constant of 14s. Base ratio analysis revealed DNA like RNA molecules in agreement with the expectation that it was produced on very many DNA templates along the bacterial chromosome. Soon afterwards, Hall and Spiegelman[39] formed artificial T2 DNA; T2 messenger RNA hybrid molecules and in several laboratories[40], hybrid molecules were subsequently formed between *E. coli* DNA and *E. coli* pulse RNA. The DNA template origin for messenger RNA was thus established beyond doubt.

The Role of Messenger RNA in Cell-Free Protein Synthesis

It was then possible to suggest why deoxyribonuclease partially inhibits amino acid incorporation in *E. coli* extracts. The messenger hypothesis prompts the idea that DNA in the extract is a template for messenger RNA. This newly made messenger then attaches to ribosomes where it serves as additional protein templates. Since deoxyribonuclease only destroys the capacity to make messenger, it has no effect upon the messenger present at the time of extract formation. Hence, no matter how high the deoxyribonuclease concentration employed, a residual fraction of synthesis will always occur. Experiments by Tissières and Hopkins[41] in our laboratories and by Berg, Chamberlain, and Wood[42] at Stanford confirmed these ideas. First it was shown that addition of DNA to extracts previously denuded of DNA significantly increased amino acid incorporation. Secondly, RNA synthesis occurs simultaneously with *in vitro* protein synthesis. This RNA has a DNA like composition, attaches to ribosomes in 10^{-2} M Mg^{++}, and physically resembles *in vivo* synthesized messenger RNA.

Furthermore, Tissières showed that addition of fractions rich in messenger RNA stimulated *in vitro* protein synthesis 2–5 fold. More striking results came from Nirenberg and Matthaei[43]. They reasoned that *in vitro* messenger destruction might be the principal cause why cell-free systems stopped synthesizing protein. If so, *preincubated extracts* deficient in natural messenger should respond more to new messenger addition. This way they became able to demonstrate a 20-fold increase in protein synthesis following addition of phenol-purified *E. coli* RNA. Like Tissières' active fraction, their stimulating fraction sedimented heterogeneously arguing against an effect due to either ribosomal or soluble RNA. More convincing support came when they next added TMV RNA to preincubated *E. coli* extracts. Again a 10–20 fold stimulation occurred. Here there could be no confusion with possible ribosomal RNA templates. Even more dramatic[44] was the effect of polyuridylic acid (like TMV RNA single stranded) addition. This specifically directed the incorporation of phenylalanine into polyphenylalanine. With this experiment (June 1961) the messenger concept became a fact. Direct proof then existed that single stranded messenger was the protein template.

Presence of Messenger RNA in Active Ribosomes

In *in vitro* systems ordinarily only 10–20% of *E. coli* ribosomes contain attached messenger RNA. This first was shown in experiments of Risebrough[45] who centrifuged extracts of T2 infected cells through a sucrose gradient. Ribosomes containing labeled messenger were found to centrifuge faster than ordinary ribosomes. Similarly, Gilbert[46] showed that these faster sedimenting ribosomes are «active», that is, able to incorporate amino acids into proteins. A fresh cell-free extract was centrifuged through a sucrose gradient. Samples along the gradient were collected and then tested for their ability to make protein. A complete parallel was found between «activity» and the presence of messenger.

Furthermore, if an extract is centrifuged *after* it has incorporated amino acids, the nascent protein chains also sediment attached to a small fraction of fast sedimenting ribosomes[45]. These ribosomes still contain messenger RNA. For when the messenger molecules are destroyed by ribonuclease (ribosomes remain intact in the presence of γ amounts of ribonuclease), the ribosomal bound nascent protein sediments as 70s ribosomes. The nascent protein is thus not attached to messenger RNA but must be directly bound to ribosomes.

Binding of sRNA to Ribosomes

Experiments by Schweet[47] and Dintzes[48] show that proteins grow by stepwise addition of individual amino acids beginning at the amino terminal end.

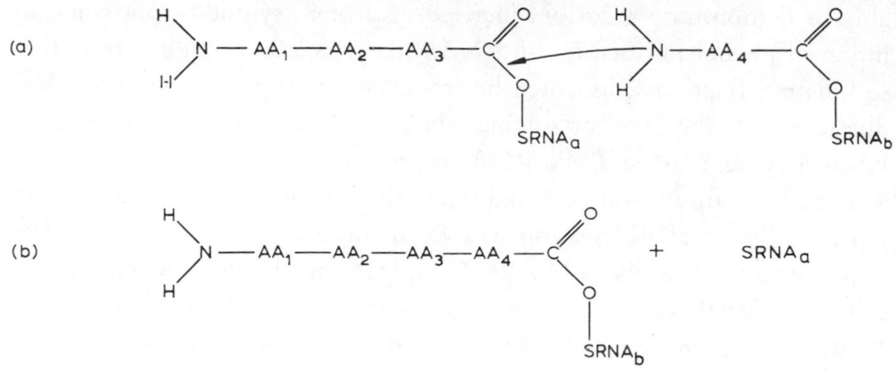

(a) $H_2N—AA_1—AA_2—AA_3—C(=O)—O—SRNA_a$ ← $H_2N—AA_4—C(=O)—O—SRNA_b$

(b) $H_2N—AA_1—AA_2—AA_3—AA_4—C(=O)—O—SRNA_b$ + $SRNA_a$

Fig. 5. Stepwise growth of a polypeptide chain. Initiation begins at the free NH_2 end with the growing point terminated by a sRNA molecule.

Since the immediate precursors are amino–acyl–sRNA molecules, their result predicts that the polypeptide chain is terminated at its carboxyl growing end by an sRNA molecule (Fig. 5). To test this scheme, we began some studies to see whether sRNA bound specifically to ribosomes. Cannon and Krug[49] first examined binding in the absence of protein synthesis. They showed that in 10^{-2} *M* Mg^{++} each 50s sub-unit of the 70s ribosome reversibly bound one sRNA molecule. The same amount of reversible binding occurs with amino–acyl–sRNA or with free sRNA and in the presence or absence of protein synthesis.

Protein synthesis, however, effects the binding observed in 10^{-4} *M* Mg^{++}. In the absence of protein synthesis no sRNA remains ribosomal bound when the Mg^{++} level is lowered from 10^{-2} *M* to 10^{-4} *M*. On the contrary, following amino acid incorporation, sRNA molecules become tightly fixed to the « stuck » 70s ribosomes, whose nascent polypeptide chains prevent easy dissociation to 30s and 50s ribosomes. One sRNA molecule appears to be attached to each stuck ribosome. Prolonged dialysis against 10^{-4} *M* Mg^{++} eventually breaks apart the stuck ribosomes. Then all the bound sRNA as well as almost all the nascent protein is seen attached to the 50s component supporting the hypothesis that these bound sRNA molecules are directly attached to nascent chains (Fig. 6). Direct proof comes from recent experiments in which Gilbert[50] used the detergent duponol to further dissociate the 50s ribosomes to their protein and RNA components. Then the nascent protein and bound sRNA remained together during both sucrose gradient centrifugation and separation on G200 Sephadex columns. Following exposure, however, to either weak alkali or to hydroxylamine, treatments known to break amino–acyl–bonds, the sRNA and nascent proteins move separately.

The significance of the reversible binding by non-active (no messenger) ribosomes is not known. Conceivably inside growing cells, all ribosomes have attached messenger and synthesize protein. Under these conditions, only those sRNA molecules corresponding to the specific messenger sequence can slip into the ribosomal cavities. But when most ribosomes lack messenger templates, as in our *in vitro* extracts, then any sRNA molecule, charged or uncharged, may fill the empty site.

All evidence suggests that covalent bonds are not involved in holding nascent chains to ribosome. Instead it seems probable that the point of firm attachment involves the terminal sRNA residue, bound by Mg^{++} dependent

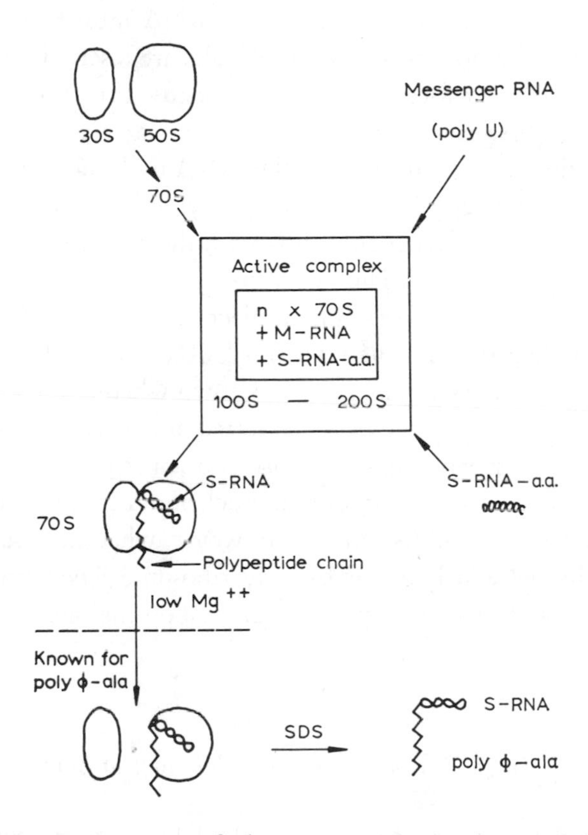

Fig. 6. Diagrammatic summary of ribosome participation in protein synthesis. (The active complex is pictured in Fig. 7.)

secondary forces to a cavity in the 50s ribosome. Extensive dialysis against 5×10^{-5} M Mg^{++} (which leaves intact 30s and 50s ribosomes) strips the nascent chains off the 50s ribosomes[50,51]. The released polypeptides sediment about 4s and if the latent ribonuclease is not activated, most likely still have terminally bound sRNA. When the Mg^{++} level is again brought to 10^{-2} M many released chains again stick to ribosomes.

Movement of the Messenger Template over the Ribosomal Surface

At any given time, each functioning ribosome thus contains only one nascent chain. As elongation proceeds, the NH_3-terminal end moves away from the point of peptide bond formation and conceivably may assume much of its final three-dimensional configuration before the terminal amino acids are added to the carboxyl end. The messenger RNA must be so attached that only the correct amino–acyl–sRNA molecules are inserted into position for possible peptide bond formation. This demands formation of specific hydrogen bonds (base-pairs?) between the messenger template and several (most likely three) nucleotides along the sRNA molecule. Then, in the presence of the necessary enzymes, the amino-acyl linkage to the then terminal sRNA breaks and a peptide bond forms with the correctly placed incoming amino-acyl-RNA (Fig. 5). This must create an energetically unfavorable environment for the now free sRNA molecule, causing it to be ejected from the sRNA binding site. The new terminal sRNA then moves into this site completing a cycle of synthesis. It is not known whether the messenger template remains attached to the newly inserted amino-acyl-sRNA. But if so, the messenger necessarily moves the correct distance over the ribosomal surface to place its next group of specific nucleotides in position to correctly select the next amino acid. No matter, however, what the mechanism is, the messenger tape necessarily moves over the ribosome. They cannot remain in static orientation if there is only one specific ribosomal site for peptide bond formation.

Attachment of Single Messenger RNA Molecules to Several Ribosomes

Addition of the synthetic messenger poly U to extracts containing predominantly 70s ribosomes creates new active ribosomes which sediment in

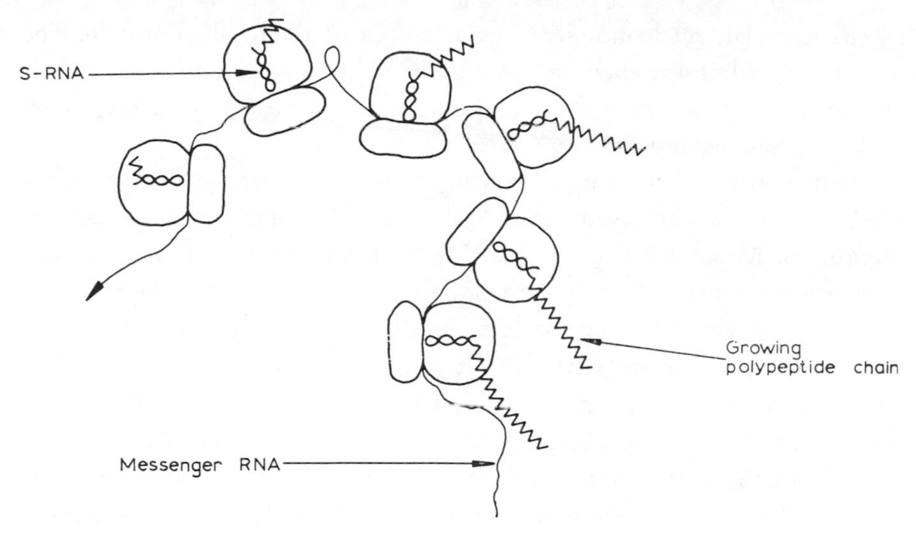

Fig. 7. Messenger RNA attachment to several ribosomes. (This illustration is schematic since the site of messenger attachment to ribosomes is not known.)

the 150–200s region[52]. Fixation of a single poly U molecule (mol. wt. = 100,000) to a 70s ribosome (mol. wt. = 3×10^6) should not significantly increase ribosomal sedimentation. Nor is it likely that a very large number of poly U molecules have combined with individual ribosomes. In these experiments, the molar ratio of fixed poly U to 70s ribosomes was less than $\frac{1}{5}$. Instead, the only plausible explanation involves formation of ribosomal aggregates attached to single poly U molecules. The 300 nucleotides in a poly U molecule of mol. wt. $\sim 10^5$ will have a contour length of about 1000 Å if the average internucleotide distance is 3.4 Å. Simultaneous attachment is thus possible to groups of 4–8 ribosomes (diameter $\sim$ 200 Å) depending upon the way the messenger passes over (through) the ribosomal surface. This estimate agrees well with the average aggregate size suggested by the sedimentation rate of the « active » complexes. Sedimentation of extracts *after* incorporation reveals most polyphenylalanine attached to the rapidly sedimenting « active » ribosomes.

Single messenger molecules thus most likely move simultaneously over the surfaces of several ribosomes, functioning on each as protein templates (Fig. 7). A progression of increasingly long polypeptide chains should be attached to successive ribosomes depending upon the fraction of the messenger tape to which they were exposed. When all the messenger has moved across the site of synthesis, some mechanism, perhaps itself triggered by a specific template nucleotide sequence must release the finished protein. The now vacant ribosome then becomes competent to receive the free end of another (or perhaps even the same) messenger molecule and start a new cycle of protein synthesis.

The realization that a single messenger molecule attaches to many ribosomes resolves a bothersome paradox which accompanied the messenger hypothesis. About 2–4% of *E. coli* RNA is messenger[40,53]. Its average sedimentation constant of 14s[54] suggests an average molecular weight about 500,000. This value may be too low since it is very difficult to completely prevent all enzymatic degradation. There thus must be *at least* 6–8 70s ribosomes for every messenger molecule. It was very difficult to believe that only 10–20% of the ribosomes functioned at a given moment. For, under a variety of conditions, the rate of protein synthesis is proportional to ribosome concentration[55]. Instead, it seems much more likely that, *in vivo*, almost all ribosomes are active. During the preparation of cell extracts, however, many ribosomes may lose their messenger and become inactive. If true, we may expect that use of more gentle techniques to break open *E. coli* cells will reveal larger fractions of fast-sedimenting active material. Already there are reports[56] that over 50% of mammalian reticulocyte ribosomes exist as aggregates of 5–6 80s particles. Furthermore, it is these aggregated ribosomes which make protein, both *in vivo* and *in vitro*.

Template Lifetime

Under the above scheme a messenger molecule might function indefinitely. On the contrary, however, the unstable bacterial templates function on the average only 10–20 times. This fact comes from experiments done in Levinthal's laboratory[57] where new messenger synthesis was blocked by addition of the antibiotic antinomycin D. Preexisting messenger (*Bacillus subtilus* growing with a 60 minute generation time) then broke down with a half-life of 2 minutes. Correspondingly, protein synthesis ceased at the expected rate. A mechanism(s) must thus exist to specifically degrade messenger molecules. Several enzymes (polynucleotide phosphorylase and a K^+ dependent diesterase) which rapidly degrade free messenger are active in bacterial cell extracts[58]. They function, however, much less efficiently when the messenger is attached to ribosomes[59]. Conceivably, a random choice exists whether the free forward-moving end of a messenger tape attaches to a vacant ribosome,

56. A. Gierer, *J. Mol. Biol.*, 6 (1963) 148; J. R. Warner, P. M. Knopf, and A. Rich, *Natl. Acad. Sci. U.S.*, 49 (1963) 122.
57. C. Levinthal, A. Keynan, and A. Higa, *Proc. Natl. Acad. Sci. U.S.*, 48 (1962) 1
58. H. Sekiguchi and S. S. Cohen, *J. Biol. Chem.*, 238 (1963) 349; D. Schlessinger P. F. Spahr, *J. Biol. Chem.*, 238 (1963) 6.
59. R. Gesteland and J. D. Watson, (will be published, 1963).

or is enzymatically degraded. If so, this important decision is settled by a chance event unrelated to the biological need for specific messengers.

Conclusion

We can now have considerable confidence that the broad features of protein synthesis are understood. RNA's involvement is very much more complicated than imagined in 1953. There is not one functional RNA. Instead, protein synthesis demands the ordered interaction of three classes of RNA – ribosomal, soluble, and messenger. Many important aspects, however, remain unanswered. For instance, there is no theoretical framework for the ribosomal sub-units nor, for that matter, do we understand the functional significance of ribosomal RNA. Most satisfying is the realization that all the steps in protein replication will be shown to involve well-understood chemical forces. As yet we do not know all the details. For example, are the DNA base-pairs involved in messenger RNA selection of the corresponding amino-acyl-sRNA? With luck, this will soon be known. We should thus have every expectation that future progress in understanding selective protein synthesis (and its consequences for embryology) will have a similar well-defined and, when understood, easy-to-comprehend chemical basis.

Acknowledgment

I have been very fortunate in having the collaboration of many able students and colleagues. The Ph.D. thesis work of Dr. C. G. Kurland, Dr. David Schlessinger and Dr. Robert Risebrough established many ideas reported here. Equally significant have been experiments by Drs. Kimiko Asano, Michael Cannon, Walter Gilbert, François Gros, Françoise Gros, Johns Hopkins, Masayasu Nomura, Pierre François Spahr, Alfred Tissières, and Jean-Pierre Waller. The visit of François Gros in the spring of 1960 was crucial in focusing attention on messenger RNA. Most importantly, I wish to mention my lengthy and still continuing successful collaboration with Alfred Tissières. Since 1960, I have the good fortune to also work closely with Walter Gilbert.

1. L. Pauling and R. B. Corey, *Proc. Natl. Acad. Sci. U.S.*, 37 (1951) 235.
2. J. D. Watson and F. H. C. Crick, *Nature*, 171 (1953) 737.
3. J. D. Watson and F. H. C. Crick, *Nature*, 171 (1953) 964.
4. P. Jordan, *Physik. Z.*, 39 (1938) 711.
The reader is also referred to the discussion of possible implications of long-range forces in biology, by H. J. Muller in his 1946 Pilgrim Trust Lecture, *Proc. Roy. Soc. London*, B (1947).
5. A sample of Pauling's views is found in his note with M. Delbrück, *Science*, 92 (1940) 77.
6. J. Brachet, *Arch. Biol. Liege*, 53 (1942) 207; T. Caspersson, *Naturwiss.*, 29 (1941) 33.
7. H. Borsook, C. L. Deasy, A. J. Haagen-Smit, G. Keighley, and P. H. Lowy, *J. Biol. Chem.*, 187 (1950) 839; and T. Hultin, *Exp. Cell Res.*, 1 (1950) 376.
8. J. W. Littlefield, E. B. Keller, J. Gross, and P. C. Zamecnik, *J. Biol. Chem.*, 217 (1955) 111; V. G. Allfrey, M. M. Daly, and A. E. Mirsky, *J. Gen. Physiol.*, 37 (1953) 157.
9. A. Rich and J. D. Watson, *Nature*, 173 (1954) 995; A. Rich and J. D. Watson, *Proc. Natl. Acad. Sci. U.S.*, 40 (1954) 759.
10. F. H. C. Crick and J. D. Watson, *Nature*, 177 (1956) 473; F. H. C. Crick and J. D. Watson, *Ciba Foundation Symposium*, «*The Nature of Viruses*», 1957.
11. D. L. D. Caspar, *Nature*, 177 (1956) 475.
12. M. L. Peterman and M. G. Hamilton, *J. Biol. Chem.*, 224 (1957) 725; P. O. Tso, J. Bonner, and J. Vinograd, *J. Biophys. Biochem. Cytol.*, 2 (1956) 725.
13. H. K. Schachman, A. B. Pardee, and R. Y. Stanier, *Arch. Biochem. Biophys.*, 38 (1952) 245.
14. E. T. Bolton, B. H. Hoyer, and D. B. Ritter, *Microsomal Particles and Protein Synthesis*, Pergamon Press, New York, 1958, p. 18.
15. A. Tissières and J. D. Watson, *Nature*, 182 (1958) 778; A. Tissières, J. D. Watson, D. Schlessinger, and B. R. Hollingworth, *J. Mol. Biol.*, 1 (1959) 221; C. E. Hall and H. S. Slayeter, *J. Mol. Biol.*, 1 (1959) 329; H. E. Huxley and G. Zubay, *J. Mol. Biol.*, 2 (1960) 10.
16. H. Lamfrom and E. R. Glowacki, *J. Mol. Biol.*, 5 (1962) 97; P. O. Tso and J. Vinograd, *Biochim. Biophys. Acta*, 49 (1961) 113.
17. B. Hall and P. Doty, *J. Mol. Biol.*, 1 (1959) 111, U.Z. Littauer and H. Eisenberger, *Biochim. Biophys. Acta*, 32 (1959) 320; S. M. Timasheff, A. Brown, J. S. Colter, and M. Davies, *Biochim. Biophys. Acta*, 27 (1958) 662.
18. C. G. Kurland, *J. Mol. Biol.*, 2 (1960) 83.
19. J. P. Waller and J. I. Harris, *Proc. Natl. Acad. Sci. U.S.*, 47 (1961) 18.
20. P. F. Spahr, *J. Mol. Biol.*, 4 (1962) 395.
21. D. Elson, *Biochim. Biophys. Acta*, 27 (1958) 216 and 36 (1959) 372.
22. P. F. Spahr and B. R. Hollingworth, *J. Biol. Chem.*, 236 (1961) 823.
23. J. W. Littlefield, E. B. Keller, J. Gross, P. C. Zamecnik, *J. Biol. Chem.*, 217 (1955) 111; J. W. Littlefield and E. B. Keller, *J. Biol. Chem.*, 224 (1957) 13; P. C. Zamecnik and E. B. Keller, *J. Biol. Chem.*, 209 (1954) 337; E. B. Keller and P. C. Zamecnik, *J. Biol. Chem.*, 221 (1956) 45.

24. M. B. Hoagland, P. C. Zamecnik, and M. L. Stephenson, *Biochim. Biophys. Acta*, 24 (1957) 215.
25. M. B. Hoagland, M. L. Stephenson, J. F. Scott, L. I. Hecht, and P. C. Zamecnik, *J. Biol. Chem.*, 231 (1958) 241.
26. F. H. C. Crick, *Symp. Soc. Exptl. Biol.*, 12 (1958) 138.
27. P. Berg and E. J. Ofengand, *Proc. Natl. Acad. Sci. U.S.*, 44 (1958) 78.
28. A. Tissières, D. Schlessinger, and F. Gros, *Proc. Natl. Acad. Sci. U.S.*, 46 (1960) 1450.
29. F. Gros and D. Schlessinger, unpublished experiments.
30. C. I. Davern and M. Meselson, *J. Mol. Biol.*, 2 (1960) 153.
31. M. Riley, A. Pardee, F. Jacob, and J. Monod, *J. Mol. Biol.*, 2 (1960) 216.
32. S. Naono and F. Gros, *Compt. Rend.*, 250 (1960) 3889.
33. A. D. Hershey, J. Dixon, and M. Chase, *J. Gen. Physiol.*, 36 (1953) 777.
34. E. Volkin and L. Astrachan, *Virology*, 2 (1956) 149.
35. M. Nomura, B. D. Hall, and S. Spiegelman, *J. Mol. Biol.*, 2 (1960) 306
36. F. Jacob and J. Monod, *J. Mol. Biol.*, 3 (1961) 318.
37. S. Brenner, F. Jacob, and M. Meselson, *Nature*, 190 (1961) 576.
38. F. Gros, H. Hiatt, W. Gilbert, C. G. Kurland, R. W. Risebrough, a
son, *Nature*, 190 (1961) 581.
39. B. D. Hall and S. Spiegelman, *Proc. Natl. Acad. Sci. U.S.*, 47 (1
40. M. Hayashi and S. Spiegelman, *Proc. Natl. Acad. Sci. U.S.*, 4
Gros, W. Gilbert, H. Hiatt, G. Attardi, P. F. Spahr, and J. D.
Harbor Symp. Quant. Biol., 26 (1961).
41. A. Tissières and J. W. Hopkins, *Proc. Natl. Acad. Sci. U.S.*
42. M. Chamberlin and P. Berg, *Proc. Natl. Acad. Sci. U.S.*,
Wood and P. Berg, *Proc. Natl. Acad. Sci. U.S.*, 48 (196
43. M. W. Nirenberg and J. H. Matthaei, *Biochem. Bioph*
404.
44. M. W. Nirenberg and J. H. Matthaei, *Proc. Natl. Ac*
45. R. W. Risebrough, A. Tissières, and J. D. Watson
(1962) 430.
46. W. Gilbert, *J. Mol. Biol.*, 6 (1963) 374.
47. J. Bishop, J. Leahy, and R. Schweet, *Proc. Na*
48. H. Dintzes, *Proc. Natl. Acad. Sci. U.S.*, 47 (1
49. M. Cannon, R. Krug, and W. Gilbert, *J. M*
50. W. Gilbert, *J. Mol. Biol.*, 6 (1963) 389.
51. D. Schlessinger and Françoise Gros, *J. M*
52. S. H. Barondes, M. W. Nirenberg, *Sci*
Lipmann, *Proc. Natl. Acad. Sci. U.S.*,
(1963) 374.
53. S. S. Cohen, H. D. Barner, and J.
54. R. Monier, S. Naono, D. Hayes, F
K. Asano, unpublished experime
55. O. Maaløe, *Cold Spring Harbor S*
D. Fraenkel, *Cold Spring Harb*

Biography

James Dewey Watson was born in Chicago, Ill., on April 6th, 1928, as the only son of James D. Watson, a businessman, and Jean Mitchell. His father's ancestors were originally of English descent and had lived in the midwest for several generations. His mother's father was a Scottish-born taylor married to a daughter of Irish immigrants who arrived in the United States about 1840. Young Watson's entire boyhood was spent in Chicago where he attended for eight years Horace Mann Grammar School and for two years South Shore High School. He then received a tuition scholarship to the University of Chicago, and in the summer of 1943 entered their experimental four-year college.

In 1947, he received a B.Sc. degree in Zoology. During these years his boyhood interest in bird-watching had matured into a serious desire to learn genetics. This became possible when he received a Fellowship for graduate study in Zoology at Indiana University in Bloomington, where he received his Ph.D. degree in Zoology in 1950. At Indiana, he was deeply influenced both by the geneticists H. J. Muller and T. M. Sonneborn, and by S.E.Luria, the Italian-born microbiologist then on the staff of Indiana's Bacteriology Department. Watson's Ph.D. thesis, done under Luria's able guidance, was a study of the effect of hard X-rays on bacteriophage multiplication.

From September 1950 to September 1951 he spent his first postdoctoral year in Copenhagen as a Merck Fellow of the National Research Council. Part of the year was spent with the biochemist Herman Kalckar, the remainder with the microbiologist Ole Maaløe. Again he worked with bacterial viruses, attempting to study the fate of DNA of infecting virus particles. During the spring of 1951, he went with Kalckar to the Zoological Station at Naples. There at a Symposium, late in May, he met Maurice Wilkins and saw for the first time the X-ray diffraction pattern of crystalline DNA. This greatly stimulated him to change the direction of his research toward the structural chemistry of nucleic acids and proteins. Fortunately this proved possible when Luria, in early August 1951, arranged with John Kendrew for him to work at the Cavendish Laboratory, where he started work in early October 1952.

He soon met Crick and discovered their common interest in solving the DNA structure. They thought it should be possible to correctly guess its structure, given both the experimental evidence at King's College plus careful examination of the possible stereochemical configurations of polynucleotide chains. Their first serious effort, in the late fall of 1951, was unsatisfactory. Their second effort based upon more experimental evidence and better appreciation of the nucleic acid literature, resulted, early in March 1953, in the proposal of the complementary double-helical configuration.

At the same time, he was experimentally investigating the structure of TMV, using X-ray diffraction techniques. His object was to see if its chemical sub-units, earlier revealed by the elegant experiments of Schramm, were helically arranged. This objective was achieved in late June 1952, when use of the Cavendish's newly constructed rotating anode X-ray tubes allowed an unambiguous demonstration of the helical construction of the virus.

From 1953 to 1955, Watson was at the California Institute of Technology as Senior Research Fellow in Biology. There he collaborated with Alexander Rich in X-ray diffraction studies of RNA. In 1955–1956 he was back in the Cavendish, again working with Crick. During this visit they published several papers on the general principles of virus construction.

Since the fall of 1956, he has been a member of the Harvard Biology Department, first as Assistant Professor, then in 1958 as an Associate Professor, and as Professor since 1961. During this interval, his major research interest has been the role of RNA in protein synthesis. Among his collaborators during this period were the Swiss biochemist Alfred Tissières and the French biochemist François Gros. Much experimental evidence supporting the messenger RNA concept was accumulated. His present principal collaborator is the theoretical physicist Walter Gilbert who, as Watson expressed it, «has recently learned the excitement of experimental molecular biology».

The honours that have to come to Watson include: the John Collins Warren Prize of the Massachusetts General Hospital, with Crick in 1959; the Eli Lilly Award in Biochemistry in the same year; the Lasker Award, with Crick and Wilkins in 1960; the Research Corporation Prize, with Crick in 1962; membership of the American Academy of Arts and Sciences and the National Academy of Sciences, and Foreign membership of the Danish Academy of Arts and Sciences. He is also a consultant to the President's Scientific Advisory Committee.

Watson is unmarried. His recreations are bird-watching and walking.

FRANCIS H. C. CRICK

On the Genetic Code

December 11, 1962

Part of the work covered by the Nobel citation, that on the structure and replication of DNA, has been described by Wilkins in his Nobel Lecture this year. The ideas put forward by Watson and myself on the replication of DNA have also been mentioned by Kornberg in his Nobel Lecture in 1959, covering his brilliant researches on the enzymatic synthesis of DNA in the test tube. I shall discuss here the present state of a related problem in information transfer in living material - that of the genetic code - which has long interested me, and on which my colleagues and I, among many others, have recently been doing some experimental work.

It now seems certain that the amino acid sequence of any protein is determined by the sequence of bases in some region of a particular nucleic acid molecule. Twenty different kinds of amino acid are commonly found in protein, and four main kinds of base occur in nucleic acid. The genetic code describes the way in which a sequence of twenty or more things is determined by a sequence of four things of a different type.

It is hardly necessary to stress the biological importance of the problem. It seems likely that most if not all the genetic information in any organism is carried by nucleic acid - usually by DNA, although certain small viruses use RNA as their genetic material. It is probable that much of this information is used to determine the amino acid sequence of the proteins of that organism. (Whether the genetic information has any other major function we do not yet know.) This idea is expressed by the classic slogan of Beadle: « one gene -one enzyme », or in the more sophisticated but cumbersome terminology of today: « one cistron-one polypeptide chain ».

It is one of the more striking generalizations of biochemistry - which surprisingly is hardly ever mentioned in the biochemical text-books - that the twenty amino acids and the four bases, are, with minor reservations, the same throughout Nature. As far as I am aware the presently accepted set of twenty amino acids was first drawn up by Watson and myself in the summer of 1953 in response to a letter of Gamow's.

In this lecture I shall not deal with the intimate technical details of the problem, if only for the reason that I have recently written such a review[1] which will appear shortly. Nor shall I deal with the biochemical details of messenger RNA and protein synthesis, as Watson has already spoken about these. Rather I shall ask certain general questions about the genetic code and ask how far we can now answer them.

Let us assume that the genetic code is a simple one and ask how many bases code for one amino acid? This can hardly be done by a pair of bases, as from four different things we can only form $4 \times 4 = 16$ different pairs, whereas we need at least twenty and probably one or two more to act as spaces or for other purposes. However, triplets of bases would give us 64 possibilities. It is convenient to have a word for a set of bases which codes one amino acid and I shall use the word « codon » for this.

This brings us to our first question. Do codons overlap? In other words, as we read along the genetic message do we find a base which is a member of two or more codons? It now seems fairly certain that codons do *not* overlap. If they did, the change of a single base, due to mutation, should alter two or more (adjacent) amino acids, whereas the typical change is to a single amino acid, both in the case of the « spontaneous » mutations, such as occur in the abnormal human haemoglobin or in chemically induced mutations, such as those produced by the action of nitrous acid and other chemicals on tobacco mosaic virus[2]. In all probability, therefore, codons do not overlap.

This leads us to the next problem. How is the base sequence, divided into codons? There is nothing in the backbone of the nucleic acid, which is perfectly regular, to show us how to group the bases into codons. If, for example, all the codons are triplets, then in addition to the correct reading of the message, there are two *in*correct readings which we shall obtain if we do not start the grouping into sets of three at the right place. My colleagues and I[3] have recently obtained experimental evidence that each section of the genetic message is indeed read from a fixed point, probably from one end. This fits in very well with the experimental evidence, most clearly shown in the work of Dintzis[4] that the amino acids are assembled into the polypeptide chain in a linear order, starting at the amino end of the chain.

This leads us to the next general question: the size of the codon. How many bases are there in any one codon? The same experiments to which I have just referred[3] strongly suggest that all (or almost all) codons consist of a triplet of bases, though a small multiple of three, such as six or nine, is not completely ruled out by our data. We were led to this conclusion by the study of mutations in the A and B cistrons of the r_{II} locus of bacteriophage

T4. These mutations are believed to be due to the addition or subtraction of one or more bases from the genetic message. They are typically produced by acridines, and cannot be reversed by mutagens which merely change one base into another. Moreover these mutations almost always render the gene completely inactive, rather than partly so.

By testing such mutants in pairs we can assign them all without exception to one of two classes which we call + and −. For simplicity one can think of the + class as having one extra base at some point or other in the genetic message and the − class as having one too few. The crucial experiment is to put together, by genetic recombination, three mutants of the same type into one gene. That is, either(+ with + with +) or (− with − with −). Whereas a single + or a pair of them (+ with +) makes the gene completely inactive, a set of three, suitably chosen, has some activity. Detailed examination of these results show that they are exactly what we should expect if the message were read in triplets starting from one end.

We are sometimes asked what the result would be if we put four +'s in one gene. To answer this my colleagues have recently put together not merely four but six +'s. Such a combination is active as expected on our theory, although sets of four or five of them are not. We have also gone a long way to explaining the production of « minutes » as they are called. That is, combinations in which the gene is working at very low efficiency. Our detailed results fit the hypothesis that in some cases when the mechanism comes to a triplet which does not stand for an amino acid (called a « nonsense » triplet) it very occasionally makes a slip and reads, say, only two bases instead of the usual three. These results also enable us to tie down the direction of reading of the genetic message, which in this case is from left to right, as the r_{II} region is conventionally drawn. We plan to write up a detailed technical account of all this work shortly. A final proof of our ideas can only be obtained by detailed studies on the alterations produced in the amino acid sequence of a protein by mutations of the type discussed here.

One further conclusion of a general nature is suggested by our results. It would appear that the number of nonsense triplets is rather low, since we only occasionally come across them. However this conclusion is less secure than our other deductions about the general nature of the genetic code.

It has not yet been shown directly that the genetic message is co-linear with its product. That is, that one end of the gene codes for the amino end of the polypeptide chain and the other for the carboxyl end, and that as one proceeds along the gene one comes in turn to the codons in between in the linear order in which the amino acids are found in the polypeptide chain. This seems highly likely, especially as it has been shown that in several systems mutations affecting the same amino acid are extremely near together on the genetic map. The experimental proof of the co-linearity of a gene and the polypeptide chain it produces may be confidently expected within the next year or so.

There is one further general question about the genetic code which we can ask at this point. Is the code universal, that is, the same in all organisms? Preliminary evidence suggests that it may well be. For example something very like rabbit haemoglobin can be synthesized using a cell-free system, part of which comes from rabbit reticulocytes and part from *Escherichia coli*[5]. This would not be very probable if the code were very different in these two organisms. However as we shall see it is now possible to test the universality of the code by more direct experiments.

In a cell in which DNA is the genetic material it is not believed that DNA itself controls protein synthesis directly. As Watson has described, it is believed that the base sequence of the DNA – probably of only one of its chains – is copied onto RNA, and that this special RNA then acts as the genetic messenger and directs the actual process of joining up the amino acids into polypeptide chains. The breakthrough in the coding problem has come from the discovery, made by Nirenberg and Matthaei[6], that one can use synthetic RNA for this purpose. In particular they found that polyuridylic acid – an RNA in which every base is uracil – will promote the synthesis of polyphenylalanine when added to a cell-free system which was already known to synthesize polypeptide chains. Thus one codon for phenylalanine appears to be the sequence UUU (where U stands for uracil: in the same way we shall use A, G, and C for adenine, guanine, and cytosine respectively). This discovery has opened the way to a rapid although somewhat confused attack on the genetic code.

It would not be appropriate to review this work in detail here. I have discussed critically the earlier work in the review mentioned previously[1] but such is the pace of work in this field that more recent experiments have already made it out of date to some extent. However, some general conclusions can safely be drawn.

The technique mainly used so far, both by Nirenberg and his colleagues[6] and by Ochoa and his group[7], has been to synthesize enzymatically «random» polymers of two or three of the four bases. For example, a polynucleotide, which I shall call poly (U,C), having about equal amounts of uracil and cytosine in (presumably) random order will increase the incorporation of the amino acids phenylalanine, serine, leucine, and proline, and possibly threonine. By using polymers of different composition and assuming a triplet code one can deduce limited information about the composition of certain triplets.

From such work it appears that, with minor reservations, each polynucleotide incorporates a characteristic set of amino acids. Moreover the four bases appear quite distinct in their effects. A comparison between the triplets tentatively deduced by these methods with the *changes* in amino acid sequence produced by mutation shows a fair measure of agreement. Moreover the incorporation requires the same components needed for protein synthesis, and is inhibited by the same inhibitors. Thus the system is most unlikely to be a complete artefact and is very probably closely related to genuine protein synthesis.

As to the actual triplets so far proposed it was first thought that possibly every triplet had to include uracil, but this was neither plausible on theoretical grounds nor supported by the actual experimental evidence. The first direct evidence that this was not so was obtained by my colleagues Bretscher and Grunberg-Manago[8], who showed that a poly (C,A) would stimulate the incorporation of several amino acids. Recently other workers[9,10] have reported further evidence of this sort for other polynucleotides not containing uracil. It now seems very likely that many of the 64 triplets, possibly most of them, may code one amino acid or another, and that in general several distinct triplets may code one amino acid. In particular a very elegant experiment[11] suggests that both (UUC) and (UUG) code leucine (the brackets imply that the order within the triplets is not yet known). This general idea is supported by several indirect lines of evidence which cannot be detailed here. Unfortunately it makes the unambiguous determination of triplets by these methods much more difficult than would be the case if there were only one triplet for each amino acid. Moreover, it is not possible by using polynucleotides of « random » sequence to determine the *order* of bases in a triplet. A start has been made to construct polynucleotides whose exact sequence is known at one end, but the results obtained so far are suggestive rather than conclusive[12]. It seems likely however from this and other unpublished evidence that the amino end of the polypeptide chain corresponds to the « right-hand » end of the polynucleotide chain – that is, the one with the 2′, 3′ hydroxyls on the sugar.

It seems virtually certain that a single chain of RNA can act as messenger

RNA, since poly U is a single chain without secondary structure. If poly A is added to poly U, to form a double or triple helix, the combination is inactive. Moreover there is preliminary evidence[9] which suggests that secondary structure within a polynucleotide inhibits the power to stimulate protein synthesis.

It has yet to be shown by direct biochemical methods, as opposed to the indirect genetic evidence mentioned earlier, that the code is indeed a triplet code.

Attempts have been made from a study of the changes produced by mutation to obtain the relative order of the bases within various triplets, but my own view is that these are premature until there is more extensive and more reliable data on the composition of the triplets.

Evidence presented by several groups[8,9,11] suggest that poly U stimulates both the incorporation of phenylalanine and also a lesser amount of leucine. The meaning of this observation is unclear, but it raises the unfortunate possibility of ambiguous triplets; that is, triplets which may code more than one amino acid. However one would certainly expect such triplets to be in a minority.

It would seem likely, then, that most of the sixty-four possible triplets will be grouped into twenty groups. The balance of evidence both from the cell-free system and from the study of mutation, suggests that this does not occur at random, and that triplets coding the same amino acid may well be rather similar. This raises the main theoretical problem now outstanding. Can this grouping be deduced from theoretical postulates? Unfortunately, it is not difficult to see how it might have arisen at an extremely early stage in evolution by random mutations, so that the particular code we have may perhaps be the result of a series of historical accidents. This point is of more than abstract interest. If the code does indeed have some logical foundation then it is legitimate to consider all the evidence, both good and bad, in any attempt to deduce it. The same is not true if the codons have no simple logical connection. In that case, it makes little sense to guess a codon. The important thing is to provide enough evidence to prove each codon independently. It is not yet clear what evidence can safely be accepted as establishing a codon. What is clear is that most of the experimental evidence so far presented falls short of proof in almost all cases.

In spite of the uncertainty of much of the experimental data there are certain codes which have been suggested in the past which we can now reject with some degree of confidence.

or is enzymatically degraded. If so, this important decision is settled by a chance event unrelated to the biological need for specific messengers.

Conclusion

We can now have considerable confidence that the broad features of protein synthesis are understood. RNA's involvement is very much more complicated than imagined in 1953. There is not one functional RNA. Instead, protein synthesis demands the ordered interaction of three classes of RNA – ribosomal, soluble, and messenger. Many important aspects, however, remain unanswered. For instance, there is no theoretical framework for the ribosomal sub-units nor, for that matter, do we understand the functional significance of ribosomal RNA. Most satisfying is the realization that all the steps in protein replication will be shown to involve well-understood chemical forces. As yet we do not know all the details. For example, are the DNA base-pairs involved in messenger RNA selection of the corresponding amino-acyl-sRNA? With luck, this will soon be known. We should thus have every expectation that future progress in understanding selective protein synthesis (and its consequences for embryology) will have a similar well-defined and, when understood, easy-to-comprehend chemical basis.

Acknowledgment

I have been very fortunate in having the collaboration of many able students and colleagues. The Ph.D. thesis work of Dr. C. G. Kurland, Dr. David Schlessinger and Dr. Robert Risebrough established many ideas reported here. Equally significant have been experiments by Drs. Kimiko Asano, Michael Cannon, Walter Gilbert, François Gros, Françoise Gros, Johns Hopkins, Masayasu Nomura, Pierre François Spahr, Alfred Tissières, and Jean-Pierre Waller. The visit of François Gros in the spring of 1960 was crucial in focusing attention on messenger RNA. Most importantly, I wish to mention my lengthy and still continuing successful collaboration with Alfred Tissières. Since 1960, I have the good fortune to also work closely with Walter Gilbert.

1. L. Pauling and R. B. Corey, *Proc. Natl. Acad. Sci. U.S.*, 37 (1951) 235.
2. J. D. Watson and F. H. C. Crick, *Nature*, 171 (1953) 737.
3. J. D. Watson and F. H. C. Crick, *Nature*, 171 (1953) 964.
4. P. Jordan, *Physik. Z.*, 39 (1938) 711.
The reader is also referred to the discussion of possible implications of long-range forces in biology, by H. J. Muller in his 1946 Pilgrim Trust Lecture, *Proc. Roy. Soc. London*, B (1947).
5. A sample of Pauling's views is found in his note with M. Delbrück, *Science*, 92 (1940) 77.
6. J. Brachet, *Arch. Biol. Liege*, 53 (1942) 207; T. Caspersson, *Naturwiss.*, 29 (1941) 33.
7. H. Borsook, C. L. Deasy, A. J. Haagen-Smit, G. Keighley, and P. H. Lowy, *J. Biol. Chem.*, 187 (1950) 839; and T. Hultin, *Exp. Cell Res.*, 1 (1950) 376.
8. J. W. Littlefield, E. B. Keller, J. Gross, and P. C. Zamecnik, *J. Biol. Chem.*, 217 (1955) 111; V. G. Allfrey, M. M. Daly, and A. E. Mirsky, *J. Gen. Physiol.*, 37 (1953) 157.
9. A. Rich and J. D. Watson, *Nature*, 173 (1954) 995; A. Rich and J. D. Watson, *Proc. Natl. Acad. Sci. U.S.*, 40 (1954) 759.
10. F. H. C. Crick and J. D. Watson, *Nature*, 177 (1956) 473; F. H. C. Crick and J. D. Watson, *Ciba Foundation Symposium, « The Nature of Viruses »*, 1957.
11. D. L. D. Caspar, *Nature*, 177 (1956) 475.
12. M. L. Peterman and M. G. Hamilton, *J. Biol. Chem.*, 224 (1957) 725; P. O. Tso, J. Bonner, and J. Vinograd, *J. Biophys. Biochem. Cytol.*, 2 (1956) 725.
13. H. K. Schachman, A. B. Pardee, and R. Y. Stanier, *Arch. Biochem. Biophys.*, 38 (1952) 245.
14. E. T. Bolton, B. H. Hoyer, and D. B. Ritter, *Microsomal Particles and Protein Synthesis*, Pergamon Press, New York, 1958, p. 18.
15. A. Tissières and J. D. Watson, *Nature*, 182 (1958) 778; A. Tissières, J. D. Watson, D. Schlessinger, and B. R. Hollingworth, *J. Mol. Biol.*, 1 (1959) 221; C. E. Hall and H. S. Slayeter, *J. Mol. Biol.*, 1 (1959) 329; H. E. Huxley and G. Zubay, *J. Mol. Biol.*, 2 (1960) 10.
16. H. Lamfrom and E. R. Glowacki, *J. Mol. Biol.*, 5 (1962) 97; P. O. Tso and J. Vinograd, *Biochim. Biophys. Acta*, 49 (1961) 113.
17. B. Hall and P. Doty, *J. Mol. Biol.*, 1 (1959) 111, U.Z. Littauer and H. Eisenberger, *Biochim. Biophys. Acta*, 32 (1959) 320; S. M. Timasheff, A. Brown, J. S. Colter, and M. Davies, *Biochim. Biophys. Acta*, 27 (1958) 662.
18. C. G. Kurland, *J. Mol. Biol.*, 2 (1960) 83.
19. J. P. Waller and J. I. Harris, *Proc. Natl. Acad. Sci. U.S.*, 47 (1961) 18.
20. P. F. Spahr, *J. Mol. Biol.*, 4 (1962) 395.
21. D. Elson, *Biochim. Biophys. Acta*, 27 (1958) 216 and 36 (1959) 372.
22. P. F. Spahr and B. R. Hollingworth, *J. Biol. Chem.*, 236 (1961) 823.
23. J. W. Littlefield, E. B. Keller, J. Gross, P. C. Zamecnik, *J. Biol. Chem.*, 217 (1955) 111; J. W. Littlefield and E. B. Keller, *J. Biol. Chem.*, 224 (1957) 13; P. C. Zamecnik and E. B. Keller, *J. Biol. Chem.*, 209 (1954) 337; E. B. Keller and P. C. Zamecnik, *J. Biol. Chem.*, 221 (1956) 45.

24. M. B. Hoagland, P. C. Zamecnik, and M. L. Stephenson, *Biochim. Biophys. Acta*, 24 (1957) 215.
25. M. B. Hoagland, M. L. Stephenson, J. F. Scott, L. I. Hecht, and P. C. Zamecnik, *J. Biol. Chem.*, 231 (1958) 241.
26. F. H. C. Crick, *Symp. Soc. Exptl. Biol.*, 12 (1958) 138.
27. P. Berg and E. J. Ofengand, *Proc. Natl. Acad. Sci. U.S.*, 44 (1958) 78.
28. A. Tissières, D. Schlessinger, and F. Gros, *Proc. Natl. Acad. Sci. U.S.*, 46 (1960) 1450.
29. F. Gros and D. Schlessinger, unpublished experiments.
30. C. I. Davern and M. Meselson, *J. Mol. Biol.*, 2 (1960) 153.
31. M. Riley, A. Pardee, F. Jacob, and J. Monod, *J. Mol. Biol.*, 2 (1960) 216.
32. S. Naono and F. Gros, *Compt. Rend.*, 250 (1960) 3889.
33. A. D. Hershey, J. Dixon, and M. Chase, *J. Gen. Physiol.*, 36 (1953) 777.
34. E. Volkin and L. Astrachan, *Virology*, 2 (1956) 149.
35. M. Nomura, B. D. Hall, and S. Spiegelman, *J. Mol. Biol.*, 2 (1960) 306.
36. F. Jacob and J. Monod, *J. Mol. Biol.*, 3 (1961) 318.
37. S. Brenner, F. Jacob, and M. Meselson, *Nature*, 190 (1961) 576.
38. F. Gros, H. Hiatt, W. Gilbert, C. G. Kurland, R. W. Risebrough, and J. D. Watson, *Nature*, 190 (1961) 581.
39. B. D. Hall and S. Spiegelman, *Proc. Natl. Acad. Sci. U.S.*, 47 (1961) 137.
40. M. Hayashi and S. Spiegelman, *Proc. Natl. Acad. Sci. U.S.*, 47 (1961) 1564; F. Gros, W. Gilbert, H. Hiatt, G. Attardi, P. F. Spahr, and J. D. Watson, *Cold Spring Harbor Symp. Quant. Biol.*, 26 (1961).
41. A. Tissières and J. W. Hopkins, *Proc. Natl. Acad. Sci. U.S.*, 47 (1961) 2015.
42. M. Chamberlin and P. Berg, *Proc. Natl. Acad. Sci. U.S.*, 48 (1962) 81 and W. B. Wood and P. Berg, *Proc. Natl. Acad. Sci. U.S.*, 48 (1962) 94.
43. M. W. Nirenberg and J. H. Matthaei, *Biochem. Biophys. Res. Commun.*, 4 (1961) 404.
44. M. W. Nirenberg and J. H. Matthaei, *Proc. Natl. Acad. Sci. U.S.*, 47 (1961) 1588.
45. R. W. Risebrough, A. Tissières, and J. D. Watson, *Proc. Natl. Acad. Sci. U.S.*, 48 (1962) 430.
46. W. Gilbert, *J. Mol. Biol.*, 6 (1963) 374.
47. J. Bishop, J. Leahy, and R. Schweet, *Proc. Natl. Acad. Sci. U.S.*, 46 (1960) 1030.
48. H. Dintzes, *Proc. Natl. Acad. Sci. U.S.*, 47 (1961) 247.
49. M. Cannon, R. Krug, and W. Gilbert, *J. Mol. Biol.*, 7 (1963) 360.
50. W. Gilbert, *J. Mol. Biol.*, 6 (1963) 389.
51. D. Schlessinger and Françoise Gros, *J. Mol. Biol.*, 7 (1963) 350.
52. S. H. Barondes, M. W. Nirenberg, *Science*, 138 (1962) 813; G. J. Spyrides and F. Lipmann, *Proc. Natl. Acad. Sci. U.S.*, 48 (1962) 1977; W. Gilbert, *J. Mol. Biol.*, 6 (1963) 374.
53. S. S. Cohen, H. D. Barner, and J. Lichtenstein, *J. Biol. Chem.*, 236 (1961) 1448.
54. R. Monier, S. Naono, D. Hayes, F. Hayes, and F. Gros, *J. Mol. Biol.*, 5 (1962) 311; K. Asano, unpublished experiments (1962).
55. O. Maaløe, *Cold Spring Harbor Symp. Quant. Biol.*, 26 (1961) 45; F. C. Neihardt and D. Fraenkel, *Cold Spring Harbor Symp. Quant. Biol.*, 26 (1961) 63.

56. A. Gierer, *J. Mol. Biol.*, 6 (1963) 148; J. R. Warner, P. M. Knopf, and A. Rich, *Proc. Natl. Acad. Sci. U.S.*, 49 (1963) 122.
57. C. Levinthal, A. Keynan, and A. Higa, *Proc. Natl. Acad. Sci. U.S.*, 48 (1962) 1631.
58. H. Sekiguchi and S. S. Cohen, *J. Biol. Chem.*, 238 (1963) 349; D. Schlessinger and P. F. Spahr, *J. Biol. Chem.*, 238 (1963) 6.
59. R. Gesteland and J. D. Watson, (will be published, 1963).

Biography

James Dewey Watson was born in Chicago, Ill., on April 6th, 1928, as the only son of James D. Watson, a businessman, and Jean Mitchell. His father's ancestors were originally of English descent and had lived in the midwest for several generations. His mother's father was a Scottish-born taylor married to a daughter of Irish immigrants who arrived in the United States about 1840. Young Watson's entire boyhood was spent in Chicago where he attended for eight years Horace Mann Grammar School and for two years South Shore High School. He then received a tuition scholarship to the University of Chicago, and in the summer of 1943 entered their experimental four-year college.

In 1947, he received a B.Sc. degree in Zoology. During these years his boyhood interest in bird-watching had matured into a serious desire to learn genetics. This became possible when he received a Fellowship for graduate study in Zoology at Indiana University in Bloomington, where he received his Ph.D. degree in Zoology in 1950. At Indiana, he was deeply influenced both by the geneticists H. J. Muller and T. M. Sonneborn, and by S.E.Luria, the Italian-born microbiologist then on the staff of Indiana's Bacteriology Department. Watson's Ph.D. thesis, done under Luria's able guidance, was a study of the effect of hard X-rays on bacteriophage multiplication.

From September 1950 to September 1951 he spent his first postdoctoral year in Copenhagen as a Merck Fellow of the National Research Council. Part of the year was spent with the biochemist Herman Kalckar, the remainder with the microbiologist Ole Maaløe. Again he worked with bacterial viruses, attempting to study the fate of DNA of infecting virus particles. During the spring of 1951, he went with Kalckar to the Zoological Station at Naples. There at a Symposium, late in May, he met Maurice Wilkins and saw for the first time the X-ray diffraction pattern of crystalline DNA. This greatly stimulated him to change the direction of his research toward the structural chemistry of nucleic acids and proteins. Fortunately this proved possible when Luria, in early August 1951, arranged with John Kendrew for him to work at the Cavendish Laboratory, where he started work in early October 1952.

He soon met Crick and discovered their common interest in solving the DNA structure. They thought it should be possible to correctly guess its structure, given both the experimental evidence at King's College plus careful examination of the possible stereochemical configurations of polynucleotide chains. Their first serious effort, in the late fall of 1951, was unsatisfactory. Their second effort based upon more experimental evidence and better appreciation of the nucleic acid literature, resulted, early in March 1953, in the proposal of the complementary double-helical configuration.

At the same time, he was experimentally investigating the structure of TMV, using X-ray diffraction techniques. His object was to see if its chemical sub-units, earlier revealed by the elegant experiments of Schramm, were helically arranged. This objective was achieved in late June 1952, when use of the Cavendish's newly constructed rotating anode X-ray tubes allowed an unambiguous demonstration of the helical construction of the virus.

From 1953 to 1955, Watson was at the California Institute of Technology as Senior Research Fellow in Biology. There he collaborated with Alexander Rich in X-ray diffraction studies of RNA. In 1955–1956 he was back in the Cavendish, again working with Crick. During this visit they published several papers on the general principles of virus construction.

Since the fall of 1956, he has been a member of the Harvard Biology Department, first as Assistant Professor, then in 1958 as an Associate Professor, and as Professor since 1961. During this interval, his major research interest has been the role of RNA in protein synthesis. Among his collaborators during this period were the Swiss biochemist Alfred Tissières and the French biochemist François Gros. Much experimental evidence supporting the messenger RNA concept was accumulated. His present principal collaborator is the theoretical physicist Walter Gilbert who, as Watson expressed it, « has recently learned the excitement of experimental molecular biology ».

The honours that have to come to Watson include: the John Collins Warren Prize of the Massachusetts General Hospital, with Crick in 1959; the Eli Lilly Award in Biochemistry in the same year; the Lasker Award, with Crick and Wilkins in 1960; the Research Corporation Prize, with Crick in 1962; membership of the American Academy of Arts and Sciences and the National Academy of Sciences, and Foreign membership of the Danish Academy of Arts and Sciences. He is also a consultant to the President's Scientific Advisory Committee.

Watson is unmarried. His recreations are bird-watching and walking.

FRANCIS H. C. CRICK

On the Genetic Code

December 11, 1962

Part of the work covered by the Nobel citation, that on the structure and replication of DNA, has been described by Wilkins in his Nobel Lecture this year. The ideas put forward by Watson and myself on the replication of DNA have also been mentioned by Kornberg in his Nobel Lecture in 1959, covering his brilliant researches on the enzymatic synthesis of DNA in the test tube. I shall discuss here the present state of a related problem in information transfer in living material – that of the genetic code – which has long interested me, and on which my colleagues and I, among many others, have recently been doing some experimental work.

It now seems certain that the amino acid sequence of any protein is determined by the sequence of bases in some region of a particular nucleic acid molecule. Twenty different kinds of amino acid are commonly found in protein, and four main kinds of base occur in nucleic acid. The genetic code describes the way in which a sequence of twenty or more things is determined by a sequence of four things of a different type.

It is hardly necessary to stress the biological importance of the problem. It seems likely that most if not all the genetic information in any organism is carried by nucleic acid – usually by DNA, although certain small viruses use RNA as their genetic material. It is probable that much of this information is used to determine the amino acid sequence of the proteins of that organism. (Whether the genetic information has any other major function we do not yet know.) This idea is expressed by the classic slogan of Beadle: « one gene –one enzyme », or in the more sophisticated but cumbersome terminology of today: « one cistron–one polypeptide chain ».

It is one of the more striking generalizations of biochemistry – which surprisingly is hardly ever mentioned in the biochemical text-books – that the twenty amino acids and the four bases, are, with minor reservations, the same throughout Nature. As far as I am aware the presently accepted set of twenty amino acids was first drawn up by Watson and myself in the summer of 1953 in response to a letter of Gamow's.

In this lecture I shall not deal with the intimate technical details of the problem, if only for the reason that I have recently written such a review[1] which will appear shortly. Nor shall I deal with the biochemical details of messenger RNA and protein synthesis, as Watson has already spoken about these. Rather I shall ask certain general questions about the genetic code and ask how far we can now answer them.

Let us assume that the genetic code is a simple one and ask how many bases code for one amino acid? This can hardly be done by a pair of bases, as from four different things we can only form $4 \times 4 = 16$ different pairs, whereas we need at least twenty and probably one or two more to act as spaces or for other purposes. However, triplets of bases would give us 64 possibilities. It is convenient to have a word for a set of bases which codes one amino acid and I shall use the word «codon» for this.

This brings us to our first question. Do codons overlap? In other words, as we read along the genetic message do we find a base which is a member of two or more codons? It now seems fairly certain that codons do *not* overlap. If they did, the change of a single base, due to mutation, should alter two or more (adjacent) amino acids, whereas the typical change is to a single amino acid, both in the case of the «spontaneous» mutations, such as occur in the abnormal human haemoglobin or in chemically induced mutations, such as those produced by the action of nitrous acid and other chemicals on tobacco mosaic virus[2]. In all probability, therefore, codons do not overlap.

This leads us to the next problem. How is the base sequence, divided into codons? There is nothing in the backbone of the nucleic acid, which is perfectly regular, to show us how to group the bases into codons. If, for example, all the codons are triplets, then in addition to the correct reading of the message, there are two *in*correct readings which we shall obtain if we do not start the grouping into sets of three at the right place. My colleagues and I[3] have recently obtained experimental evidence that each section of the genetic message is indeed read from a fixed point, probably from one end. This fits in very well with the experimental evidence, most clearly shown in the work of Dintzis[4] that the amino acids are assembled into the polypeptide chain in a linear order, starting at the amino end of the chain.

This leads us to the next general question: the size of the codon. How many bases are there in any one codon? The same experiments to which I have just referred[3] strongly suggest that all (or almost all) codons consist of a triplet of bases, though a small multiple of three, such as six or nine, is not completely ruled out by our data. We were led to this conclusion by the study of mutations in the A and B cistrons of the r_{II} locus of bacteriophage

T4. These mutations are believed to be due to the addition or subtraction of one or more bases from the genetic message. They are typically produced by acridines, and cannot be reversed by mutagens which merely change one base into another. Moreover these mutations almost always render the gene completely inactive, rather than partly so.

By testing such mutants in pairs we can assign them all without exception to one of two classes which we call + and −. For simplicity one can think of the + class as having one extra base at some point or other in the genetic message and the − class as having one too few. The crucial experiment is to put together, by genetic recombination, three mutants of the same type into one gene. That is, either (+ with + with +) or (− with − with −). Whereas a single + or a pair of them (+ with +) makes the gene completely inactive, a set of three, suitably chosen, has some activity. Detailed examination of these results show that they are exactly what we should expect if the message were read in triplets starting from one end.

We are sometimes asked what the result would be if we put four +'s in one gene. To answer this my colleagues have recently put together not merely four but six +'s. Such a combination is active as expected on our theory, although sets of four or five of them are not. We have also gone a long way to explaining the production of « minutes » as they are called. That is, combinations in which the gene is working at very low efficiency. Our detailed results fit the hypothesis that in some cases when the mechanism comes to a triplet which does not stand for an amino acid (called a « nonsense » triplet) it very occasionally makes a slip and reads, say, only two bases instead of the usual three. These results also enable us to tie down the direction of reading of the genetic message, which in this case is from left to right, as the r_{II} region is conventionally drawn. We plan to write up a detailed technical account of all this work shortly. A final proof of our ideas can only be obtained by detailed studies on the alterations produced in the amino acid sequence of a protein by mutations of the type discussed here.

One further conclusion of a general nature is suggested by our results. It would appear that the number of nonsense triplets is rather low, since we only occasionally come across them. However this conclusion is less secure than our other deductions about the general nature of the genetic code.

It has not yet been shown directly that the genetic message is co-linear with its product. That is, that one end of the gene codes for the amino end of the polypeptide chain and the other for the carboxyl end, and that as one proceeds along the gene one comes in turn to the codons in between in the linear order in which the amino acids are found in the polypeptide chain. This seems highly likely, especially as it has been shown that in several systems mutations affecting the same amino acid are extremely near together on the genetic map. The experimental proof of the co-linearity of a gene and the polypeptide chain it produces may be confidently expected within the next year or so.

There is one further general question about the genetic code which we can ask at this point. Is the code universal, that is, the same in all organisms? Preliminary evidence suggests that it may well be. For example something very like rabbit haemoglobin can be synthesized using a cell-free system, part of which comes from rabbit reticulocytes and part from *Escherichia coli*[5]. This would not be very probable if the code were very different in these two organisms. However as we shall see it is now possible to test the universality of the code by more direct experiments.

In a cell in which DNA is the genetic material it is not believed that DNA itself controls protein synthesis directly. As Watson has described, it is believed that the base sequence of the DNA – probably of only one of its chains – is copied onto RNA, and that this special RNA then acts as the genetic messenger and directs the actual process of joining up the amino acids into polypeptide chains. The breakthrough in the coding problem has come from the discovery, made by Nirenberg and Matthaei[6], that one can use synthetic RNA for this purpose. In particular they found that polyuridylic acid – an RNA in which every base is uracil – will promote the synthesis of polyphenylalanine when added to a cell-free system which was already known to synthesize polypeptide chains. Thus one codon for phenylalanine appears to be the sequence UUU (where U stands for uracil: in the same way we shall use A, G, and C for adenine, guanine, and cytosine respectively). This discovery has opened the way to a rapid although somewhat confused attack on the genetic code.

It would not be appropriate to review this work in detail here. I have discussed critically the earlier work in the review mentioned previously[1] but such is the pace of work in this field that more recent experiments have already made it out of date to some extent. However, some general conclusions can safely be drawn.

The technique mainly used so far, both by Nirenberg and his colleagues[6] and by Ochoa and his group[7], has been to synthesize enzymatically « random » polymers of two or three of the four bases. For example, a polynucleotide, which I shall call poly (U,C), having about equal amounts of uracil and cytosine in (presumably) random order will increase the incorporation of the amino acids phenylalanine, serine, leucine, and proline, and possibly threonine. By using polymers of different composition and assuming a triplet code one can deduce limited information about the composition of certain triplets.

From such work it appears that, with minor reservations, each polynucleotide incorporates a characteristic set of amino acids. Moreover the four bases appear quite distinct in their effects. A comparison between the triplets tentatively deduced by these methods with the *changes* in amino acid sequence produced by mutation shows a fair measure of agreement. Moreover the incorporation requires the same components needed for protein synthesis, and is inhibited by the same inhibitors. Thus the system is most unlikely to be a complete artefact and is very probably closely related to genuine protein synthesis.

As to the actual triplets so far proposed it was first thought that possibly every triplet had to include uracil, but this was neither plausible on theoretical grounds nor supported by the actual experimental evidence. The first direct evidence that this was not so was obtained by my colleagues Bretscher and Grunberg-Manago[8], who showed that a poly (C,A) would stimulate the incorporation of several amino acids. Recently other workers[9,10] have reported further evidence of this sort for other polynucleotides not containing uracil. It now seems very likely that many of the 64 triplets, possibly most of them, may code one amino acid or another, and that in general several distinct triplets may code one amino acid. In particular a very elegant experiment[11] suggests that both (UUC) and (UUG) code leucine (the brackets imply that the order within the triplets is not yet known). This general idea is supported by several indirect lines of evidence which cannot be detailed here. Unfortunately it makes the unambiguous determination of triplets by these methods much more difficult than would be the case if there were only one triplet for each amino acid. Moreover, it is not possible by using polynucleotides of « random » sequence to determine the *order* of bases in a triplet. A start has been made to construct polynucleotides whose exact sequence is known at one end, but the results obtained so far are suggestive rather than conclusive[12]. It seems likely however from this and other unpublished evidence that the amino end of the polypeptide chain corresponds to the « right-hand » end of the polynucleotide chain – that is, the one with the 2′, 3′ hydroxyls on the sugar.

It seems virtually certain that a single chain of RNA can act as messenger

RNA, since poly U is a single chain without secondary structure. If poly A is added to poly U, to form a double or triple helix, the combination is inactive. Moreover there is preliminary evidence[9] which suggests that secondary structure within a polynucleotide inhibits the power to stimulate protein synthesis.

It has yet to be shown by direct biochemical methods, as opposed to the indirect genetic evidence mentioned earlier, that the code is indeed a triplet code.

Attempts have been made from a study of the changes produced by mutation to obtain the relative order of the bases within various triplets, but my own view is that these are premature until there is more extensive and more reliable data on the composition of the triplets.

Evidence presented by several groups[8,9,11] suggest that poly U stimulates both the incorporation of phenylalanine and also a lesser amount of leucine. The meaning of this observation is unclear, but it raises the unfortunate possibility of ambiguous triplets; that is, triplets which may code more than one amino acid. However one would certainly expect such triplets to be in a minority.

It would seem likely, then, that most of the sixty-four possible triplets will be grouped into twenty groups. The balance of evidence both from the cell-free system and from the study of mutation, suggests that this does not occur at random, and that triplets coding the same amino acid may well be rather similar. This raises the main theoretical problem now outstanding. Can this grouping be deduced from theoretical postulates? Unfortunately, it is not difficult to see how it might have arisen at an extremely early stage in evolution by random mutations, so that the particular code we have may perhaps be the result of a series of historical accidents. This point is of more than abstract interest. If the code does indeed have some logical foundation then it is legitimate to consider all the evidence, both good and bad, in any attempt to deduce it. The same is not true if the codons have no simple logical connection. In that case, it makes little sense to guess a codon. The important thing is to provide enough evidence to prove each codon independently. It is not yet clear what evidence can safely be accepted as establishing a codon. What is clear is that most of the experimental evidence so far presented falls short of proof in almost all cases.

In spite of the uncertainty of much of the experimental data there are certain codes which have been suggested in the past which we can now reject with some degree of confidence.

Comma-less triplet codes
All such codes are unlikely, not only because of the genetic evidence but also because of the detailed results from the cell-free system.

Two-letter or three-letter codes
For example a code in which A is equivalent to O, and G to U. As already stated, the results from the cell-free system rule out all such codes.

The combination triplet code
In this code all permutations of a given combination code the same amino acid. The experimental results can only be made to fit such a code by very special pleading.

Complementary codes
There are several classes of these. Consider a certain triplet in relation to the triplet which is complementary to it on the other chain of the double helix. The second triplet may be considered either as being read in the same direction as the first, or in the opposite direction. Thus if the first triplet is UCC, we consider it in relation to either AGG or (reading in the opposite direction) GGA.

It has been suggested that if a triplet stands for an amino acid its complement necessarily stands for the same amino acids, or, alternatively in another class of codes, that its complement will stand for no amino acid, i.e. be nonsense.

It has recently been shown by Ochoa's group that poly A stimulates the incorporation of lysine[10]. Thus presumably AAA codes lysine. However since UUU codes phenylalanine these facts rule out all the above codes. It is also found that poly (U,G) incorporates quite different amino acids from poly (A,C). Similarly poly (U,C) differs from poly (A,G)[9,10]. Thus there is little chance that any of this class of theories will prove correct. Moreover they are all, in my opinion, unlikely for general theoretical reasons.

A start has already been made, using the same polynucleotides in cell-free systems from different species, to see if the code is the same in all organisms. Eventually it should be relatively easy to discover in this way if the code is universal, and, if not, how it differs from organism to organism. The preliminary results presented so far disclose no clear difference between *E. coli* and mammals, which is encouraging[10,13].

At the present time, therefore, the genetic code appears to have the following general properties:

(1) Most if not all codons consist of three (adjacent) bases.
(2) Adjacent codons do not overlap.
(3) The message is read in the correct groups of three by starting at some fixed point.
(4) The code sequence in the gene is co-linear with the amino acid sequence, the polypeptide chain being synthesized sequentially from the amino end.
(5) In general more than one triplet codes each amino acid.
(6) It is not certain that some triplets may not code more than one amino acid, i.e. they may be ambiguous.
(7) Triplets which code for the same amino acid are probably rather similar.
(8) It is not known whether there is any general rule which groups such codons together, or whether the grouping is mainly the result of historical accident.
(9) The number of triplets which do not code an amino acid is probably small.
(10) Certain codes proposed earlier, such as comma-less codes, two- or three-letter codes, the combination code, and various transposible codes are all unlikely to be correct.
(11) The code in different organisms is probably similar. It may be the same in all organisms but this is not yet known.

Finally one should add that in spite of the great complexity of protein synthesis and in spite of the considerable technical difficulties in synthesizing polynucleotides with defined sequences it is not unreasonable to hope that all these points will be clarified in the near future, and that the genetic code will be completely established on a sound experimental basis within a few years.

The references have been kept to a minimum. A more complete set will be found in the first reference.

1. F. H. C. Crick in *Progress in Nucleic Acid Research*, J. N. Davidson and Waldo E. Cohn (Eds.), Academic Press Inc., New York (in the press).
2. H. G. Wittmann, *Z. Vererbungslehre*, 93 (1962) 491.
   A. Tsugita, *J. Mol. Biol.*, 5 (1962) 284, 293.
3. F. H. C. Crick, L. Barnett, S. Brenner, and R. J. Watts-Tobin, *Nature*, 192 (1961) 1227.
4. M. A. Naughton and Howard M. Dintzis, *Proc. Natl. Acad. Sci. U.S.*, 48 (1962) 1822.
5. G. von Ehrenstein and F. Lipmann, *Proc. Natl. Acad. Sci. U.S.*, 47 (1961) 941.
6. J. H. Matthaei and M. W. Nirenberg, *Proc. Natl. Acad. Sci. U.S.*, 47 (1961) 1580.
   M. W. Nirenberg and J. H. Matthaei, *Proc. Natl. Acad. Sci. U.S.*, 47 (1961) 1588.
   M. W. Nirenberg, J. H. Matthaei, and O. W. Jones, *Proc. Natl. Acad. Sci. U.S.*, 48 (1962) 104.
   J. H. Matthaei, O. W. Jones, R. G. Martin, and M. W. Nirenberg, *Proc. Natl. Acad. Sci. U.S.*, 48 (1962) 666.
7. P. Lengyel, J. F. Speyer, and S. Ochoa, *Proc. Natl. Acad. Sci. U.S.*, 47 (1961) 1936.
   J. F. Speyer, P. Lengyel, C. Basilio, and S. Ochoa, *Proc. Natl. Acad. Sci. U.S.*, 48 (1962) 63.
   P. Lengyel, J. F. Speyer, C. Basilio, and S. Ochoa, *Proc. Natl. Acad. Sci. U.S.*, 48 (1962) 282.
   J. F. Speyer, P. Lengyel, C. Basilio, and S. Ochoa, *Proc. Natl. Acad. Sci. U.S.*, 48 (1962) 441.
   C. Basilio, A. J. Wahba, P. Lengyel, J. F. Speyer, and S. Ochoa, *Proc. Natl. Acad. Sci. U.S.*, 48 (1962) 613.
8. M. S. Bretscher and M. Grunberg-Manago, *Nature*, 195 (1962) 283.
9. O. W. Jones and M. W. Nirenberg, *Proc. Natl. Acad. Sci. U.S.*, 48 (1962) 2115.
10. R. S. Gardner, A. J. Wahba, C. Basilio, R. S. Miller, P. Lengyel, and J. F. Speyer, *Proc. Natl. Acad. Sci. U.S.*, 48 (1962) 2087.
11. B. Weisblum, S. Benzer, and R. W. Holley, *Proc. Natl. Acad. Sci. U.S.*, 48 (1962) 1449.
12. A. J. Wahba, C. Basilio, J. F. Speyer, P. Lengyel, R. S. Miller, and S. Ochoa, *Proc. Natl. Acad. Sci. U.S.*, 48 (1962) 1683.
13. H. R. V. Arnstein, R. A. Cox, and J. A. Hunt, *Nature*, 194 (1962) 1042.
    E. S. Maxwell, *Proc. Natl. Acad. Sci. U.S.*, 48 (1962) 1639.
    I. B. Weinstein and A. N. Schechter, *Proc. Natl. Acad. Sci. U.S.*, 48 (1962) 1686.

Biography

Francis Harry Compton Crick was born on June 8th, 1916, at Northampton, England, being the elder child of Harry Crick and Annie Elizabeth Wilkins. He has one brother, A. F. Crick, who is a doctor in New Zealand.

Crick was educated at Northampton Grammar School and Mill Hill School, London. He studied physics at University College, London, obtained a B.Sc. in 1937, and started research for a Ph.D. under Prof. E. N. da C. Andrade, but this was interrupted by the outbreak of war in 1939. During the war he worked as a scientist for the British Admiralty, mainly in connection with magnetic and acoustic mines. He left the Admiralty in 1947 to study biology.

Supported by a studentship from the Medical Research Council and with some financial help from his family, Crick went to Cambridge and worked at the Strangeways Research Laboratory. In 1949 he joined the Medical Research Council Unit headed by M. F. Perutz of which he has been a member ever since. This Unit was for many years housed in the Cavendish Laboratory Cambridge, but in 1962 moved into a large new building – the Medical Research Council Laboratory of Molecular Biology – on the New Hospital site. He became a research student for the second time in 1950, being accepted as a member of Caius College, Cambridge, and obtained a Ph.D. in 1954 on a thesis entitled « X-ray diffraction: polypeptides and proteins ».

During the academic year 1953–1954 Crick was on leave of absence at the Protein Structure Project of the Brooklyn Polytechnic in Brooklyn, New York. He has also lectured at Harvard, as a Visiting Professor, on two occasions, and has visited other laboratories in the States for short periods.

In 1947 Crick knew no biology and practically no organic chemistry or crystallography, so that much of the next few years was spent in learning the elements of these subjects. During this period, together with W. Cochran and V. Vand he worked out the general theory of X-ray diffraction by a helix, and at the same time as L. Pauling and R. B. Corey, suggested that the α-keratin pattern was due to α-helices coiled round each other.

A critical influence in Crick's career was his friendship, beginning in 1951, with J. D. Watson, then a young man of 23, leading in 1953 to the proposal of the double-helical structure for DNA and the replication scheme. Crick and Watson subsequently suggested a general theory for the structure of small viruses.

Crick in collaboration with A. Rich has proposed structures for polyglycine II and collagen and (with A. Rich, D. R. Davies, and J. D. Watson) a structure for polyadenylic acid.

In recent years Crick, in collaboration with S. Brenner, has concentrated more on biochemistry and genetics leading to ideas about protein synthesis (the « adaptor hypothesis »), and the genetic code, and in particular to work on acridine-type mutants.

Crick was made an F.R.S. in 1959. He was awarded the Prix Charles Leopold Meyer of the French Academy of Sciences in 1961, and the Award of Merit of the Gairdner Foundation in 1962. Together with J. D. Watson he was a Warren Triennial Prize Lecturer in 1959 and received a Research Corporation Award in 1962. With J. D. Watson and M. H. F. Wilkins he was presented with a Lasker Foundation Award in 1960. In 1962 he was elected a Foreign Honorary Member of the American Academy of Arts and Sciences, and a Fellow of University College, London. He was a Fellow of Churchill College, Cambridge, in 1960–1961, and is now a non-resident Fellow of the Salk Institute for Biological Studies, San Diego, California.

In 1940 Crick married Ruth Doreen Dodd. Their son, Michael F. C. Crick is a scientist. They were divorced in 1947. In 1949 Crick married Odile Speed. They have two daughters, Gabrielle A. Crick and Jacqueline M. T. Crick. The family lives in a house appropriately called « The Golden Helix », in which Crick likes to find his recreation in conversation with his friends.

1965

François Jacob
André Lwoff
Jacques Monod

François Jacob

Genetics of the Bacterial Cell

December 11, 1965

If I find myself here today, sharing with André Lwoff and Jacques Monod this very great honor which is being bestowed upon us, it is undoubtedly because, when I entered research in 1950, I was fortunate enough to arrive at the right place at the right time. At the right place, because there, in the attics of the Pasteur Institute, a new discipline was emerging in an atmosphere of enthusiasm, lucid criticism, nonconformism, and friendship. At the right time, because then biology was bubbling with activity, changing its ways of thinking, discovering in microorganisms a new and simple material, and drawing closer to physics and chemistry. A rare moment, in which ignorance could become a virtue.

Lysogeny and Bacterial Conjugation

The laboratory of André Lwoff was traversed by a long corridor where everyone would meet for endless discussions of experiments and hypotheses. At one end of the corridor, Jacques Monod's group was adding β-galactosides to bacterial cultures to initiate the biosynthesis of β-galactosidase; at the other end, André Lwoff and his collaborators were dousing cultures of lysogenic bacteria with ultraviolet light, having just discovered that means of initiating the biosynthesis of bacteriophage. Each was therefore «inducing» in his own way, convinced that the two phenomena had nothing in common, save a word.

Having come to prepare a doctoral thesis with André Lwoff, I was assigned the study of lysogeny in *Pseudomonas pyocyanea*. Thus I conscientiously set out to irradiate this organism. However, it soon became apparent that the problem of lysogeny was primarily that of the relationship between the bacterium and bacteriophage, in other words, a matter of genetics.

The genetics of bacteria and bacteriophages was born just 10 years earlier, with a paper by Luria and Delbrück[1]. It had continued to grow with the investigations of Lederberg and Tatum[2], Delbrück and Bailey[3], and Hershey[4].

But this young science had already produced many surprises for biologists. The most important was the demonstration by Avery, MacLeod, and McCarty[5], and later by Hershey and Chase[6], that genetic specificity is carried by DNA. For the first time, it became possible to give some chemical and physical meaning to the old biological concepts of heredity, variation, and evolution. Such a molecular interpretation of genetic phenomena is exactly what was provided in the structure of DNA proposed by Watson and Crick[7].

Another surprise was the realization that their rapid growth rate, their ability to adapt to many different media, and the variety of their mechanisms of genetic transfer make bacteria and viruses objects of choice for studying the functions and reproduction of the cell. The work of Beadle and Tatum[8], Lederberg[9], and Benzer[10] had shown that with a little imagination one can exert on a population of microorganisms such a selective pressure as to isolate, almost at will, individuals in which a particular function has been altered by mutation. Indeed, one of the most effective ways of determining the normal mechanisms of the cell is to explore abnormalities in suitably selected monsters.

The first attempts to analyze lysogeny genetically, intended to determine the location of the prophage in the bacterial cell, were carried out in 1952 by E. and J. Lederberg[11] and by Wollman[12]. Certain crosses between lysogenic and nonlysogenic bacteria suggested some linkage between the lysogenic character – determined by the λ prophage of *Escherichia coli* – and other characters controlled by bacterial genes. However, other crosses gave anomalous results. In fact, the answer obtained from these experiments could hardly be decisive, since the mechanism of conjugation was not understood at the time.

It was with the intention of continuing this study under somewhat different conditions that I began to work with Elie Wollman; very soon our collaboration became a particularly close and friendly one. We wanted in the first place to understand the anomalies observed in crosses between lysogenic and nonlysogenic bacteria, in particular the fact that the lysogenic character was not transmitted to recombinants except when carried by the female. To study this problem, we used a mutant male bacterium recently isolated by William Hayes and named *Hfr*, because, in crosses with females, it produced recombinants with high frequency[13]. Upon crossing such lysogenic *Hfr* males to nonlysogenic females, we were surprised to find out that the zygotes formed by more than half the males happened to lyse and produce phage[14]. This phenomenon, termed *zygotic induction*, showed that the equilibrium between the prophage and the bacterium is maintained by some regulatory system present in the cytoplasm of a lysogenic bacterium but absent from a nonlysogenic one. Moreover, it showed that a genetic character transferred by the male can be expressed in the zygote without being integrated into the chromosome of the female bacterium. It thus became possible in bacterial conjugation to distinguish experimentally between transfer of genetic material and recombinational event.

In order to bring the analysis of conjugation down from the level of the population to the level of the individual bacterial pair, we had to understand how the genetic material of the male is transferred to the female. In particular, one could try to interrupt conjugation after various times in order to find out when the transfer takes place. Elie Wollman had the somewhat startling idea of interrupting conjugation by placing a mating mixture in one of those blenders which ordinarily find service in the kitchen. It turned out that the shearing forces generated in the blender separate males from females; the male chromosome is broken during transit, but the chromosomal fragment that has already penetrated the female can express its potentialities and undergo recombination. In this way, it could be shown that, after pairing, the male slowly injects its chromosome into the female. This injection follows a strict schedule and, with any particular strain, the injection always starts at the same point[15]. Marvellous organism, in which conjugal bliss can last for nearly three times the life-span of the individual!

With this system, it became relatively easy to analyze the genetic constitution of *E. coli*, to show that bacterial characters are arranged on a single linkage group, termed the bacterial chromosome, and to map them, not only by classical genetic methods, but also by physical and chemical measurements. Moreover, two new insights emerged from this study. First, the bacterial chromosome turned out to be a closed, or circular, structure. Second, it was not as fixed a unit as one might have believed: other genetic elements, termed *episomes* (for example, a phage chromosome or a sex factor), can be added to or subtracted from it[16]. These properties happened to be of great value in subsequent studies of the bacterial cell and of its functioning.

Expression of the Genetic Material: The Messenger

In addition to its interest for the analysis of strictly genetic phenomena, bacterial conjugation proved particularly well adapted for the analysis of cellular functions, since it provided a means of transferring to an entire population a given gene at a given time. The phenotypic effects produced by the sudden appearance of a new gene in the recipient bacterium are then manifested without the accompanying complications that occur in higher organisms as a result of morphogenesis and cellular differentiation.

At his end of the corridor, Jacques Monod had reached the conclusion that further progress in the understanding of enzymatic induction required genetic analysis. Two types of mutations which altered the induced biosynthesis of β-galactosidase were known at that time. One type abolished the capacity to produce an active protein. The other changed the inducible character of enzyme synthesis so that it became *constitutive*, that is, able to proceed even in the absence of a β-galactoside inducer. How are these genes expressed? What is the relationship between the genetic determinants revealed by these mutations? What do the «inducible» or «constitutive» characters result from? Many of these questions could be experimentally approached through bacterial conjugation. By using male and female bacteria of suitable genotypes, one could transfer the desired allele of a given gene into a bacterium, and then study the conditions of enzyme synthesis in the zygote.

Such experiments were carried out in collaboration with Arthur Pardee, who had come to spend a year at the Pasteur Institute[17]. They led to two new concepts. The first was relevant to the mechanism of induction itself. Transfer into a constitutive bacterium of the genetic determinant for inducibility of the enzyme by β-galactosides resulted in formation of transitory diploids, heterozygous for the characters «inducible / constitutive». Obviously, the phenotype of such zygotes should permit a choice among the different hypotheses then proposed to explain induction. The experiments showed that the «inducible» allele can express itself independently of the gene controlling the synthesis of the enzyme; and that it is dominant over the «constitutive» allele. This result revealed the existence of a special gene which controls induction by forming a cytoplasmic product that inhibits synthesis of the enzyme in the absence of inducer. As we shall see later, this finding changed existing notions about the mechanism of induction and made possible a genetic analysis of the systems which regulate the rates of protein synthesis.

The second observation concerned the functioning of the genetic material. By transferring the gene that governs the structure of a protein into a bacterium which lacks it, one can determine the conditions under which this gene is expressed in the zygote. Here again, different predictions could be made, depending on the nature of the mechanisms postulated for information transfer in the formation of proteins. From kinetic analysis of protein synthesis, one could expect information concerning the primary gene product, the time required for its synthesis, and its mode of action. The experiments showed that, once transferred into a bacterium, and before genetic recombination has occurred, the gene controlling the structure of a protein can begin to function without detectable delay, producing protein at the maximal rate.

This was quite a surprising observation, for it was inconsistent with the notions that were prevalent at the time. Gene expression was then usually believed to consist in the accumulation of stable structures in the cytoplasm, probably the RNA of ribosomes, which were assumed to serve as templates specifying protein structures (see ref. 18). Such a scheme, which can be summarized by the aphorism «one gene–one ribosome–one enzyme», was hardly compatible with an immediate protein synthesis at maximal rate.

Further study of this problem required the withdrawal of a gene from a bacterium in order to examine the consequences of this withdrawal on the synthesis of the corresponding protein: stable templates, if present, should permit residual synthesis. But, although conjugation made it easy to inject a particular gene, the extraction of a particular gene from a whole bacterial population appeared to be an impossible operation. What could be done, however, was to transfer a sègment of chromosome heavily labeled with ^{32}P and then destroy the gene under study by ^{32}P decay. This delicate experiment was carried out by Monica Riley in the laboratory of Arthur Pardee; it showed unambiguously that capacity to produce the protein does not survive destruction of the gene[19].

The answer was clear: gene expression cannot proceed through formation of stable templates. About the same time, the genetic and kinetic analyses of induction further strengthened this belief. Induction appeared to take place almost instantaneously and to act on structures which often specified several proteins, not merely a single one. This finding was likewise inconsistent with the theories then current, since it did not fit with the observed homogeneity of the ribosomes.

By virtue of their stability, their homogeneity, and their base composition, the two known species of RNA did not fulfill the requirements for cytoplasmic templates. Since the notion that synthesis of proteins could occur directly on DNA was incompatible with the cytoplasmic localization of the ribosomes and their role in this synthesis, only one possible hypothesis remained: it was necessary to postulate the existence of a third species of RNA, the *messenger*, a short-lived molecule charged with transmission of genetic information to the cytoplasm[20]. According to this hypothesis, the ribosomes are non-specific structures, which function as machines to translate the nucleic language, carried by the messenger, into the peptidic language, with the aid of the transfer RNA's. In other words, the synthesis of a protein must be a two-step process: the deoxyribonucleotide sequence of DNA is first *transcribed* into messenger, the primary gene product; this messenger binds to the ribosomes, bringing them a specific «program», and the nucleotide sequence of the messenger is then *translated* into the amino acid sequence. Despite the objections raised against it, this messenger hypothesis possessed two main virtues in our eyes: on the one hand, it allowed a coherent interpretation of a number of known facts which had, until then, remained isolated or incompatible; on the other hand, it led to some precise experimental predictions.

In fact, even before it had appeared in print, the messenger hypothesis received two experimental confirmations, Sydney Brenner and I had decided to spend the month of June 1960 with M. Meselson, hunting for the messenger in the laboratory of Max Delbrück at the California Institute of Technology. The best candidate for the role of messenger seemed to us to be the RNA detected by Hershey[21] and later by Volkin and Astrachan[22] in bacteria infected with T2 phage. Thanks to the extraordinary intellectual and experimental agility of Sydney Brenner, we were able to show, within a few weeks, that the RNA formed by the phage associates with ribosomes synthesized wholly *before* infection, to produce on them phage proteins. The same ribosomes can thus make either phage or bacterial proteins, depending on the messenger with which they associate. Accordingly, it is the messenger which brings to the ribosomes a specific program for synthesis[23].

At this time, another member of our group at the Pasteur Institute, François Gros, had gone to spend several months at Harvard in the laboratory of J. D. Watson. With their collaborators, they rapidly succeeded in demonstrating the existence of a messenger fraction in the RNA of growing bacteria, and in establishing its principal properties[24]. The course of events which led to the recognition and isolation of the messenger has been described by J. D. Watson[25].

Genetic Activity and Its Regulation: The Operon

Experiments on genetic transfer by conjugation not only led to a revision of the concepts on the mechanisms of information transfer which occur in protein synthesis; they also made it possible to analyze the regulation of this synthesis.

The most striking observation that emerged from the study of phage production by lysogenic bacteria and of induction of β-galactosidase synthesis was the extraordinary degree of analogy between the two systems. Despite the obvious differences between the production of a virus and that of an enzyme, the evidence showed that in both cases protein synthesis is subject to a double genetic determinism: on the one hand, by *structural genes*, which specify the configuration of the peptide chains; on the other hand, by *regulatory genes*, which control the expression of these structural genes. In both cases, the properties of mutants showed that the effect of a regulatory gene consists in inhibiting the expression of the structural genes, by forming a cytoplasmic product which was called the *repressor*. In both cases, the induction of synthesis (whether of phage or of enzyme) seemed to result from a similar process: an inhibition of the inhibitor. Thus, to our surprise, these two phenomena, studied at opposite ends of the corridor, appeared to share a common fundamental mechanism. It should be emphasized that this analogy was invaluable to us. In biology, each material has its own virtues and is of particular value for a certain kind of experimental investigation. The combination of two systems significantly increased our means of analysis.

The existence of a specific inhibitor, the repressor, had an immediate corollary: the protein-forming apparatus must contain a site on which the repressor acts in order to block synthesis. The repressor itself could be regarded as a chemical signal emitted by the regulatory gene. The signal must have a receptor. The receptor had to be specific, hence genetically determined, and hence accessible to mutation. In a system that permits induced biosynthesis of an enzyme, any mutation damaging one element of the emitter–receptor system which inhibits the synthesis should result in constitutive enzyme production. Consequently it seemed difficult to distinguish mutations affecting the emitter from those affecting the receptor, until we realized that the distinction should be relatively easy to make in a diploid. This point can be illustrated by a simple analogy. Let us consider a house in which the opening of each of two doors is controlled by a little radio receiver. Let us suppose, furthermore, that somewhere in the vicinity there exist two transmitters, each sending out the same signal, which prevents the opening of the doors. If one of these transmitters is damaged, the other continues to send out signals and the doors remain closed: the damaged transmitter can be considered as «recessive» with respect to the normal one. On the other hand, if one of the receivers is damaged, it no longer responds to the inhibitory signal, and the door which it controls (but only that one) opens. The damaged receiver is thus «dominant» over the normal one, but the lesion is manifested only by the door which it controls: the effect is *cis* and not *trans*, in genetic terminology[26].

Thus it should be possible in principle, by the use of diploid bacteria, to distinguish, among constitutive mutations, those due to the regulatory gene from those due to the receptor. In fact, phage mutants corresponding to one or the other of these types had been known for a long time, although their nature became clear only in the light of this scheme. The existence of such mutations in phages encouraged us to search for analogous bacterial mutations affecting the enzymes of the lactose system. For that purpose, however, diploid bacteria were required. Although conjugation allowed formation of transitory diploids, their production was tricky and their analysis complicated. However, certain observations which had recently been made by Edward Adelberg led to the idea that the sexual episome F, which governs conjugation in *E. coli*, might under certain conditions incorporate and subsequently replicate a small fragment of the bacterial chromosome[27]. By using a series of strains in which the sexual episome was inserted at various points along the bacterial chromosome, we succeeded in isolating episomes which had incorporated a neighboring chromosomal fragment. Bacteria harboring such an episome become stable diploids for a small genetic segment, so that it is easy to make all possible combinations of alleles in this segment[28].

Having thus constructed the requisite genetic tool for our analysis, we set out to isolate under different conditions a whole series of mutants constitutive for the lactose system, in order to subject them to functional analysis. These mutants proved to belong to two quite distinct groups, which possessed the predicted properties for the transmitter and the receiver, respectively.

Many of these mutations were found to be «recessive» with respect to the wild-type allele. They allowed a definition of the transmitter, that is, of the regulatory gene. Some of these mutations possessed characteristic properties which led to the indirect identification of the repressor, the product of the regulatory gene[29]. They are discussed in greater detail in Jacques Monod's lecture.

In the second group, the mutations turned out to be «dominant» over the wild-type allele, and only those genes which were located on the same chromosome, that is, in *cis* position, were expressed constitutively. With these mutations, it was possible to define the receptor of the repressor, termed the *operator*[30].

The study of these mutants led, furthermore, to the notion that in bacteria the genetic material is organized into units of activity called *operons*, which are often more complex than the gene considered as the unit of function. In fact, the lactose system of *E. coli* contains three known proteins, and the three genes governing their structure are adjacent to one another on a small segment of the chromosome with the operator at one end (Fig. 1). Constitutive mutations, whether due to the alteration of the regulatory gene or of the operator, always display the remarkable property of being pleiotropic; that is, they affect simultaneously, and to the same extent, the production of the three proteins. The regulatory circuit therefore had to act on one integral structure containing the information which specifies the amino acid sequences of the three proteins. This structure could only be either the DNA itself or a messenger common to the three genes. This idea was further supported by the properties observed in mutations affecting the structural genes of the lactose system. Whereas some of these mutations obey Beadle and Tatum's «one gene-one enzyme» rule in the sense that they abolish only one of the three biochemical activities, others violate this rule by affecting the expression of several genes at a time[31, 32].

The notion of the operon, a grouping of adjacent structural genes controlled by a common operator, explained why the genes controlling the enzymes of the same biochemical pathway tend to remain clustered in bacteria, as observed by Demereč and Hartman[33]. Similarly, it accounted for the coordinate production of enzymes already found in certain biochemical pathways[34]. Although at first the operon concept was based exclusively on genetic criteria, it now includes biochemical criteria as well. There are, in fact, a number of experimental arguments, both genetic[32, 35] and biochemical[36], in support of the inference that an operon produces a single messenger, which binds to ribosomes to form the series of peptide chains determined by the different structural genes of the operon.

We can therefore envision the activity of the genome of *E. coli* as follows. The expression of the genetic material requires a continuous flow of unstable messengers which dictate to the ribosomal machinery the specificity of the proteins to be made. The genetic material consists of operons containing one or more genes, each operon giving rise to one messenger. The production of messenger by the operon is, in one way or another, inhibited by regulatory loops composed of three elements: regulatory gene, repressor, operator. Specific metabolites intervene at the level of these loops to play their role as signals: in inducible systems, to inactivate the repressor and hence allow production of messenger and ultimately of proteins; in repressible systems, to activate the repressor, and hence inhibit production of messenger and of proteins. Ac-

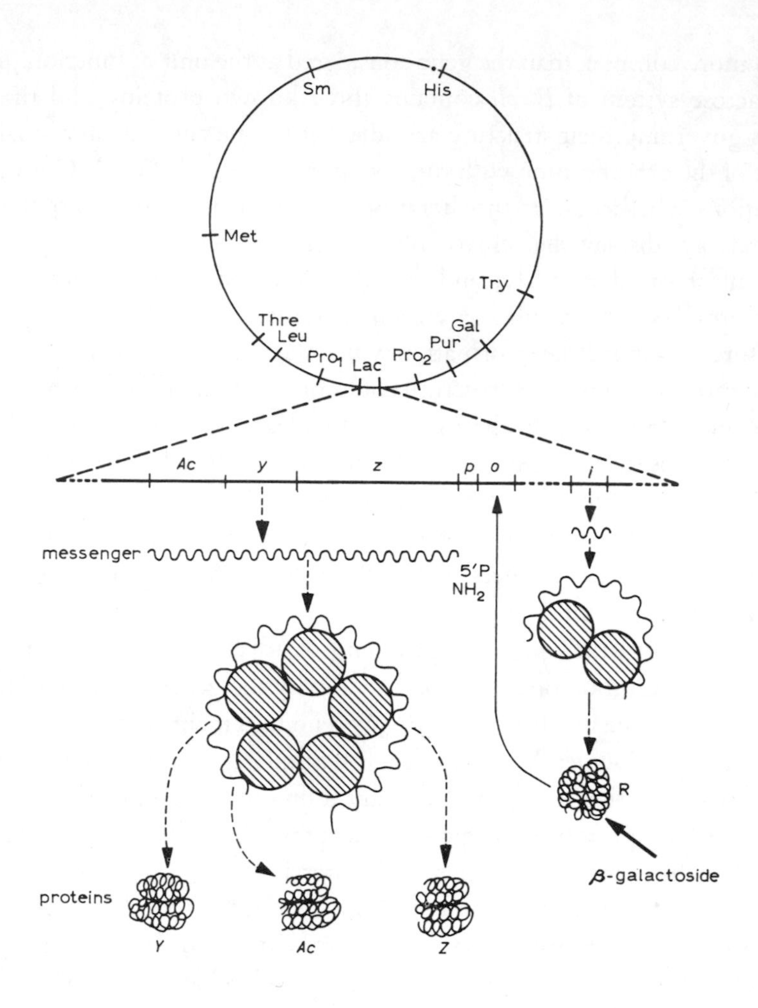

Fig. 1. The lactose region of *Escherichia coli*. The circle represents the *E. coli* chromosome and shows the position of the lactose region (*Lac*) among other markers. An enlargement of the lactose region is shown below. *i*, Regulatory gene; *o*, operator; *p*, promotor; *z*, structural gene for β-galactosidase; *y*, structural gene for β-galactoside permease; *Ac*, structural gene for β-galactoside transacetylase. The structural genes probably synthesize a single messenger (of which the 5′-phosphate end is most likely at the operator end) which associates with the ribosomes to form a polysome where the different peptide chains (of which the amino end probably corresponds to the operator end) are synthesized. The regulatory gene produces a specific repressor which, acting at the level of the operator, blocks the production of the messenger and hence of the proteins. The β-galactoside inducers act on the repressor to inactivate it, thus permitting the production of the messenger and consequently of the proteins determined by the operon.

cording to this scheme, only a fraction of the genes of the cell can be expressed at any moment, while the others remain repressed. The network of specific, genetically determined circuits selects at any given time the segments of DNA that are to be transcribed into messenger and consequently translated into proteins, as a function of the chemical signals coming from the cytoplasm and from the environment.

From the beginning, the conception of the genetic material as being formed of juxtaposed operons whose activity is regulated by a single operator site entailed a precise experimental prediction: a chromosomal rearrangement which would separate some structural genes from their operator and link them to a different operon controlled by its own operator should place the activity of these structural genes under a new regulatory control. But for a while, although some genes could be separated from their operators as a result of certain mutations, they became reattached to an unidentified region of the chromosome and consequently subject to an unknown system of regulation which remained beyond our experimental reach[37, 38].

Only recently has it become possible to obtain a fusion of the lactose operon of *E. coli* with another known operon, by using bacteria which were diploid for the chosen region[39]. At present, we know only a limited number of genes on the bacterial chromosome and a still more limited number of genes whose activity can be modified by the action of external metabolites. Any deletion which fuses two of these regions is likely to be relatively large and therefore to include a gene whose product is required for growth or division; it would thus be lethal in a haploid bacterium. In diploid bacteria, on the other hand, it has been possible to isolate a series of deletions covering about 50 to 80 genes; at one end, these deletions terminate in the gene controlling the structure of β-galactosidase, and at the other, in different regions of the chromosome. Some terminate in one of the two cistrons belonging to a purine operon, while leaving the other cistron intact (Fig. 2).

In these mutants, synthesis of the two proteins of the lactose region determined by the two genes left intact by the deletion is no longer inducible by β-galactosides. Such a result could be predicted since the deletion has destroyed both of the elements (regulatory gene and operator) responsible for the specific regulation of the lactose system. *But this synthesis has become repressible by the addition of purines.* It is thus clear that, in the deletion, the fragment of the lactose operon and the fragment of the purine operon have been fused to form a new operon which, from all appearances, produces a single messenger containing the genetic information for the synthesis of the proteins

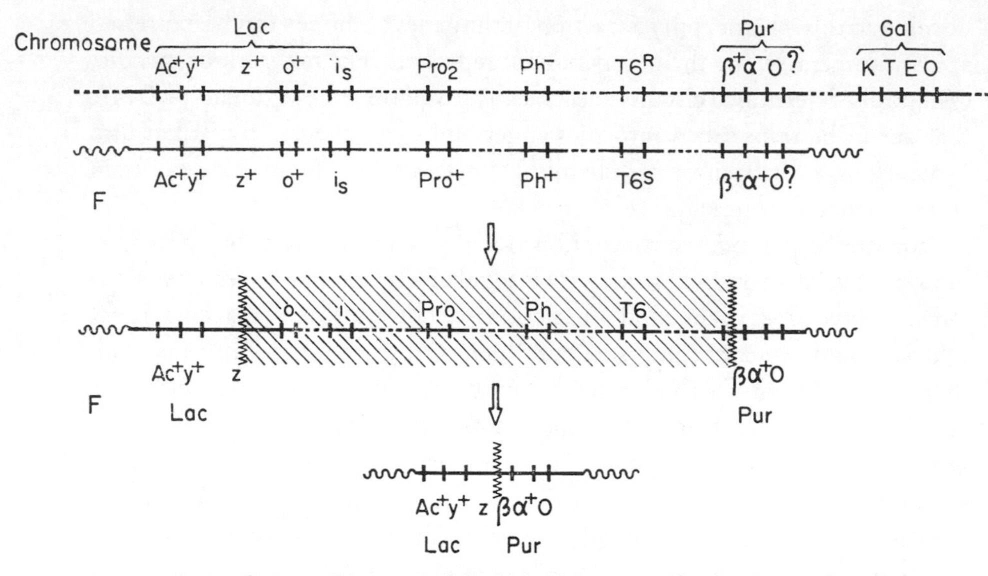

Fig. 2. Deletion fusing a fragment of the lactose operon and a fragment of a purine operon in *E. coli*. The upper part of the diagram represents the original heterozygous diploid structure of a bacterium containing a sexual episome which has incorporated an important chromosomal fragment by recombination. In the central part of the diagram, the hatched region shows the zone eliminated by a deletion which occurred in the episome. The lower part of the diagram shows the structure formed as a result of the deletion. The latter connected a terminal fragment of the *z* gene (determining β-galactosidase) with an initial fragment of the *Pur* β gene (determining an enzyme of purine biosynthesis). A new operon is thus formed from the *Pur* α gene (determining one of the proteins of purine biosynthesis), a structure formed by part of the *Pur* β gene and part of the *z* gene (probably producing a hybrid peptide chain consisting at the amino end of a *Pur* β sequence and at the carboxyl end of a sequence from *z*), the *y* gene (determining β-galactoside permease), and the *Ac* gene (determining β-galactoside transacetylase). The expression of the operon is repressed by purines, presumably at the level of a purine operator, which is itself sensitive to a repressor specifically activated by purines[39].

involved both in the biosynthesis of purines and in the utilization of lactose. But the system which determines the regulation of this messenger must be the operator of the purine region, sensitive to a repressor activated by purines.

In the same way, deletions fusing the lactose operon with the operon controlling tryptophan biosynthesis have recently been isolated[40]. The expression of the genes of the lactose operon which are still intact has consequently become repressible by tryptophan.

The type of regulation imposed on the expression of the genes belonging to a given operon thus depends exclusively on the operator, that is, in some way, on the nucleotide sequence located at the proximal extremity of the operon. In this manner, both the nature of the metabolites on which regulation depends and the inducible or repressible character of the regulation are determined by the respective *positions* of the genes along the chromosome and, more particularly, by their *association* with a particular operator segment. Obviously, it is the associations most favorable to the organism that are selected.

The presence of units of activity and regulation constituted by polycistronic operons implies the existence of a double system of punctuation in the nucleic text. One type of punctuation must permit the «slicing» of the long DNA duplex into «sections of transcription» corresponding to operons: this must serve as the point of recognition for the RNA polymerase, to show it not only where to start and finish the transcription of an operon, but also which strand of DNA is to be transcribed. Under certain conditions, one can obtain transpositions of the lactose operon into a region of the chromosome different from the normal region, and these insertions can be oriented in either one direction or the other[40]. It should be noted that, in case of an inversion, the 3′,5′ polarity of the DNA strands requires that the sequence to be transcribed into messenger must change with respect not only to direction, but also to strand (Fig. 3). For the insertions obtained, the lactose operon seems to be equally well expressed whether or not there has been an inversion. Conse-

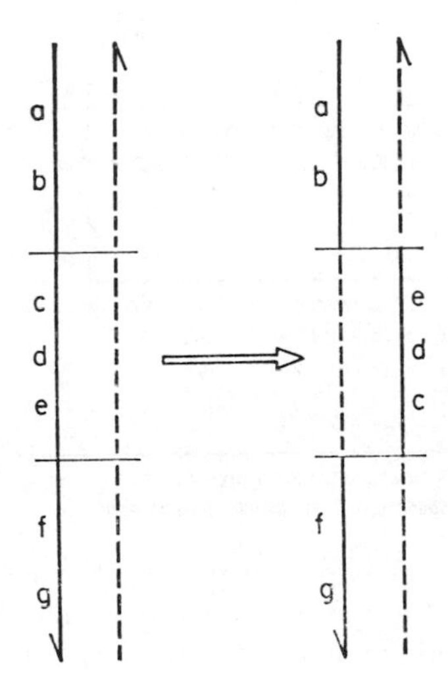

Fig. 3. Schematic diagram showing a change of polarity, and hence of strand, in an inversion.

quently, it must be assumed that (*i*) all the genetic information of *E. coli* is not necessarily contained in the same strand of DNA; (*ii*) a genetic signal ought to indicate the start of the operon, as well as the direction of transcription; and (*iii*) another signal ought to mark the end of the operon. If two operons of opposite orientations occur in juxtaposition, in the absence of a signal «end of transcription», the transcription of one operon could eventually proceed along the other, in the DNA strand which is not normally supposed to be transcribed.

The second punctuation mark of the nucleic text must, at the time of translation, allow the «slicing» of the messenger into the various peptide chains corresponding to the respective genes of the operon. This punctuation serves as a signal to the translation system (ribosomes, tRNA, and so on) to delimit the amino terminal and the carboxy terminal ends of each peptide chain.

In the lactose system of *E. coli*, the analysis of a series of deletions shows that the operator is situated outside the first known structural gene of the operon (refs. 41,38), from which it appears to be separated by a region called the *promotor*, which is indispensable to the expression of the entire operon[37]. The promotor probably corresponds to one of the punctuation marks, either for transcription or for translation. There are reasons to believe that the operator itself is not translated into a peptide chain, but we still do not know whether it is transcribed into messenger and whether the repressor acts at the level of the messenger or of the DNA itself (Fig. 4). It is not possible to discuss in detail

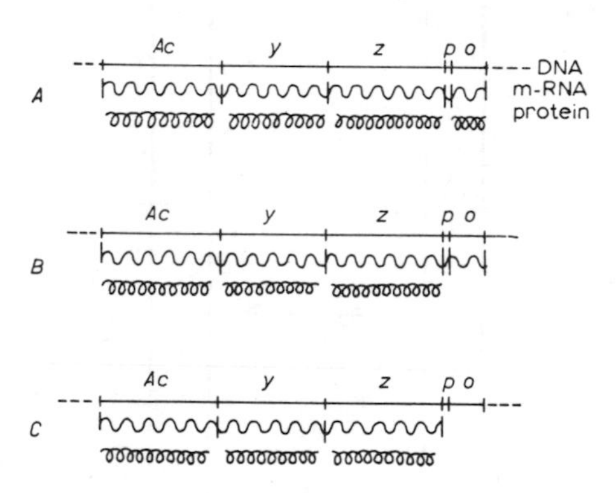

Fig. 4. The three possible models for the functioning of an operator. *A*, The operator (*o*) is transcribed and translated into protein. *B*, The operator is transcribed but not translated. *C*, The operator is neither transcribed nor translated. Obviously, depending on whether or not the operator is translated or is transcribed, the repressor could act at the level of the protein, of the messenger, or of the DNA itself.

here the experimental arguments or the hypotheses[20, 32, 42] concerning the site of action of the repressor. However, the ensemble of results recently obtained in various laboratories[43] indicates that synthesis of the messenger from its 5′-phosphate terminus, like that of the first peptide chain from its amino terminus, both begin at the operator end of the operon. The simplest hypothesis compatible with the results of genetic analysis, particularly with the study of deletions covering different segments of the operator region, is that the promotor represents the punctuation of transcription, providing the signal for RNA polymerase to start the synthesis of the messenger for this operon on one of the two DNA strands. If this is correct, the operator is not transcribed into messenger and repression can be exerted only at the level of DNA. This is the interpretation that now seems the most plausible to the geneticist; but it is clear that, as usual, the last word will belong to the chemist.

Organization of Genetic Material in the Bacterial Cell: The Replicon

Genetic analysis had revealed the logic of the circuits involved in the regulation of protein synthesis by showing that these circuits are composed of elements which can be connected in different ways to respond to the needs of the cell. It seemed reasonable to assume that analogous circuits, constructed on similar principles and using similar elements, might participate in other aspects of cellular regulation, and in particular to direct the replication of DNA in coordination with cellular division.

The systems involved in the replication of DNA seem to function more subtly in the cell than when isolated in the test tube. There are a number of experimental arguments in favor of a semiconservative mechanism of replication, as predicted by Watson and Crick from their model. Thanks to the work of Kornberg[44] and his collaborators an enzyme is known which can polymerize deoxyribonucleotides in the order dictated by the sequence of a piece of DNA serving as template. However, if a fragment of bacterial DNA is transferred into a recipient bacterium by transformation or incomplete conjugation, this fragment is incapable of replicating by itself. It can replicate only when integrated by recombination with one of the genetic structures in the host bacterium.

In bacteria, the DNA is organized into much simpler units than those observed in the cells of higher organisms. The essential information for the growth and division of the bacterium is carried by a single element, the so-called «bacterial chromosome». In addition, other nonessential elements, the «episomes», may be introduced into the bacterial cell[16].

Many different kinds of work have revealed that the best known of these genetic elements, the chromosome, behaves genetically, structurally, and biochemically as a single, integrated element. It seems to consist of one double-stranded chain of DNA, very probably closed or circular[45]. Replication appears to start at a fixed point on the molecule, and to continue regularly until the point of departure is reached again[45,46]. Under normal growth conditions, a new round of replication cannot begin until completion of the previous one[47].

Although other bacterial genetic elements are less well understood, their properties seem to be analogous. The genetic equipment of a bacterium can thus be considered to consist of distinct structures, each containing a «molecule» of DNA which is circular and of variable length.

Together with Sydney Brenner, we have tried to explain the regulation of DNA synthesis by means of circuits resembling those involved in the control of protein synthesis[48]. We have been led to postulate that each genetic element constitutes a unit of replication or *replicon*, which determines a circuit controlling its own replication in coordination with cell division. This hypothesis carries with it three distinct predictions.

(*1*) If each element contains some genetic determinants controlling its own replication, it should be possible to isolate mutants in which the regulatory circuit is impaired. In fact, for each of the three elements examined–bacterial chromosome, sexual episome, and phage–mutations can be obtained which abolish replication of the mutated element but not of others[49]. The nature and properties of these mutations suggest that they modify a diffusible product which normally acts on a punctuation mark of the replicon, that is, on a particular nucleotide sequence, to permit the start of replication. Once the reaction is initiated, the system replicates the entire sequence attached to this punctuation.

Here again, a genetically determined regulatory loop appears to operate. But, whereas in the synthesis of proteins the regulation seems to be *negative* or repressive, in the synthesis of DNA the available evidence indicates that the regulation involves a *positive* element, that is, one which acts on the DNA to trigger replication.

(*2*) The circumstances of bacterial conjugation can be best interpreted by assuming that the sexual episome is attached to the bacterial membrane near the zone through which the male chromosome passes during mating. Furthermore, to explain, during bacterial growth, the segregation of DNA after replication and the distribution of the two DNA copies in the two bacteria formed by cellular division, the simplest hypothesis consists in assuming that all cellular replicons are attached to the bacterial membrane. It is the synthesis of the membrane between the points of attachment of the two DNA copies that would insure their normal segregation.

The latter prediction appears to be confirmed by Antoinette Ryter's electron microscopic study of the «nuclear bodies» in *Bacillus subtilis*[50]. Each of these nuclear bodies seems to be attached to a «mesosome», a structure formed by an invagination of the membrane (Figs. 5 and 6). Furthermore, by staining the membrane with a tellurium salt, it can be shown that membrane synthesis does not take place uniformly over the bacterial surface, but rather in particular zones close to the attachment sites of the nuclear bodies. It is thus the growth of the membrane which seems to bring about the segregation of the DNA elements formed by replication. Finally, François Cuzin has succeeded in demonstrating that two independent replicons, such as the chromosome and the sex factor, do not segregate independently but remain associated during bacterial multiplication[51]; presumably, each of these structures is linked to the same element, possibly the same fragment of membrane. This fragment which might remain intact during growth and bacterial division would thus constitute the actual unit of segregation, analogous to a chromosome in a higher organism.

(3) To explain the coordination between the replication of DNA and the growth and subsequent division of the bacterial cell, it must be assumed that DNA replication and its regulation occur in the membrane. This seems to be indicated by certain phenomena observed during bacterial conjugation: presumably a surface reaction which occurs while the male and female come in contact triggers, in some way, a round of replication in the male. One of the structures thus synthesized remains in the male, while the other is progressively driven, during its formation, into the female[52]. Furthermore, the idea that DNA synthesis occurs in the membrane has recently received some biochemical support[53].

Thus one comes to envision the genetic equipment of bacteria as formed of circular DNA «molecules» constituting independent units of replication. These units would be associated with one membrane element, which would coordinate their replication with cell growth by means of regulatory circuits (Fig. 7). The basic genetic information would be contained in the longest of these units, but supplementary information could be added by the fixation of

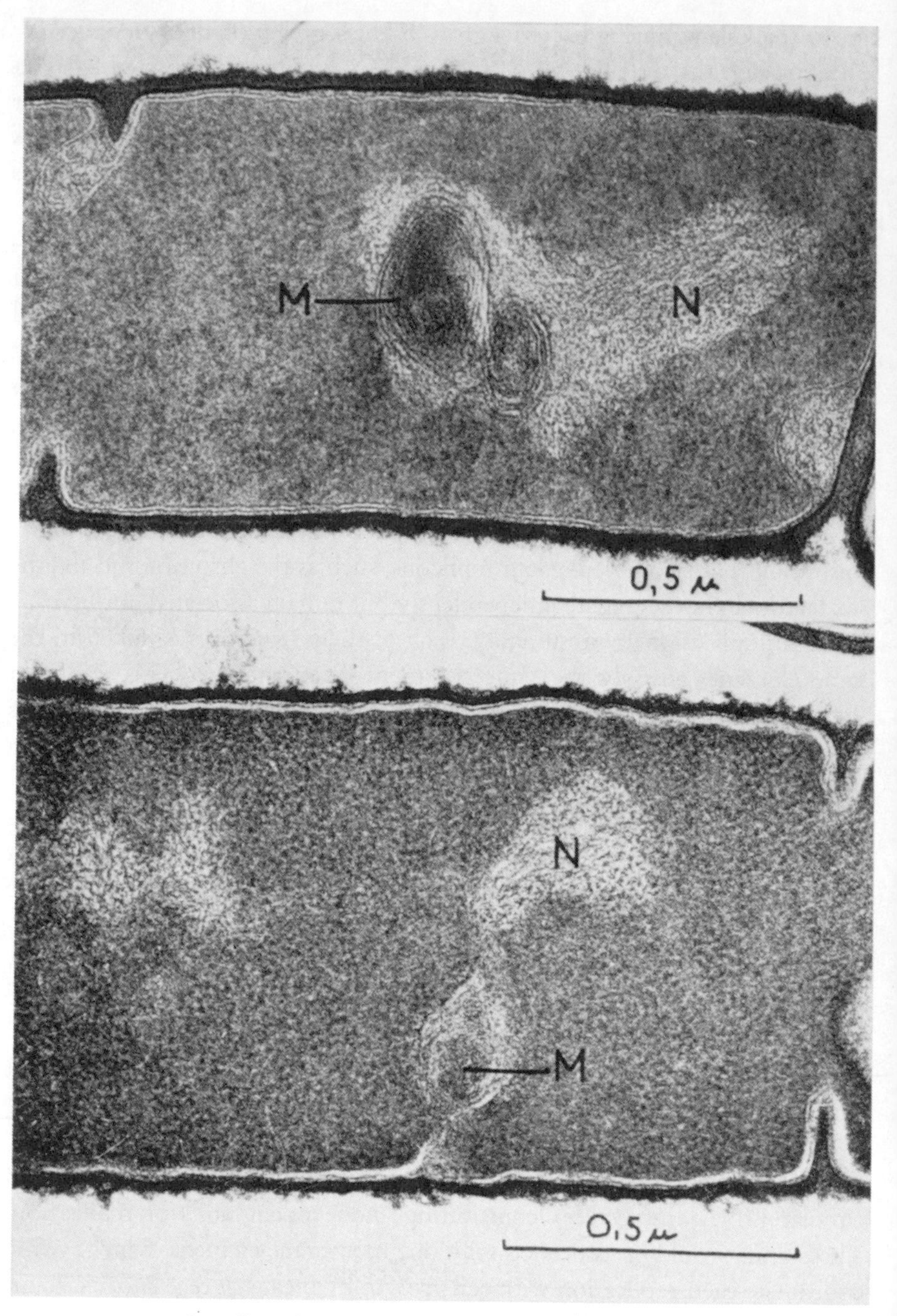

Fig. 5. Sections of *Bacillus subtilis*. The nuclear body (N) is bound to the membrane by means of a mesosome (M)[50].

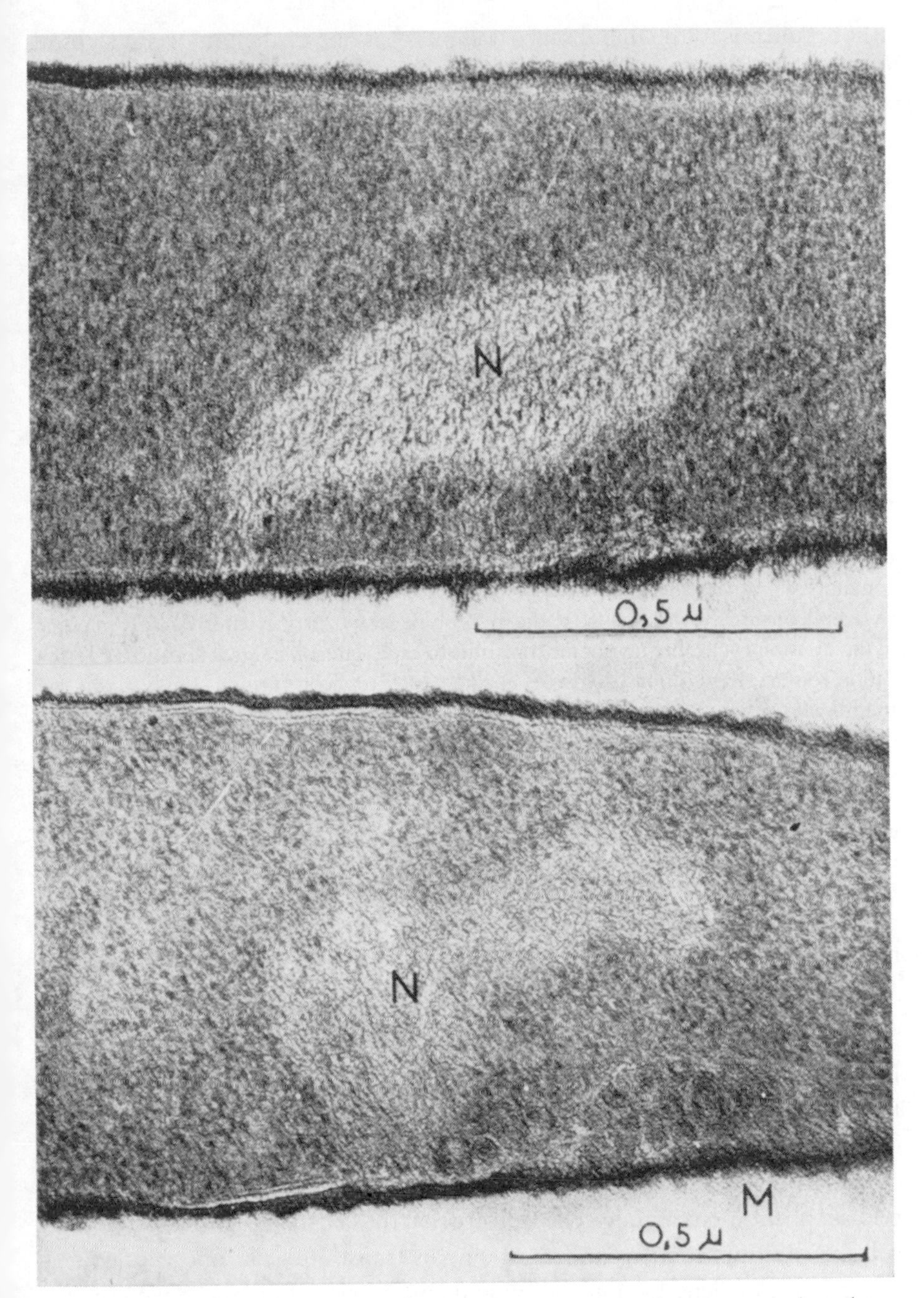

Fig. 6. Sections of *B. subtilis* placed for 30 min in 0.5 *M* sucrose. The mesosomes have been pushed out of the cytoplasm. During their retraction, they pull with them the nuclear bodies (N), which then appear bound directly to the membrane[50].

other replicons to the membrane. It seems that one of the important steps in the passage from the cellular organization of procaryotes to that of eucaryotes involves invaginations of membranous structures followed by differentiation into specialized organelles (mitochondria, genetic apparatus) whose functions were originally carried out by the bacterial membrane.

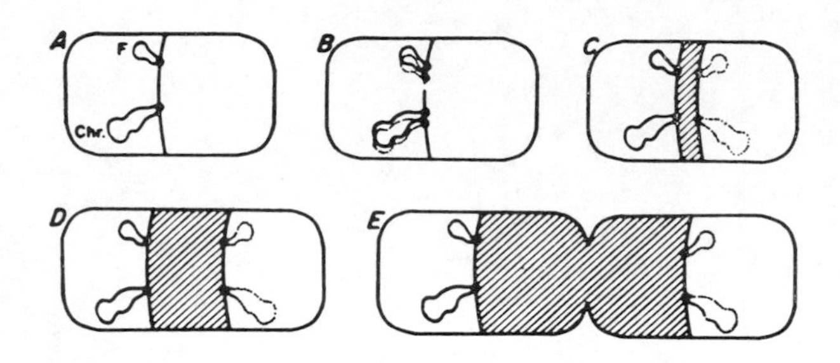

Fig. 7. Schematic diagram of DNA replication in bacteria. The bacterium represented contains two independent units: the «chromosome» and the sexual episome F. These two replicons are shown attached to the membrane at two distinct stites. At a given point in the division cycle, the membrane transmits to each replicon a signal permitting replication to start. Replication progresses linearly, each replicon turning slowly across the membrane in which the enzymatic complex responsible for replication is thought to occur. Two copies of each replicon are thus formed and are assumed to be attached to the membrane side by side. Membrane synthesis is considered to occur between these regions to which the two copies of each replicon are attached, thus drawing them to either side, with the septum forming in the median region. No new round of replication is allowed until the membrane, having returned to its original state following cell division transmits a new signal. The process is simplified in the sense that: (*i*) since the replication of DNA occurs one cycle ahead of division, bacteria generally have from two to four nuclear bodies (and not from one to two); and (*ii*) each step is considered to be completed before the start of the next one[48].

It is remarkable that the study of the bacterial cell leads us to attribute to the membrane such an important role in the coordination of growth and cell division, since a similar conclusion has been reached on the basis of the study of cells from higher organisms. In such cells, the process of morphogenesis and contact phenomena among cells indicate that the cell surface must also control cellular multiplication by means of signals transmitted, in one way or another, to the nucleus. Despite the evident complexity of such cells in comparison with bacteria, it must be assumed that evolution has conserved a system of molecular communication between cell surface and DNA.

The two chemical activities of DNA, transcription, the copying of a single strand into a ribonucleotide sequence, and replication, the copying of both strands into deoxyribonucleotide sequences, are controlled by a network of specific molecular interactions determined by the genes. The message inscribed in the genetic material thus contains not only the plans for the architecture of the cell, but also a program to coordinate the synthetic processes, as well as the means of insuring its execution.

Perhaps one of the most important contributions of microbial genetics is the definitive answer given to the old problem of the interaction between genes and cytoplasm, between heredity and environment. Although the demonstration that acquired characters cannot be inherited had already been made by classical genetics, the explanation for this fact is now provided by the nature of the nucleic message and of the genetic code. It is clear, on the other hand, that the expression of the genetic material is subject to external influences. Ten years ago, it still seemed possible that, in certain processes such as the induced biosynthesis of enzymes or of antibodies, the presence of specific compounds could modify the synthesis of proteins, mold their configurations, and hence alter their properties. The environment seemed to exercise an *instructive* action on the genes, to use Lederberg's expression[54], and thus to modulate the sense of the genetic text. What the study of regulatory circuits has shown is that the compounds in question serve only as simple stimuli: they act as signals to initiate a synthesis whose mechanism and final product remain entirely determined by the nucleotide sequence of the DNA. If the nucleic message may be compared with the text of a book, the regulatory network determines which pages are to be read at any given time. In the expression of the nucleic message, as well as in its reproduction, adaptation results from an *elective* rather than an instructive effect of the environment.

Of course genetic analysis can do no more than indicate the existence of regulatory circuits. A chemical analysis, which should disclose the specific molecular interactions, remains to be made. No repressor has yet been isolated, and the nature of the complexes that it can form with an operator or a metabolite remains obscure. We do not know how molecules find each other, recognize each other, and combine to constitute the regulatory network or to form such cellular superstructures as a membrane, a mitochondrion, or a chromosome. We do not know how molecules transmit the signals which modify the activity of their neighbors. What is clear, however, is that the problems to be solved by cellular biology and genetics in the years to come tend increasingly to merge with those in which biochemistry and physical chemistry are involved.

Acknowledgment

The work summarized here could not have been pursued without the help of so many collaborators that it is impossible for me to cite them without risking the injustice of unintentional omissions.

This work has been supported by successive grants from the Centre National de la Recherche Scientifique, the Délégation Générale à la Recherche Scientifique et Technique, the Commissariat à l'Energie Atomique, the Fondation pour la Recherche médicale française, the Jane Coffin Childs Memorial Fund, and the National Science Foundation.

1. S.E. Luria and M. Delbrück, *Genetics*, 28 (1943) 499.
2. J. Lederberg and E.L. Tatum, *Nature*, 158 (1946) 558.
3. M. Delbrück and W.T. Bailey Jr., *Cold Spring Harbor Symp. Quant. Biol.*, 11 (1946) 33.
4. A.D. Hershey, *Cold Spring Harbor Symp. Quant. Biol.*, 11 (1946) 67.
5. O.T. Avery, C.M. MacLeod, M. McCarty, *J. Exptl. Med.*, 79 (1944) 137.
6. A.D. Hershey and M. Chase, *J. Gen. Physiol.*, 36 (1952) 39.
7. J.D. Watson and F.H.C. Crick, *Nature*, 171 (1953) 737.
8. G.W. Beadle and E.L. Tatum, *Proc. Natl. Acad. Sci. (U.S.)*, 27 (1941) 499.
9. J. Lederberg, in J.H. Comroe Jr. (Ed.), *Methods in Medical Research*, Vol. 3, Year Book Publishers, Chicago, 1950, p. 5.
10. S. Benzer, in W.D. McElroy and B. Glass (Eds.), *The Chemical Basis of Heredity*, Johns Hopkins, Baltimore, 1957, p. 70.
11. E.M. Lederberg and J. Lederberg, *Genetics*, 38 (1953) 51.
12. E.L. Wollman, *Ann. Inst. Pasteur*, 84 (1953) 281.
13. W. Hayes, *Cold Spring Harbor Symp. Quant. Biol.*, 18 (1953) 75.
14. F. Jacob and E.L. Wollman, *Compt. Rend.*, 239 (1954) 455.
15. E.L. Wollman and F. Jacob, *Compt. Rend.*, 240 (1955) 2449.
16. F. Jacob and E.L. Wollman, *Sexuality and the Genetics of Bacteria*, Academic Press, New York, 1961.
17. A.B. Pardee, F. Jacob and J. Monod, *J. Mol. Biol.*, 1 (1959) 165.
18. T. Caspersson, *Chromosoma*, 1 (1940) 562; J. Brachet, *Enzymologia*, 10 (1941) 87; F.H.C. Crick, *Symp. Soc. Exptl. Biol.*, 12 (1958) 138; R.B. Roberts, *Microsomal Particles and Protein Synthesis*, Pergamon, Oxford, 1958; A. Tissières, J.D. Watson, D.

Schlesinger and B.R.Hollingworth, *J.Mol.Biol.*, 1 (1959) 221; C.I.Davern and M.Meselson, *J.Mol.Biol.*, 2 (1960) 153.
19. M.Riley, A.B.Pardee, F.Jacob and J.Monod, *J.Mol.Biol.*, 2 (1960) 216.
20. F.Jacob and J.Monod, *J.Mol.Biol.*, 3 (1961) 318.
21. A.D.Hershey, J.Dixon and M.Chase, *J.Gen.Physiol.*, 36 (1953) 777.
22. E.Volkin and L.Astrachan, *Virology*, 2 (1956) 149.
23. S.Brenner, F.Jacob and M.Meselson, *Nature*, 190 (1961) 576.
24. F.Gros, W.Gilbert, H.Hiatt, C.G.Kurland, R.W.Risebrough and J.D.Watson, *Nature*, 190 (1961) 581.
25. J.D.Watson, The involvement of RNA in the synthesis of proteins, in *Nobel Lectures, Physiology or Medicine, 1942–1962*, Elsevier, Amsterdam, 1964, p.785.
26. G.Pontecorvo, *Advan.Enzymol.*, 13 (1952) 121.
27. E.A.Adelberg and S.N.Burns, *J.Bacteriol.*, 79 (1960) 321.
28. F.Jacob and E.A.Adelberg, *Compt.Rend.*, 249 (1959) 189.
29. C.Willson, D.Perrin, M.Cohn, F.Jacob and J.Monod., *J.Mol.Biol.*, 8 (1964) 582; S.Bourgeois, M.Cohn and L.Orgel, *J.Mol.Biol.*, 14 (1965) 300.
30. F.Jacob, D.Perrin, C.Sanchez and J.Monod, *Compt.Rend.*, 250 (1960) 1727.
31. F.Jacob and J.Monod, *Cold Spring Harbor Symp.Quant.Biol.*, 26 (1961) 193; N. Franklin and S.E.Luria, *Virology*, 15 (1961) 299.
32. B.N.Ames and P.Hartman, *Cold Spring Harbor Symp.Quant.Biol.*, 28 (1963) 349.
33. M.Demereč and P.Hartman, *Ann.Rev.Microbiol.*, 13 (1959) 377.
34. B.N.Ames and B.Gary, *Proc.Natl.Acad.Sci.(U.S.)*, 45 (1959) 1453.
35. J.R.Beckwith, *Structure and Function of the Genetic Material*, Akademie Verlag, Berlin, 1964, p.119.
36. G.Attardi, S.Naono, J.Rouvière, F.Jacob and F.Gros, *Cold Spring Harbor Symp. Quant.Biol.*, 28 (1963) 363; B.Guttman and A.Novick, *Cold Spring Harbor Symp. Quant.Biol.*, 28 (1963) 373; R.G.Martin, *Cold Spring Harbor Symp.Quant.Biol.*, 28 (1963) 357; S.Spiegelman and M.Hayashi, *Cold Spring Harbor Symp.Quant.Biol.*, 28 (1963) 161.
37. B.N.Ames, P.Hartman and F.Jacob, *J.Mol.Biol.*, 7 (1963) 23; A.Matsushiro, S. Kido, J.Ito, K.Sato and F.Imamoto, *Biochem.Biophys.Res.Commun.*, 9 (1962) 204.
38. F.Jacob, A.Ullmann and J.Monod, *Compt.Rend.*, 258 (1964) 3125.
39. F.Jacob, A.Ullmann and J.Monod, *J.Mol.Biol.*, 13 (1965) 704.
40. J.R.Beckwith, E.Signer and W.Epstein, *Cold Spring Harbor Symp.Quant.Biol.*, 23 (1966) 393.
41. J.R.Beckwith, *J.Mol.Biol.*, 8 (1964) 427.
42. L.Szilard, *Proc.Natl.Acad.Sci.(U.S.)*, 46 (1960) 277; G.S.Stent, *Science*, 144 (1964) 816; W.K.Maas and E.McFall, *Ann.Rev.Microbiol.*, 18 (1964) 95.
43. R.L.Somerville and C.Yanofsky, *J.Mol.Biol.*, 8 (1964) 616; G.Streisinger, *Mendel Symposium on the Mutational Process*, Prague, 1965; H.Bremer, M.W.Konrad, K. Gaines and G.S.Stent, *J.Mol.Biol.*, 13 (1965) 540; F.Imamoto, N.Morikawa and K.Sato, *J.Mol.Biol.*, 13 (1965) 169; U.Maitra and J.Hurwitz, *Proc.Natl.Acad.Sci. (U.S.)*, 54 (1965) 815; M.Salas, M.A.Smith, W.M.Stanley Jr., A.J.Wahba and S.Ochoa, *J.Biol.Chem.*, 240 (1965) 3988; R.E.Thach, M.A.Cecere, T.A.Sunarajan and P.Doty, *Proc.Natl.Acad.Sci.(U.S.)*, 54 (1965) 1167.

44. A. Kornberg, The biologic synthesis of deoxyribonucleic acid, in *Nobel Lectures, Physiology or Medicine, 1942–1962*, Elsevier, Amsterdam, 1964, p. 665.
45. J. Cairns, *J. Mol. Biol.*, 6 (1963) 208.
46. N. Sueoka and H. Yoshikawa, *Cold Spring Harbor Symp. Quant. Biol.*, 28 (1963) 47; F. Bonhoeffer and A. Gierer, *J. Mol. Biol.*, 7 (1963) 534.
47. O. Maaløe, *Cold Spring Harbor Symp. Quant. Biol.*, 26 (1961) 45; R. H. Pritchard and K. G. Lark, *J. Mol. Biol.*, 9 (1964) 288.
48. F. Jacob and S. Brenner, *Compt. Rend.*, 256 (1963) 298; F. Jacob, S. Brenner and F. Cuzin, *Cold Spring Harbor Symp. Quant. Biol.*, 28 (1963) 329.
49. M. Kohiyama, H. Lamfrom, S. Brenner and F. Jacob, *Compt. Rend.*, 257 (1963) 1979; F. Cuzin and F. Jacob, *Proc. Intern. Congr. Genet. 11th, The Hague 1962*, Vol. 1, 1963, p. 40; F. Jacob, C. Fuerst and E. L. Wollman, *Ann. Inst. Pasteur*, 93 (1957) 724.
50. A. Ryter and F. Jacob, *Ann. Inst. Pasteur*, 107 (1964) 384.
51. F. Cuzin and F. Jacob, *Compt. Rend.*, 260 (1965) 5411.
52. J. D. Gross and L. G. Caro, *Science*, 150 (1965) 1679; M. Ptashne, *J. Mol. Biol.*, 11 (1965) 829; A. A. Blinkova, S. E. Bresler and V. A. Lanzov, *Z. Vererbungsl.*, 96 (1965) 267.
53. A. T. Ganesan and J. Lederberg, *Biochem. Biophys. Res. Commun.*, 18 (1965) 824.
54. J. Lederberg, *J. Cellular Comp. Physiol.*, Suppl. 1, 52 (1958) 398.

Biography

François Jacob was born in June 1920 in Nancy (France). He was the only son of Simon Jacob and Thérèse Franck. After attending the Lycée Carnot in Paris, he began studying medicine at the Faculty of Paris, with the intention of becoming a surgeon. These studies were interrupted by the war. In June 1940, when in his second year of medicine, he left France and joined the Free French Forces in London. He was sent to Africa as a medical officer and saw action in Fezzan, Libya, Tripolitania and Tunisia, where he was wounded. He was posted to the Second Armoured Division, and was severely wounded in Normandy, in August 1944. He remained in the hospital for seven months, and was awarded the Croix de la Libération, the highest French military decoration of this war.

After the war, François Jacob completed his medical studies and submitted his doctoral thesis in Paris in 1947. He was unable to practise surgery on account of his injuries, and worked in various fields before turning to biology. He obtained a science degree in 1951, and then a doctorate in science in 1954 at the Sorbonne, with a thesis on «Lysogenic bacteria and the provirus concept».

In 1950, Francois Jacob joined the Institut Pasteur under Dr. André Lwoff. He was appointed Laboratory Director in 1956, then in 1960 Head of the Department of Cell Genetics, recently created at the Institut Pasteur. In 1964 he was appointed Professor at the College de France, where a chair of Cell Genetics was created for him.

The work of François Jacob has dealt mainly with the genetic mechanisms existing in bacteria and bacteriophages, and with the biochemical effects of mutations. He first studied the properties of lysogenic bacteria and demonstrated their «immunity», *i.e.* the existence of a mechanism inhibiting the activity of genes in the prophage as in infective particles of the same type. In 1954 he began a long and fruitful collaboration with Elie Wollman, in an attempt to establish the nature of the relationships between the prophage and genetic material of the bacterium. This study led to a definition of the mechanism of bacterial conjugation, and also enabled an analysis of the genetic apparatus of the bacterial cell. From this work there emerged a whole series of new concepts, such as the oriented process of genetic transfer from the male to the female, the circularity of the bacterial chromosome or the episome concept. The whole of this work was summarized in a book *Sexuality and the Genetics of Bacteria*.

In 1958 the remarkable analogy revealed by genetic analysis of lysogeny and that of the induced biosynthesis of β-galactosidase led François Jacob, with Jacques Monod, to study the mechanisms responsible for the transfer of genetic information as well as the regulatory pathways which, in the bacterial cell, adjust the activity and synthesis of macromolecules. Following this analysis, Jacob and Monod proposed a series of new concepts, those of messenger RNA, regulator genes, operons and allosteric proteins.

In 1963, together with Sydney Brenner, François Jacob put forward the «replicon» hypothesis to account for certain aspects of cell division in bacteria. Since then, he has devoted his attention to the genetic analysis of the mechanisms of cell division. In 1970 he began to study cultured mammalian cells, particularly certain aspects of their genetic properties.

In 1970, François Jacob published a book *La logique du vivant, une Histoire de l'Hérédité*, in which, beginning with the 16th century, he traces the stages in the study of living beings that have led up to molecular biology.

François Jacob has been awarded a number of French scientific prizes, notably the Charles Leopold Mayer prize by the Académie des Sciences (1692). He is a foreign member of the Danish Royal Academy of Arts and Sciences (1962), the American Academy of Arts and Sciences (1964), the National Academy of Sciences of the United States (1969), and the American Philosophical Society (1969). He has received honorary degrees from several universities. He was invited to give a Harvey Lecture (New York, 1958) and the Dunham Lectures (Harvard, 1964).

In 1947 François Jacob married the pianist Lise Bloch. They have four children: Pierre (born in 1949), who has become a philosopher, Laurent and Odile (born in 1952) and Henri (born in 1954), who are still undifferentiated.

André Lwoff

Interaction among Virus, Cell and Organism

December 11, 1965

An organism is an integrated system of interdependent structures and functions. An organism is constituted of cells, and a cell consists of molecules which must work in harmony. Each molecule must know what the others are doing. Each one must be capable of receiving messages and must be sufficiently disciplined to obey. You are familiar with the laws which control regulation. You know how our ideas have developed and how the most harmonious and sound of them have been fused into a conceptual whole which is the very foundation of biology and confers on it its unity.

For the philosopher, order is the entirety of repetitions manifested, in the form of types or of laws, by perceived objects. Order is an intelligible relation. For the biologist, order is a sequence in space and time. However, according to Plato, all things arise out of their opposites. Order was born of the original disorder, and the long evolution responsible for the present biological order necessarily had to engender disorder.

An organism is a molecular society, and biological order is a kind of social order. Social order is opposed to revolution, which is an abrupt change of order, and to anarchy, which is the absence of order.

I am presenting here today both revolution and anarchy, for which I am fortunately not the only one responsible. However, anarchy cannot survive and prosper except in an ordered society, and revolution becomes sooner or later the new order. Viruses have not failed to follow the general law. They are strict parasites which, born of disorder, have created a very remarkable new order to ensure their own perpetuation.

For very many years, a group of eminent researchers have devoted their activity to the study of viral order. My own work simply prolongs a long chain of discoveries and ideas. I intend to discuss certain aspects of the relations between virus and cell and between virus and organism, and specifically the interaction between viral and cellular metabolism. I shall attempt to trace the development and evolution of the concepts, their ontogeny and phylogeny.

In this development one man has played a decisive role. By the logic of his thinking, by the rigor of his method, and in the selection of his followers, Max Delbrück has profoundly influenced the evolution of contemporary virology and of molecular biology. One of his followers, Hershey, in 1952, was responsible for a fundamental discovery: bacteriophages reproduce themselves from their sole genetic material. This strange property, which seemed to be a singularity of the bacteriophage, quickly became a general property of the viruses and even more than a property, a characteristic. Actually, any organized particle reproducing itself only from its own genetic material is and can only be a virus. Thus, thanks to Hershey, the category «virus» could be separated from the category «microbe», so that it became possible to distinguish the viruses by their essential difference, that is, to define them. This was also the discovery which has governed all interpretation of data with regard to the various aspects of viral development and all of the evolution of fundamental virology.

A molecule of nucleic acid can reproduce itself, express its potentialities, and give rise to virions, only within a cell. Here it finds whatever it lacks: enzymes, building blocks, a source of energy, and ribosomes. The virus is necessarily an intracellular parasite.

The genetic material of a virus has thus entered the cell. The cellular and viral molecules will confront each other, and the fate of the two partners will be decided. Two extreme cases may present themselves. Either the virus will multiply in the cell or else the cell will enslave the virus. Quite naturally, investigation was first directed toward the total war, which offers greater attraction for the combative intellect than peaceful coexistence.

When the genetic material of a highly virulent bacteriophage penetrates a bacterium, the bacterial chromosome is disintegrated and the bacterium consequently becomes incapable of producing messengers and bacterial proteins. The DNA of the bacteriophage synthesizes its own messengers with the aid of ribonucleotides synthesized by its host and of the enzymes of its host. The messengers of the bacteriophage will establish themselves on the bacterial ribosomes. With the aid of the activated transfer RNA and of the bacterial enzymes, the proteins of the bacteriophage are synthesized. Some of these proteins are enzymes necessary, for example, for the manufacture of specific constituents of the phages such as 5-hydroxymethylcytosine. Others are enzymes necessary for the replication of the bacteriophage DNA. Still others are structural proteins of the virion. One of the last to be formed is endolysine, which destroys the wall of the bacterium, provokes its rupture, and ensures the liberation of the virions. When we examine the kinetics of production of the various proteins, we find that each is formed in a given period of the evolutionary cycle. Everything takes place as if a system of sequential repression and derepression was acting.

As far as we know, the bacteriophage itself controls its own regulation. The bacterium infected by a virulent bacteriophage has become a virus factory which cannot be stopped except by its own disintegration. A bacterium has no control over the development of a virulent bacteriophage. But this is an extreme case. The relations between virus and bacterium do not always have this dramatic character.

As a matter of fact bacteriophages exist which do not kill all the bacteria which they infect. Some infected bacteria survive and perpetuate the ability to produce bacteriophages. These are lysogenic bacteria. Their investigation has profoundly modified our ideas on the relations between cell and virus. As so often happens, the hypotheses and theoretical concepts preceded the facts. Let us therefore begin with the theory.

In 1923, Duggar and Armstrong proposed the bold idea that viruses are not small bacteria but rebellious genes that have escaped from the chains of coordination. In 1925 and again in 1928 this idea was taken up and developed by Eugène Wollman. To him the transmission of properties from one bacterium to another was the result of the transmission by the external environment of certain genes endowed with a relative stability, and the viruses were compared to lethal genes. Today we know that the viruses and the chromosomes of their host cell may have important nucleotide sequences in common. It would be difficult to assume that such common structural characteristics could be the result of chance. Many virologists believe that viruses originated by mutation from cellular elements, that is, from normal structures. The virus, this element of disorder, arose from cellular order. Plato is justified, and the ideas of Duggar and Armstrong and those of Eugène Wollman now seem prophetic visions.

These ideas were bitterly opposed for a long time. In papers published between 1925 and 1940, very often a passion shows through whose violence astonishes us. The scientific discussions frequently recall the invectives of the heroes of Homer. It is my impression that the scientific mind today is much better prepared for the acceptance of new ideas, and we must also say that new ideas are in general firmly grounded in experimental data.

Let us return to the past and attempt to determine how our knowledge and our ideas on viruses and lysogeny, as well as our concepts of the relations between cell and virus, have evolved.

Already in 1915, Twort had thought that the bacteriophagy might be due to a virus. This was also the opinion of d'Hérelle: bacteriophages are viruses which kill the bacteria. Lysogeny came to confuse the bacteriologists. D'Hérelle at first denied the lysogeny. Later, he became convinced that he had discovered it. None of this is important, but the bacteria producing bacteriophages posed a curious problem.

Jules Bordet wrote in 1925: «The faculty for producing bacteriophages is incorporated in the heredity of the lysogenic bacteria. It is inherent in the normal physiology of the bacterium.» Nevertheless, it is of interest to note that the great immunologist did not conceive that heredity might be linked to a structure. For Bordet, heredity was the perpetuation of an individual physiology. The bacteriophage is not a materialized hereditary property, and Bordet affirmed in 1931: «The invisible virus of d'Hérelle does not exist. The intense lytic activity represents a pathological exaggeration of a normal function of the bacterium.» It seems strange to us today that such an eminent mind could have conceived of specific functions independent of any specific structure.

In 1929 lysogeny underwent a revival. Sir MacFarlane Burnet and his collaborator Margot MacKie began to investigate lysogenic Salmonellas. These Australian authors noted that only 0.1 percent of the bacteria contain bacteriophages. However, since all possess the property of producing bacteriophages, they must contain a specific *Anlage* coordinated in the hereditary constitution of the bacterium. The bacteriophage is «liberated» only if the bacterium is «activated». In the thinking of Burnet, this liberation probably corresponds to an unmasking, for he wrote in 1934: «We are forced to admit that each lysogenic bacterium encloses one or several particles of a bacteriophage which multiply by binary division with the bacterium.»

Everybody at that time believed that viruses were small microbes. A small microbe would necessarily have to reproduce itself by division. What, then, was the significance of this noninfectious phase? It might not represent anything important. We know of many organisms, specifically the protozoa, which go through a noninfectious stage during their cycle of evolution. Burnet had discovered the noninfectious phase of the bacteriophage in the lysogenic bacterium. In 1937 Eugène and Elisabeth Wollman noted that immediately after infection the bacteriophages passes through a noninfectious stage. This was confirmed in 1948 by Doermann, a follower of Max Delbrück, who for the first time methodically investigated the complete cycle of a bacteriophage.

However, Wollman understood that the bacteriophage particle, the virion, is not the direct descendant of the infecting particle. An infectious and a non-infectious phase necessarily would have to alternate. In a nonlysogenic bacterium, this alternation should take place in each bacterial cycle. In each division, each lysogenic bacterium *should* liberate *one* bacteriophage.

In 1938, Northrop started with the idea that bacteriophages are proteins and decided on a parallel investigation of the kinetics of the production of enzymes and that of the production of bacteriophages in a lysogenic bacterium. He concluded from his experiments that the bacteriophage, like the enzymes, is produced during the normal growth of the bacterium. At this point an important remark is necessary. Whether, in a bacterial population, one bacterium in a hundred produces one hundred bacteriophages or whether each of the bacteria produces one, the overall kinetics will remain the same. In 1949, the new school of American virologists, to which virology owes so much, condemned lysogeny. In nature, no problems exist but only solutions. The solution, the lysogenic bacterium, was enslaved as a typing tool for the identification of the bacterial families. Like those wisps of cloud that a breath of wind dispels, the problem was blown away from the temple of science and a smell of sulfur was left floating in the air. Lysogeny had become a heresy.

However, a few heretics survived – and among them Jacques Monod, who played a decisive role in my decision to return to the problem of lysogeny. I decided to operate with individual bacteria.

Here I must make a confession. I was led to this decision because I do not like either mathematics or statistics. I began my career as a protozoologist. I like to see things, not calculate probabilities.

Consequently, I took a lysogenic bacterium and immersed it in a drop of culture medium. The bacterium divided, the daughters were separated, and at each division a specimen was taken from the medium. One bacterium thus divided 19 times without liberating bacteriophages, and the daughter bacteria were still lysogenic.

When we subject lysogenic bacteria to lysis, we note that they do not enclose any bacteriophage. Lysogeny is consequently perpetuated in a noninfectious form. We were then in 1950 – and Hershey's discovery dates from 1952. However, I did not like the idea that noninfectious virions might exist. The noninfectious phase should be something different from a virion. The term «prophage» was therefore proposed, and it seemed that the world eagerly awaited its coming. In spite of its French origin, the Greek word was rapidly and unanimously adopted.

By giving a name to an unknown particle, we confer on it the dignity of a problem. The problem of the prophage had been posed, and now the history of lysogeny began again.

The prophage and the bacteria live in equilibrium. However, in a large population of lysogenic bacteria, we always find bacteriophages. How and why? Should we consider the problem as statisticians? Should we calculate the probability that a bacterium will produce bacteriophages within a given time? Should we content ourselves with a formula which would express the state of health of the population in terms of Greek symbols? I have already said that I do not have a statistical soul, that my mind tends to the concrete, and that I like to observe because I like to see. Accordingly, I again observed isolated bacteria. Some of them multiplied normally. Others multiplied for a time and then the descendants underwent lysis. And each of the bacteria which were lysed liberated bacteriophages. All this happened as if, in some drops of the medium, the development of the bacteriophage had been induced. This was my conclusion and I published it, to my regret. I now had to show that induction was not a fanciful hypothesis but a reality.

With Louis Siminovitch and Niels Kjeldgaard we went to work on an enterprise that was hard and discouraging because it seemed fruitless for a long time. After a year of effort, our faith was finally vindicated. Bacteria were irradiated by ultraviolet radiation. For an interval of 45 minutes, they continued to grow, but then they began to undergo lysis. In the process, each of them liberated some hundred bacteriophages. Induction had been discovered. And since induction affected 99.9 percent of the population, any statistical analysis was obviated. I was saved.

It thus appeared that the development of the prophage into bacteriophage is a mortal disease. The prophage is a potentially lethal factor. Irradiation forces it to express its potentialities.

For a long time, it was believed that such lethal agents as ultraviolet radiation or X-rays kill the cell because they destroy an essential structure. This concept seemed perfectly natural. It was in harmony with concepts in regard to death. After all, it is simple to consider death as the result of the suppression of some indispensable function. However, because every theory is a generalization, the risk in a theoretical concept increases with the fraction of truth it contains. Biological theories explain the various phenomena of life either in terms of disappearance of structure or of function, or in terms of the development of new structures. Our minds are tuned to a mode that we might call positive or negative. It is apparently difficult to make the transition from one to the other or to realize that the two modes are not necessarily incompatible.

Radiations sometimes kill by provoking alterations in or disappearance of structures. Sometimes, too, it permits a potentially lethal gene to express or to effect new syntheses, whether this concerns a bacterial protein or a virus, and thus to engender disease and death. Radiation may trigger lethal syntheses.

Nevertheless, induction was only a stage in our knowledge of the lysogenic bacteria. Induction, like the prophage, raised a whole series of new problems. Their investigation quickly surpasses the specific cases of the bacteriophage and of lysogeny and merged with the fundamental problems of molecular biology.

First of all, what is the nature of prophage? The use of radioactive molecules showed that the prophage is a deoxyribonucleic acid. It is the genetic material of the bacteriophage, a conclusion in harmony with the discoveries of Hershey.

Next, where is the prophage located? The discovery of sexualtity in *Escherichia coli* and the investigation of the bacteriophage lambda made it possible to answer this question, and in a general way. The prophage is attached to the bacterial chromosome. It is localized on the chromosome at a well-defined point, the receptor, which is unique and specific for each type of bacteriophage.

A temperate bacteriophage infects a nonlysogenic bacterium under conditions in which the bacterium will survive. The genetic material of the bacteriophage then penetrates the cytoplasm, explores the bacterial chromosome, recognizes the receptor, and pairs with it. Recognition and pairing can only be the consequence of structural homology – that is, of common nucleotide sequences. The DNA of the temperate bacteriophage is a circular–that is, closed – structure. The sector of the bacteriophage that is homologous with the bacterial chromosome opens, the bacterial chromosome also opens, and thus the genetic material of the bacteriophage is inserted, like a bacterial gene, into the bacterial chromosome. It becomes an integral part of the chromosome and behaves as if it were a bacterial gene. It will be reproduced by the system of enzymes which reproduces the chromosome of the bacteria. It happens sometimes that the prophage, when it detaches, wins out by taking with it some bacterial genes. These bacterial genes will be reproduced by the enzymes which provide for the autonomous multiplication of the bacteriophage.

Several years ago, in 1953, it occurred to me that the properties and activity of a molecule or of a particle might not be dependent only on its structure but also on its geographic situation, and I wrote: «The position is the fourth dimension of the prophage». My friends chided me for this formula by saying that it was devoid of meaning, and at the time they were perhaps right. I still believe that, under its somewhat esoteric and fanciful aspect, it has a profound significance.

In a bacterium, the DNA–RNA polymerase synthesizes RNA on a DNA matrix and not on an RNA matrix. However, *in vitro*, the same enzyme is able to utilize RNA as matrix. It is likely that, in a bacterium, the DNA–RNA polymerase is at the locus of its activity and not where it would have an opportunity to engage in actions reproved by molecular morals. In a normal cell, each molecule is at the place where it should be and not elsewhere, and that is why each of them does what it should do and not something else. We must then ask whether certain diseases of cellular metabolism are not provoked by molecular incursion into foreign territory. Molecular societies obey the same laws as more complex societies.

Let us go back to the lysogenic bacteria. Now then, the prophage is reproduced by bacterial enzymes. Why does the lysogenic bacterium not produce bacteriophages? We believe today that at least one of the genes of the prophage expresses itself and produces a repressor. This repressor attaches itself to an operator gene and blocks the expression of the structural gene that determines the formation of the enzymes necessary for autonomous reproduction of the bacteriophage. A lysogenic bacterium produces virions if it is derepressed, and here we are again entangled in the problem of induction.

In addition to such physical agents as ultraviolet and various other kinds of radiation, we know of many chemical inducers, such as the organic peroxides, ethyleneimines, and mitomycin. All inducers have in common the property of disturbing the metabolism of nucleic acids. According to Goldthwait and Jacob, the final product of the change might be a derivative of adenine. The product will attach itself to the active repressor and thus bring about an allosteric modification. The active repressor will become an inactive aporepressor. An operon being derepressed, a structural gene can express itself, and a new enzyme is produced which assures the autonomous multiplication of the viral genetic material and permits the expression of all of the genes that regulate viral structures.

The vegetative phase takes its course, virions are formed, and the bacterium explodes and dies.

Thus the inducing agents act by inactivating the repressor. Now, then, the repressor is responsible for immunity. This is why, under the action of inducers, the immunity of the lysogenic bacteria to the superinfective homologous phage is lost.

When a nonlysogenic bacterium is infected by a temperate phage, it will either undergo lysis or become lysogenic. The conditions of the environment here decide the evolution of the genetic material of the bacteriophage; that is, they decide the fate of the bacterium. In order to be effective, these conditions must begin to operate within 7 minutes after infection. The fate of the bacterium–virus system manifestly depends on whether a repressor or the key enzyme responsible for autonomous multiplication is formed first.

The repressor is produced by a regulator gene and acts on an operator gene. It is obvious that either gene is susceptible to mutation. A regulator gene, under the influence of mutation, will give a repressor incapable of inhibiting a given operator. An operator, as a result of mutation, may become insensitive to a given repressor. In the last analysis, the fate of a bacterium infected by a bacteriophage thus depends on the genetic constitution of the bacteriophage, on the genetic constitution of the bacterium, and on the metabolism of the bacterium which is in turn controlled by the environment. Moreover, the genetic material of the bacteriophage may confer on the bacterium not only the power of producing bacteriophages in the absence of infection but also other properties, such as the capability of synthesizing a toxin like diphtheria toxin or the synthesis of a new antigen which will modify the structure of the bacterial wall.

Be that as it may, through the infection the lysogenic bacterium has become a new organism, a cell–virus system whose fate will depend on the bacterial metabolism, which itself depends on the environment.

Any valid proposition, however singular it may appear, is necessarily the particular expression of a general law. Since generalization is one of the most productive heuristic methods, we shall attempt to express the relations between bacteriophage and bacterium in such a way that the generality on which they depend will be included in the expression. Here, then, is this general expression: the course of the viral cycle is dependent on allosteric proteins whose structure and activity are controlled by the metabolism of the host cell.

This general proposition shares at the same time both in the strictness of a law and the weakness of a hypothesis. We shall now have to abandon the heights of theoretical conception and descend into the underworld of disease.

We know that herpetic infections are frequently latent. The infected individual does not present any symptoms of disease. However, under the influence of a great number of factors, the disease erupts. The variety of the effectors is astonishing, as will be seen from the following list of those promoting the outbreak of herpes:

Local hyperpyrexia
Artificial fever
Febrile disorders (malaria, pneumonia, brucellosis, typhoid fever)
Local ultraviolet radiation
Hormone treatments
Menstruation
Unbalanced diet
Leukemia
Administration of proteins foreign to the system
Anaphylactic shock
Lesions of the Gasserian ganglion
Section of the trigeminal nerve
Emotion

During the latent infections not only are there no symptoms, but it is not possible to detect the virus. We do not know in which form it is present, and to say that the virus is masked simply masks our ignorance. When the lesions develop, the virus appears in abundance. Actually, it is the viral multiplication which is responsible for the disease. During the latent infection, the viral cycle is blocked. The agents which trigger the disease thus induce the viral development of which the disease is the consequence. We can therefore assume that all these agents, in spite of their diversity, provoke by different mechanisms the same modification of cellular metabolism, and precisely that modification which will be responsible for triggering viral multiplication.

And here we become entangled in a new hypothesis, according to which the development of an animal virus is controlled in a positive or negative man-
same modification of cellular metabolism, and precisely that modification which will be responsible for triggering viral multiplication.

Some experimental data will be welcome. Guanidine inhibits the development of certain viruses, and in particular that of the poliovirus. At concentrations which are inhibiting for the virus, however, guanidine does not observably influence metabolism and cellular growth. Guanidine consequently is a specific inhibitor of the poliovirus. How does it act? Does it act at the level of the nucleic acid, as some seem to think? Sensitivity to guanidine may disappear upon mutation. A structural gene is a sequence of several hundred nucleotides, and a point mutation is the substitution of one nucleotide for another. It is difficult to conceive that the presence, at some particular locus, of a nucleotide already abundantly represented along a long chain could modify the properties of the chain so that, for example, a difference of temperature of one-tenth of one degree or the presence of 0.0002 *M* guanidine would notably affect the structure and function of the molecule.

Let us therefore assume that guanidine does not act directly on the nucleic acid. The hypothesis proposed by us a few years ago is the following. Guanidine, like temperature, affects the tertiary or quaternary structure of a protein. Today we would state that it is responsible for an allosteric modification.

What is this protein?

In the presence of guanidine, viral RNA is not synthesized, and it has been believed that guanidine acts in some manner on the viral RNA-replicase. This was a logical conclusion. However, we became aware that methionine and choline neutralize the inhibiting effects of guanidine. A number of experiments have led us to believe that the guanidine must block the activity of a virus-determined transmethylase. The simplest hypothesis is that this enzyme methylates the viral RNA.

The DNA of the polyoma virus contains 5-methylcytosine, and so does that of bacteriophage lambda. Methionine intervenes in the modification induced by the host of this bacteriophage. However, we do not know the physiological significance of such methylation. Investigation of the poliovirus has afforded an indication that methylation in certain cases may well control the course of the viral cycle. Such methylation would be effected by a virus-determined enzyme which is sentitive to guanidine and to cellular metabolites possessing a guanyl group. Thus the evolution of viral proteins, like the evolution of proteins in general, should terminate in the development of sites capable of accepting specific effectors, inhibitors and anti-inhibitors, which are cellular metabolites. I should like to draw attention to this conclusion.

A cell becomes cancerous under the action of a virus. The virus has introduced into the normal cell its genetic material, which brings with it new functions, and these functions are the cause of the malignancy. It is reasonable to assume that a viral protein carries the phenotypical responsibility of the malign transormation.

If the functions of the oncogenic viruses, like the functions of other viruses, depend on specific effectors, we may hope some day to convert a malignant cell into a phenotypically normal one.

This leads us to remark on methodology. It would seem that we have so far been occupied in finding substances which specifically kill the malignant cell in the culture to the exclusion of normal cells or which specifically prevent the malignant cell from multiplying. The experiments are generally made in environments which may contain anti-effectors, as is the case for the couple methionine/guanidine. A change in methodology might perhaps be profitable.

There is also another obvious theoretical possibility. Instead of attempting to repress the viral functions, we might attempt to intensify them in such a manner that the virus whose cycle is blocked develops and kills the host cell.

The search for specific effectors of the viral functions and of the viral development is empirical at the moment. Such research must be developed and extended. Our ignorance of the nature of the factors which govern the relations among oncogenic virus and the cells should not incline us to pessimism but should instead be a stimulant. We should declare war on oncogenic viruses and carry it to victory.

The experimental data and concepts discussed here encompass a vast field. There are very many who have made important contributions to this domain. It would have been impossible to do each of them justice within a lecture of 30 minutes. I have mentioned some names, and my selection has necessarily been arbitrary. I would have liked to and I should cite, among others, T. F. Anderson, L. Astrachan and E. Volkin, L. Barksdale, G. Bertani, A. Campbell, S. S. Cohen, V. J. Freeman, N. B. Groman, L. M. Kozloff, S. Lederberg, S. E. Luria, F. W. Putnam, G. Stent, Elie Wollman, and N. D. Zinder.

The bibliography concerning bacteriophages will be found in the excellent book by G. Stent, *Molecular Biology of Bacterial Viruses* (Freeman, San Francisco, 1963) and the no less excellent treatise of W. Hayes. *The Genetics of Bacteria and Their Viruses* (Blackwell, Oxford; Wiley, New York, 1964). The data concerning the effectors of the development of animal viruses are discussed in A. Lwoff, «The specific effectors of viral development» (The First Keilin Memorial Lecture), *Biochem. J.*, *96*, (1965) 289–301.

Biography

André Michel Lwoff was born on 8 May 1902 in Ainay-le-Château (Allier). He joined the Institut Pasteur at the age of 19. He had graduated in science and had done one year of medicine. Lwoff completed his studies while working in the laboratory. In 1921, he had the good fortune to study under a very great microbiologist, Edouard Chatton; Lwoff remained his colleague for seventeen years. It was through him that Lwoff joined the Institut Pasteur in the laboratory of Félix Mesnil. His first investigations were on the parasitic ciliates, their developmental cycle, and morphogenesis. Later, he worked on the problems involved in the nutrition of protozoans. André Lwoff obtained his M.D. in 1927 and his Ph.D. in 1932.

In 1932–1933 a grant from the Rockefeller Foundation enabled him to spend a year in Heidelberg in the laboratory of Otto Meyerhof. He studied haematin – a growth factor for the flagellates – the specificity of protohaematin, its quantitative effect on growth, and the part it played in the respiratory catalyst system.

Then in 1936, again with the aid of a grant from the Rockefeller Foundation, Lwoff and his wife spent seven months in Cambridge in the laboratory of David Keilin; factor V, which is required by *Haemophilus influenzae*, was identified with cozymase and its physiological role for the bacterium was defined.

There were many other investigations on growth factors for flagellates and ciliates with regard to growth factors, loss of function, and physiological development until the time when Lwoff began working on the problem of lysogenic bacteria.

Dr. Lwoff was appointed Head of the Department at the Institut Pasteur in 1938, and Professor of Microbiology at the Science Faculty in Paris in 1959.

The observation of isolated bacteria led him to the conclusion that lysogenic bacteria did not secrete bacteriophages, that the production of bacteriophages led to the death of the bacterium, and above all that this production must be induced by external factors. It was this hypothesis which, together with Louis Siminovitch and Niels Kjeldgaard, led Lwoff to discover the inductive action of ultraviolet irradiation (1950).

In 1954 Prof. Lwoff began studying poliovirus. Experiments on the relations between the temperature sensitivity of viral development and neurovirulence led him to consider the problem of viral infection. In this way it became clear that non-specific factors play an important part in the development of the primary infection. He has now begun to investigate the action mechanism of specific inhibitors of viral development.

André Lwoff has been honoured by the following prizes of the Académie des Sciences: Lallemant, Noury, Longchampt, Chaussier, Petit d'Ormoy prizes and the Charles-Leopold Mayer Foundation prize. He also received the Barbier prize from the Académie de Medécine, and the Leeuwenhoek Medal of the Royal Netherlands Academy of Science and Arts (Amsterdam, 1960), as well as the Keilin Medal of the British Biochemical Society (1964).

He is a Honorary Member of the Harvey Society (1954), of the American Society of Biological Chemists (1961), of the Society for General Microbiology (1962), and a Corresponding Member of the Botanical Society of America (1956).

He is President of the International Association of Microbiological Societies, and a Member of the International Committee for the Organization of Medical Sciences. He is a Member of the Société Zoologique de France, of the Société de Pathologie exotique, of the Société de Biologie and President of the Société des Microbiologistes de langue française. Furthermore a Honorary Member of the New York Academy of Sciences (1955), Honorary Foreign Member of the American Academy of Arts and Sciences (1958), Associate of the National Academy of Sciences of the United States of America (1955), and a Foreign Member of the Royal Society, London (1958).

He holds honorary degrees from the following universities: Chicago (D. Sc., 1959), Oxford (D. Sc., 1959), Glasgow (Doctor of Laws, 1963) and Louvain (M. D., 1966).

JACQUES MONOD

From Enzymatic Adaptation to Allosteric Transitions

December 11, 1965

One day, almost exactly 25 years ago – it was at the beginning of the bleak winter of 1940 – I entered André Lwoff's office at the Pasteur Institute. I wanted to discuss with him some of the rather surprising observations I had recently made.

I was working then at the old Sorbonne, in an ancient laboratory that opened on a gallery full of stuffed monkeys. Demobilized in August in the Free Zone after the disaster of 1940, I had succeeded in locating my family living in the Northern Zone and had resumed my work with desperate eagerness. I interrupted work from time to time only to help circulate the first clandestine tracts. I wanted to complete as quickly as possible my doctoral dissertation, which, under the strongly biometric influence of Georges Teissier, I had devoted to the study of the kinetics of bacterial growth. Having determined the constants of growth in the presence of different carbohydrates, it occurred to me that it would be interesting to determine the same constants in paired mixtures of carbohydrates. From the first experiment on, I noticed that, whereas the growth was kinetically normal in the presence of certain mixtures (that is, it exhibited a single exponential phase), two complete growth cycles could be observed in other carbohydrate mixtures, these cycles consisting of two exponential phases separated by a complete cessation of growth (Fig.1).

Lwoff, after considering this strange result for a moment, said to me, «That could have something to do with enzyme adaptation.»

«Enzyme adaptation? Never heard of it!» I said.

Lwoff's only reply was to give me a copy of the then recent work of Marjorie Stephenson, in which a chapter summarized with great insight the still few studies concerning this phenomenon, which had been discovered by Duclaux at the end of the last century. Studied by Dienert and by Went as early as 1901 and then by Euler and Josephson, it was more or less rediscovered by Karström, who should be credited with giving it a name and attracting attention to its existence. Marjorie Stephenson and her students Yudkin and

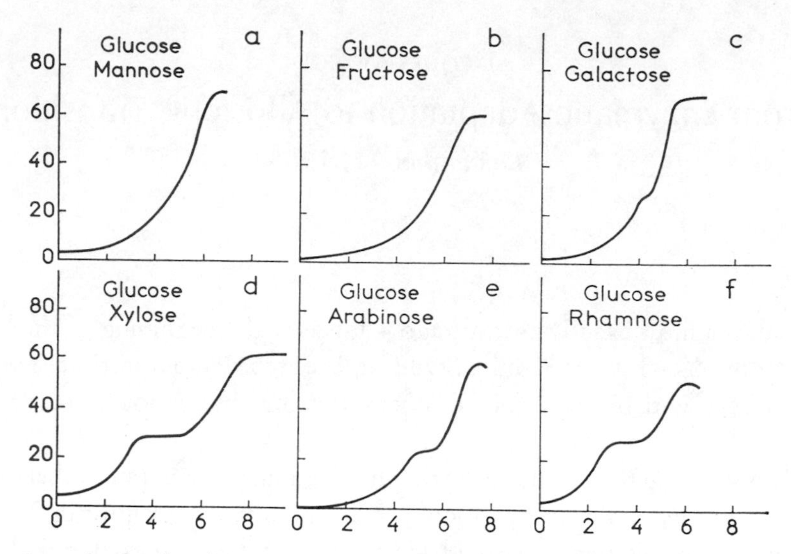

Fig. 1. Growth of *Escherichia coli* in the presence of different carbohydrate pairs serving as the only source of carbon in a synthetic medium[50].

Gale had published several papers on this subject before 1940. [See ref. 1 for a bibliography of papers published prior to 1940.]

Lwoff's intuition was correct. The phenomenon of «diauxy» that I had discovered was indeed closely related to enzyme adaptation, as my experiments, included in the second part of my doctoral dissertation, soon convinced me. It was actually a case of the «glucose effect» discovered by Dienert as early as 1900, today better known as «catabolic repression» from the studies of Magasanik[2].

The die was cast. Since that day in December 1940, all my scientific activity has been devoted to the study of this phenomenon. During the Occupation, working, at times secretly, in Lwoff's laboratory, where I was warmly received, I succeeded in carrying out some experiments that were very significant for me. I proved, for example, that agents that uncouple oxidative phosphorylation, such as 2,4-dinitrophenol, completely inhibit adaptation to lactose or other carbohydrates[3]. This suggested that «adaptation» implied an expenditure of chemical potential and therefore probably involved the true synthesis of an enzyme. With Alice Audureau, I sought to discover the still quite obscure relations between this phenomenon and the one Massini, Lewis, and others had discovered: the appearance and selection of «spontaneous» mutants (see ref. 1). Using a strain of *Escherichia coli mutabile* (to which we had given the initials ML because it had been isolated from André Lwoff's intestinal tract), we showed that an apparently spontaneous mutation was allowing these originally «lactose-negative» bacteria to become «lactose-positive». However, we proved that the original strain (*Lac*$^-$) and the mutant strain (*Lac*$^+$) did not differ from each other by the presence of a specific enzyme system, but rather by the ability to produce this system in the presence of lactose. In other words, the mutation affected a truly genetic property that became evident only in the presence of lactose[4].

There was nothing new about this; geneticists had known for a long time that certain genotypes are not always expressed. However, this mutation involved the selective control of an enzyme by a gene, and the conditions necessary for its expression seemed directly linked to the chemical activity of the system. This relation fascinated me. Influenced by my friendship with and admiration for Louis Rapkine, whom I visited frequently and at length in his laboratory, I had been tempted, even though I was poorly prepared, to study elementary biochemical mechanisms, that is, enzymology. But under the influence of another friend whom I admired, Boris Ephrussi, I was equally tempted by genetics. Thanks to him and to the Rockefeller Foundation, I had had an opportunity some years previously to visit Morgan's laboratory at the California Institute of Technology. This was a revelation for me – a revelation of genetics, at that time practically unknown in France; a revelation of what a group of scientists could be like when engaged in creative activity and sharing in a constant exchange of ideas, bold speculations, and strong criticisms. It was a revelation of personalities of great stature, such as George Beadle, Sterling Emerson, Bridges, Sturtevant, Jack Schultz, and Ephrussi, all of whom were then working in Morgan's department. Upon my return to France, I had again taken up the study of bacterial growth. But my mind remained full of the concepts of genetics and I was confident of its ability to analyze and convinced that one day these ideas would be applied to bacteria.

«Discovery» of Bacterial Genetics

Toward the end of the war, while still in the army, I discovered in an American army bookmobile several miscellaneous issues of *Genetics*, one containing the beautiful paper in which Luria and Delbrück[5] demonstrated for the first time rigorously, the spontaneous nature of certain bacterial mutants. I think I have never read a scientific article with such enthusiasm; for me, bacterial genetics was established. Several months later, I also «discovered» the paper by Avery, MacLeod, and McCarty[6] – another fundamental revelation. From then on I read avidly the first publications by the «phage-church», and when I entered Lwoff's department at the Pasteur Institute in 1945, I was tempted to abandon enzyme adaptation in order to join the church myself and work with bacteriophage. In 1946 I attended the memorable symposium at Cold Spring Harbor where Delbrück and Bailey, and Hershey, revealed their discovery of virus recombination at the same time that Lederberg and Tatum announced their discovery of bacterial sexuality[7]. In 1947 I was invited to the Growth Symposium to present a report[1] on enzyme adaptation, which had begun to arouse the interest of embryologists as well as of geneticists. Preparation of this report was to be decisive for me. In reviewing all the literature, including my own, it became clear to me that this remarkable phenomenon was almost entirely shrouded in mystery. On the other hand, by its regularity, its specificity, and by the molecular-level interaction it exhibited between a genetic determinant and a chemical determinant, it seemed of such interest and of a significance so profound that there was no longer any question as to whether I should pursue its study. But I also saw that it would be necessary to make a clean sweep and start all over again from the beginning.

The central problem posed was that of the respective roles of the inducing substrate and of the specific gene (or genes) in the formation and the structure of the enzyme. In order to understand how this problem was considered in 1946, it would be well to remember that at that time the structure of DNA was not known, little was known about the structure of proteins, and nothing was known of their biosynthesis. It was necessary to resolve the following question: Does the inducer effect total synthesis of a new protein molecule from its precursors, or is it rather a matter of the activation, conversion, or «remodeling» of one or more precursors?

This required first of all that the systems to be studied be carefully chosen and defined. With Madeleine Jolit and Anne-Marie Torriani, we isolated β-galactosidase, then the amylomaltase of *Escherichia coli*[8]. Our work was advanced greatly by the valuable collaboration of Melvin Cohn, an excellent immunologist, who knew better than I the chemistry of proteins. He knew, for example, how to operate that marvelous apparatus that had intimidated me, the «Tiselius»[9]. With Anne-Marie Torriani, he characterized β-galactosidase as an antigen[10]. Being familiar with the system, we could now study with precision the kinetics of its formation. A detailed study of the kinetics carried out in collaboration with Alvin Pappenheimer and Germaine Cohen-Bazire[11] strongly suggested that the inducing effect of the substrate entailed

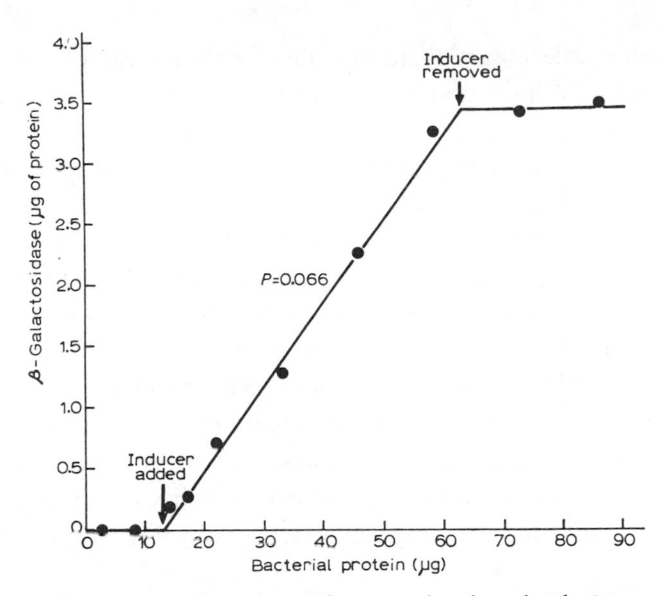

Fig. 2. Induced biosynthesis of β-galactosidase in *Escherichia coli*. The increase in enzyme activity is expressed not as a function of time but as a function of the concomitant growth of bacterial proteins. The slope of the resulting curve (P) indicates the differential rate of synthesis[11].

total biosynthesis of the protein from amino acids (Fig. 2). This interpretation seemed surprising enough at that time, but from the first, I must say, it won my firm belief. There is in science, however, quite a gap between belief and certainty. But would one ever have the patience to wait and to establish the certainty if the inner conviction were not already there?

We were to establish certainty a little later, thanks to some experiments with isotopic tracers done by Hogness, Cohn, and myself[12]. To tell the truth, the results of these labeling experiments were even more surprising in view of the ideas then current on the biosynthesis of proteins and their state within the cell. The work of Schoenheimer[13] had actually persuaded most biochemists that in an organism proteins are in herently in a «dynamic state», each molecule being perpetually destroyed and reconstructed by exchange of amino acid residues. Our experiments, however, showed that β-galactosidase is entirely stable *in vivo*, as are other bacterial proteins, under conditions of normal growth. They did not, of course, contradict the results of Schoenheimer, but very seriously questioned their interpretation and the dogma of the «dynamic state».

Be that as it may, these conclusions were invaluable to us. We knew, thenceforth, that «enzyme adaptation» actually corresponds to the total biosynthesis of a stable molecule and that, consequently, the increase of enzyme activity in the course of induction is an authentic measure of the synthesis of the specific protein.

These results took on even more significance as our system became more accessible to experiment. With Germaine Cohen-Bazire and Melvin Cohn [14,15], I was able to continue the systematic examination of a question I had repeatedly encountered: the correlations between the specificity of action of an inducible enzyme and the specificity of its induction. Pollock's pertinent observations on the induction of penicillinase by penicillin[16] made it necessary to consider this problem in a new way. We conducted a study of a large number of galactosides or their derivatives, comparing their properties as inducers, substrates, or as antagonists of the substrates of the enzyme, once more reaching a quite surprising conclusion, namely, that inductive ability is by no means a prerogative of the substrates of the enzyme, or even of the substances capable

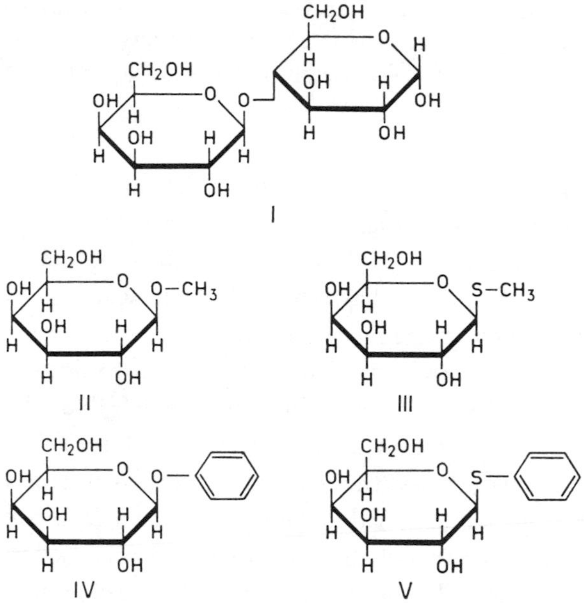

Fig. 3. Comparison of various β-galactosides as substrates and as inducers of β-galactosidase. I, Lactose: substrate of the enzyme, but deprived of inductive activity. II, Methyl-β-D-galactoside: low-affinity substrate effective inducer. III, Methyl-β-D-thiogalactoside: not hydrolyzable by the enzyme, but a powerful inducer. IV, Phenyl-β-D-galactoside: excellent enzyme substrate, high affinity, no inductive ability. V, Phenyl-β-D-thiogalactoside: no activity either as a substrate or as an inducer, but capable of acting as an antagonist of the inducer.

of forming the most stable complexes with it. For example, certain thiogalactosides, not hydrolyzed by the enzyme or used metabolically, appeared to be very powerful inducers. Certain substrates, on the other hand, were not inducers. The conclusion became obvious that the inducer did not act (as frequently assumed) either as a substrate or through combination with preformed active enzyme, but rather at the level of another specific cellular constituent that would one day have to be identified (Fig. 3).

Generalized Induction

In the course of this work, we observed a fact that seemed very significant: a certain compound, phenyl-β-D-thiogalactoside, devoid of inductive capacity, proved capable of counteracting the action of an effective inducer, such as methyl-β-D-thiogalactoside. This suggested the possibility of utilizing such «anti-induction» effects to prove a theory that we called, somewhat ambitiously, «generalized induction». From the very beginning of my research, I had been preoccupied with the problem posed by the existence, together with inducible enzymes, of «constitutive» systems; in other words (according to the then current definition), systems synthesized in the absence of any substrate or exogenous inducer, as is the case, of course, with all the enzymes of intermediate and biosynthetic metabolism. It did not seem unreasonable to suppose that the synthesis of these enzymes was controlled by their endogenous substrate, which would imply that the mechanism of induction is in reality universal. We were encouraged in this hypothesis by the work of Roger Stanier on the supposedly sequential induction of systems attacking phenolic compounds in *Pseudomonas*.

I sought, therefore, along with Germaine Cohen-Bazire, to prove that the biosynthesis of a typically «constitutive» enzyme (according to the ideas of the time), tryptophan synthetase, could be inhibited by an analogue of the presumed substrate. The reaction product seemed a good candidate for an analogue of the substrate, and we were soon able to prove that tryptophan and 5-methyltryptophan are powerful inhibitors of the biosynthesis of the enzyme. This was the first known example of a «repressible» system—discovered, it turned out, as proof of a false hypothesis[17].

I did not have, I must say, complete confidence in the ambitious theory of «generalized» induction, which soon encountered various difficulties. I was, however, encouraged by an interesting observation made by Vogel and

Davis[18] concerning another enzyme, acetylornithinase, involved in the formation of arginine. Using a mutant requiring arginine or *N*-acetylornithine, Vogel and Davis found that, when the bacteria are cultivated in the presence of arginine, they do not produce acetylornithinase, whereas when they are cultivated in the presence of *N*-acetylornithine, acetylornithinase is synthesized. Hence these authors concluded that this enzyme must be induced by its substrate, *N*-acetylornithine. When Henry Vogel was passing through Paris, I drew his attention to the fact that their very interesting observations could just as well be explained as resulting from an inhibitory effect of arginine as from an inductive effect of acetylornithine. In order to resolve this problem, it was necessary to study the biosynthesis of the enzyme in a mixture of the two metabolites. The experiment proved that it is indeed a question of an inhibiting effect rather than an inductive effect. Vogel, quite rightly, proposed the term «repression» to designate this effect and thus established «repressible» systems alongside of «inducible» systems. Later on, thanks especially to the studies of Maas, Gorini, Pardee, Magasanik, Cohen, Ames, and many others (see ref.19 for literature), the field of repressible systems was considerably extended; it is now generally accepted that practically all bacterial biosynthetic systems are controlled by such mechanisms.

Nevertheless, I remained faithful to the study of the β-galactosidase of *Escherichia coli*, knowing well that we were far from having exhausted the resources of this system. During the years spent in establishing the biochemical nature of the phenomenon, I had been able only partially to approach the question of its genetic control – enough, however, to convince me that it was extremely specific and that it justified the idea that Beadle and Tatum's postulate, «one gene–one enzyme», was applicable to inducible and degradative enzymes as well as to the enzymes of biosynthesis, which the Stanford school had principally studied. These conclusions led me to abandon an idea I had adopted as a working hypothesis – that is, that many different inducible enzymes may result from the «conversion» of a single precursor whose synthesis is controlled by a single gene; this hypothesis was also contradicted by the results of our experiments with tracers.

But genetic analysis once more encountered grave difficulties. First, the low frequency of recombination, in the systems of conjugation known at that time, did not permit fine genetic analysis. Another difficulty holding us back was the existence of some mysterious phenotypes; certain mutants («cryptic»), incapable of metabolizing the galactosides, nevertheless appeared capable of synthesizing β-galactosidase. The solution to this problem came to us by accident while we were looking for something entirely different. In 1954, when the chairmanship of the new Department of Cellular Biochemistry had just been bestowed upon me, Georges Cohen joined us, and I suggested to him, and simultaneously to Howard Rickenberg, to make use of the properties of thiogalactosides as gratuitous inducers in attempting to study their fate in inducible bacteria, employing a thiogalactoside labeled with carbon-14. We noted that the radioactivity associated with the galactoside accumulated rapidly in wild-type induced bacteria, but not in the so-called cryptic mutants. Neither did the radioactivity accumulate in wild-type bacteria not previously induced. The capacity for accumulation depended, therefore, on an inducible factor. Study of the kinetics, of the specificity of action, and of the specificity of induction of this system, as well as the comparison of various mutants, led us to the conclusion that the element responsible for this accumulation could only be a specific protein whose synthesis, governed by a gene (y) distinct from that of galactosidase (z), was induced by the galactosides at the same time as the synthesis of the enzyme. To this protein we gave the name «galactoside permease»[20,21] (Fig. 4).

The very existence of a specific protein responsible for the permeation and accumulation of galactosides was occasionally put in doubt because the evidence for it was based entirely on observations *in vivo*. Some of the researchers who did not really doubt its existence still reproached me from time to time for giving a name to a protein when it had not been isolated. This attitude reminded me of that of two traditional English gentlemen who, even if they know each other well by name and by reputation, will not speak to each other before having been formally introduced. On my part, I never for a moment doubted the existence of this protein, for our results could be interpreted in no other way. Nevertheless, I was only too happy to learn, recently, that by a recent series of experiments, Kennedy has identified *in vitro* and isolated the specific inducible protein, galactoside permease[22]. Kennedy was brilliantly successful where we had failed, for we had repeatedly sought to isolate galactoside permease *in vitro*. These efforts of ours, however, were not in vain, since they led Irving Zabin, Adam Kepes, and myself to isolate still another protein, galactoside transacetylase[23]. For several weeks we believed that this enzyme was none other than the permease itself. This was an erroneous assumption, and the physiological function of this protein is still totally unknown. It was a profitable discovery, nevertheless, because the transacetylase, determined by a gene belonging to the lactose operon, has been very useful to experimenters, if not to the bacterium itself.

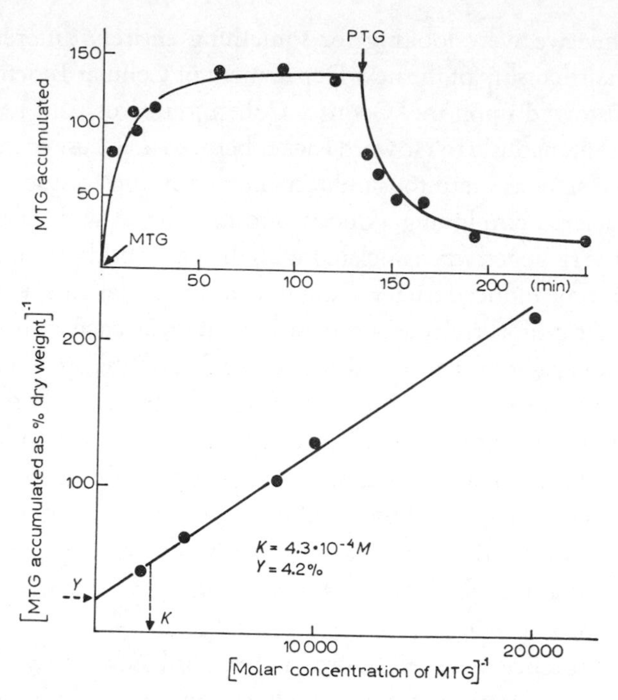

Fig. 4. Evidence for the existence of galactoside permease. (Top) Accumulation of labeled methyl-β-D-thiogalactoside (MTG) by a suspension of previously induced bacteria. Displacement of accumulated galactoside (phenyl-β-D-thiogalactoside, PTG). (Bottom) Accumulation of a galactoside in previously induced bacteria as a function of the concentration of the external galactoside. Inverse coordinates: The constants *K* and *Y* define, respectively, the constant of apparent dissociation and the constant of apparent activity of the system of accumulation[21].

The study of galactoside permease was to reveal another fact of great significance. Several years earlier, following Lederberg's work, we had isolated some «constitutive» mutants of β-galactosidase, that is, strains in which the enzyme was synthesized in the absence of any galactoside. But we now proved that the constitutive mutation has a pleiotropic effect. In these mutants, galactoside permease as well as galactosidase (and the transacetylase) were indeed simultaneously constitutive, whereas we knew on the other hand that each of the three proteins is controlled by a distinct gene. We then had to admit that a constitutive mutation, although very strongly linked to the loci governing galactosidase, galactoside permease, and transacetylase, had taken place in a gene (*i*) distinct from the other three (*z*, *y*, and *Ac*), and that the relationship of this gene to the three proteins violated the postulate of Beadle and Tatum.

New Perspectives

These investigations were given new meaning by the perspectives opened to biology around 1955. It was in 1953 that Watson and Crick, on the basis of observations made by Chargaff and Wilkins, proposed their model of the structure of DNA. From the first, in this complementary double sequence, one could see a mechanism for exact replication of the genetic material. Meanwhile, one year earlier, Sanger had described the peptide sequence of insulin, and it was also already known, from the work of Pauling and Itano[24] in particular, that a genetic mutation can cause a limited modification in the structure of a protein. In 1954, Crick and Watson[25] and Gamow[26] proposed the genetic code theory: The primary structure of proteins is determined and defined by the linear sequence of the nucleotides in DNA. Thus the profound logical intuition of Watson and Crick had allowed them to discover a structure that immediately explained, at least in principle, the two essential functions long assigned by geneticists to hereditary factors: to control its own synthesis and to control that of the nongenetic constituents. Molecular biology had been born, and I realized that, like Monsieur Jourdain, I had been doing molecular biology for a long time without knowing it.

More than ten years have elapsed since then, and the ideas whose hatching I recall here were then far from finding a uniformly enthusiastic audience. My conviction, however, had been established long before absolute certainty could be acquired. This certainty exists today, thanks to a succession of discoveries, some of them almost unhoped for, that have enriched our discipline since that time.

Once the physiological relations of galactosidase and galactoside permease were understood, and once it was proved that they depend on two distinct genetic elements while remaining subject to the same induction determinism and to the same constitutive mutations, it became imperative to analyze the corresponding genetic structures. In particular, the expression of these genes and the relations of dominance between their alleles had to be studied in detail.

Precisely at this time, the work of Jacob and Wollman[27] had clarified the mechanism of bacterial conjugation; we knew that this conjugation consists of the injection, without cytoplasmic fusion, of the chromosome of a male bacterium into a female. It was even possible to follow the kinetics of penetration of a given gene. I decided, along with Arthur Pardee and François Jacob, to use these new experimental tools to follow the «expression» of the z^+ and i^+ genes injected into a female carrying mutant alleles of these genes.

This difficult undertaking, carried out successfully thanks to the experimental talent of Arthur Pardee, brought about two remarkable and at least partially unexpected results. First, the *z* gene (which we knew to be the determinant of the structure) is expressed (by the synthesis of β-galactosidase) very fast and at maximum rate from the beginning. I will pass over the development and the consequences of this observation, which was one of the sources of the messenger theory. Second, the inducible allele of the *i* gene is dominant with respect to the constitutive allele, but this dominance is expressed very slowly. Everything seemed to indicate that this gene is responsible for the synthesis of a product that inhibits, or represses, the biosynthesis of the enzyme. This was the reason for designating the product of the gene as a «repressor» and hypothesizing that the inducer acts not by provoking the synthesis of the enzyme but by «inhibiting an inhibitor» of this synthesis[28].

The Theory of Double Bluff

Of course I had learned, like any schoolboy, that two negatives are equivalent to a positive statement, and Melvin Cohn and I, without taking it too seriously, debated this logical possibility that we called the «theory of double bluff», recalling the subtle analysis of poker by Edgar Allan Poe.

I see today, however, more clearly than ever, how blind I was in not taking this hypothesis seriously sooner, since several years earlier we had discovered that tryptophan inhibits the synthesis of tryptophan synthetase; also, the subsequent work of Vogel, Gorini, Maas, and others (cited in ref. 15) showed that repression is not due, as we had thought, to an anti-induction effect. I had always hoped that the regulation of «constitutive» and inducible systems would be explained one day by a similar mechanism. Why not suppose, then, since the existence of repressible systems and their extreme generality were now proven, that induction could be effected by an anti-repressor rather than by repression by an anti-inducer? This is precisely the thesis that Leo Szilard, while passing through Paris, happened to propose to us during a seminar. We had only recently obtained the first results of the injection experiment, and we were still not sure about its interpretation. I saw that our preliminary observations confirmed Szilard's penetrating intuition, and when he had finished his presentation, my doubts about the «theory of double bluff» had been removed and my faith established – once again a long time before I would be able to achieve certainty.

Some of the more important developments of this study, such as the discovery of operator mutants and of the operon, considered as a single coordinated expression of the genetic material, and the bases and demonstration of the messenger theory, have been presented by François Jacob in his lecture[27], and I will not pause over these, in order to return to that constituent whose existence and role had so long escaped me, the repressor. To tell the truth, I find some excuses for myself even now. It was not easy to get away completely from the quite natural idea that a structural relation, inherent in the mechanism of the phenomenon of induction, must exist between the inducer of an enzyme and the enzyme itself. And I must admit that, up until 1957, I tried to «rescue» this hypothesis, even at the price of reducing almost to nothing the «didactic» role (as Lederberg would say) of the inducer.

From now on it was necessary to reject it completely. An experiment carried out in collaboration with David Perrin and François Jacob proved, moreover, that the mechanism of induction functioned perfectly in certain mutants, producing a modified galactosidase totally lacking in affinity for galactosides[29].

What now had to be analyzed and understood were the interactions of the repressor with the inducer on the one hand, with the operator on the other. Otto Warburg said once, about cytochrome oxidase, that this protein–or presumed protein–was as inaccessible as the matter of the stars. What is to be said, then, of the repressor, which is known only by the results of its interactions? In this respect we are in a position somewhat similar to that of the police inspector who, finding a corpse with a dagger in its back, deduces that somewhere there is an assassin; but as for knowing who the assassin is, what his name is, whether he is tall or short, dark or fair, that is another matter. The police in this case, it seems, sometimes get results by sketching a composite portrait of the culprit from several clues. This is what I am going to try to do now with regard to the repressor.

First, it is necessary to assign to the assassin–I mean the repressor–two properties: the ability to recognize the inducer and the ability to recognize the operator. These recognitions are necessarily steric functions and are thus susceptible to being modified or abolished by mutation. Loss of the ability to recognize the operator would result in total derepression of the system. Every mutation that causes a shift in the structure of the repressor or the abolition of its synthesis must therefore appear «constitutive», and this is without doubt the reason for the relatively high frequency of this type of mutation.

However, if the composite portrait is correct, it can be seen that certain mutations might abolish the repressor's ability to recognize the inducer but leave unaffected its ability to recognize the operator. Such mutations should exhibit a very special phenotype. They would be noninducible (that is, lactose-negative), and in diploids they would be dominant in *cis* as well as in *trans*. Clyde Willson, David Perrin, Melvin Cohn, and I[30] were able to isolate two mutants that possessed precisely these properties, and Suzanne Bourgeois (ref. 31) has recently isolated a score of others.

In tracing this first sketch of the composite portrait, I implicitly supposed that there was only one assassin; that is, the characteristics of the system were explained by the action of a single molecular species, the repressor, produced from gene *i*. This hypothesis is not necessary *a priori*. It could be supposed, for example, that the recognition of the inducer is due to another constituent distinct from that which recognizes the operator. Then we would have to assume that these two constituents could recognize each other. Today this latter hypothesis seems to be practically ruled out by the experiments of Bourgeois, Cohn, and Orgel[31], which show, among other important results, that the mutation of type i^- (unable to recognize the operator) and the mutations of the type i^s (unable to recognize the inducer) occur in the same cistron and, from all appearances, involve the same molecule, a unique product of the regulator gene *i*.

An essential question is the chemical nature of the repressor. Inasmuch as it seems to act directly at the level of the DNA, it seemed logical to assume that it could be a polyribonucleotide whose association with a DNA sequence would take place by means of specific pairing. Although such an assumption could explain the recognition of the operator, it could not explain the recognition of the inducer, because probably only proteins are able to form a *stereospecific* complex with a small molecule. This indicates that the repressor, that is, the active product of the gene *i*, must be a protein. This theory, based until now on purely logical considerations, has just received indirect but decisive confirmation.

It should be remembered that, thanks to the work of Benzer[32], Brenner[33], and Garen[34], a quite remarkable type of mutation has been recognized, called «nonsense» mutation. This mutation, as is well known, interrupts the reading of the messenger in the polypeptide chain. But on the other hand, certain «suppressors», today well identified, are able to restore the reading of the triplets (UAG and UAA) corresponding to the nonsense mutations. The fact that a given mutation may be restored by one of the carefully catalogued suppressors provides proof that the phenotype of the corresponding mutant is due to the interruption of the synthesis of a protein. Using this principle, Bourgeois, Cohn, and Orgel[31] showed that certain constitutive mutants of the gene *i* are nonsense mutants and that, consequently, the active product of this gene is a protein.

This result, which illustrates the surprising analytical ability of modern biochemical genetics, is of utmost importance. It must be emphasized that, with respect to the suppression of a constitutive mutant (i^-), it shows that the recognition of the operator (as well as recognition of the inducer) is linked to the structure of the protein produced by the gene *i*.

The problem of the molecular mechanism that permits this protein to play the role of relay between the inducer and the operator still remains. Until now this problem has been inaccessible to direct experimentation, in that the repressor itself remains to be isolated and studied *in vitro*. However, in conclusion, I would like to explain why and how this inaccessibility was itself the source of new preoccupations that we hope will be fruitful.

First of all, is should be recalled that we had tried repeatedly, even before the existence of the repressor was demonstrated, to learn something of the mode of action of the inducer by following its tracks *in vivo* with radioactive markers. One after the other, Georges Cohen, François Gros, and Agnes Ullmann engaged in this approach, using different fractionation techniques. Some of these experiments led to some unexpected and important discoveries, such as that of galactoside permease and galactoside transacetylase. But concerning the way in which galactosides act as inducers, the results were completely negative. Nothing whatever indicated that the inductive interaction is accompanied by a chemical change, however transient, or by any kind of covalent reaction in the inducer itself. The kinetics of induction, elaborated on in the elegant work of Kepes[35, 36], also revealed that the inductive interaction is extremely rapid and completely reversible (Fig. 5).

This is quite a remarkable phenomenon, if one thinks of it, since this non-covalent, reversible stereospecific interaction – an interaction that in all probability involves only a few molecules and can involve only a very small amount of energy – triggers the complex transcription mechanism of the operon, the reading of the message, and the synthesis of three proteins, leading to the formation of several thousand peptide links. During this entire process, the inducer acts, it seems, exclusively as a chemical signal, recognized by the repressor, but without directly participating in any of the reactions which it initiates.

One would be inclined to consider such an interpretation of the inductive

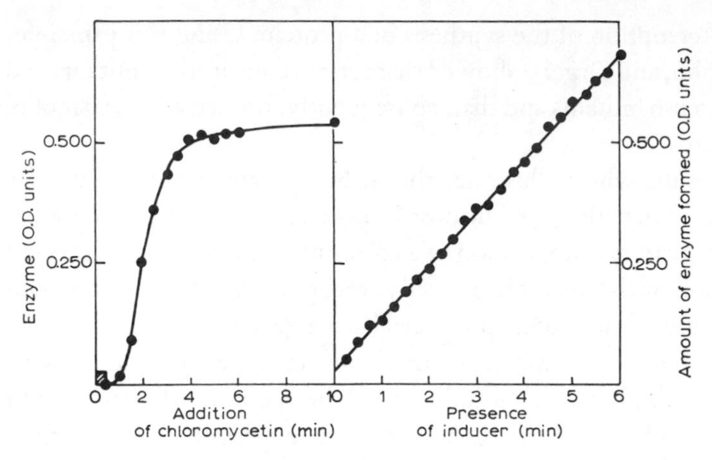

Fig. 5. Kinetics of the synthesis of galactosidase after a short period of induction. Left: Inducer added at time zero. Inducer eliminated after a time corresponding to the width of the cross-hatched rectangle. On the ordinates: accumulation of the enzyme. Right: Total amount of enzyme formed (asymptote of the curve at the left) as a function of the duration of the presence of the inducer. The linear relation obtained indicates that the inductive interaction is practically immediate and reversible[35].

interaction as highly unlikely if one did not know today of numerous examples in which similar mechanisms participate in the regulation of the activity as well as the synthesis of certain enzymes. It was as a possible model of inductive interactions that Jacob, Changeux, and I first became interested in regulatory enzymes[37]. The first example of such an enzyme was undoubtedly phosphorylase *b* from rabbit muscle; as Cori[38] and his group[39] showed, this enzyme is activated specifically by adenosine 5′-phosphate, although the nucleotide does not participate in the reaction in any way. We are indebted to Novick and Szilard[40], to Pardee[41], and to Umbarger[42] for their discovery of feedback inhibition, which regulates the metabolism of biosynthesis – their discovery led to a renewal of studies and demonstrated the extreme importance of these phenomena.

In a review that we devoted to these phenomena[43], a systematic comparison and analysis of the properties of some of the regulatory enzymes led us to conclude that, in most if not all cases, the observed effects were due to *indirect* interactions between distinct stereospecific receptors on the surface of the protein molecule, these interactions being in all likelihood transmitted by means of conformational modifications induced or stabilized at the time of the formation of a complex between the enzyme and the specific agent – hence the name «allosteric effects», by which we proposed to distinguish this partic-

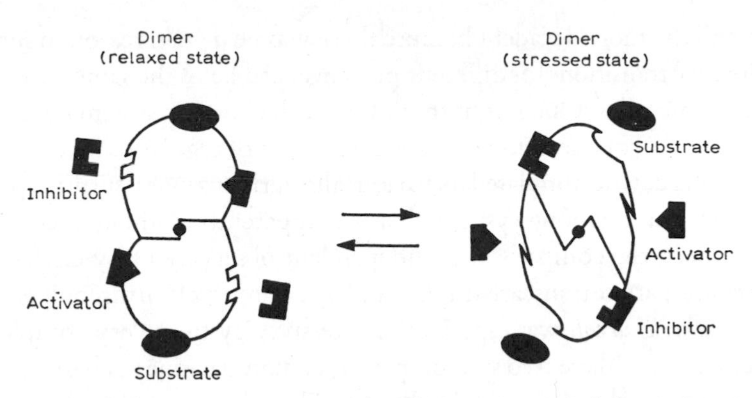

Fig. 6. Model of allosteric transition produced in a symmetrical dimer. In one of the two conformations, the protein can attach itself to the substrate as well as to the activating bond. In the other conformation, it can attach itself to the inhibiting bond.

ular class of interactions, and the term «allosteric transitions», used to designate the modification undergone by the protein (Fig. 6).

By virtue of being indirect, the allosteric interactions do not depend on the structure or the particular chemical reactivity of the ligands themselves, but entirely on the structure of the protein, which acts as a relay. This is what confers upon these effects their profound significance. The metabolism, growth, and division of a cell require, obviously, not only the operation of the principal metabolic pathways—those through which pass the necessary energy and chemical materials—but also that the activity of the various metabolic pathways be closely and precisely coordinated by a network of appropriate specific interactions. The creatin and development of such networks during the course of evolution obviously would have been impossible if only *direct* interactions at the surface of the protein had been used; such interactions would have been severely limited by chemical structure, the reactivity or lack of reactivity of metabolites among which the existence of an interaction could have been physiologically beneficial. The «invention» of indirect allosteric interactions, depending exclusively on the structure of the protein itself, that is on the genetic code, would have freed molecular evolution from this limitation[43].

The disadvantage of this concept is precisely that its ability to explain is so great that it excludes nothing, or nearly nothing; there is no physiological phenomenon so complex and mysterious that it cannot be disposed of, at least on paper, by means of a few allosteric transitions. I was very much in agreement with my friend Boris Magasanik, who remarked to me several years ago that this theory was the most decadent in biology.

It was all the more decadent because there was no *a priori* reason to suppose that allosteric transitions for different proteins need be of the same nature and obey the same rules. One might think that each allosteric system constituted a specific and unique solution to a given problem of regulation. However, as experimental data accumulated on various allosteric enzymes, surprising analogies were found among systems that had apparently nothing in common. In this respect, the comparison of independent observations by Gerhart and Pardee[44] on aspartate transcarbamylase and by Changeux[45] on threonine deaminase of *Escherichia coli* was especially impressive. By their very complexity, the interactions in these two systems presented unusual kinetic characteristics, almost paradoxical and yet quite analogous. Therefore it could not be doubted that the same basic solution to the problem of allosteric interactions had been found during evolution in both cases; it remained only for the researcher to try to discover it in his turn.

Among the properties common to these two systems, as well as to the great majority of known allosteric enzymes, the most significant seemed to us to be the fact that their saturation functions are not linear (as is the case for «classic» enzymes) but multimolecular. An example of such a pattern of saturation has been known for a long time: it is that of hemoglobin by oxygen (Fig. 7). Jeffries Wyman had noted several years earlier[46] that the symmetry of the saturation curves of hemoglobin by oxygen seemed to suggest the existence of a

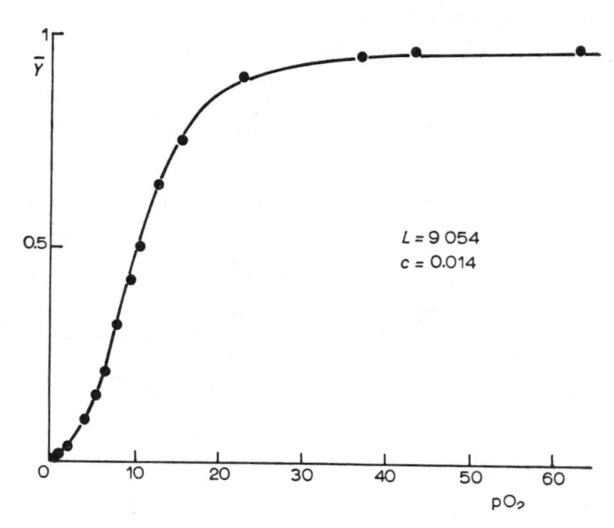

Fig. 7. Saturation of hemoglobin with oxygen. Abscissa: partial pressure of O_2. Ordinate: saturated fraction. The points correspond to experimental points[51]. The interpolation curve was calculated from a theoretical model essentially similar to that of Fig. 6.

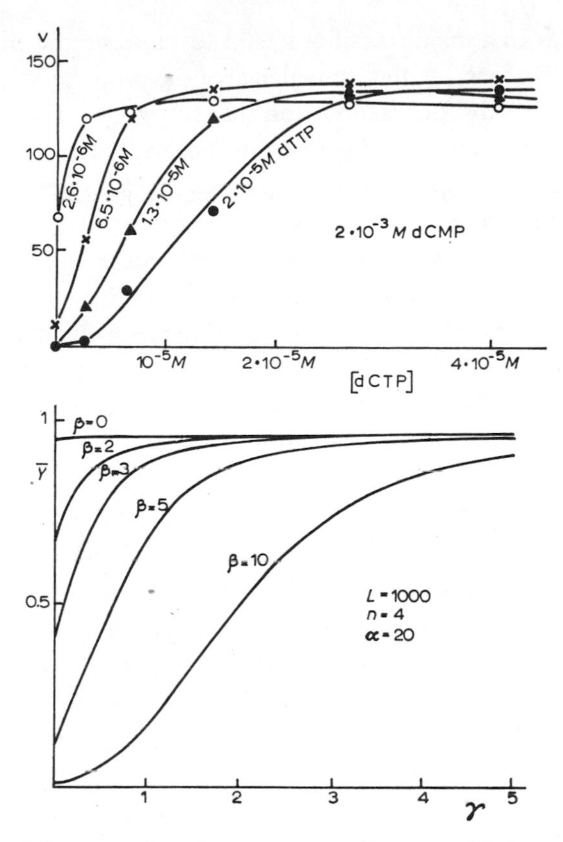

Fig. 8. Activity of deoxycytidine deaminase as a function of the concentration of the substrate (dCMP), of the activator (dCTP), and of the inhibitor (dTTP). (Top) Experimental results (from Scarano; see ref. 48). (Bottom) Theoretical curve calculated for a similar case according to the model of Monod, Wyman and Changeux[48].

structural symmetry within the protein molecule itself; this idea was brilliantly confirmed by the work of Perutz[47].

These indications encouraged us – Wyman, Changeux, and myself – to look for a physical interpretation of the allosteric interactions in terms of molecular structure. This exploration led us to study the properties of a model defined in the main by the following postulates:

(*1*) An allosteric protein is made up of several identical subunits (protomers).

(*2*) The protomers are arranged in such a way that none can be distinguished from the others; this implies that there are one or more axes of molecular symmetry.

(*3*) Two (or more) conformational states are accessible to this protein.

(4) These conformational transitions tend to preserve the molecular symmetry, or, more generally, the equivalence of the protomers[48].

We were pleasantly surprised to find that this very simple model made it possible to explain, classify, and predict most of the kinetic properties, sometimes very complex in appearance, of many allosteric systems (Figs. 7 and 8). Obviously, this model represents only a first approximation in the description of real systems. It is not likely, moreover, that it represents the only solution to the problem of regulative interactions found during evolution; certain systems seem to function according to quite different principles[49], which will also need to be clarified.

However, the ambition of molecular biology is to interpret the essential properties of organisms in terms of molecular structures. This objective has already been achieved for DNA, and it is in sight for RNA, but it still seems very remote for the proteins. The model that we have studied is interesting primarily because it proposes a functional correlation between certain elements of the molecular structure of proteins and certain of their physiologic properties, specifically those that are significant at the level of integration, of dynamic organization, of metabolism. If the proposed correlation is experimentally verified, I would see an additional reason for having confidence in the development of our discipline which, transcending its original domain, the chemistry of heredity, today is oriented toward the analysis of the more complex biological phenomena: the development of higher organisms and the operation of their networks of functional coordinations.

Acknowledgment

The research by my collaborators and myself since 1945 has been carried out entirely at the Pasteur Institute. This work has received decisive assistance from numerous institutions, in particular the Centre National de la Recherche Scientifique, the Rockefeller Foundation of New York, the National Science Foundation and the National Institutes of Health of the United States, the Jane Coffin Childs Memorial Fund, the Commissariat à l'Energie Atomique, and the Délégation Générale à la Recherche Scientifique et Technique. A donation by Mesdames Edouard de Rothschild and Bethsabée de Rothschild permitted, in large part, the establishment in 1954 of the Department of Cellular Biochemistry at the Pasteur Institute.

1. J. Monod, *Growth*, 11 (1947) 223.
2. B. Magasanik, *Mécanismes de Régulation des Activités Cellulaires chez les Microorganismes*, Centre National de la Recherche Scientifique, Paris, 1965, p.179.
3. J. Monod, *Ann. Inst. Pasteur*, 70 (1944) 381.
4. J. Monod and A. Andureau, *Ann. Inst. Pasteur*, 72 (1946) 868.
5. S. E. Luria and M. Delbrück, *Genetics*, 28 (1943) 491.
6. O. T. Avery, C. M. MacLeod and M. McCarty, *J. Exptl. Med.*, 79 (1944) 409.
7. J. Lederberg and E. L. Tatum, *Cold Spring Harbor Symp. Quant. Biol.*, 11 (1946) 113.
8. J. Monod, A. M. Torriani and J. Gribetz, *Compt. Rend.*, 227 (1948) 315; J. Monod, *Intern. Congr. Biochem., 1st, Cambridge, 1949*, Abs. Commun., p. 303; *Unités Biologiques douées de Continuité Génétique*, Centre National de la Recherche Scientifique, Paris, 1949, p.181.
9. J. Monod and M. Cohn, *Biochim. Biophys. Acta*, 7 (1951) 153.
10. M. Cohn and A. M. Torriani, *J. Immunol.*, 69 (1952) 471.
11. J. Monod, A. M. Pappenheimer and G. Cohen-Bazire, *Biochim. Biophys. Acta*, 9 (1952) 648.
12. D. S. Hogness, M. Cohn and J. Monod, *Biochim. Biophys. Acta*, 16 (1955) 99; J. Monod and M. Cohn, *Intern. Congr. Microbiol., 6th, Rome, 1953, Symp. Microbial Metabolism*, p. 42.
13. R. Schoenheimer, *The Dynamic State of Body Constituents*, Harvard Univ. Press, Cambridge, 1942.
14. J. Monod, G. Cohen-Bazire and M. Cohn, *Biochim. Biophys. Acta*, 7 (1951) 585.
15. J. Monod and M. Cohn, *Advan. Enzymol.*, 13 (1952) 67.
16. M. R. Pollock, *Brit. J. Exptl. Pathol.*, 31 (1950) 739.
17. J. Monod and G. Cohen-Bazire, *Compt. Rend.*, 236 (1953) 530; M. Cohn and J. Monod, *Adaptation in Microorganisms*, Cambridge Univ. Press, Cambridge, 1953, p.132.
18. H. J. Vogel and B. D. Davis, *Federation Proc.*, 11 (1952) 485.
19. G. N. Cohen, *Ann. Rev. Microbiol.*, 19 (1965) 105.
20. J. Monod, *Enzymes: Units of Biological Structure and Function*, Academic Press, New York, 1956, p. 7; G. N. Cohen and J. Monod, *Bacteriol. Rev.*, 21 (1957) 169.
21. H. V. Rickenberg, G. N. Cohen, G. Buttin and J. Monod, *Ann. Inst. Pasteur*, 91 (1956) 829.
22. C. F. Fox and E. P. Kennedy, *Proc. Natl. Acad. Sci. (U. S.)*, 54 (1965) 891.
23. I. Zabin, A. Kepes and J. Monod, *Biochem. Biophys. Res. Commun.*, 1 (1959) 289; *J. Biol. Chem.*, 237 (1962) 253.
24. L. Pauling, H. A. Itano, S. J. Singer and I. C. Wells, *Nature*, 166 (1950) 677.
25. J. D. Watson, The involvement of RNA in the synthesis of proteins; F. H. C. Crick, On the genetic code, in *Nobel Lectures, Physiology or Medicine, 1942–1962*, Elsevier, Amsterdam, 1964, pp. 785, 811.
26. G. Gamow, *Nature*, 173 (1954) 318.
27. F. Jacob, Genetics of the bacterial cell, in *Nobel Lectures, Physiology or Medicine, 1963–1970*, Elsevier, Amsterdam, 1972, p. 148.
28. A. B. Pardee, F. Jacob and J. Monod, *Compt. Rend.*, 246 (1958) 3125; A. B. Pardee,

F. Jacob and J. Monod, *J. Mol. Biol.*, 1 (1959) 165; F. Jacob and J. Monod, *Compt. Rend.*, 249 (1959) 1282.
29. D. Perrin, F. Jacob and J. Monod, *Compt. Rend.*, 250 (1960) 155.
30. C. Willson, D. Perrin, M. Cohn, F. Jacob and J. Monod, *J. Mol. Biol.*, 8 (1964) 582.
31. S. Bourgeois, M. Cohn and L. Orgel, *J. Mol. Biol.*, 14 (1965) 300.
32. S. Benzer and S. P. Charupe, *Proc. Natl. Acad. Sci. (U. S.)*, 48 (1962) 1114.
33. S. Brenner, A. O. W. Stretton and S. Kaplan, *Nature*, 206 (1965) 994.
34. M. G. Weigert and A. Garen, *Nature*, 206 (1965) 992.
35. A. Kepes, *Biochim. Biophys. Acta*, 40 (1960) 70.
36. A. Kepes, *Biochim. Biophys. Acta*, 76 (1963) 293; *Cold Spring Harbor Symp. Quant. Biol.*, 28 (1963) 325.
37. J. Monod and F. Jacob, *Cold Spring Harbor Symp. Quant. Biol.*, 26 (1961) 389; J. P. Changeux, *Cold Spring Harbor Symp. Quant. Biol.*, 26 (1961) 313.
38. C. F. Cori *et al.*, see references in C. F. Cori and G. T. Cori, Polysaccharide phosphorylase, in *Nobel Lectures, Physiology or Medicine, 1942–1962*, Elsevier, Amsterdam, 1964, p. 186.
39. E. Helmreich and C. F. Cori, *Proc. Natl. Acad. Sci. (U. S.)*, 51 (1964) 131.
40. A. Novick and L. Szilard, *Dynamics of Growth Process*, Princeton Univ. Press, Princeton, N. J., 1954, p. 21.
41. R. A. Yates and A. B. Pardee, *J. Biol. Chem.*, 221 (1956) 757.
42. H. E. Umbarger, *Science*, 123 (1956) 848.
43. J. Monod, J. P. Changeux and F. Jacob. *J. Mol. Biol.*, 6 (1963) 306.
44. J. C. Gerhart and A. B. Pardee, *Federation Proc.*, 20 (1961) 224; *J. Biol. Chem.*, 237 (1962) 891; *Cold Spring Harbor Symp. Quant. Biol.*, 28 (1963) 491; *Federation Proc.*, 23 (1964) 727.
45. J. P. Changeux, *Cold Spring Harbor Symp. Quant. Biol.*, 26 (1961) 313; *J. Mol. Biol.*, 4 (1962) 220; *Bull. Soc. Chim. Biol.*, 46 (1964) 927, 947, 1151; 47 (1965) 115, 267, 281.
46. D. W. Allen, K. F. Guthe and J. Wyman, *J. Biol. Chem.*, 187 (1950) 393.
47. M. F. Perutz, X-Ray analysis of haemoglobin, in *Nobel Lectures, Chemistry, 1942–1962*, Elsevier, Amsterdam, 1964, p. 653.
48. J. Monod, J. Wyman and J. P. Changeux, *J. Mol. Biol.*, 12 (1965) 88.
49. C. A. Woolfolk and E. R. Stadman, *Biochem. Biophys. Res. Commun.*, 17 (1964) 313.
50. J. Monod, *Recherches sur la Croissance des Cultures Bactériennes*, Hermann, Paris, 1941.
51. Lyster, unpublished results.

Biography

Jacques Lucien Monod was born in Paris on February 9th, 1910. In 1917 his parents settled in the South of France, where Monod spent his early years, and he therefore thinks of himself as a Southerner rather than as a Parisian. His father was a painter, something of an unusual vocation for a Huguenot family in which doctors, ministers of the Church, civil servants, and professors predominated. His mother was American, born in Milwaukee, with a father of Scottish descent – again somewhat out of the ordinary considering French bourgeois tradition at the end of the nineteenth century. His secondary education took place at the lycée de Cannes, and he owes a great deal to some of the masters under whom he was fortunate enough to study. Monod in particular recalls Monsieur Dor de la Souchère, well known as the founder and curator of the Antibes museum. Although Monod remembers nothing of the Greek grammar studied under him, the admiration which he soon developed for this highly cultured and worthy man was of the greatest spiritual benefit for him as a youngster. It is difficult to express just how much Monod owes to his father, who combined artistic sensitivity with prodigious erudition and a passionate concern for intellectual affairs. He had a positivist faith in the joint progress of science and society. It was through his father, who used to read Darwin, that Jacques Monod developed his interest in biology very early in life.

Monod came to Paris in 1928 to begin his higher education, and registered at the Faculty for a degree in Natural Sciences, not realising (as he later found out) that this course was then some twenty years or more behind contemporary biological science. It was from others, a few years senior to himself, rather than from the professional staff, that he gained his true initiation into biology. To George Teissier he owes a preference for quantitative descriptions; André Lwoff initiated him into the potentials of microbiology; to Boris Ephrussi he owes the discovery of physiological genetics, and to Louis Rapkine the concept that only chemical and molecular descriptions could provide a complete interpretation of the function of living organisms.

Monod obtained his Science Degree in 1931, and his doctorate in Natural

Sciences in 1941. After lecturing at the Faculty of Sciences in 1934, and spending some time at the California Institute of Technology on a Rockefeller grant in 1936, Monod joined the Institut Pasteur after the liberation as Laboratory Director in Lwoff's Department. He was made Director of the Cell Biochemistry Department in 1954, and in 1959 was appointed Professor of the Chemistry of Metabolism at the Sorbonne. In 1967 he became Professor at the Collège de France, and in 1971 he was appointed Director of the Institut Pasteur.

The following honours and distinctions were awarded to Professor Monod: Montyon Physiology Prize of the Académie des Sciences (Paris, 1955), Louis Rapkine Medal (London, 1958), Honorary Foreign Member of the American Academy of Arts and Sciences (1960), Chevalier de l'Ordre des Palmes Académiques (1961), Charles Léopold Mayer Prize of the Académie des Sciences (1962), Officier de la Légion d'Honneur (1963), Honorary Foreign Member of the Deutsche Akademie der Naturforscher «Leopoldina» (1965), D. Sc. *h.c.* University of Chicago (1965), Foreign Member of the Royal Society (1968), Foreign Member of the Academy National of Sciences (Washington, 1968), Foreign Member of the American Philosophical Society (1969), D. Sc. *h.c.* of the Rockefeller University (1970). His military distinctions include: Honorary Colonel of the Reserve, Chevalier de la Légion d'Honneur (military) (1945), Croix de Guerre (1945), and the Bronze Star Medal.

In 1938, Jacques Monod married Odette Bruhl, now the curator of the Guimet Museum. As an archeologist and orientalist with the most sensitive and impeccable taste, his wife brought to the marriage a culture complementary to his own. They have twin sons, Olivier and Philippe. Their father did nothing to influence them to become men of science like himself. On the contrary, he made very effort to persuade them that the realm of knowledge and ideas is not confined to the present-day connotation of the word «science». Both of them nevertheless became scientists: one a geologist, the other a physicist. These two sons gave the parents what they lacked before: two daughters, or rather daughters-in-law, and even a grand-daughter with the pretty name of Claire. The interests of Jacques Monod include almost all aspects of Arts and Sciences, his favourite recreations are music and sailing.

1968

Robert W. Holley
H. Gobind Khorana
Marshall Nirenberg

ROBERT W. HOLLEY

Alanine Transfer RNA

December 12, 1968

Work on the alanine transfer RNA actually began in 1956 in James Bonner's laboratories at the California Institute of Technology. I was on sabbatical leave from the Geneva Experiment Station of Cornell University and was studying protein synthesis. Toward the end of my leave I carried out experiments designed to detect the acceptor of activated amino acids.

At that time it was already known from the work of Hoagland, Keller and Zamecnik[1], DeMoss, Genuth and Novelli[2], and Berg and Newton[3] that amino acids are activated enzymatically to give enzyme-bound amino acyladenylates (Enz-AA-AMP, Fig. 1). It seemed likely that these amino acyladenylates would react with something, indicated as ‹X› in Fig. 1, and one product of the reaction would be AMP (adenosine 5'-monophosphate), as formulated in the second equation in Fig. 1. It seemed quite possible that such

$$\text{Enz} + \text{AA} + \text{ATP} \rightleftarrows \text{Enz-AA-AMP} + \text{pyrophosphate}$$
$$\text{Enz-AA-AMP} + \langle X \rangle \rightleftarrows \text{Enz} + \text{AA-}\langle X \rangle + \text{AMP}$$

Fig. 1. Schematic representation of amino acid activation.

a reaction would be reversible, and if so, it might be possible to detect the overall back reaction as an incorporation of radioactive AMP into ATP that required amino acids. Using this approach, an alanine-dependent incorporation of AMP into ATP was found in the «pH 5 enzyme» prepared from the low molecular weight, «soluble» fraction of rat liver homogenate. Of greatest interest was the finding that the AMP incorporation was inhibited by ribonuclease[4]. Subsequently, this alanine-dependent AMP incorporation system was reconstructed by combining a partially purified alanine-activating enzyme with low molecular weight RNA prepared from rat liver «pH 5 enzyme»[5].

In the meantime it was shown by Hoagland *et al.*[6], and by Ogata and Nohara[7] that radioactive amino acids became bound to a low molecular weight RNA in a rat-liver «pH 5 enzyme» preparation. The RNA was referred to as «soluble RNA» and is now known as «transfer RNA»[8]. Thus it became clear that the acceptor of activated amino acids was a low molecular weight RNA.

The work of Zachau, Acs and Lipmann[9], and Hecht, Stephenson and Zamecnik[10] showed that all of the activated amino acids became attached to a terminal adenosine residue in transfer RNA. Since different amino acids did not compete for the same attachment site[6,11], it seemed likely that different transfer RNA's were serving as acceptors for the different amino acids.

For a chemist, the existence of amino acid-specific, low molecular weight RNA's was very intriguing. It seemed possible that these RNA's might be small enough to permit detailed structural studies. This would be of great interest because it is the nucleotide sequences of nucleic acids that provide specificity and enable nucleic acids to carry out their many vital functions.

Isolation of yeast alanine transfer RNA

When transfer RNA's are extracted from cells, a mixture is obtained that contains at least one transfer RNA for each of the 20 different amino acids involved in protein synthesis. For detailed structural analysis, a highly purified transfer RNA was needed; therefore, in 1958, at the U. S. Plant, Soil and Nutrition Laboratory, a U. S. Department of Agriculture Laboratory at Cornell University, we set out to try to isolate an individual transfer RNA for chemical study.

Our first problem was to find a fractionation technique that was applicable to transfer RNA's. Various procedures were investigated, and the Craig countercurrent distribution technique[12] was found to be promising. In collaboration with J. Apgar, B. P. Doctor and S. H. Merrill[13,14], the countercurrent distribution procedure was developed, over a period of four years, into the first generally applicable method for the fractionation of transfer RNA's. Fig. 2 shows the results of a countercurrent distribution of bulk yeast transfer RNA. Yeast transfer RNA was used because it is readily obtained in large quantity[15]. By repeated countercurrent distribution of the most active fractions obtained in Fig. 2, three of the transfer RNA's, the alanine, tyrosine, and valine RNA's, were obtained in a relatively homogeneous form and essentially free of activity as acceptors of other amino acids[14]. The results with the alanine RNA are shown in Fig. 3. The excellent correlation between the experimental curves and the calculated theoretical distribution curve, shown in Fig. 3, encouraged us to believe that the RNA was pure enough for structural analysis. Nevertheless, to undertake structural work was a gamble, since there was the possibility that the preparation might not be pure, or that it

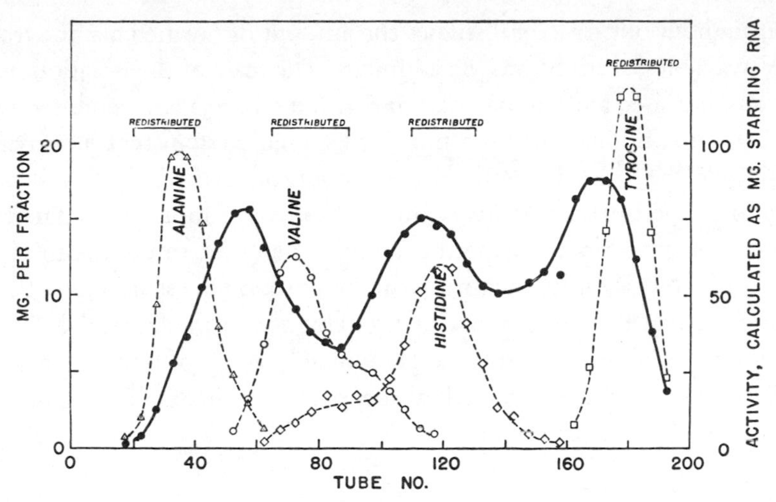

Fig. 2. 200-Transfer countercurrent distribution of 500 mg of bulk yeast transfer RNA (from ref. 14).

might be a mixture of different molecular species, all of which accepted alanine. Since attempts to fractionate the material further were unsuccessful, there seemed no alternative but to gamble a few years of work on the problem hoping that the material was sufficiently pure for structural analysis. If the starting material was impure, we could expect that attempts at structural analysis would lead to hopeless confusion. Fortunately this was not the outcome.

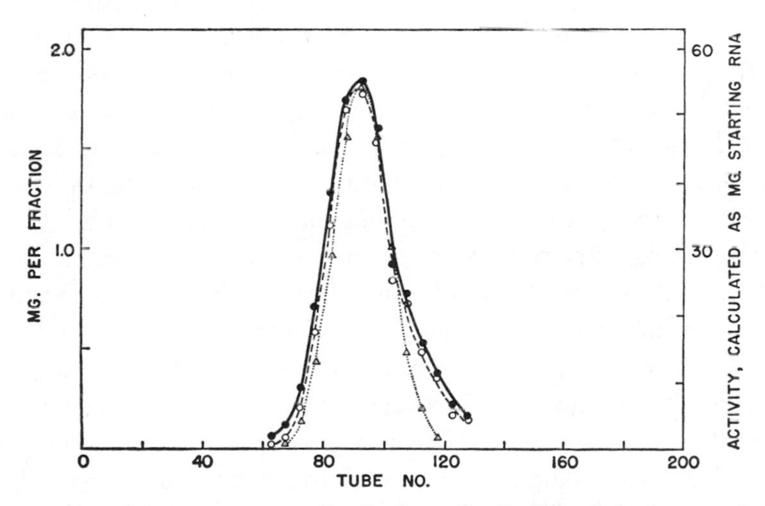

Fig. 3. 875-Transfer countercurrent distribution of redistributed alanine transfer RNA (from ref. 14).

Throughout our structural studies the amount of purified alanine transfer RNA available for study was very limited. The scale of the isolation procedure was increased tenfold over that in Fig. 3, by using a large countercurrent apparatus in combination sith a modified solvent system that increased the solubility of the RNA. Nevertheless, the supply of purified RNA available for individual experiments was always limited to tens of milligrams. Therefore, to the extent possible, the experiments were designed to use the minimum amount of RNA. During the three years of work on the structure of the alanine transfer RNA, we used a total of 1 g of highly purified material. This was isolated in our laboratories from approximately 200 g of bulk yeast transfer RNA, which in turn was obtained by phenol extraction of approximately 140 kg of commercial baker's yeast.

Cleavage of the RNA into small fragments

The evidence obtained in preliminary analyses indicated that the alanine transfer RNA molecule consisted of a single chain of approximately 80 nucleotide residues[16]. Therefore, in principle, structural analysis required the identification of the nucleotide residues and the determination of their sequences. Formally, the problem was analogous to determination of the sequence of approximately 80 letters in a sentence.

The experimental approach that was used involved cleavage of the polynucleotide chain into small fragments, identification of the small fragments, and then reconstruction of the original nucleotide sequence by determining the order in which the small fragments occurred in the RNA molecule. In terms of the analogy of a sentence, the approach was equivalent to breaking a sentence into words, identifying the words, and reconstructing the sequence of the letters in the sentence by determining the order of the words.

Briefly, the experiments were carried out as follows. Pancreatic ribonuclease was used to cleave the RNA chain next to pyrimidine nucleotides, to give one set of fragments in which each fragment ended in a pyrimidine nucleotide such as cytidylic acid (C–) or uridylic acid (U–). Then, takadiastase ribonuclease Tl, the enzyme discovered by Sato-Asano and Egami[17], was used, separately, to cleave the RNA chain specifically at guanylic acid (G–) residues. This gave a different set of small fragments. The individual small fragments were isolated by ion-exchange chromatography, followed by paper electrophoresis or rechromatography, where necessary. Fig. 4 shows

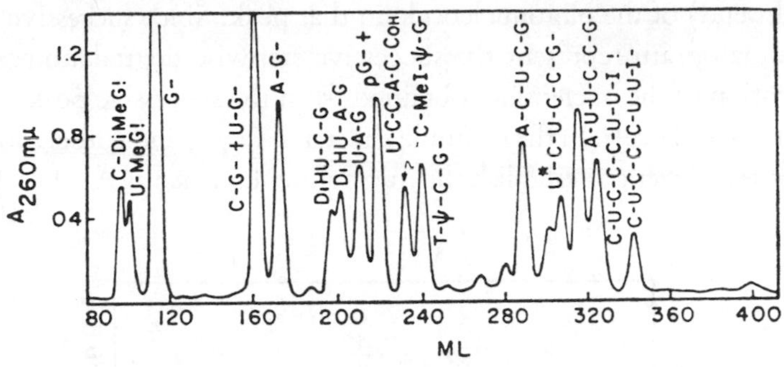

Fig. 4. Separation of ribonuclease Tl digest fragments of the alanine transfer RNA by chromatography on DEAE-cellulose. (For abbreviations see Table 1) (From refs. 23 and 31)

the pattern obtained after chromatography of the ribonuclease Tl digest on a diethylaminoethylcellulose column, using Tomlinson and Tener's procedures[18] with modifications by M. Marquisee and J. Apgar. Under these conditions, almost all of the different fragments obtained in the digest are separated. Each of the separated fragments was hydrolyzed with alkali, and the component mononucleotides were identified by chromatographic and electrophoretic properties and spectra. This was sufficient to determine the sequence of each of the dinucleotides, because the position of attack by each ribonuclease was known. Additional information was needed to establish the nucleotide sequences of the trinucleotides and larger oligonucleotides.

New methods of sequence determination were required in the identification of several of the larger oligonucleotides. One new method that was especially useful is outlined in Figs. 5 and 6. As indicated in Fig. 5, partial digestion of an oligonucleotide with snake venom phosphodiesterase gives a mixture of degradation products. A chromatographic pattern obtained from such a partial digest is shown in Fig. 6. Alkaline hydrolysis of the material recovered from each peak gives a nucleoside, which arises from the 3'-terminal residue

$$\text{A-U-U-C-C-G} \xrightarrow[\text{phosphodiesterase}]{\text{partial digestion with snake venom}} \begin{array}{l} \text{A-U-U-C-C} + \\ \text{A-U-U-C} + \\ \text{A-U-U} + \\ \text{A-U} + \\ \text{A} + \text{mononucleotides} \end{array}$$

Fig. 5. Partial digestion of an oligonucleotide with snake venom phosphodiesterase.

(the right end) of the oligonucleotide in that peak. Since successive peaks in the chromatogram represent the successive stepwise degradation products, identification of the nucleosides obtained from the successive peaks gives the nucleotide sequence[19]. In the example shown in Figs. 5 and 6, the information obtained is sufficient to establish the nucleotide sequence as A–U–U–C–C–G–.

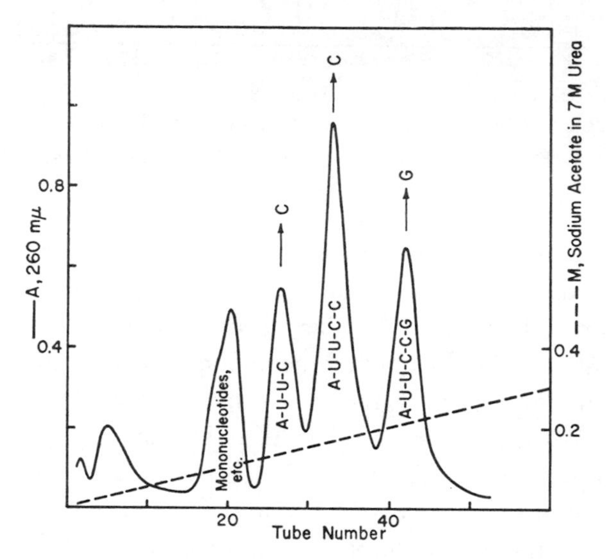

Fig. 6. Chromatographic pattern of partial digest of A–U–U–C–C–G. Recovery of the oligonucleotide from each peak followed by alkaline hydrolysis gives the nucleoside indicated (from ref. 32).

Tables 1 and 2 list the identities of the fragments obtained in the pancreatic ribonuclease and ribonuclease Tl digest, respectively[20]. Determination of the structures of all of the fragments required approximately 2.5 years. It was the work of J. T. Madison and A. Zamir, with assistance in the identification of the nucleotides from G. Everett. Especially time-consuming were the determination of the structures of the larger oligonucleotides and the identification of unusual nucleotides, particularly 1-methylinosinic acid (MeI–) and 5,6-dihydrouridylic acid (DiHU–)[21]. (The latter nucleotide had never been found in a natural nucleic acid. It does not absorb light at 260 mμ, and, as a consequence, it is invisible in the usual procedures for detecting nucleotides.)

The presence of distinctive end groups (a free 5′-phosphate group (p) at the left end of the RNA molecule, as the structure is conventionally written, and a free 3′-hydroxyl group ($_{OH}$) at the right end) established that the left end of

Table 1

Fragments obtained by complete digestion of alanine RNA with pancreatic ribonuclease[a]

C_{OH}[b]	MeG–G–C–
13 C–	A–G–C–
Ψ–	A–G–DiHU–
6 U–	G–A–U–
A–C–	I–G–C–
MeI–Ψ–	G–G–T–
DiMeG–C–	G–G–DiHU–
2 G–C–	G–G–A–C–
4 G–U–	pG–G–G–C–
G–G–G–A–G–A–G–U*–	

[a] Abbreviations: p and – are used interchangeably to represent a phosphate residue; A, adenosine 3′-phosphate; C, cytidine 3′-phosphate; C_{OH}, cytidine (with free 3′-hydroxyl group emphasized); DiHU, 5,6-dihydrouridine 3′-phosphate; DiMeG, N^2-dimethyl guanosine 3′-phosphate; I, inosine 3′-phosphate; MeG, 1-methylguanosine 3′-phosphate; MeI, 1-methylinosine 3′-phosphate; Ψ, pseudouridine 3′-phosphate; T, ribothymidine 3′-phosphate; U, uridine 3′-phosphate; U*, a mixture of U, and DiHU; p!, 2′,3′-cyclic phosphate, for example: Ip!, inosine 2′,3′-cyclic phosphate.

[b] The presence of a free 3′-hydroxyl group on this fragment indicates that cytidine occupies the terminal position in the purified alanine RNA. This establishes that the terminal adenylic acid residue is missing, as it is from most transfer RNA's isolated from commercial baker's yeast. A terminal adenylic acid residue is replaced under assay conditions before the amino acid is attached.

Table 2

Fragments obtained by complete digestion of alanine RNA with takadiastase ribonuclease Tl

9 G–	DiHU–C–G–
pG–	DiHU–A–G–
C–DiMeGp!	C–MeI–Ψ–G–
U–MeGp!	T–Ψ–C–G–
4 C–G–	A–C–U–C–G–
2 A–G–	U–C–C–A–C–C_{OH}[a]
U–G–	U*–C–U–C–C–G–
U–A–G–	A–U–U–C–C–G–
C–U–C–C–C–U–U–I–	

[a] See Table 1, second footnote.

the alanine transfer RNA molecule has the structure pG–G–G–C–, and the right end the structure U–C–C–A–C–C–A_{OH}.

The presence of the unusual nucleotides, and also of certain unique sequences, gave a number of overlaps between the two sets of sequences shown in Tables 1 and 2. For example, there is only one I– in the molecule, and this is found in the sequence C–U–C–C–C–U–U–I– in the ribonuclease Tl digest and in the sequence I–G–C– in the pancreatic ribonuclease digest. These two sequences must overlap, and the overall sequence must be C–U–C–C–C–U–U–I–G–C–.

All of the information in Tables 1 and 2 is summarized in Table 3, in which the sequences are listed in such a way that all the nucleotides in the alanine RNA are accounted for in 16 sequences that total 77 nucleotide residues.

Table 3

Sequences that account for the nucleotide residues in the alanine RNA[a]

pG–G–G–C–, G–U–G–, U–MeG–G–C–, G–C–,
G–U–A–G–, DiHU–C–G–, G–DiHU–A–G–,
C–G–, C–DiMeG–, C–U–C–C–C–U–U–I–G–C–,
. . . .
MeI–Ψ–, G–G–G–A–G–A–G–U*–C–U–C–C–G–,
G–T–Ψ–C–G–, A–U–U–C–C–G–, G–A–C–U–C–G–,
U–C–C–A–C–C–A_{OH}

[a] Dotted line separates sequences present in one half of the molecule from those present in the other half.

Cleavage of the RNA into large fragments

Once the 16 sequences shown in Table 3 were known, with the positions of the two end sequences established, the structural problem became one of determining the positions of the 14 intermediate sequences. This was done by isolating a number of large fragments from the RNA. In a crucial experiment it was found by J. R. Penswick that very brief treatment of the RNA with ribonuclease Tl at 0 °C in the presence of magnesium ion splits the molecule at one position[22]. The two halves of the molecule could be separated by chromatography (Fig. 7). Subsequent digestion of the separated half molecules with ribonuclease Tl, followed by chromatographic analysis (Fig. 8), established that the sequences listed above the dotted line in Table 3 were present in the left half of the molecule and the remaining sequences were present in the right half[22].

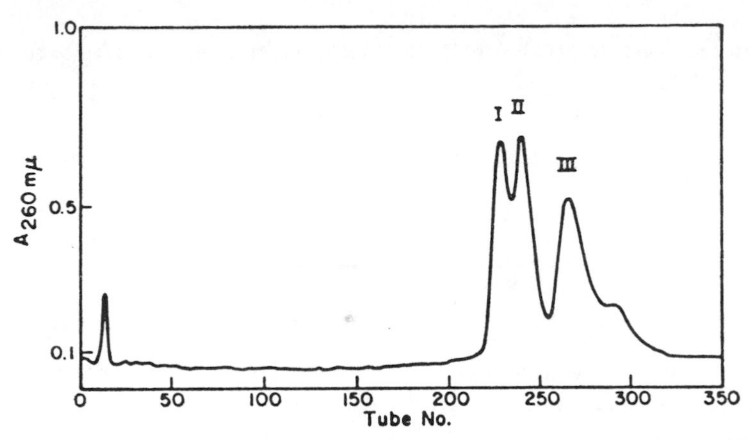

Fig. 7. Chromatographic separation of two large fragments (I and II) obtained by very limited digestion of the alanine transfer RNA with ribonuclease Tl at 0° in the presence of magnesium ion (from ref. 22).

Using somewhat more vigorous, but still limited treatment of the RNA with ribonuclease Tl, we then obtained, with J. Apgar, a number of additional large fragments. To determine the structures of the large fragments, each large fragment was degraded completely with ribonuclease Tl, the digest was chromatographed to give two or more of the ribonuclease Tl peaks already identified in Fig. 4, and these known sequences were put together, one after

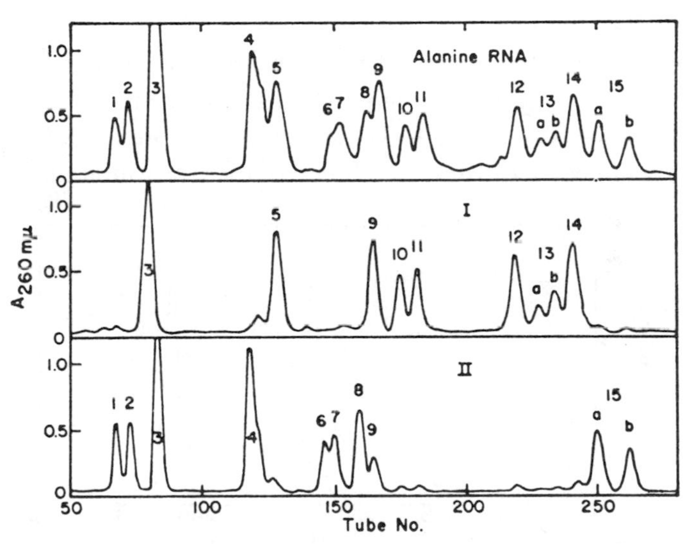

Fig. 8. Chromatography of ribonuclease Tl digests of the alanine transfer RNA and the large fragments I and II from Fig. 7 (from ref. 22).

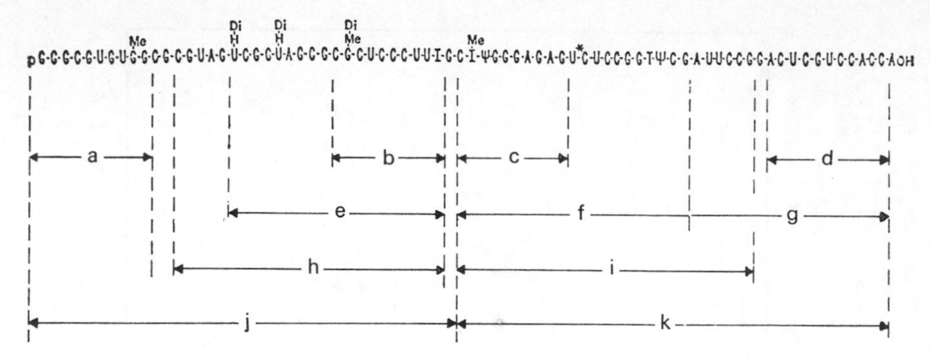

Fig. 9. The nucleotide sequence of the yeast alanine transfer RNA is shown at the top. Large oligonucleotide fragments isolated and used to reconstruct the complete sequence are indicated below (from ref. 25).

another, until the complete nucleotide sequence of the large fragment was known. The sequences that were determined are indicated in Fig. 9 by the letters *a* to *k*[23,24].

The approach used in reconstructing the long sequences can be illustrated by considering two fragments in detail.

The chromatographic analysis of a complete ribonuclease Tl digest of fragment *d* is shown in Fig. 10. The presence of U–C–C–A–C–C_{OH} indicates that fragment d is from the right end, the 3′-end, of the molecule. Therefore, the A–C–U–C–G– sequence must be to the left of this, and the sequence of d is known[23].

The chromatographic analysis of a ribonuclease Tl digest of fragment a is shown in Fig. 11. The analysis indicates that fragment a is composed of U–MeG–, 3 G–, C–G–, U–G–, and pG–. The presence of pG– establishes that fragment a is from the left end of the RNA molecule. Since it is

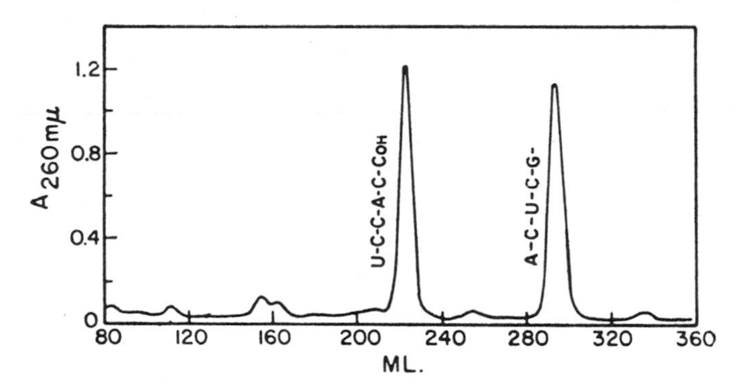

Fig. 10. Chromatography of complete ribonuclease Tl digest of fragment d (from ref. 23).

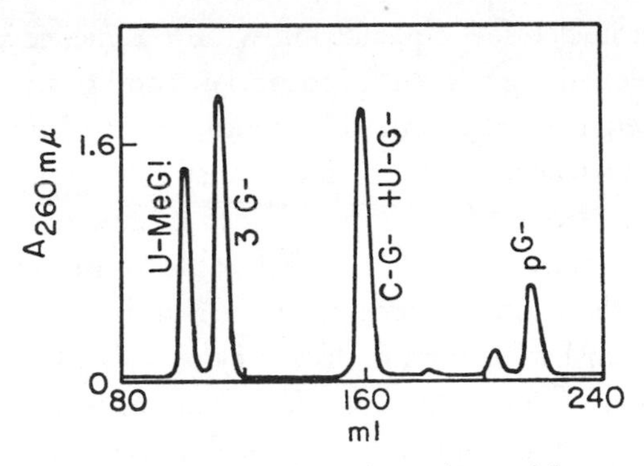

Fig.11. Chromatography of complete ribonuclease Tl digest of fragment a (from refs. 23 and 31).

already known (Tables 1 and 3) that the terminal sequence at the left end of the RNA is pG–G–G–C–, the positions of two of the three G–'s and the C–G– are known, and the terminal five nucleotides must be pG–G–G–C–G–. The positions of the U–G–, U–MeG–, and G– are established by the following information. It is known (Table 3) that the U–MeG– is present in the RNA in the sequence U–MeG–G–C. Since there is only one C in fragment a, and its position has already been established, fragment a must terminate before the C of the U–MeG–G–C– sequence. Therefore, the U–G– must be to the left of the U–MeG–, and the structure of fragment a can be represented as pG–G–G–C–G–··· U–G···U–MeG–, with one G– remaining to be placed. If the G– is placed to the left or the right of the U–G– in this structure, there will be G–G–U– sequence in the RNA. Since any G–G–U– sequence would appear in a pancreatic ribonuclease digest, and such a sequence is not found (Table 1), the remaining G– must be to the right of the MeG–, and the sequence of fragment a is pG–G–G–C–G–U–G–U–MeG–G [23].

The structural proofs for the other large fragments were carried out in a similar fashion[24]. Some of these proofs were straightforward; others were difficult. Eventually, the analyses of the large fragments furnished sufficient information to establish the sequences of the halves of the RNA molecule (fragments j and k in Fig. 9). Since the terminal sequences of the RNA were already known, the halves could be joined in only one way, to give the I–G–C– sequence which was known to be present in the RNA (Table 1), and the complete nucleotide sequence of the yeast alanine transfer RNA is that shown in Fig. 9[25].

This is the first nucleotide sequence known for a nucleic acid. Also, it can be said that the sequence gives, with appropriate modifications for DNA, the first nucleotide sequence of a gene. This would be the sequence of the gene that determines the structure of the alanine transfer RNA in yeast cells.

It was, of course, tremendously satisfying to be able to solve each experimental problem as it arose, and eventually be able to complete the nucleotide sequence. The satisfaction was increased by the fact that we were able to work with the alanine transfer RNA from discovery to isolation to structural analysis. In these times of highly competitive research, few scientists have the satisfaction of carrying through a research problem that takes 9 years. Without minimizing the pleasure of receiving awards and prizes, I think it is true that the greatest satisfaction for a scientist comes from carrying a major piece of research to a successful conclusion.

Three-dimensional structure

When a problem is solved, one's attention turns to other problems. With the complete nucleotide sequence of the alanine transfer RNA established, we became concerned with other questions about the alanine RNA structure. One question of particular interest has to do with the interaction of the transfer RNA with a messenger RNA. Speculation suggests that the three-dimensional structure of a transfer RNA, in the presence of the magnesium ion under conditions suitable for protein synthesis, should have the coding triplet of nucleotides, the anticodon, exposed in a way that will permit it to interact with a triplet of nucleotides, the codon, in the messenger RNA[26]. The sequence that constitutes the anticodon in the alanine transfer RNA is the sequence I–G–C, present in the middle of the molecule and including the linkage that is so sensitive to attack by ribonuclease Tl. One arrangement of the RNA chain, suggested by E. B. Keller and by Penswick, has the I–G–C–sequence in an exposed position and also has very interesting symmetry. This «cloverleaf» arrangement is shown in Fig. 12[25]. In drawing this arrangement, it was assumed that there would be Watson–Crick-type pairing of A to U and G to C in the double-stranded regions and the unpaired regions would form loops as suggested by Fresco, Alberts and Doty[27]. The strongest evidence for the «cloverleaf» arrangement of the secondary structure of transfer RNA's comes from the finding that all of the transfer RNA sequences that have been determined since 1965 fit the same type of base-pairing arrange-

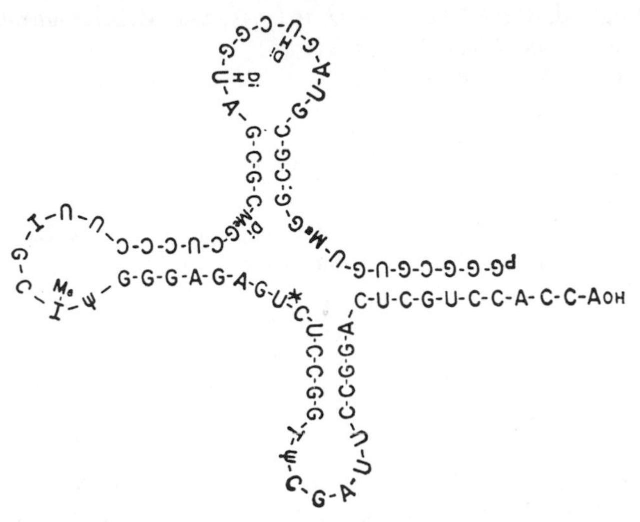

Fig.12. Suggested secondary structure of the alanine transfer RNA (from ref. 25).

ment. There are now 12 of these sequences and they have come from structural studies in many different laboratories[28]. In all instances the anticodon sequence is found at the same position in the middle loop. The «cloverleaf» arrangement can be only a partial description of the three-dimensional structural. This is clear from chemical and enzymatic studies, which indicate that the molecule is folded in some way[29]. However, details of the folding are not clear. Some further information can no doubt be obtained by chemical and enzymatic probing, but it seems likely that proof of the three-dimensional structure of a transfer RNA will wait for X-ray analysis[30].

That then is our story of the alanine transfer RNA. It all followed quite naturally from taking a sabattical leave. I strongly recommend sabbatical leaves.

Acknowledgment

Our work was made possible by financial assistance from the National Science Foundation. Fellowships and grant support from the National Institutes of Health are also gratefully acknowledged. Special thanks are due Dr. W. H. Allaway, Director of the U. S. Plant, Soil and Nutrition Laboratory, for his confidence and encouragement throughout the sequence determination.

1. M. B. Hoagland, *Biochim. Biophys. Acta*, 16 (1955) 288; M. B. Hoagland, E. B. Keller and P. C. Zamecnik, *J. Biol. Chem.*, 218 (1956) 345.
2. J. A. DeMoss, S. M. Genuth and G. D. Novelli, *Proc. Natl. Acad. Sci (U. S.)*, 42 (1956) 325.
3. P. Berg and G. Newton, *Federation Proc.*, 15 (1956) 219.
4. R. W. Holley, *J. Am. Chem. Soc.*, 79 (1957) 658.
5. R. W. Holley and J. Goldstein, *J. Biol. Chem.*, 234 (1959) 1765.
6. M. B. Hoagland, P. C. Zamecnik and M. L. Stephenson, *Biochim. Biophys. Acta*, 24 (1957) 215; M. B. Hoagland, M. L. Stephenson, J. F. Scott, L. I. Hecht and P. C. Zamecnik, *J. Biol. Chem.*, 231 (1958) 241.
7. K. Ogata and H. Nohara, *Biochim. Biophys. Acta*, 25 (1957) 659.
8. E. A. Allen, E. Glassman and R. Schweet, *J. Biol. Chem.*, 235 (1960) 1068.
9. H. G. Zachau, G. Acs and F. Lipmann, *Proc. Natl. Acad. Sci. (U. S.)*, 44 (1958) 885.
10. L. I. Hecht, M. L. Stephenson and P. C. Zamecnik, *Proc. Natl. Acad. Sci. (U. S.)*, 45 (1959) 505.
11. P. Berg and E. J. Ofengand, *Proc. Natl. Acad. Sci. (U. S.)*, 44 (1958) 78; R. S. Schweet, F. P. Bovard, E. Allen and E. Glassman, *Proc. Natl. Acad. Sci. (U. S.)*, 44 (1958) 173.
12. L. C. Craig and D. Craig, in A. Weissberger (Ed.), *Technique of Organic Chemistry*, Vol. 3, Part 1, 2nd Edn., Interscience, New York, 1956, p. 149.
13. B. P. Doctor, J. Apgar and R. W. Holley, *J. Biol. Chem.*, 236 (1961) 1117.
14. J. Apgar, R. W. Holley and S. H. Merrill, *J. Biol. Chem.*, 237 (1962) 796.
15. R. Monier, M. L. Stephenson and P. C. Zamecnik, *Biochim. Biophys. Acta*, 43 (1960) 1; R. W. Holley, *Biochem. Biophys. Res. Commun.*, 10 (1963) 186.
16. R. W. Holley, J. Apgar, S. H. Merrill and P. L. Zubkoff, *J. Am. Chem. Soc.*, 83 (1961) 4861.
17. K. Sato-Asano and F. Egami, *Nature*, 185 (1960) 462.
18. R. V. Tomlinson and G. M. Tener, *J. Am. Chem. Soc.*, 84 (1962) 2644; *Biochemistry*, 2 (1963) 697.
19. R. W. Holley, J. T. Madison and A. Zamir, *Biochem. Biophys. Res. Commun.*, 17 (1964) 389.
20. R. W. Holley, G. A. Everett, J. T. Madison and A. Zamir, *J. Biol. Chem.*, 240 (1965) 2122.
21. J. T. Madison and R. W. Holley, *Biochem. Biophys. Res. Commun.*, 18 (1965) 153.
22. J. R. Penswick and R. W. Holley, *Proc. Natl. Acad. Sci. (U. S.)*, 53 (1965) 543.
23. J. Apgar, G. A. Everett and R. W. Holley, *Proc. Natl. Acad. Sci. (U. S.)*, 53 (1965) 546.
24. J. Apgar, G. A. Everett and R. W. Holley, *J. Biol. Chem.*, 241 (1966) 1206.
25. R. W. Holley, J. Apgar, G. A. Everett, J. T. Madison, M. Marquisee, S. H. Merrill, J. R. Penswick and A. Zamir, *Science*, 147 (1965) 1462.
26. M. R. Bernfield and M. W. Nirenberg, *Science*, 147 (1965) 479.
27. J. R. Fresco, B. M. Alberts and P. Doty, *Nature*, 188 (1960) 98.
28. H. G. Zachau, D. Dütting and H. Feldman, *Z. Physiol. Chem.*, 347 (1966) 212; J. T. Madison, G. A. Everett and H. Kung, *Science*, 153 (1966) 531; U. L. RajBhandary, S. H. Chang, A. Stuart, R. D. Faulkner, R. M. Hoskinson and H. G. Khorana, *Proc. Natl. Acad. Sci. (U. S.)*, 57 (1967) 751; A. A. Baev, T. V. Vekstern, A. D. Mirzabekov, A. I. Krutilina, L. Li and V. D. Axelrod, *Mol. Biol.*, 1 (1967) 754; H. M. Goodman,

J. Abelson, A. Landy, S. Brenner and J. D. Smith, *Nature*, 217 (1968) 1019; S. K. Dube, K. A. Marcker, B. F. C. Clark and S. Cory, *Nature*, 218 (1968) 232; S. Takemura, T. Miqutani and M. Miyazaki, *Biochem. J.*, 63 (1968) 277; M. Staehelin, H. Rogg, B. C. Baguley, T. Ginsberg and W. Wehrli, *Nature*, 219 (1968) 1363.

29. J. A. Nelson, S. C. Ristow and R. W. Holley, *Biochim. Biophys. Acta*, 149 (1967) 590.
30. B. F. C. Clark, B. P. Doctor, K. C. Holmes, A. Kug, K. A. Marcker, S. J. Morris and H. H. Paradies, *Nature*, 219 (1968) 1222.
31. R. W. Holley, *J. Am. Med. Ass.*, 194 (1965) 868.
32. R. W. Holley, *Progr. Nucl. Acid. Res. Mol. Biol.*, 8 (1968) 37.

Biography

Robert W. Holley was born in Urbana, Illinois, on January 28th, 1922, one of four sons of Charles and Viola Holley. His parents were both educators. He attended public schools in Illinois, California and Idaho, and graduated from Urbana High School in 1938. He studied chemistry at the University of Illinois and received his B. A. degree in 1942. Graduate work was at Cornell University, where the Ph. D. degree in organic chemistry, with Professor Alfred T. Blomquist, was awarded in 1947. Graduate work was interrupted during the war. He spent two years, 1944–1946, with Professor Vincent du Vigneaud at Cornell University Medical College, where he participated in the first chemical synthesis of penicillin.

After completing the Ph. D. degree, Holley spent 1947–1948 as an American Chemical Society Postdoctoral Fellow with Professor Carl M. Stevens at Washington State University. He then returned to Cornell University as Assistant Professor of Organic Chemistry at the Geneva Experiment Station in 1948. He was Associate Professor there from 1950–1957. During a sabbatical year, 1955–1956, he was a Guggenheim Memorial Fellow in the Division of Biology at the California Institute of Technology. In 1958, he returned to Ithaca, New York, as a Research Chemist at the U. S. Plant, Soil and Nutrition Laboratory, a U. S. Department of Agriculture Laboratory on the Cornell University campus. He had an appointment in the University throughout this period and became Professor of Biochemistry in 1962. He rejoined the faculty of Cornell University full time in 1964 as Professor of Biochemistry and Molecular Biology, and was Chairman of the Department from 1965 to 1966. The following year, 1966–1967, was spent at the Salk Institute for Biological Studies and the Scripps Clinic and Research Foundation in La Jolla, California, as a National Science Foundation Postdoctoral Fellow. In 1968, though maintaining an affiliation with Cornell University, he joined the permanent staff of the Salk Institute, where he is a Resident Fellow and an American Cancer Society Professor of Melocular Biology. He is also an Adjunct Professor at the University of California at San Diego.

Holley's training as a chemist did not alter his basic interest in living things.

This interest has influenced his choice of research, which began with the organic chemistry of natural products. There followed a gradual drift toward more biological subjects, with work on amino acids and peptides, and eventually work on the biosynthesis of proteins. During the latter, the alanine transfer RNA was discovered. The following 10 years were spent working with this RNA, first concentrating on the isolation of the RNA, and then working on the determination of the structure of the RNA. The nucleotide sequence was completed at the end of 1964. It was for this work that the Nobel prize was awarded. More recently, his work has been concerned with factors that control cell division in mammalian cells.

Holley is a member of the National Academy of Sciences, the American Academy of Arts and Sciences, the American Association for the Advancement of Science, The American Society of Biological Chemists and the American Chemical Society. He received the Albert Lasker Award in Basic Medical Research in 1965, the Distinguished Service Award of the U.S. Department of Agriculture in 1965, and the U.S. Steel Foundation Award in Molecular Biology of the National Academy of Sciences in 1967.

Holley was married to Ann Dworkin in 1945. They have one son, Frederick. Mrs. Holley's professional interests are concerned with the teaching of mathematics. The three of them especially enjoy the ocean and the mountains.

H. GOBIND KHORANA

Nucleic Acid Synthesis in the Study of the Genetic Code

December 12, 1968

1. Introduction

Recent progress in the understanding of the genetic code is the result of the efforts of a large number of workers professing a variety of scientific disciplines. Therefore, I feel it to be appropriate that I attempt a brief review of the main steps in the development of the subject before discussing our own contribution which throughout has been very much a group effort. I should also like to recall that a review of the status of the problem of the genetic code up to 1962 was presented by Crick in his Nobel lecture[1].

While it is always difficult, perhaps impossible, to determine or clearly define the starting point in any area of science, the idea that genes make proteins was an important step and this concept was brought into sharp focus by the specific one gene-one enzyme hypothesis of Beadle and Tatum[2]. The field of biochemical genetics was thus born. The next step was taken when it was established that genes are nucleic acids. The transformation experiments of Avery and coworkers[3] followed by the bacteriophage experiments of Hershey and Chase[4] established this for DNA and the work with TMV-RNA a few years later established the same for RNA[5,6]. By the early 1950's it was, therefore, clear that genes are nucleic acids and that nucleic acids direct protein synthesis, the direct involvement of RNA in this process being suggested by the early work of Caspersson[7] and of Brachet[8]. It was important at this stage to know more about the chemistry of the nucleic acids and, indeed, the accelerated pace of discovery that soon followed, was largely because of work at the chemical and biochemical level in the field of nucleic acids.

The structural chemistry of the nucleic acids, which developed over a period of some seventy years in many countries, progressed step-by-step from the chemistry of the constituent purines, pyrimidines and the sugar moieties, to work on the nucleosides and then onto the nucleotides. A distinct climax was reached in 1952 with the elucidation of the internucleotidic linkage in nucleic acids by Brown and Todd and their coworkers[9]. (It was my good fortune to be associated with Professor, now Lord, Todd's laboratory before the start of our own work in the nucleotide field). Shortly thereafter, the Watson–Crick structure[10] for DNA was proposed, which focused attention, in particular, on the biological meaning of its physical structure. It is also about this time that the hypothesis that a linear sequence of nucleotides in DNA specifies the linear sequence of amino acids in proteins was born. A few years later, the enzymology of DNA got into its stride with the work of Kornberg and his coworkers[11]: their discovery and characterization of the enzyme DNA polymerase was a major triumph of modern enzymology and the methods developed distinctly aided the characterization, a few years later, of DNA-dependent RNA polymerase[12-16]. The discovery of this enzyme clarified the manner by which information in DNA is transcribed into an RNA, which we now equate with messenger RNA[17-21]. The last biochemical landmark to be introduced in the development of a cell-free amino acid incorporating system. Work on this really began with efforts to understand the biosynthesis of the peptide bond. The subject has a long history but critical progress began to be made in the early fifties. One thinks, in particular, of the pioneering work of Zamecnik and Hoagland[22], of Lipmann[23], of Berg[24] and in regard to the bacterial system that of Watson's laboratory[25], of Berg[26], and of the important refinement made in 1961 by Matthaei and Nirenberg[27].

With the knowledge of the chemical structures of the nucleic acids, the two major tasks which faced the chemists were those of synthesis and sequential analysis. Chemical synthesis of short-chain oligonucleotides began to be a preoccupation in my laboratory. The types of problems that one faced were: (*1*) activation of the phosphomonoester group of a mononucleotide so as to phosphorylate the hydroxyl group of another nucleoside or nucleotide; (*2*) design of suitable protecting groups for the various functional groups (primary and secondary hydroxyl groups in the sugar rings, amino groups in the purine and pyrimidine rings, phosphoryl dissociation in the phosphomonoester group); (*3*) development of methods for the polymerization of mononucleotides and for the separation and characterization of the resulting polynucleotides, and (*4*) evaluation of approaches to the stepwise synthesis of polynucleotides of specific sequences.

While even at present, organo-chemical methods demand further investigation and refinement, nevertheless, synthesis of short chains of deoxyribopolynucleotides with predetermined and fully controlled sequences became possible in the early sixties. In addition, unambiguous synthesis of short ribo-oligonucleotides containing strictly the $3' \rightarrow 5'$-internucleotidic linkages also became feasible. A discussion of the chemical aspects of these problems is outside the scope of the present lecture and reviews given elsewhere[28-33] should be consulted. The following review will be restricted to that part of the synthetic work which bears on the problem of the genetic code and attention will be focused in the main on the biochemical experiments made possible by the synthetic polynucleotides.

2. *Polynucleotide Synthesis in Relation to the Genetic Code*

A few words about the experimental development of the coding problem are now appropriate. In the fifties, possible rules governing the genetic code engaged the attention of many theoreticians, Gamow[34] being the first to speculate on the possible relation between DNA and protein structure. However, until 1961 the only experimental approach was that of direct correlation of the sequence of a nucleic acid with that of a protein specified by it. It was hoped to do this either chemically, for example, by working with the coat protein of a virus and its RNA, or by mutagenic techniques. An ingenious application of the ‹frameshift mutation› idea was, indeed, that of Crick and coworkers, who correctly deduced several of the fundamental properties of the genetic code[1]. These approaches, however, offered little immediate hope of getting directly at the coding problem.

The discovery which introduced a direct experimental attack on the genetic code was that of Matthaei and Nirenberg[27] who observed that polyuridylate directs the synthesis of polyphenylalanine in the bacterial cell-free amino acid incorporating system. The aim now was to use synthetic polynucleotides of defined composition as messengers in the *in vitro* system. A great deal, in fact, was learned during the years 1961–1963, both in the laboratories of Ochoa[35]

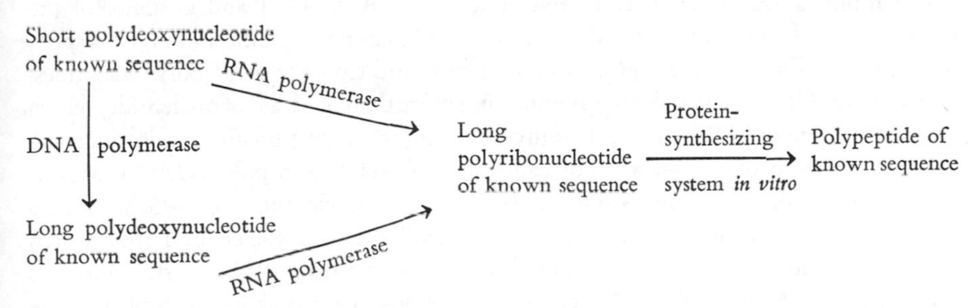

Scheme 1. Proposed reaction sequence for the preparation of high-molecular-weight RNA messengers and the subsequent *in vitro* synthesis of polypeptides of known amino acid sequences.

and of Nirenberg and their coworkers[36], about the overall nucleotide composition of the coding units by using polynucleotides made by the agency of the enzyme, polynucleotide phosphorylase.

(*a*) *Short chains of deoxyribopolynucleotides containing repeating nucleotide sequences**

The hope in my laboratory was to prepare ribopolynucleotide messengers of completely defined nucleotide sequences. However, chemical methodology permitted at this time the synthesis of oligonucleotides containing but a few ribonucleotide units (see below for trinucleotide syntheses). In the deoxyribonucleotide series, chemical synthesis was a little more advanced and the synthesis of longer, but still short, chains containing ten to fifteen nucleotide units was feasible. Therefore, it was decided to study the RNA-transcribing enzyme with the hope that this enzyme might use short chemically synthesized deoxyribopolynucleotides as templates in the manner that it uses biologically functional DNA.

The initial experiments with the RNA polymerase[37] were, in fact, a follow-up of the observation first made by Hurwitz and coworkers[38] that a mixture of chemically synthesized thymidine oligonucleotides served as a template for the synthesis of ribo-polyadenylate. Our aim was to obtain a ribopolynucleotidic product matching in chain length the deoxypolynucleotide used as the template. However, analysis showed that irrespective of the size of the short deoxypolynucleotide template, the RNA product was always much longer; it contained invariably more than 100 nucleotide units. At first sight, the results appeared to be discouraging in that we were losing control on the exact

* The abbreviations used are as follows. The letters, A, C, G, T and U stand for the nucleosides *or* the nucleotides of adenine, cytosine, guanine, thymine and uracil respectively; the prefixes d and r represent deoxyribose and ribose series of polynucleotides, respectively. All the polynucleotides containing more than one kind of nucleotide, which are used in the text, in Tables and in Figures, have strictly repeating nucleotide sequences and the repeating unit is shown. For example, poly-rUG and poly-rUAG represent polymers in which the dinucleotide, U–G, and the trinucleotide, U–A–G, sequences repeat. Polythymidylate containing eleven nucleotide units in the chain if abbreviated to dT_{11} and the hepta-deoxyadenylate if abbreviated to dA_7. tRNA or sRNA stands for transfer RNA; met-$tRNA^{met}$ stands for the non-formylatable species of methionine-specific tRNA which has been charged with the amino acid and fmet-$tRNA^{met}$ stands for the formylatable species of methionine-specific tRNA which has been charged with methionine and the latter residue subsequently formylated.

size of the product, despite our having carefully defined the size of the oligothymidylate template. However, it soon became apparent that this ‹slipping› or reiterative copying on the part of the enzyme could be a highly useful device to amplify the messages contained in the short chemically-synthesized polynucleotides. In a further study, attention was paid to understand a little better the conditions for the ‹amplification› to occur[39].

Some months later I visited Kornberg's laboratory (this was one of the many pilgrimages that I have made to this great laboratory) and started a few experiments with the DNA polymerase and here, again, very short synthetic oligonucleotides containing alternating A and T units induced the enzyme to bring about extensive synthesis of the previously characterized high molecular weight dAT polymer[40]. These encouraging results led to a generalized scheme (Scheme 1) for the *in vitro* studies of the coding problem. The amplification to produce DNA or RNA products was conceived at this time to be a general behaviour of the polymerases so long as there was a repeating pattern of nucleotide sequences (homopolynucleotides, repeating di- or tri-nucleotides) in the chemically-synthesized deoxypolynucleotide templates. Everything from this point on went remarkably well and the period starting with the spring of 1963 and ending with 1967 was a period, essentially, of uninterrupted success in work devoted to the genetic code.

The decision to synthesize deoxyribopolynucleotides with repeating nucleotide sequences was fortunate for another reason. The cell-free protein-synthesizing system, being a crude bacterial extract, undoubtedly contained powerful nucleases and peptidases. The use of messengers with completely defined but strictly repeating nucleotide sequences could be expected to give unequivocal answers despite (*1*) the exo- and/or endo-nucleolytic damages to the synthetic messengers and (*2*) the corresponding activities of the proteolytic enzymes on the polypeptide products.

The actual choice of nucleotide combinations in the deoxyribopolynucleotides to be synthesized was influenced by the then available knowledge that, at least in the cell-free protein-synthesizing system, the messenger RNA appears to be used in the single-stranded configuration. In fact, the only DNA containing more than one type of nucleotide, whose sequence was completely known, was the above-mentioned poly-dAT. Although RNA polymerase nicely produces from it the expected poly-rAU containing the two bases in strictly alternating sequence, this product, because of self-complementarity, has a tight double-stranded structure and elicits no response from the ribosomes in the cell-free system. It was, therefore, clear that those combinations of nucleotides which would lead to overwhelming base-pairing in the polynucleotides should be avoided[30-32].

All of the chemical syntheses relevant to the genetic code which were carried out are shown in Table I. First, we made the two sets of polynucleotides shown on the left, which contained repeating dinucleotide sequences: one set contains the hexamer of the dinucleotide with alternating thymidylate and guanylate residues and the hexamer of the dinucleotide with alternating adenylate and cytidylate residues; the second set consists of the hexamer of alternating thymidylate and cytidylate residues and the hexamer of alternating adenylate and guanylate residues[41]. This work was then extended to

Table I

Synthetic deoxyribopolynucleotides with repeating nucleotide sequences

Repeating dinucleotide sequences		*Repeating trinucleotide sequences*				*Repeating tetranucleotide sequences*	
d[TC]$_6$ d[AG]$_6$	d[TG]$_6$ d[AC]$_6$	d[TTC]$_4$ d[AAG]$_4$	d[CCT]$_{3-5}$ d[GGA]$_{3-5}$	d[TAC]$_{4-6}$ d[TAG]$_{4-6}$	d[CCA]$_{3-5}$ d[GGT]$_{3-5}$	d[TTAC]$_4$ d[GTAA]$_2$	d[TCTA]$_3$ d[TAGA]$_2$
		d[TTG]$_{4-6}$ d[CAA]$_{4-6}$	d[CGA]$_{3-5}$ d[CGT]$_{35-}$	d[ATC]$_{3-5}$ d[ATG]$_{3-5}$			

polynucleotides with repeating trinucleotide sequences. There is a theoretical maximum of ten such sets that can contain more than one nucleotide base and we prepared seven such sets[42-46]. Shown also in Table I are two sets of polymers with repeating tetranucleotide sequences[47,48]. Two additional considerations for the selection of the nucleotide sequences in them are: (*1*) they contain in every fourth place the chain-terminating codons and (*2*) this class of polymers can be used to prove the direction of reading of the messenger RNA[49-51]. Two general points about all the synthetic polynucleotides shown in Table I may be noted. The first point is that every set comprises two polynucleotides which are complementary in the *antiparallel* Watson–Crick base-pairing sense. A set of repeating trinucleotide polymers, which was complementary in the *parallel* sense, was found to be unacceptable to the DNA polymerase[52]. The second point is that it *was* necessary to synthesize segments corresponding to both strands of the DNA-polymer eventually desired (see below). DNA polymerase failed to bring about polymerization reactions when given only one of the segments of a set as a template.

(b) *Double-stranded DNA-like polymers with repeating nucleotide sequences*

In a part of the early work, short-chain deoxyribopolynucleotides with repeating sequences, for example, $(TTC)_3$ and $(TC)_5$, were directly used as templates for the RNA polymerase of *Escherichia coli*. While these experiments were successful[31,53], further work soon showed that use of the DNA polymerase as the first ‹amplification› device was preferable by far. Therefore, a major concern, following the chemical syntheses of the templates, was the study of the DNA polymerase and characterization of the DNA-like products produced by it in response to the short templates.

Scheme 2 lists the four types of reactions which have been elicited from the DNA polymerase, including the use of short homopolynucleotides. As seen, in reaction 1, a mixture of dT_{11} and dA_7 caused the extensive polymerization of dATP and dTTP to give a DNA-like polymer containing polyadenylate and polythymidylate in the two strands. In reaction 2, a mixture of the two short-chain polynucleotides with repeating dinucleotide sequences directed the extensive synthesis of a double-stranded DNA-like polymer containing exactly the sequences present in the short-chain templates[54,55]. In further work, similar reactions were demonstrated with short-chain templates containing repeating tri- as well as tetra-nucleotide sequences[52,56]. Characterization of the high molecular weight DNA-like polymers was accomplished by a variety of methods. The techniques used included nearest-neighbor analysis, electron microscopy (in a part of the early work[55]), sedimentation velocity and banding in alkaline cesium chloride density gradients[57].

Many of the features of the DNA-polymerase catalyzed reactions are truly remarkable. Thus: (*1*) in all the reactions studied (Scheme 2) the enzyme shows complete fidelity in the reproduction of sequences; (*2*) the synthesis is exstensive, 50–200-fold, and the products are of high molecular weight (300000 to over 1000000); (*3*) the enzyme thus amplifies and multiplies the information created by chemical methods; (*4*) finally, from the standpoint of an organic chemist, the most satisfying aspect is that the DNA polymers thus made can be used repeatedly for further production of the same polymers. It is unnecessary to go back to the time-consuming chemical synthesis for obtaining the templates again. DNA polymerase assures the continuity of these sequences.

Table II catalogues the different kinds of polymers which have so far been prepared and characterized. Thus, we have three classes of polymers: two

$$dT_{11} + dA_7 + \begin{Bmatrix} dTTP \\ dATP \end{Bmatrix} \rightarrow \text{poly-dA:dT} \quad (1)$$

$$d[TG]_6 + d[AC]_6 + \begin{Bmatrix} dTTP \\ dATP \\ dCTP \\ dGTP \end{Bmatrix} \rightarrow \text{poly-dTG:dCA} \quad (2)$$

$$d[TTC]_4 + d[AAG]_3 + \begin{Bmatrix} dTTP \\ dATP \\ dCTP \\ dGTP \end{Bmatrix} \rightarrow \text{poly-dTTC:dGAA} \quad (3)$$

$$d[TATC]_3 + d[TAGA]_2 + \begin{Bmatrix} dTTP \\ dATP \\ dCTP \\ dGTP \end{Bmatrix} \rightarrow \text{poly-dTATC:dGATA} \quad (4)$$

Scheme 2. Types of reactions catalyzed by DNA polymerase. All of the DNA-like polymers are written so that the colon separates the two complementary strands. The complementary sequences in the individual strands are written so that antiparallel base-pairing is evident.

Table II

DNA-like polymers with repeating nucleotide sequences

Repeating dinucleotide sequences	*Repeating trinucleotide sequences*	*Repeating tetranucleotide sequences*
Poly-dTC:GA	Poly-dTTC:GGA	Poly-dTTAC:GTAA
Poly-dTG:GA	Poly-dTTG:CAA	Poly-dTATC:GATA
	Poly-dTAC:GTA	
	Poly-dATC:GAT	

double-stranded polymers with repeating dinucleotide sequences, four polymers with repeating trinucleotide sequences and two polymers with repeating tetranucleotide sequences.

(c) *Single-stranded ribo-polynucleotides with repeating nucleotide sequences*

The next step was the transcription of the DNA-like polymers by means of RNA polymerase to form single-stranded ribo-polynucleotides. The prin-

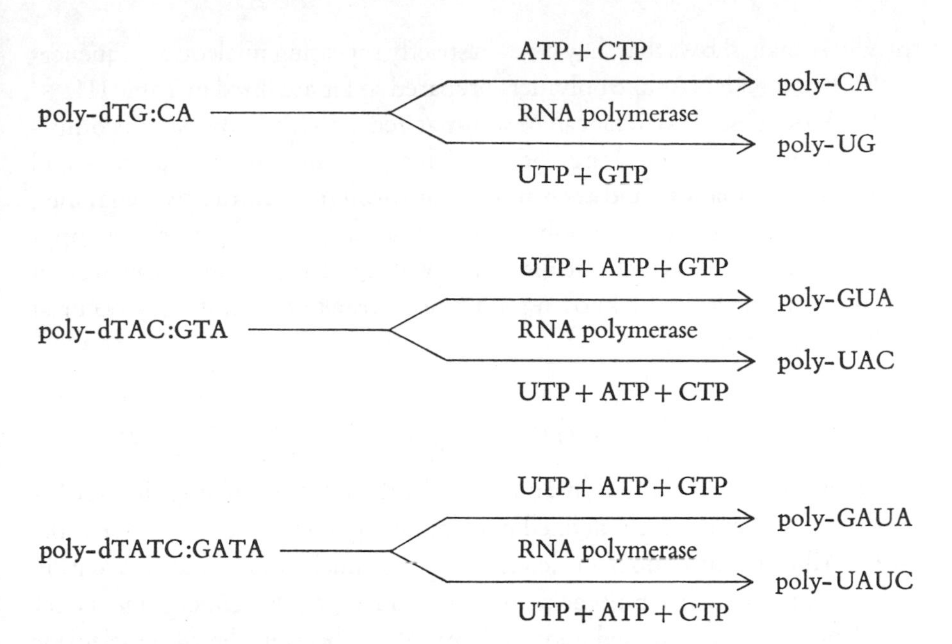

Scheme 3. The preparation of single-stranded ribopolynucleotides from DNA-like polymers containing repeating nucleotide sequences.

ciple used throughout is illustrated in Scheme 3. All of the DNA-like polymers contain two, or a maximum of three, different bases in individual strands. It is therefore possible, by giving the nucleoside triphosphates required for copying only one strand, to restrict the action of RNA polymerase to that strand. This is the case for all of the polymers, examples of which are shown in Scheme 3. Nearest-neighbor frequency analysis of all of the RNA

Table III

Synthetic ribopolynucleotides with repeating nucleotide sequences

Repeating dinucleotide sequences	*Repeating trinucleotide sequences*	*Repeating tetranucleotide sequences*
Poly-rUG	Poly-rUAC	Poly-rUAAG
Poly-rAC	Poly-rGUA	Poly-rUAGA
Poly-rUC	Poly-rAUC	Poly-rUCUA
Poly-rAG	Poly-rGAU	Poly-rUUAC
	Poly-rUUG	
	Poly-rCAA	
	Poly-rUUC	
	Poly-rGAA	

products again shows that they contain strictly repeating nucleotide sequences [58-60]. The total RNA-like polymers prepared so far are listed in Table III.

The work described so far can be summarized as follows. By using a combination of purely chemical methods, which are required to produce new and specified information, and then following through with the two enzymes, DNA polymerase and RNA polymerase, which are beautifully precise copying machines, we have at our disposal a variety of high-molecular-weight ribo-polynucleotides of known sequences. Mistake levels, if they occur at all, are insignificant.

(*d*) *Chemical synthesis of the sixty-four possible ribotrinucleotides*

At about the time that the above methods for the synthesis of long ribo-polynucleotides of completely defined nucleotide sequences were developed, the use of ribo-trinucleotides in determining the nucleotide sequences within codons for different amino acids was introduced by Nirenberg and Leder (see below). As mentioned above, chemical methods in the ribonucleotide field developed in this laboratory had previously resulted in general methods for the synthesis of the ribotrinucleotides. In view of the importance of these oligonucleotides in work on the genetic code, all the 64 trinucleotides derivable from the four common ribonucleotides, A, C, U, and G, were unambiguously synthesized and characterized. A separate report[61] should be consulted for the details of the chemical principles used in these syntheses. The use of the trinucleotides in work on the codon assignments for different amino acids is reviewed below.

3. *Polypeptide Synthesis in vitro and the Genetic Code*

(*a*) *Cell-free polypeptide synthesis using polynucleotides with repeating sequences*

Polymers with repeating dinucleotide sequences, $(AB)_n$, contain two triplets, ABA and BAB, in alternating sequence. Assuming three-letter, non-overlapping properties of the code, such polymers should direct incorporations of two amino acids in strictly alternating sequence. Repeating trinucleotide polymers, $(ABC)_n$, contain three repeating triplets depending upon the starting point. These are: ABC, BCA, and CAB. Here one would predict that one amino acid should be incorporated at a time to form a homopolypeptide chain, and a maximum of three such chains should result. Similar considerations for polynucleotides with repeating tetranucleotide sequences, $(ABCD)_n$, show that *in vitro* polypeptide synthesis should give products containing repeating tetrapeptide sequences, irrespective of the starting point in the reading of the messengers. All these predictions have been fully borne out experimentally without a single exception. The results with the three classes of polymers may be reviewed as follows.

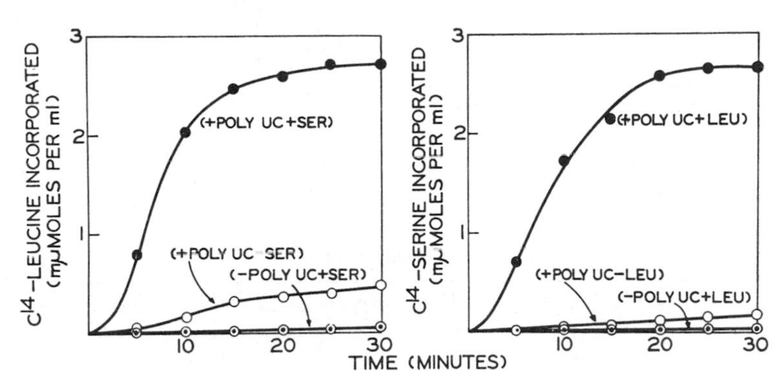

Fig. 1. Characteristics of the incorporation of [^{14}C]serine and [^{14}C]leucine into polypeptide in the presence of poly-UC.

Shown in Fig. 1 is an example of the type of results obtained with ribopolynucleotides containing two nucleotides in alternating sequence. Three features of the amino acid incorporations shown in Fig. 1, and which are common for all the messengers of this class, are (*1*) incorporation of only two amino acids is observed; (*2*) incorporation of one of these amino acids is dependent on the presence of the second amino acid, and, (*3*) the incorporations of the two amino acids are equimolar. All these features suggest that in these reactions copolypeptides containing two amino acids in alternating sequence are being produced. This has been demonstrated by extensive analysis of all the four series of polypeptidic products which are listed in Table IV[58,62].

Table V shows the results obtained with repeating trinucleotide polymers. Thus, these polymers, have as a rule given three homopolypeptides[63,64] and it should now be emphasized that this was because in all the work with the cell-free system artificially high Mg^{2+} ions concentration was used and, therefore, polypeptide chains could initiate without a proper signal. Two polymers, poly-rUAG and poly-rAUG, were exceptions in that they stimulated the incorporation of only two amino acids[64]. These polymers contain each a

Table IV

Cell-free copolypeptide syntheses using messengers containing repeating dinucleotide sequences
(System, *Escherichia coli* B)

Polynucleotides	*Copolypeptides with 2 amino acids in alternating sequence*
Poly-UC	$(\text{ser–leu})_n$
Poly-AG	$(\text{arg–glu})_n$
Poly-UG	$(\text{val–cys})_n$
Poly-AC	$(\text{thr–his})_n$

Table V

Cell-free homopolypeptide syntheses using messengers containing repeating trinucleotide sequences
(System, *Escherichia coli* B)

Polynucleotide	*Homopolypeptides of single amino acids*
Poly-UUC	phe, ser, leu
Poly-AAG	lys, glu, arg
Poly-UUG	cys, leu, val
Poly-CCA	gln, thr, asn
Poly-GUA	val, ser, (chain-terminator)
Poly-UAC	tyr, thr, leu
Poly-AUC	ileu, ser, his
Poly-GAU	met, asp, (chain-terminator)

chain-terminating triplet; UAG is the well-known amber triplet, and UGA is also now known to be a chain-terminating triplet.

Finally, as seen in Table VI, repeating tetranucleotide polymers, in fact, direct amino acid incorporations such that products containing repeating tetrapeptide sequences are formed except when chain-terminating triplets are present. This has been proved by analysis for the two products shown in Table VI[50]. This analysis of the repeating tetrapeptide sequences proves independently that the direction of reading of the messenger is from the 5'- to the 3'-end. This result is in agreement with that from a number of other laboratories[65,66]. The last two polynucleotides shown contain in every fourth place the chain-terminating triplets, UAG and UAA; for this reason, they fail to give any continuous peptides, but the formation of tripeptides has been demonstrated[51].

Table VI

Cell-free polypeptide syntheses using messengers containing repeating tetranucleotide sequences
(System, *Escherichia coli* B)

Polynucleotide	*Polypeptide*
Poly-UAUC	$(\text{tyr–leu–ser–ileu})_n$
Poly-UUAC	$(\text{leu–leu–thr–tyr})_n$
Poly-GUAA	di-and tri-peptides
Poly-AUAG	di-and tri-peptides

The results summarized above lead to the following general conclusions: (*1*) DNA does, in fact, specify the sequence of amino acids in proteins and this information is relayed through an RNA. (This was the first time that a direct sequence correlation between DNA and a protein had been established.) (*2*) All the results prove the 3-letter and non-overlapping properties of the code. (*3*) Finally, information on codon assignments can also be derived from these results.

(b) Codon assignments – The structure of the code

For this large question of codon assignments, however, unless one does a large number of polypeptide syntheses, the experiments reviewed above do not individually provide unique assignments to the codons. For example, of the two codons, UCU and CUC, which stand for serine and leucine, it is not possible to say which stands for which amino acid. Now the code has, in fact, been derived by a combination of the results obtained by the use of the binding technique developed by Nirenberg and Leder[67] and the work with the repeating polymers reviewed above. In Nirenberg's technique, one looks for the stimulation of the binding of different aminoacyl-tRNA's to ribosomes in the presence of specific trinucleotides. An example of its use is shown in Fig. 2, where the question of which of the three sequence isomers, AAG, AGA, GAA, which codes for lysine, is investigated. One measures the binding of [^{14}C]lysyl-tRNA to ribosomes in the presence of increasing amounts of these trinucleotides. As seen in Fig. 2 the binding is specifically induced by AAG. The other trinucleotide which also promotes a strong binding is AAA and the experiment using this is also included in Fig. 2. The trinucleotides AAG and AAA are, therefore, the codons for lysine. This technique has been used extensively in Nirenberg's laboratory, and my own colleagues have also tested

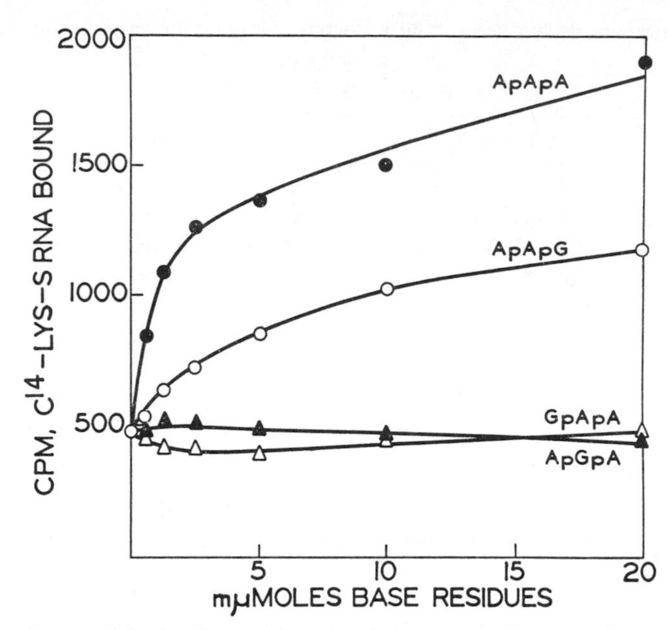

Fig. 2. Stimulation of the binding of [^{14}C]lysyl-tRNA to ribosomes by trinucleotides.

all of the 64 synthetic trinucleotides in this type of analysis. While extremely useful, the technique has not proved to be completely reliable. Often the effects are very small and there are cases where certain trinucleotides stimulate the binding of unexpected tRNA's. Conversely, there are cases where authentic trinucleotide codons do not give any binding. As already mentioned, most of the code actually has been worked out by using this technique in combination with the results from the repeating polymers and often by using evidence from a number of *in vivo* experiments.

The structure of the code which has emerged is shown in Table VII. This is by now a familiar method of presentation[65]. There is a box in the left-hand column for each base as the first letter; within each box in the right-hand column is shown each one of the four bases as the third letter; and in the middle are four columns, one for each base as the second letter. (For use of the Table see the legend). Only a few general observations may be made. (*1*) The code as shown is for the micro-organism *Escherichia coli* B, but probably will hold essentially for other organisms as well, although detailed and systematic checking in other systems (plants and animals) remains to be carried out. (*2*) There are entries for all of the sixty-four trinucleotides (there is no absolute nonsense). The code is highly degenerate in a semi-systematic way. Most of the degeneracy pertains to the third letter, where all of the four bases may

Table VII

The abbreviations for amino acids are standard. C.T. stands for chain termination, *i.e.*, the trinucleotide sequence does not stand for any amino acid but probably signals the end of protein chain formation. C.I. stands as a signal for chain initiation in protein synthesis. The method of presentation used in this Table follows the conventional way of writing of trinucleotides: thus, the first letter (base) of the trinucleotide is on the left (the 5′-end) and the third letter (the 3′-end) is to the right of the middle (second) base. The use of the Table for derivation of codons for different amino acids is exemplified as follows: codons for the amino acid, PHE, are UUU and UUC; condons for the amino acid, ALA, are GCU, GCC, GCA and GCG.

THE GENETIC CODE					
1ST LETTER	2ND LETTER				3RD LETTER
	U	C	A	G	
U	PHE	SER	TYR	CYS	U
	PHE	SER	TYR	CYS	C
	LEU	SER	C.T.	C.T.	A
	LEU	SER	C.T.	TRY	G
C	LEU	PRO	HIS	ARG	U
	LEU	PRO	HIS	ARG	C
	LEU	PRO	GLN	ARG	A
	LEU	PRO	GLN	ARG	G
A	ILEU	THR	ASN	SER	U
	ILEU	THR	ASN	SER	C
	ILEU	THR	LYS	ARG	A
	MET (C.I.)	THR	LYS	ARG	G
G	VAL	ALA	ASP	GLY	U
	VAL	ALA	ASP	GLY	C
	VAL	ALA	GLU	GLY	A
	VAL (C.I.)	ALA	GLU	GLY	G

stand for the same amino acid or where the two purine bases may stand for one amino acid and the two pyrimidines may stand for another amino acid. An exception is the box with the first letter A and the second letter U. Here, AUU, AUC and AUA represent isoleucine while the fourth codon, AUG, stands for methionine. Three amino acids show additional degeneracy in positions other than the third letter: thus, leucine and arginine are degenerate in the first letter while serine is unique in changing its position with regard to, both, the first and the second letters. (*3*) While the code is now generally accepted to be essentially universal, it should not be inferred that all organisms use the same codons for protein synthesis. What the universality means is that a trinucleotide codon does not change its meaning from one organism to the next. After all, there is very great divergence in the DNA composition of diverse organisms and they therefore probably use different codons for the same amino acid to varying extents. (*4*) The codons AUG and GUG, which stand respectively for methionine and valine, are also used as signals for initiation of polypeptide chain synthesis (see also a later section for initiation of protein synthesis). (*5*) There are three trinucleotides, UAA, UAG and UGA, which cause termination of polypeptide chain growth. It is not clear which ones are used naturally and under what circumstances a particular one is used. More recent work (see the lecture by M. W. Nirenberg, p. 372) indicates that there may be protein factors which have specificity for the different termination codons.

Finally, it should be emphasized that large portions of the code have been derived or confirmed by the prolonged and intensive studies of Yanofsky and coworkers, by the studies of Streisinger and coworkers, by Whitmann and by Tsugita and others (for comprehensive accounts of these studies see ref. 65).

4. Transfer RNA Structures: The Anticodons and Codon Recognition

The eludication of the nucleotide sequence of yeast alanine tRNA by Holley and coworkers[68] has been followed by similar work on a number of other tRNA's. At present some six yeast tRNA's, four *E. coli* tRNA's, one rat-liver tRNA, and one wheat-germ tRNA, have been sequenced and it is likely that the structures of many more will be known in the near future. Dr. RajBhandary and coworkers[69] have determined the primary structure of yeast phenylalanine tRNA and this structure is shown in Fig. 3, the usual cloverleaf model being used. In fact, a common feature of all the tRNA's, whose primary structures are known, is that they all can adopt the cloverleaf secondary structure. As discussed in detail by RajBhandary and coworkers[69] and by others, there is a remarkable overall similarity in regard to many important physical features between the different tRNA's. It is not my intention here to dwell in detail on the broad and exciting subject of tRNA structure and its biological function. The following paragraphs will be confined to those aspects where particularly relevant information has been forthcoming from the work of my own colleagues in Madison.

The first general question that one can ask is: How are the trinucleotide codons recognized by the protein-synthesizing apparatus? The first impor-

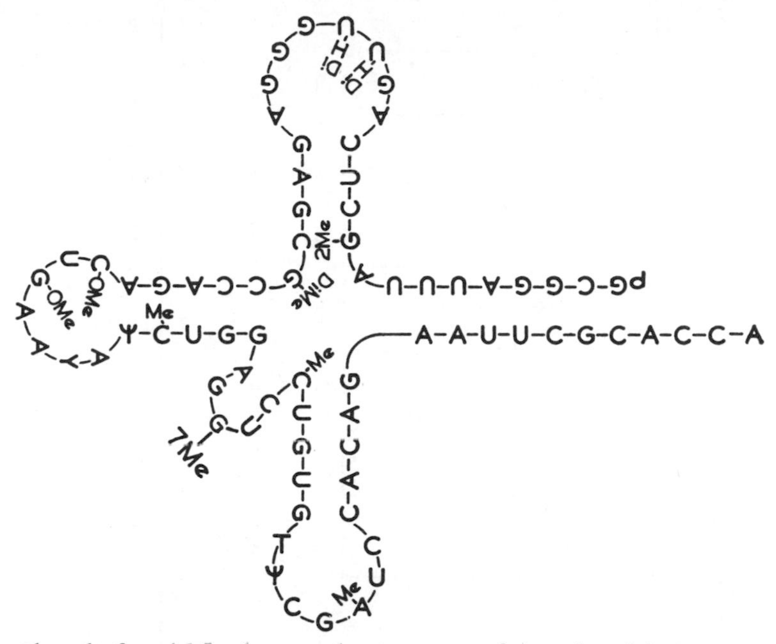

Fig. 3. Cloverleaf model for the secondary structure of yeast phenylalanine tRNA. The modifications in the minor bases are evident from the abbreviations shown against them. *Ψ* is pseudouridine while the nucleoside, Y, next to the presumed anticodon, G–A–A, is as yet unidentified.
OMe

tant advance here was the concept of an adapter molecule, which now is clearly seen to be a tRNA molecule. The experiments of Benzer, Lipmann and their coworkers[70] brought an elegant confirmation of this hypothesis. The next question is: What is the evidence that the recognition of codons, in fact, involves nucleotide–nucleotide interaction by virtue of base-pairing? If this is so, then one might expect to find in the primary structure of an amino acid specific tRNA three contiguous nucleotide units, ‹complementary› to the established codons for the particular amino acid. Indeed, the most encouraging common feature of all the tRNA's, whose primary sequence is known, turns out to be that they all contain an identical looped-out region in which such trinucleotide sequences are present. In Table VIII are assembled many of the known anticodons and the established codons for different amino acids. It can be seen that in every case the inferred anticodons show antiparallel Watson–Crick base-pairing with the codons. Moreover, the findings (*1*) that a single nucleotide change in the anticodon of *E. coli* $tRNA^{tyr}$ brings about a change in the coding properties of the tRNA[71] and (*2*) that an oligonucleotide

Table VIII

Codon–anticodon pairing[a] as derived from established codons for certain amino acids and from the primary structures of tRNA's for the corresponding amino acids

Amino acid	ala	tyr	tyr	amber codon	phe	val	ser	met
	→	→	→	→	→	→	→	→
Codons	G C U	U A U	U A U	U A G	U U U	G U U	U C U	A U G
	C	C	C		C	C	C	
	A					A	A	
Anticodons[a]	C G I	A Ψ G	A U G[b]	A U C	A A G-OMe	C A I	A G I	U A C[c]
	←	←	←	←	←	←	←	←
(*tRNA source*)	yeast	yeast	*E. coli*	*E. coli* (tyr-suppressor tRNA)	yeast and wheat germ	yeast	yeast and rat liver	*E. coli*

[a] Codon–anticodon pairing takes place in the antiparallel direction. Thus, the anticodons, as distinct from the conventional way of writing oligonucleotides, are written in the reverse direction.

[b] The structure of this, evidently a derivative of G, is as yet unknown.

[c] The structure of this, a derivative of C, is as yet unknown.

fragment of *E. coli* tRNA^{fmet} containing the anticodon sequence binds to ribosomes specifically in response to the codons for formyl-methionine[72] give us confidence that the concept of an anticodon consisting of three adjacent nucleotides in all tRNA's is correct.

Another important aspect of the biological function of tRNA's deserves comment. Can one tRNA molecule recognize more than one codon? At the present time we believe that this is often the case for the third letter of the codons and this occurs by a certain amount of «wobble» on the tRNA molecule[73], thus permitting base-pairs additional to the standard Watson-Crick base-pairs. One such case of multiple recognition of codons by one tRNA has been proven[74]. Thus phenylalanine tRNA, which we have available in a pure state in our laboratory and of which we know the anticodon to be 2′-*O*-methyl-GAA, can recognize both UUU and UUC which are the established codons for phenylalanine. This has been done by actual polyphenylalanine synthesis using precharged phenylalanyl-tRNA and the two polymers (*1*) polyuridylate and (*2*) poly-UUC which contains a repeating trinucleotide sequence (Fig. 4). There are other possibilities for multiple recognition. For example, it appears that inosine in the first position may recognize U, A and C[73]. Support for this pattern of multiple recognition has also been provided

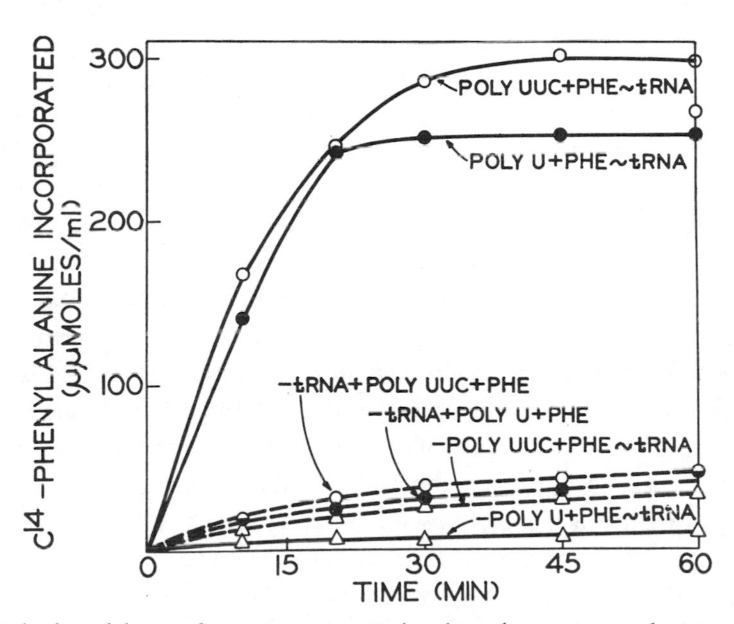

Fig. 4. Polyphenylalanine formation using *Escherichia coli* protein-synthesizing system and purified yeast phe-tRNA. Messengers are poly-U and poly-UUC.

[75,76]. Possible biological implications of multiple codon recognition by tRNA molecules have been discussed elsewhere[75].

Transfer RNA's are a unique class of molecules in the biological realm. They clearly have to perform a variety of functions. There is a good deal of evidence to suggest that in addition to a common secondary structure, these molecules possess a tertiary structure[65]. Further, a very plausible and attractive model for the anticodon loop has been put forward[77]. The very recent success in several laboratories[78–83] in obtaining crystals of tRNA's signifies in all probability a new era in the study of tRNA structure and function. The progress here would be very exciting not only for deepening our understanding of the mechanism of protein synthesis but also because of the possibility that a good part of the evolution of the genetic code is synonymous with the evolution of tRNA molecules.

5. *Further Aspects of the Code and of Protein Synthesis*

(*a*) *Initiation of protein synthesis*

As far as the initiation of protein synthesis in *E. coli* is concerned, a surge of activity occurred with the discovery of formylmethionyl-RNA by Marcker and Sanger[65]. It is now generally believed that formylmethionine (fmet) as carried by a particular species of methionine-specific tRNA is the initiation signal in protein synthesis. As mentioned above, in most of the work on the codon assignments using synthetic polynucleotides as messengers, artificially high Mg^{2+} ion concentrations were used. Under those conditions, the need for specific initiation of polypeptide chain synthesis is obviated. However, the requirement for the latter can be introduced by lowering the Mg^{2+} ion concentration to about 4–5 m*M* (compared with 10–15 m*M* used in earlier work). It is then found that prompt response in the cell-free system is elicited by only those messengers which contain codons that can recognize fmet-$tRNA^{fmet}$. Now, the peptide synthesis starts with fmet at the amino terminus. Once again, time does not permit a complete account of the work reported from different laboratories on this subject. My attention will be restricted to those experiments from my own laboratory which (*1*) permit derivation of codons involved in chain initiation in *E. coli*[84] and (*2*) shed a little light on the role of the ribosomal subunits in protein synthesis[85].

Poly-rAUG, as mentioned above, directs polymethionine synthesis. When this experiment is carried out at 5 m*M* Mg^{2+} ion concentration, the results shown in Fig. 5 are obtained. Thus, synthesis proceeds with a lag and poorly, when only met-tRNA^{met} is provided. Addition of fmet-RNA^{fmet} gives a dramatic acceleration of the rate of polymethionine synthesis. It is therefore concluded that fmet-tRNA^{fmet} is required for initiation and met-tRNA^{met} is required for chain propagation.

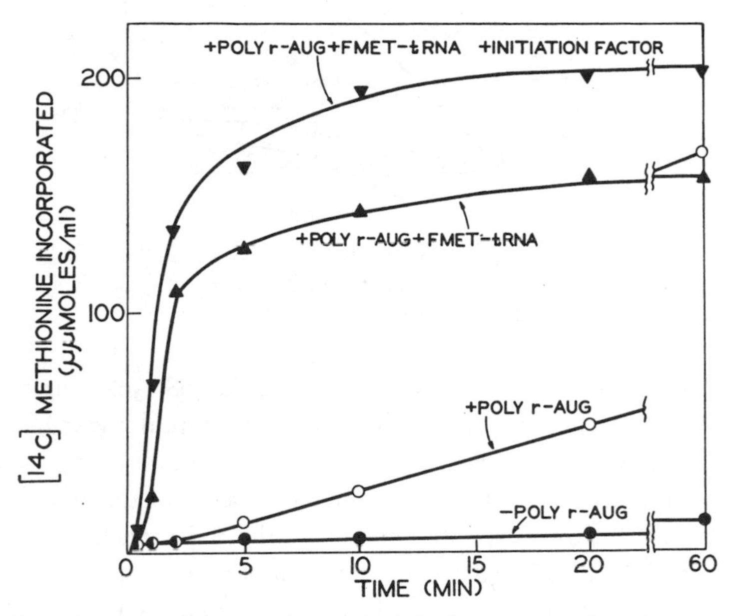

Fig. 5. Polymethionine synthesis in the presence of poly-rAUG and [^{14}C]met-tRNA^{met}. The effects obtained on supplementing the system with fmet-tRNA^{fmet} and initiation factors are shown.

Similarly, at 5 m*M* Mg^{2} ion concentration, poly-rUG-directed synthesis of val–cys copolypeptide (Fig. 6) requires the presence of fmet-tRNA^{fmet} and there is a striking effect following the addition of protein fractions which have been designated initiation factors[65]. Analysis of the terminal sequence of the polypeptidic product formed showed that fmet was present at the amino end and it was followed by cys and then by val.

From the above results, it is concluded that AUG and GUG are the codons for initiation in *E. coli*[84]. An intriguing point here is degeneracy in the first letter.

That the 70S ribosomal particle from *E. coli* can be split to 30S and 50S subunits was already evident in the late fifties. However, the significance of the two subunits in protein synthesis has remained obscure until recently. Gentle lysis of *E. coli* cells was recently found to yield mainly the 30S and 50S sub-

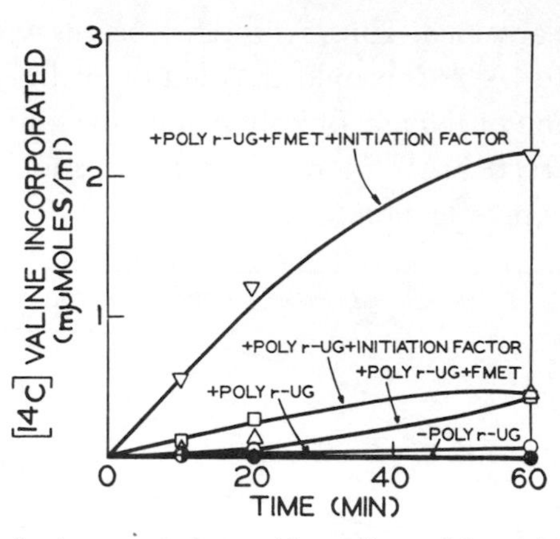

Fig. 6. The synthesis of val–cys copolypeptide as directed by poly-rUG. The synthesis was carried out at 5 mM Mg^{2+} ion concentration. The effects of fmet-$tRNA^{fmet}$ and of the initiation factors are shown.

units and this finding further suggested a role for the 30S–50S couples in protein synthesis[86]. It should be added that prior biochemical investigations had indicated at least two binding sites on the 70S ribosomes. More recently, Nomura and coworkers[87] showed that in the presence of the viral f2-RNA, fmet-$tRNA^{fmet}$ showed specific binding to the 30S particles, whereas the noninitiator tRNA's showed no binding to the 30S particles.

A further study of the above-described polymethionine synthesis as directed by poly-rAUG gave results[85] which can be summarized as follows. (*1*) The 30S particles bind fmet-$tRNA^{fmet}$ in the presence of poly-rAUG at 5 mM Mg^{2+} ion concentration. (*2*) The noninitiator rRNA, met-$tRNA^{met}$ is bound only after the addition of 50S particles to the 30S particles. (*3*) It was possible to demonstrate the synthesis of the dipeptide formylmethionylmethionine (fmet-met) by stepwise formation of the appropriate complex containing all the components. Thus fmet-$tRNA^{fmet}$ was bound to the 30S particles in the presence of poly-rAUG and the ‹initiation factors›. The complex was isolated by centrifugation, supplemented first with 50S particles and then with met-$tRNA^{met}$. The resulting complex, containing now 30S particles, poly-rAUG, fmet-$tRNA^{fmet}$, 50S particles and met-$tRNA^{met}$, was again isolated by centrifugation. This complex when supplemented with the S-100 supernatant fraction gave the dipeptide fmet-met. Therefore, it is clear that both fmet-$tRNA^{fmet}$ and met-$tRNA^{met}$ were being bound simultaneously to 30S + 50S ribosomal particles. The results provide direct evidence for the presence of two tRNA binding sites on 70S ribosomes. Furthermore, the picture of the role of the ribosomal subunits which emerges from this work is that the primary event in the initiation of protein synthesis is the binding of the initiator tRNA to 30S ribosome + messenger RNA complex. The resulting initiation complex is then joined by the 50S particles and is now able to accept another aminoacyl-tRNA so that a peptide bond may be formed. Nomura and coworkers[87] arrived at the same conclusion from their work and the above results support their conclusions. Several other laboratories have subsequently obtained similar results.

As mentioned above, certain protein factors that can be released from the ribosomes are required for the initiation of protein synthesis. These factors, the general chemistry of the ribosomal proteins and the ribosomal subunits themselves are all areas which are currently the subject of investigation in many laboratories. Very recently, striking progress has been made in Nomura's laboratory on the reconstitution of the 30S subunit. These and related studies are rapidly opening up new approaches to a deeper understanding of the mechanism of protein synthesis.

(b) *Missense suppression: tRNA involvement*

Another application of the ribopolynucleotide messengers with repeating nucleotide sequences was in the study of the mechanism of genetic suppression (missense to sense). From the work of Yanofsky and his coworkers it is known that many mutants of *E. coli* can only make a defective protein A of tryptophan synthetase. In one case, mutant A-78, one glycine residue in the A protein is replaced by cysteine. A suppressed mutant (A-78-Su-78) restores, to a small extent, the original glycine in place of cysteine. Using the cell-free protein-synthesizing system from *E. coli* B, it was shown[88] that this system when supplemented with the tRNA from the strain A-78-Su-78, incorporates [^{14}C]glycine in the presence of valine under the direction of poly-r-UG (Fig. 7). As reviewed above, the latter polymer normally directs the synthesis of valine–cysteine copolypeptide. Valine–glycine copolypeptide formed *specifically* in the presence of tRNA from A-78-Su-78 strain was thoroughly characterized[88]. Similarly, Carbon, Berg and Yanofsky[89] showed that another missense suppressor of glycine to arginine mutation in protein A also acted at the level of tRNA. Previously, tRNA had been shown to be responsible for suppression of an *amber* codon in a bacteriophage RNA[90,91]. As already described above, in one case amber suppression has now been shown to be due to a single nucleotide change in the anticodon of a tRNA[71].

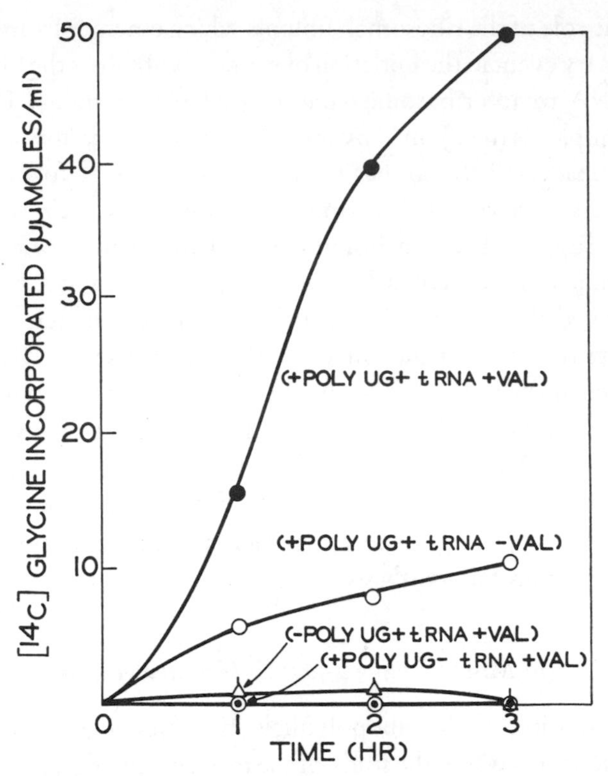

Fig. 7. The incorporation of [^{14}C]glycine into valine–glycine copolypeptide as stimulated by poly-rUG in the presence of tRNA from A-78-Su-78 strain of *Escherichia coli*.

(*c*) *Translation of single-stranded DNA-like polymers*

Recently the striking finding was reported[92] that in the presence of amino-glycoside antibiotics such as neomycin B, denatured DNA stimulated the incorporation of amino acids in the bacterial cell-free protein-synthesizing system. A further study of these observations using single-stranded deoxyribopolynucleotides with defined nucleotide sequences, poly-dTG, poly-dAC and poly-dT, gave most encouraging results[93]. Thus, the response from the DNA-like polymers was excellent and, surprisingly, the mistakes were very rare and small. For example, poly-dCA directed the synthesis of thr–his copolypeptide and no other amino acid was incorporated. In addition to providing a further opportunity for the study of ribosome function, these results may have important practical applications in future work. As pointed out elsewhere[94], it is not inconceivable that the laboratory synthesis of specific proteins will be carried out using nucleic acid templates. For this purpose, protected trinucleotides representing different codons will be made in quantity and on a commercial basis, and these will be used in the synthesis of nucleic acid templates for proteins, the approach offering flexibility and selectivity in amino acid substitutions at the template level.

6. Conclusion

While clarity in some of the detailed aspects of the genetic code is still lacking, it has been a most satisfying experience in the lives of many of us, who have worked on the problems, to see complete agreement reached in regard to its general structure. Evidence coming from a variety of techniques, genetic and biochemical, from *in vivo* and *in vitro* experiments, has furnished the codon assignments reviewed above. It is unlikely that any of the assignments would be revised. However, much remains to be done at chemical and biochemical level to obtain an adequate understanding of the very elaborate protein-synthesizing system. Nevertheless, the problem of the genetic code at least in the restricted one-dimensional sense (the linear correlation of the nucleotide sequence of polynucleotides with that of the amino acid sequence of polypeptides) would appear to have been solved. It may be hoped that this knowledge would serve as a basis for further work in molecular and developmental biology.

Acknowledgements

I wish to emphasize again that the work which has formed the content of this lecture has been so much a collaborative effort. I am deeply indebted to a very large number of devoted colleagues, chemists and biochemists, with whom I had the good fortune to be happily associated.

Work and progress in science becomes more and more interdependent: this certainly has been true in work on the genetic code. Many of the great scientists, who influenced directly or indirectly the work herein reviewed, have been mentioned in the text. I wish to make a personal acknowledgement to one more scientist. Fortunately, I was accepted by Professor V. Prelog of the Eidgenössische Technische Hochschule, Zürich, as a postdoctoral student.

The association with this great scientist and human being influenced immeasurably my thought and philosophy towards science, work and effort.

The work had its start at the British Columbia Research Council, Vancouver, Canada, where it was made possible by the encouragement of Dr. Gordon M. Shrum (now Chancellor of Simon Fraser University, B.C.) and with the financial support from the National Research Council of Canada. In more recent years, generous support has been received from the National Cancer Institute of the National Institutes of Health, U.S. Public Health Service, the National Science Foundation, Washington, the Life Insurance Medical Research Fund and the Graduate School of the University of Wisconsin.

1. F.H.C. Crick, in On the genetic code, in *Nobel Lectures, Physiology or Medicine 1942–1962*, Elsevier, Amsterdam, 1964, p. 811.
2. G.W. Beadle and E.L. Tatum, *Proc. Natl. Acad. Sci. (U.S.)*, 27 (1941) 499.
3. O.T. Avery, C.M. Macleod and M. McCarty, *J. Exptl. Med.*, 79 (1944) 137.
4. A.D. Hersey and M. Chase, *J. Gen. Physiol.*, 36 (1952) 39.
5. A. Gierer and G. Schramm, *Nature*, 177 (1956) 702.
6. H. Fraenkel-Conrat, B. Singer and R.C. Williams, *Biochim. Biophys. Acta*, 25 (1957) 87.
7. T. Caspersson, *Naturwissenschaften*, 28 (1941) 33.
8. J. Brachet, *Arch. Biol. (Liège)*, 53 (1942) 207.
9. D.M. Brown and A.R. Todd, in E. Chargaff and J.N. Davidson (Eds.), *Nucleic Acids*, Vol. 1, Academic Press, New York, 1955, p. 409.
10. J.D. Watson and F.H.C. Crick, *Nature*, 171 (1953) 737.
11. I.R. Lehman, M.J. Bessman, E.S. Simms and A. Kornberg, *J. Biol. Chem.*, 233 (1958) 163.
12. S.B. Weiss, *Proc. Natl. Acad. Sci. (U.S.)*, 46 (1960) 1020.
13. A. Stevens, *J. Biol. Chem.*, 236 (1961) PC43.
14. J. Hurwitz, J.J. Furth, M. Anders and A. Evans, *J. Biol. Chem.*, 237 (1962) 3752.
15. M. Chamberlin and P. Berg, *Proc. Natl. Acad. Sci. (U.S.)*, 48 (1962) 81.
16. D.P. Burma, H. Kroger, S. Ochoa, R.C. Warner and J.D. Weill, *Proc. Natl. Acad. Sci. (U.S.)*, 47 (1961) 749.
17. A.D. Hershey, J. Dixon and M. Chase, *J. Gen. Physiol.*, 36 (1953) 777.
18. E. Volkin and L. Astrachan, *Virology*, 2 (1956) 149.
19. F. Jacob and J. Monod, *J. Mol. Biol.*, 3 (1961) 318.
20. S. Brenner, F. Jacob and M. Meselson, *Nature*, 190 (1961) 576.
21. F. Gros, W. Gilbert, H. Hiatt, C.G. Kurland, R.W. Risebrough and J.D. Watson, *Nature*, 190 (1961) 581.
22. P.C. Zamecnik, *Harvey Lectures*, Ser. 54 (1958–1959) 256.
23. F. Lipmann, W.C. Hulsmann, G. Hartmann, Hans G. Boman and George Acs, *J. Cell. Comp. Physiol.*, 54, Suppl. 1 (1959) 75.

24. P.Berg, *Ann.Rev.Biochem.*, 30 (1961) 293.
25. J.D.Watson, in The involvement of RNA in the synthesis of proteins, in *Nobel Lectures, Physiology or Medicine 1942–1962*, Elsevier, Amsterdam, 1964, p.785.
26. W.B.Wood and P.Berg, *Proc.Natl.Acad.Sci. (U.S.)*, 48 (1962) 94.
27. J.H.Matthaei and M.W.Nirenberg, *Proc.Natl.Acad.Sci. (U.S.)*, 47 (1961) 1580.
28. H.G.Khorana, *Federation Proc.*, 19 (1960) 931.
29. H.G.Khorana, *Some Recent Developments in the Chemistry of Phosphate Esters of Biological Interest*, Wiley, New York, 1961.
30. H.G.Khorana, T.M.Jacob, M.W.Moon, S.A.Narang and E.Ohtsuka, *J.Am. Chem.Soc.*, 87 (1965) 2954.
31. H.G.Khorana, *Federation Proc.*, 24 (1965) 1473.
32. H.G.Khorana, in *Proc. 7th.Intern.Congr.Biochem., Tokyo, August 19–25, 1967*, 1968, p.17
33. H.G.Khorana, on *Proc.Plenary Lectures of IUPAC Symp.Natural Products, London*, 1968.
34. G.Gamow, *Nature*, 173 (1954) 318.
35. S.Ochoa, *Federation Proc.*, 22 (1963) 62.
36. M.W.Nirenberg, J.H.Matthaei, O.W.Jones, R.G.Martin and S.H.Barondes, *Federation Proc.*, 22 (1963) 55.
37. A.Falaschi, J.Adler and H.G.Khorana, *J.Biol.Chem.*, 238 (1963) 3080.
38. J.Hurwitz, J.J.Furth, M.Anders and A.Evans, *J.Biol.Chem.*, 237 (1962) 3752.
39. B.D.Mehrotra and H.G.Khorana, *J.Biol.Chem.*, 240 (1965) 1750.
40. A.Kornberg, L.L.Bertsch, J.F.Jackson and H.G.Khorana, *Proc. Natl. Acad. Sci. (U.S.)*, 51 (1964) 315.
41. E.Ohtsuka, M.W.Moon and H.G.Khorana, *J.Am.Chem. Soc.*, 87 (1965) 2956.
42. T.M.Jacob and H.G.Khorana, *J.Am.Chem.Soc.*, 87 (1965) 2971.
43. S.A.Nrang and H.G.Khorana, *J.Am.Chem.Soc.*, 87 (1965) 2981.
44. S.A.Narang, T.M.Jacob and H.G.Khorana, *J.Am.Chem.Soc.*, 87 (1965) 2988.
45. S.A.Narang, T.M.Jacob and H.G.Khorana, *J.Am.Chem.Soc.*, 89 (1967) 2158.
46. S.A.Narang, T.M.Jacob and H.G.Khorana, *J.Am.Chem.Soc.*, 89 (1967) 2167.
47. H.Kössel, H.Büchi and H.G.Khorana, *J.Am.Chem.Soc.*, 89 (1967) 2185.
48. E.Ohtsuka and H.G.Khorana, *J.Am.Chem.Soc.*, 89 (1967) 2195.
49. H.G.Khorana, H.Büchi, T.M.Jacob, H.Kössel, S.A.Narang and E.Ohtsuka, *J. Am.Chem.Soc.*, 89 (1967) 2154.
50. H.Kössel, A.R.Morgan and H.G.Khorana, *J.Mol.Biol.*, 26 (1967) 449.
51. H.Kössel, *Biochim.Biophys.Acta*, 157 (1968) 91.
52. Robert D.Wells, T.M.Jacob, S.A.Narang and H.G.Khorana, *J.Mol.Biol.*, 27 (1967) 237.
53. S.Nishimura, T.M.Jacob and H.G.Khorana, *Proc.Natl.Acad.Sci. (U.S.)*, 52 (1964) 1494.
54. C.Byrd, E.Ohtsuka, M.W.Moon and H.G.Khorana, *Proc.Natl.Acad.Sci. (U.S.)*, 53 (1965) 79.
55. R.D.Wells, E.Ohtsuka and H.G.Khorana, *J.Mol.Biol.*, 14 (1965) 221.
56. R.D.Wells, H.Büchli, H.Kóssel, E.Ohtsuka and H.G.Khorana, *J.Mol.Biol.*, 27 (1967) 265.

57. R.D.Wells and J.E.Blair, *J.Mol.Biol.*, 27 (1967) 273.
58. S.Nishimura, D.S.Jones and H.G.Khorana, *J.Mol. Biol.*, 13 (1965) 302.
59. H.G.Khorana, in *Genetic Elements*, Federation of European Societies for Biological Chemists, April 1966, p.209.
60. Unpublished work of A.R.Morgan.
61. R.Lohrmann, D.Söll, H.Hayatsu, E.Ohtsuka and H.G.Khorana, *J.Am.Chem.Soc.*, 88 (1966) 819.
62. D.S.Jones, S.Nishimura and H.G.Khorana, *J.Mol.Biol.*, 16 (1966) 454.
63. S.Nishimura, D.S.Jones, E.Ohtsuka, H.Hayatsu, T.M.Jacob and H.G.Khorana, *J.Mol.Biol.*, 13 (1965) 283.
64. A.R.Morgan, R.D.Wells and H.G.Khorana, *Proc.Natl.Acad.Sci.* (*U.S.*), 56 (1966) 1899.
65. For references, see The Genetic Code, *Cold Spring Harbor Symp.Quant.Biol.*, 31 (1966).
66. H.Lamfrom, C.S.McLaughlin and A.Sarabhai, *J.Mol.Biol.*, 22 (1966) 359.
67. M.W.Nirenberg and P.Leder, *Science*, 145 (1964) 1399.
68. R.W.Holley, J.Apgar, G.A.Everett, J.T.Madison, M.Marquisee, S.H.Merrill, J.R.Penswick and A.Zamir, *Science*, 147 (1965) 1462.
69. U.L.RajBhandary, S.H.Chang, A.Stuart, R.D.Faulkner, R.M.Hoskinson and H.G.Khorana, *Proc.Natl.Acad.Sci.* (*U.S.*), 57 (1967) 751.
70. F.Chapeville, F.Lipmann, G.von Ehrenstein, B.Weisblum, W.J.Ray Sr. and S. Benzer, *Proc.Natl.Acad.Sci.* (*U.S.*), 48 (1962) 1086.
71. H.M.Goodman, J.Abelson, A.Landy, S.Brenner and J.D.Smith, *Nature*, 217 (1968) 1019.
72. B.F.C.Clark, S.K.Dube and K.A.Marcker, *Nature*, 219 (1968) 484.
73. F.H.C.Crick, *J.Mol.Biol.*, 19 (1966) 548.
74. D.Söll and U.L.RajBhandary, *J.Mol.Biol.*, 29 (1967) 97.
75. D.Söll, J.Cherayil, D.S.Jones, R.D.Faulkner, A.Hampel, R.M.Bock and H.G. Khorana, p.51 in ref.65.
76. D.Söll, J.D.Cherayil and R.M.Bock, *J.Mol.Biol.*, 29 (1967) 97.
77. W.Fuller and A.Hodgson, *Nature*, 215 (1967) 817.
78. B.F.C.Clark, B.P.Doctor, K.C.Holmes, A.Klug, K.A.Marcker, S.J.Morris and H.H.Paradies, *Nature*, 219 (1968) 1222.
79. S.Kim and A.Rich, *Science*, 162 (1968) 1381.
80. A.Hampel, M.Labanauskas, P.G.Connors, L.Kirkegaard, U.L.RajBhandary, P. B.Sigler and R.M.Bock, *Science*, 162 (1968) 1384.
81. F.Cramer, F.v.d.Haar, W.Saenger and E.Schlimme, *Angew. Chem.*, 80 (1968) 969.
82. H.H.Paradies, *FEBS Letters*, 2 (1968) 112.
83. J.R.Fresco, R.D.Blake and R.Langridge, *Nature*, 220 (1968) 1285.
84. H.P.Ghosh, D.Söll and H.G.Khorana, *J.Mol.Biol.*, 25 (1967) 275.
85. H.P.Ghosh and H.G.Khorana, *Proc.Natl.Acad.Sci.* (*U.S.*), 58 (1967) 2455.
86. G.Mangiarotti and D.Schlessinger, *J.Mol.Biol.*, 20 (1966) 123; 29 (1967) 395; D.Schlessinger, G.Mangiarotti and D.Apirion, *Proc.Natl.Acad.Sci.* (*U.S.*), 58 (1967) 1782.

87. M. Nomura and C. V. Lowry, *Proc. Natl. Acad. Sci. (U. S.)*, 58 (1967) 946; M. Nomura, C. V. Lowry and C. Guthrie, *ibid.*, 58 (1967) 1487.
88. N. K. Gupta and H. G. Khorana, *Proc. Natl. Acad. Sci. (U. S.)*, 56 (1966) 772.
89. J. Carbon, P. Berg and C. Yanofsky, *Proc. Natl. Acad. Sci. (U. S.)*, 56 (1966) 764.
90. M. R. Capecchi and G. N. Gussin, *Science*, 149 (1965) 417.
91. D. L. Engelhardt, R. Webster, R. C. Wilhelm and N. D. Zinder, *Proc. Natl. Acad. Sci. (U. S.)*, 54 (1965) 1791.
92. B. J. McCarthy and J. J. Holland, *Proc. Natl. Acad. Sci. (U. S.)*, 54 (1965) 880.
93. A. R. Morgan, R. D. Wells and H. G. Khorana, *J. Mol. Biol.*, 26 (1967) 477.
94. A. Kumar and H. G. Khorana, *J. Am. Chem. Soc.*, in the press.

Biography

Har Gobind Khorana was born of Hindu parents in Raipur, a little village in Punjab, which is now part of West Pakistan. The correct date of his birth is not known; that shown in documents is January 9th, 1922. He is the youngest of a family of one daughter and four sons. His father was a «patwari», a village agricultural taxation clerk in the British Indian system of government. Although poor, his father was dedicated to educating his children and they were practically the only literate family in the village inhabited by about 100 people.

Har Gobind Khorana attended D.A.V. High School in Multan (now West Punjab); Ratan Lal, one of his teachers, influenced him greatly during that period. Later, he studied at the Punjab University in Lahore where he obtained an M. Sc. degree. Mahan Singh, a great teacher and accurate experimentalist, was his supervisor.

Khorana lived in India until 1945, when the award of a Government of India Fellowship made it possible for him to go to England and he studied for a Ph. D. degree at the University of Liverpool. Roger J. S. Beer supervised his research, and, in addition, looked after him diligently. It was the introduction of Khorana to Western civilization and culture.

Khorana spent a postdoctoral year (1948–1949) at the Eidgenössische Technische Hochschule in Zürich with Professor Vladimir Prelog. The association with Professor Prelog molded immeasurably his thought and philosophy towards science, work, and effort.

After a brief period in India in the fall of 1949, Khorana returned to England where he obtained a fellowship to work with Dr. (now Professor) G. W. Kenner and Professor (now Lord) A. R. Todd. He stayed in Cambridge from 1950 till 1952. Again, this stay proved to be of decisive value to Khorana. Interest in both proteins and nucleic acids took root at that time.

A job offer in 1952 from Dr. Gordon M. Shrum of British Columbia (now Chancellor of Simon Fraser University, British Columbia) took him to Vancouver. The British Columbia Research Council offered at that time very little by way of facilities, but there was «all the freedom in the world», to use Dr. Shrum's words, to do what the researcher liked to do. During the following years, with Dr. Shrum's inspiration and encouragement and frequent help and scientific counsel from Dr. Jack Campbell (now Head of the Department of Microbiology at the University of British Columbia), a group began to work in the field of biologically interesting phosphate esters and nucleic acids. Among the many devoted and loyal colleagues of this period, there should, in particular, be mention of Dr. Gordon M. Tener (now a Professor in the Biochemistry Department of the University of British Columbia), who contributed much to the spiritual and intellectual well-being of the group.

In 1960 Khorana moved to the Institute for Enzyme Research at the University of Wisconsin. He became a naturalized citizen of the United States. As of the fall of 1970 Khorana has been Alfred P. Sloan Professor of Biology and Chemistry at the Massachusetts Institute of Technology.

Har Gobind Khorana was married in 1952 to Esther Elizabeth Sibler, who is of Swiss origin. Esther brought a consistent sense of purpose into his life at a time when, after six years' absence from the country of his birth, Khorana felt out of place everywhere and at home nowhere. They have three children: Julia Elizabeth (born May 4th, 1953), Emily Anne (born October 18th, 1954), and Dave Roy (born July 26th, 1958).

MARSHALL NIRENBERG

The Genetic Code

December 12, 1968

Genetic memory resides in specific molecules of nucleic acid. The information is encoded in the form of a linear sequence of bases of 4 varieties that corresponds to sequences of 20 varieties of amino acids in protein. The translation from nucleic acid to protein proceeds in a sequential fashion according to a systematic code with relatively simple rules. Each unit of nucleic acid defines the species of molecule to be selected, its position relative to the previous molecule selected, and the time of the event relative to the previous event. The nucleic acid therefore functions both as a template for other molecules and as a biological clock.

The information is encoded and decoded in the form of a one-dimensional string. The polypeptide translation product then folds upon itself in a specific manner predetermined by the amino acid sequence, forming a complex, three-dimensional protein.

The Concept of a Gene–Protein Code

The advances in biochemical genetics are due to the efforts of investigators from virtually every field of science. Among the milestones are the identification of DNA as the genetic material by Avery, MacLeod, and McCarty[1], the «one gene–one enzyme» concept of Beadle and Tatum[2], and the pioneering experiments of Brachet[3] and Caspersson[4] on the relation of RNA to protein synthesis. In addition, the puzzle of protein synthesis was unraveled, bit by bit, and *in vitro* systems for protein and nucleic acid synthesis were developed.

The concept of a simple code relating the base sequence of nucleic acid to the amino acid sequence of protein originated on three independent occasions during the early 1950's. Intuition was spectacular when considered in the context of the information then available. Caldwell and Hinshelwood[5] suggested that RNA is composed of five kinds of units, the four bases and ribose phosphate, and that two adjacent units in RNA correspond to one amino acid in protein. Dounce[6] proposed that three adjacent bases in RNA correspond to one amino acid in protein. In addition, the concepts of polarity of translation and activation of amino acids were formulated in considerable detail. Dounce's conviction that templates are required for the synthesis of protein originated during his Ph.D. oral examination when he was asked by James Sumner to consider the problem of how proteins synthesize other proteins.

Concurrently, George Gamow[7] suggested that a double-strand of DNA contains binding sites for amino acids, each site defined by one base-pair and adjacent non-complementary bases on opposite strands of DNA. Gamow conceived the idea upon reading the article by Watson and Crick on the pairing of bases in DNA[8]. Other speculations concerning the nature of the code were advanced by many investigators during the latter part of the 1950's (*cf.* the recent review of Woese[9]).

Although the concept that RNA is a template for protein was well established, direct biochemical evidence was lacking. However, Hershey's[10] finding that a fraction of RNA is rapidly synthesized and then degraded in *E. coli* infected with T2 bacteriophage, and the demonstration by Volkin and Astrachan[11] that the composition of this RNA fraction resembles phage DNA rather than *E. coli* DNA were exciting, because the data suggested that the unstable RNA fraction might function as templates for the synthesis of phage protein.

I plunged into the problems of protein synthesis after I had obtained postdoctoral training. My graduate studies were in biochemistry under the guidance of James Hogg; I obtained postdoctoral training with DeWitt Stetten and so with William Jakoby at the National Institutes of Health. Then I joined Gordon Tompkins' department and began to study the steps that relate DNA, RNA, and protein. The training in enzymology and the stimulating enviroment greatly influenced the future course of my work.

Extensive studies on the mechanism of protein synthesis had yielded much information and it seemed likely then that it would be possible within the coming decade to obtain the synthesis of an enzyme in cell extracts. Since a system of this kind would provide many opportunities to study questions pertaining to the flow of information from nucleic acid to protein, I decided to work on the cell-free synthesis of penicillinase. Pollock and his colleagues[12,13] had obtained much information on the regulation of penicillinase synthesis *in vivo*, and had shown that the molecular weight of the enzyme is relatively low, and that the enzyme lacks cysteine. It seemed likely that one might selectively inhibit the synthesis of proteins that require cysteine and at the same time stimulate penicillinase synthesis *in vitro* by the addition of nucleic acid templates to cell extracts.

During the next 2 years I studied the properties of the system, particularly the effect of reaction conditions, nucleic acids and other factors upon the rate of cell-free protein synthesis. During this period results of great interest dealing with protein synthesis in *E. coli* extracts were reported by Lamborg and Zamecnik[14]; Tissieres, Schlessinger and Gros[15], and others. Tissieres, Schlessinger and Gros[15]; Kameyama and Novelli[16]; and Nisman and Fukuhara[17] reported that DNAase inhibited *in vitro* amino acid incorporation into protein. I had also observed this phenomenon and was greatly interested in it because the results strongly suggested that the cell-free synthesis of protein was dependent, ultimately, upon DNA templates.

Heinrich Matthaei then joined me in these studies. We soon showed that RNA prepared from ribosomes stimulates amino acid incorporation into protein[18]. However, amino acids were incorporated into protein rapidly without added RNA[19], so RNA-dependent protein synthesis was difficult to detect. This problem was solved, as shown in Fig.1, by incubating *E. coli* extracts with the components required for protein synthesis and DNAase in order to reduce the level of endogenous RNA templates. After a brief incubation period, the synthesis of protein stops and further protein synthesis is then dependent upon the addition of template RNA. Transfer RNA does not replace template RNA.

A rapid assay was devised based on the filtration of [^{14}C]protein precipitates that reduced the time required for each experiment about four-fold. Preparations of RNA from many sources were obtained to determine the specificity and activity of each RNA preparation as templates for protein synthesis. RNA from yeast, ribosomes, and from tobacco mosaic virus were found to be highly active in stimulating the incorporation into protein of every species of amino acid tested. In contrast, poly-U stimulated phenalanine incorporation into protein rather specifically, and the product was shown to be polyphenylalanine. Single-stranded poly-U was an active template for phenylalanine incorporation, but double- or triple-stranded poly-U · poly-A helices did not serve as templates for protein synthesis[18].

These results showed that RNA is a template for protein, that residues of U in poly-U correspond to phenylalanine in protein, and that the translation of mRNA is affected by both the primary and the secondary structure of the RNA.

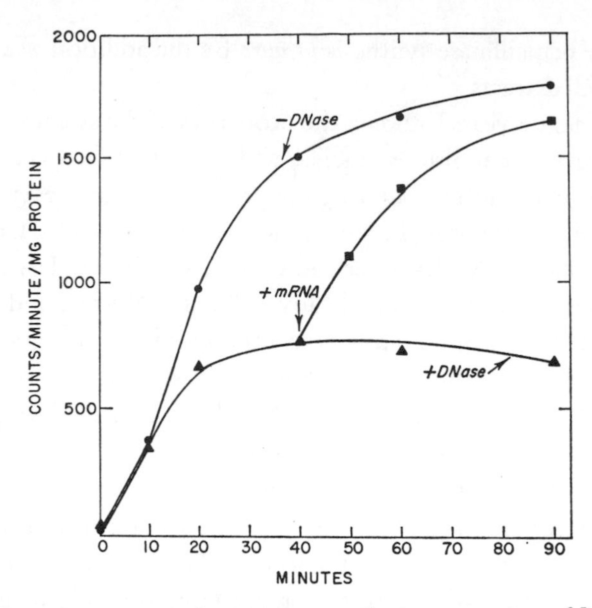

Fig.1. The effect of DNAase and mRNA upon the incorporation of [^{14}C]valine into protein in *E. coli* extracts. The symbols represent the following: ●, no addition; ▲, 10 μg DNAase added per ml of reaction; ■, 10 μg DNAase and 0.5 mg of an mRNA fraction added per ml of reaction.

In 1961, the role of tRNA was still controversial. Most investigators assumed that tRNA participated in the synthesis of protein, but direct proof that tRNA is required for this process was lacking. Lipmann and Nathans generously gave us a purified preparation of transfer enzymes and we found that Phe-tRNA is an *obligatory* intermediate in polyphenylalanine synthesis and that transfer enzymes and GTP are also required for the synthesis of this polypeptide[20].

Base Composition of Codons

The genetic code was deciphered in two experimental phases over a period of approximately six years. During the first phase, the base composition of codons and the general nature of the code were explored by directing cell-free protein synthesis with randomly-ordered RNA templates containing different combinations of bases. Such polymers were synthesized with the aid of polynucleotide phosphorylase that had been discovered by Grunberg-Manago, Ortiz and Ochoa[21].

A summary of data obtained by Ochoa and associates[23] and by ourselves[24] is shown in Table I. Only polynucleotides containing the minimum species of bases required to stimulate an amino acid into protein are shown. Poly-U, poly-C, and poly-A stimulate the incorporation into protein of phenylalanine, proline, and lysine, respectively. No template activity was detected with poly-G. In later studies Maxine Singer, Bill Jones, and I showed that poly-(U,G) preparations rich in G contain a high degree of secondary structure in solution and do not serve as templates for protein synthesis[22].

Poly-(U,C), poly-(C,G), and poly-(A,G) are templates for 2 additional amino acids per polynucleotide, whereas poly-(U,A), poly-(U,G), and poly-(C,A) are templates for 4 additional amino acids per polynucleotides. Each polynucleotide composed of 3 species of bases is a template for 10 or more amino acids.

Table I

Minimum species of bases required for mRNA codons

The specificity of randomly ordered polynucleotide templates in stimulating amino acid incorporation into protein into *E. coli* extracts is shown. Only the minimum species of bases necessary for template activity are shown, so many amino acids responding to polymers composed of two or more kinds of bases are omitted.

Polynucleotides	*Amino acids*			
U	PHE			
C	PRO			
A	LYS			
G	—			
UC	LEU	SER		
UA	LEU	TYR	ILE	ASN
UG	LEU	VAL	CYS	TRP
CA	HIS	THR	GLN	ASN
CG	ARG	ALA		
AG	ARG	GLU		
UAG	ASP	MET		
CAG	ASP	SER		

Randomly-ordered polynucleotides composed of 1, 2, 3, or 4 kinds of bases contain 1, 8, 27, and 64 kinds of triplets, respectively. The relative abundance of each kind of triplet can be calculated easily if the base-ratio of a randomly-ordered polynucleotide is known. One can derive both the *kinds* of bases that correspond to an amino acid and the *number* of bases of each kind, because the amount of each species of amino acid that is incorporated into protein due to the addition of a polynucleotide preparation and the base-ratio of the polynucleotide can be determined experimentally. In this manner the base compositions of approximately 50 codons were assigned to amino acids[23,24]. The results showed that multiple codons can correspond to the same amino acid; hence the code is highly degenerate. In most cases synonym codons differ by only one base; therefore, it was assumed that the non-variable bases occupy the same relative positions within each synonym word. By means of genetic studies, Crick, Barnett, Brenner and Watts-Tobin[25] showed that the code is a triplet code, and the biochemical studies confirmed this conclusion. Analysis of the coat protein of mutant strains of tobacco mosaic virus provided evidence that triplets in mRNA are translated in a non-overlapping fashion, because the replacement of one base by another in mRNA usually results in only one amino acid replacement in protein[26].

Base Sequence of Codons

Although base compositions of codons were determined, the *order* of bases within codons was not known. We investigated many potential methods for determining base sequence of codons. A clue to the solution of the problem stemmed from the important finding by Arlinghaus, Favelukes and Schweet[27] and by Kaji and Kaji[28] that Phe-tRNA attaches to ribosomes in response to poly-U prior to peptide bond formation. Perhaps trinucleotides or hexanucleotides of known base sequence would also stimulate binding of AA-tRNA to ribosomes. To test this possibility, Philip Leder and I devised a rapid method for separating ribosomal-bound AA-tRNA from unbound AA-tRNA that depends upon the selective retention of the ribosomal intermediate by discs of cellulose nitrate and then found that trinucleotides function as specific templates for AA-tRNA binding to ribosomes[29]. As shown in Table II, the trinucleotide, AAA, stimulates Lys-tRNA binding to ribosomes and is as active a template for Lys-tRNA as the tetra-or penta-nucleotide. The doublet, AA, has no effect upon Lys-tRNA binding; hence, 3 *sequential* bases in mRNA correspond to 1 amino acid in protein.

This experimental approach provided a relatively simple means of determining base sequence of codons. Fractionation of poly-(U,G) digests yielded 3 trinucleotides, GUU, UGU, and UUG, which were shown to be codons for valine, cysteine, and leucine, respectively[75].

Trinucleotide synthesis proved to be our major experimental problem. At

Table II

[14C]Lys-tRNA binding to ribosomes

The effect of oligo A preparations upon the binding of *E. coli* Lys-tRNA to ribosomes. The assay for AA-tRNA binding to ribosomes is described elsewhere[29]. Each 50 μl reaction contained 0.4 mμmoles of oligonucleotide as specified; 7.0 $\mu\mu$moles of [14C]-Lys-tRNA (0.150 A^{260} units); 1.1 A^{260} units of *E. ocli* ribosomes; 0.05 *M* Tris acetate, pH 7.2; 0.03 *M* magnesium acetate; and 0.05 *M* potassium chloride. In the absence of oligo A, 0.49 $\mu\mu$moles of [14C]Lys-tRNA bound to ribosomes; this amount has been subtracted each value shown above.

Addition	*[14C]Lys-tRNA bound due to oligo A ($\mu\mu$moles)*
ApA	0.01
ApApA	1.92
ApApApA	1.92
ApApApApA	1.92
ApApApApApA	2.71

that time only 20 to 30 of the 64 trinucleotides were described in the literature. Philip Leder began to explore possible enzymatic methods for synthesizing trinucleotides and we sought the advice of Leon Heppel and Maxine Singer. Throughout the course of our studies on the code Heppel and Singer advised us on problems pertaining to nucleic acids. Each visit to their laboratories became, for me, something akin to a pilgrimage to Delphi. The major difference was that the advice from either oracle was invariably clear and accurate.

Marianne Grunberg-Manago was visiting the National Institutes of Health for a few days, and both she and Maxine Singer joined Philip Leder in studying oligonucleotides synthesis catalyzed by primer-dependent polynucleotide phosphorylase (Fig. 2). Eventually, conditions for oligonucleotide synthesis were found by Leder, Singer and Brimacombe[30] and by Thatch and Doty[31].

$$\text{ApG} + \text{UDP} \underset{}{\overset{\text{Polynucleotide phosphorylase, } Mg^{2+}}{\rightleftharpoons}} \text{ApGpU} + \text{ApG(pU)}_n + \text{P} \quad (1)$$

$$\text{ApG} + \text{Uridine-2',3'-cyclic phosphate} \overset{\text{RNAase A}}{\rightleftharpoons} \text{UpApG} + \text{(UP)}_n\text{ApG} \quad (2)$$

Fig. 2. Trinucleotide synthesis catalyzed by polynucleotide phosphorylase and by pancreatic RNAase A are shown in reactions 1 and 2, respectively.

Heppel suggested another synthetic method that he, Whitfield and Markham[32] had discovered that depends upon the ability of pancreatic RNAase A to catalyze the synthesis of oligonucleotides from pyrimidine 2′,3′-cyclic phosphates and mono- or oligo-nucleotide accepter moieties. Merton Bernfield studied various aspects of the reaction and synthesized many trinucleotides with this enzyme[33-35].

In a remarkable series of studies over many years, Khorana and his associates established chemical methods for oligo- and poly-nucleotide synthesis[38]. They were able to synthesize the 64 trinucleotides by chemical methods whereas enzymatic methods were used in our laboratory.

Codon-base sequences were established both by stimulating the binding of AA-tRNA to ribosomes with trinucleotides of known sequence[36,37] and by stimulating *in vitro* protein synthesis with polyribonucleotides containing repeating doublet, triplet, or tetramers of known sequence as described by Khorana in the accompanying article.

THE GENETIC CODE

Codon	△	○	Amino acid	Codon	△	○	Amino acid	Codon	△	○	Amino acid	Codon	△	○	Amino acid
UUU	△	○	PHE	UCU	△	○	SER	UAU	△	○	TYR	UGU	△	○	CYS
UUC	△	○		UCC	△	○		UAC	△	○		UGC	△		
UUA		○	LEU	UCA	△			UAA	△		TERM	UGA	△		TERM
UUG	△	○		UCG	△	○		UAG	△			UGG	△	○	TRP
CUU	△	○	LEU	CCU	△	○	PRO	CAU	△	○	HIS	CGU	△	○	ARG
CUC	△	○		CCC	△	○		CAC	△	○		CGC	△	○	
CUA	△			CCA	△	○		CAA	△	○	GLN	CGA	△	○	
CUG	△			CCG	△	○		CAG	△			CGG	△		
AUU	△	○	ILE	ACU	△	○	THR	AAU	△	○	ASN	AGU	△		SER
AUC	△	○		ACC	△	○		AAC	△	○		AGC	△	○	
AUA	△	○		ACA	△	○		AAA	△	○	LYS	AGA	△	○	ARG
AUG	△	○	MET	ACG	△			AAG	△			AGG	△		
GUU	△	○	VAL	GCU	△	○	ALA	GAU	△	○	ASP	GGU	△	○	GLY
GUC	△			GCC	△	○		GAC	△	○		GGC	△	○	
GUA	△	○		GCA	△	○		GAA	△	○	GLU	GGA	△	○	
GUG	△			GCG	△			GAG	△			GGG	△		

△ BASE SEQUENCE. [AA-tRNA-TRINUCLEOTIDE-RIBOSOME] COMPLEX
○ BASE COMPOSITION. RNA TEMPLATES FOR PROTEIN SYNTHESIS

Fig. 3. The symbols represent the following: △, base sequences of mRNA codons determined by stimulating binding of *E. coli* AA-tRNA to *E. coli* ribosomes with trinucleotide templates; ○, base compositions of mRNA codons determined by stimulating the incorporation of amino acids into protein with randomly-ordered polynucleotide templates in extracts of *E. coli*. TERM corresponds to terminator codons (terminator- and initiator-codons are shown in Table III).

The genetic code is shown in Fig.3. Most triplets correspond to amino acids. Codons for the same amino acid usually differ only in the base occupying the third position of the triplet. Therefore, synonym codons are *systematically* related to one another. Five patterns of codon degeneracy are found, each pattern determined by the kinds of bases that occupy the third positions of synonym triplets. The third base of each degenerate triplet is shown below; the dashes correspond to the first and second bases of each triplet.

— — G	(1)
— — U	(2)
— — C	
— — A	(3)
— — G	
— — U	(4)
— — C	
— — A	
— — U	(5)
— — C	
— — A	
— — G	

The last pattern (discussed in a later section) corresponds to the sum of two patterns.

Results with trinucleotides confirm 43 of the 50 base compositions of codons that were estimated previously on the basis of studies with randomly-ordered polynucleotides and the cell-free protein synthesizing system.

From 1 to 6 codons may correspond to one amino acid, depending upon the amino acid in question. One consequence of systematic degeneracy is that the replacement of one base by another in DNA often does not result in the replacement of one amino acid by another in protein. Many mutations, therefore, are silent ones. The code appears to be arranged so that effects of base replacements in DNA, or erroneous translations of bases in mRNA, often are minimized. Amino acid replacements in protein that occur due to the replacement of one base by another in nucleic acid can be read in Fig. 3 by moving horizontally or vertically from the amino acid in question, but not diagonally

Punctuation

Punctuation of transcription and translation is illustrated schematically in Figs. 4 and 5. RNA polymerase attaches to specific site(s) on DNA and thereby selects the strand of DNA to be transcribed, the direction of transcription, and

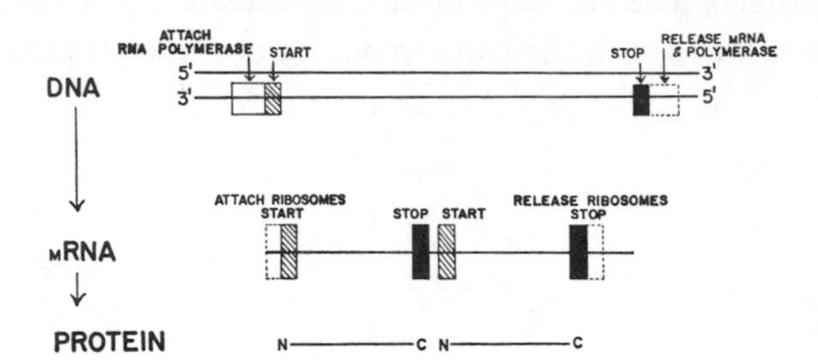

Fig. 4. The punctuation of transcription and translation is illustrated diagramatically. Ribosomal subunits attach to mRNA near the 5′-terminus of the mRNA and are released near the 3′-terminus of the mRNA. Speculations are indicated by the dotted lines. N and C represent the N- and C-terminal amino acid residues of protein, respectively.

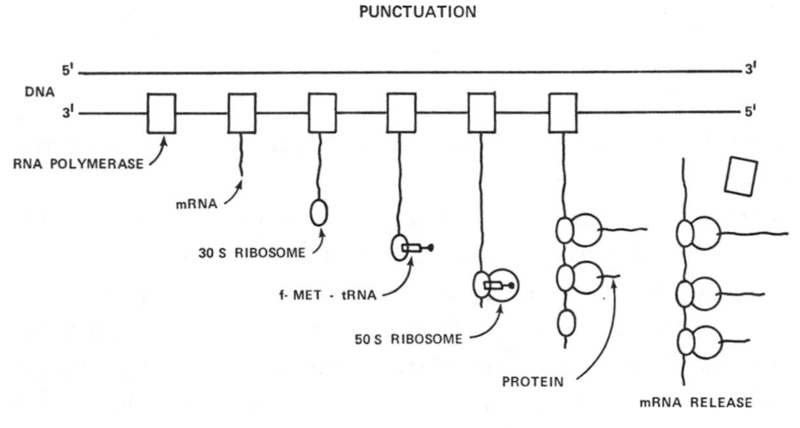

Fig. 5. Diagrammatic illustration of early steps of protein synthesis.

the first base to be transcribed. Many questions remain to be answered about the initiation of RNA synthesis.

The direction of mRNA synthesis is opposite to that of the DNA strand being read. The first base to be incorporated into the nascent mRNA chain is the 5′-terminus of the mRNA, the last base is the 3′-terminus. Similarly, the RNA template is translated during protein synthesis starting at or near the 5′-terminus of the RNA and proceeding three bases at a time, sequentially, toward the 3′-terminus of the RNA. Therefore, mRNA is synthesized and then translated with the same polarity. The first amino acid corresponds to the N-terminus of the peptide chain; the C-terminal amino acid is the last amino acid incorporated.

Protein synthesis is initiated in *E. coli* by a unique species of tRNA, *N*-formyl-$tRNA_f$, discovered by Marcker and Sanger[39]. A 30S ribosomal particle attaches to the nascent chain of mRNA near its 5'-terminus before the mRNA detaches from the DNA template. At least three non-dializable factors and GTP are required for the initiation of protein synthesis. The reactions have not been clarified fully; however, the available evidence suggests that one factor (F3) participates in the attachment of a 30S ribosomal subunit to a nascent chain of mRNA and that other factors (F1 and F2) and GTP are required for the binding of *N*-formyl-met-tRNA to the 30S ribosome-mRNA complex in response to an initiator codon (AUG or GUG[62,63]). The 50S ribosomal subunit then attaches to the 30S ribosomal complex before the next codon is recognized by AA-tRNA. *N*-Formyl-met-tRNA thus selects the first codon to be translated and phases the translation of subsequent codons.

Another species of tRNA from *E. coli*, Met-$tRNA_m$, does not accept formyl moieties, responds only to AUG, and corresponds to methionine at internal positions in protein.

The pattern of degeneracy observed with *N*-formyl-met-tRNA differs from the patterns observed with other species of AA-tRNA because initiator codons have alternate first bases rather than alternate third bases.

Each triplet can occur in three structural forms: as 5'-terminal-, 3'-terminal-, or internal-codons. Substituents attached to ribose hydroxyl groups of codons can influence codon template properties profoundly. The relation between codon structure and template activity was investigated by my colleague, Fritz Rottman[83] (Fig. 6). Relative template activities of oligo-U preparations, at limiting oligonucleotide concentrations, are as follows: p-5'-UpUpU, > UpUpU, > CH_3O-p-5'-UpUpU, > UpUpU-3'-p, > UpUpU-3'-p-OCH_3 > UpUpU-2',3'-cyclic phosphate. Trimers with (2',5') phosphodiester linkages, (2',5')-UpUpU and (2',5')-ApApA, do not serve as templates for Phe- or Lys-tRNA, respectively. The relative template efficiencies of oligo-A preparations are as follows: p-5'-ApApA > ApApA > ApApA-3'-p > ApApA-2'-p. Ikehara and Ohtsuka[70] showed that N^6-DiMeApApA does not stimulate Lys-tRNA binding to ribosomes; whereas the tubercidin (7-deazaadenosine) analog, TupApA, serves as a template for Lys-tRNA.

RNA polymerase catalyzes the synthesis of mRNA with 5'-terminal triphosphate. Also, many enzymes have been described that catalyze the transfer

Relative template activity of substituted oligonucleotides

Oligonucleotide	Template activity† relative to UpUpU
p-5′-UpUpU	510
UpUpU	100
CH_3O-pUpUpU	74
UpUpU-3′-p	48
UpUpUp-OCH_3	18
UpUpU-2′,3′-cyclic p	17
(2′-5′)-UpUpU	0
Oligodeoxy T‡	0

Oligonucleotide	Template activity† relative to ApApA
p-5′-ApApA	181
ApApA	100
ApApA-3′-p	57
ApApA-2′-p	15
(2′-5′)-ApApA	0
Oligodeoxy A‡	0

Fig. 6. Relative template activities† of substituted oligonucleotides are approximations obtained by comparing the amount of AA-tRNA bound to ribosomes in the presence of limiting concentrations of oligonucleotides compared to either UpUpU for [^{14}C]Phe-tRNA or ApApA, for [^{14}C]Lys-tRNA (each designated at 100%). The data are from Rottman and Nirenberg[83].

of molecules to or from hydroxyl groups of nucleic acids. It is possible, therefore, that certain modifications of ribose or deoxyribose hydroxyl groups of nucleic acids provide a means of regulating the rate of transcription or translation.

Since mRNA and protein synthesis are transiently coupled *via* the formation of a (DNA-mRNA-ribosome) intermediate, it is possible that the synthesis of certain species of mRNA may be regulated selectively by events at the level of the ribosome[76-79].

Termination

The first evidence for «nonsense» codons was reported in 1962 by Benzer and Champe[40]. They obtained a mutant of bacteriophage T4 with a deletion spanning part of the A gene and part of the contiguous B gene of the rII region. Presumably, the remaining segments of gene A and B are joined and thus form one gene. Nevertheless, a functional B gene product was found. However, a second mutation that mapped in the A gene resulted in the loss of a functional B gene product. These results suggested that a «sense» codon is converted by mutation to a «nonsense» codon that cannot be read; hence subsequent regions of the gene also are not read. Sarabhai, Stretton, Brenner and Bolle[41] then showed that «nonsense» mutations at various sites within the gene for the head protein of bacteriophage T4 determine the chain length of the corresponding polypeptide. These dramatic results showed that «nonsense» codons correspond to the termination of protein synthesis. Additional evidence obtained by Brenner[42,43] and by Garen[44,45] and their colleagues showed that 3 codons, UAA, UAG and UGA, correspond to the termination of protein synthesis (also *cf.* the recent review of Garen[46]).

The mechanism of peptide-chain termination was investigated by stimulating cell-free protein synthesis with randomly-ordered polynucleotides[47-49], oligonucleotides[50] and polynucleotides[51,52] of known sequence, and viral RNA[53,54]. Capecchi showed that the release of peptides from ribosomes is dependent upon both a release factor and a terminator codon [53]. The codons, UAA, UAG, and UGA, do not stimulate binding of AA-tRNA to ribosomes (although mutant strains of bacteria have been found that contain species of AA-tRNA that respond to terminator codons).

Recently my colleagues, Caskey, Tompkins, and Scolnick[55,56] found that the process of termination can be studied with trinucleotides. Incubation of terminator trinucleotides and the release factor with the [*N*-formyl-Met-tRNA-AUG-ribosomal] complex results in the release of free *N*-formyl-methionine from the ribosomal intermediate. The release factor of *E. coli* then was separated into two components that correspond to different sets of codons: R1, active with UAA or UAG; and R2, active with UAA or UGA. It is clear, therefore, that terminator codons are recognized by specific molecules. The simplest hypothesis is that R1 and R2 interact with terminator codons on ribosomes; however, the codon recognition step and the mechanism of termination have not been clarified thus far.

As shown in Table III, the pattern of codon degeneracy found with R1 (UAA and UAG) resembles that found with some species of AA-tRNA; *i.e.*, A is equivalent to G at the 3rd position of codons. However, the degeneracy pattern found with R2 (UAA and UGA) is different from that of AA-tRNA because A and G are equivalent at the 2nd but not at the 3rd position of triplets.

Table III

Codons corresponding to the initiation or termination of protein synthesis in *E. coli* are shown. Release factors 1 and 2 are required for termination with the codons indicated, but it is not known whether they interact directly with terminator codons.

Initiation (*N*-formyl-Met-tRNA)	AUG or GUG
Termination (release factor-1)	UAA or UAG
Termination (release factor-2)	UAA or UGA

Redundancy

By 1962, studies with randomly-ordered RNA templates had shown that the code is extensively degenerate and that synonym codons often differ by only one base. It was assumed that the nonvariable bases occupy the same relative positions within synonym triplets. A systematic form of degeneracy seemed probable because often U was equivalent to C, and A was equivalent to G. Attempts were made to deduce the rules governing degeneracy from the available data on base compositions of codons and amino acid replacements in protein[57,58].

Two species of Leu-tRNA were found that respond to different mRNA codons[59]. However, further work was required to determine whether one species of tRNA responds only to one codon, or to 2 or more codons.

As the order of bases within codons was established, it became abundantly clear that synonym codons are systematically related to one another. As discussed earlier, alternate bases occupy the third position of synonym triplets. Since only a few kinds of degeneracy patterns were found for the 20 amino acids, it seemed likely that correspondingly few codon recognition mechanisms were operative[60].

Evidence that one molecule of AA-tRNA can respond to two kinds of codons was provided by the demonstration that most molecules of Phe-tRNA respond both to UUU and to UUC[61]. Further evidence was obtained by determining the specificity of purified tRNA fractions for trinucleotide codons. The results showed that a purified species of tRNA responds either to 1, 2 or 3 codons[62-67,37]. A summary of our studies with purified fractions of tRNA from *E. coli* is shown in Table IV. Four, possibly 5, kinds of synonym codon sets were found, as shown below. The third base of each synonym triplet is shown; the dashes represent the first and second bases of each triplet.

Table IV

Codons recognized by species of *E. coli* AA-tRNA

Aminoacyl-tRNA preparations from *E. coli* were fractionated by reverse phase column chromatography and their response to trinucleotide templates was determined[67]. Additional results have been obtained by Khorana and his colleagues[64–66]. A dash, —, represents Leu-tRNA fractions that do not respond to trinucleotide codons. Numerals within parentheses indicate the number of redundant peaks of AA-tRNA found.

Amino acid	*AA-tRNA*		*Species*		
	1	*2*	*3*	*4*	*5*
			mRNA codons		
LEU	CUU CUC	CUA CUG	CUG	UUG	—(2)
SER	UCU UCC	UCA UCG(2)	UCG	AGU AGC	
ARG	CGU CGC CGA	AGA AGG	CGG		
ALA	GCU GCC	GCA GCG			
VAL	GUU GUC	GUA GUG			
TRP	UGG				
MET	AUG				
ILE	AUU AUC(2)				
PHE	UUU UUC(2)				
TYR	UAU UAC(2)				
CYS	UGU UGC(3)				
HIS	CAU CAC				
LYS	AAA AAG(3)				
GLU	GAA GAG				

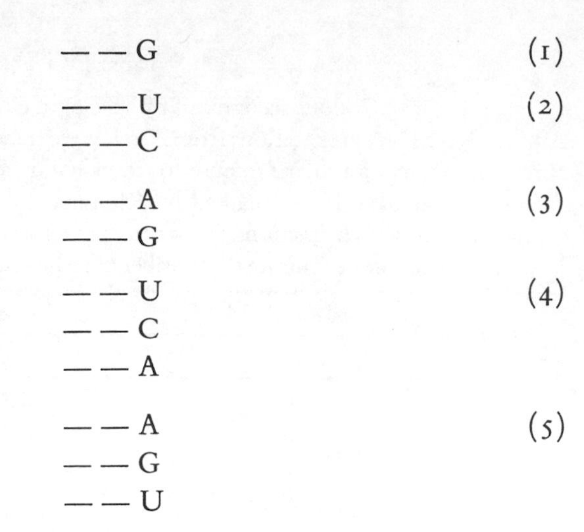

The fifth pattern of degeneracy was found with *E. coli* Ser-tRNA (possibly also with Val-tRNA) but has not been found thus far with AA-tRNA from other organisms.

The number of words, or sets of words, in the code corresponds to the number of tRNA anticodons rather than the number of amino acids. Since multiple species of tRNA for the same amino acid often respond to different sets of codons, the tRNA code consists of more word-sets than the amino acid code.

Redundant fractions of AA-tRNA for the same amino acid were found that differ in chromatographic mobility but respond similarly to codons. Such AA-tRNA fractions may be products of the same gene that have been altered in different ways by enzymes *in vivo* or perhaps have been altered *in vitro* during the fractionation procedure. Alternatively, redundant AA-tRNA fractions may be products of different genes.

Crick suggested that codon degeneracy is due to the formation of alternate base pairs between a base in a tRNA anticodon and alternate bases occupying the third positions of synonym mRNA anticodons[68]. Presumably, the first and second bases of mRNA codons form antiparallel, Watson–Crick base-pairs with corresponding bases in the tRNA anticodon. Alternate base-pairs proposed by Crick are shown in Table V; U in the tRNA anticodon pairs alternately with A or G occupying the third position of synonym mRNA codons; C pairs with G; G pairs with C or U; and I pairs with U, C, or A.

The elucidation of the base sequence of Ala-tRNA from yeast by Holley *et al.*[69] provided an opportunity to relate the base sequence of the tRNA anti-

Table V

Alternate base-pairing

Alternate base pairing between a base in a tRNA anticodon, shown in the left hand column, and the base(s) in the third position of synonym mRNA codons. Relationships are antiparallel «wobble» hydrogen bonds suggested by Crick[68].

tRNA Anticodon	*mRNA Codons (3rd base)*
U	A G
C	G
G	C U
I	U C A

codon with the mRNA codons. Holley generously gave us a preparation of tRNA^{Ala} of known sequence and of high purity and Philip Leder and I and, concurrently, Söll *et al.*[65] found that the Ala-tRNA responds to GCU, GCC, and GCA. The results confirmed Holley's prediction that the sequence, IGC, serves as the tRNA^{Ala} anticodon. Inosine in the anticodon, therefore, pairs alternately with U, C, or A, in the third position of the mRNA codons. The base sequences of other species of tRNA have been defined and in every case, codon-anticodon relationships are in accord with wobble base-pairing.

Universality

The results of many studies suggest that different forms of life use essentially the same genetic language. However, the fidelity of codon translation can change quite dramatically due to alterations that affect components required for protein synthesis. Thus cells sometimes differ in the specificity of codon translation.

Richard Marshall, Thomas Caskey, and I studied the responses of bacterial, amphibian, and mammalian AA-tRNA (*E. coli*, *Xenopus laevis*, and guinea pig liver, respectively) to trinucleotide codons. Almost identical translations of nucleotide sequences to amino acids were found with bacterial, amphibian, and mammalian AA-tRNA[71]. However, *E. coli* AA-tRNA preparations do not respond appreciably to certain codons that are active templates with metazoan AA-tRNA.

Our interest in species-dependent variation in codon recognition was stimulated by the possibility that such phenomena might serve as regulators of cell differentiation. Therefore, AA-tRNA preparations were fractionated by column chromatography and responses of tRNA fractions to trinucleotide codons were determined[67]. A summary of our results is shown in Fig. 7.

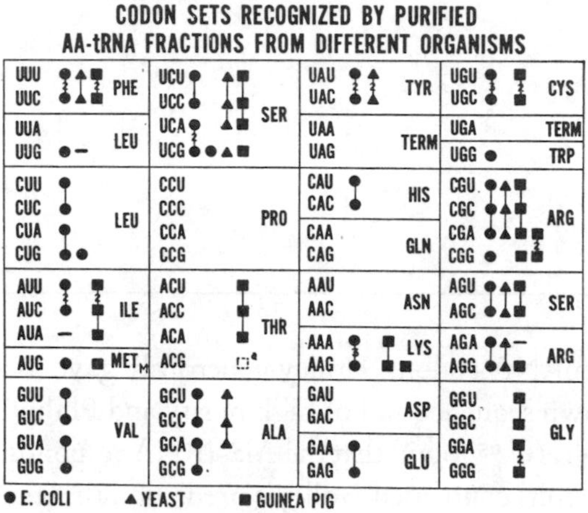

CODON SETS RECOGNIZED BY PURIFIED AA-tRNA FRACTIONS FROM DIFFERENT ORGANISMS

UUU UUC — PHE	UCU UCC UCA UCG — SER	UAU UAC — TYR	UGU UGC — CYS
UUA UUG — LEU		UAA UAG — TERM	UGA — TERM; UGG — TRP
CUU CUC CUA CUG — LEU	CCU CCC CCA CCG — PRO	CAU CAC — HIS; CAA CAG — GLN	CGU CGC CGA CGG — ARG
AUU AUC AUA — ILE; AUG — MET, MET$_M$	ACU ACC ACA ACG — THR	AAU AAC — ASN; AAA AAG — LYS	AGU AGC — SER; AGA AGG — ARG
GUU GUC GUA GUG — VAL	GCU GCC GCA GCG — ALA	GAU GAC — ASP; GAA GAG — GLU	GGU GGC GGA GGG — GLY

● E. COLI ▲ YEAST ■ GUINEA PIG

Fig. 7. A summary of the results obtained in this laboratory with purified aminoacyl-tRNA fractions[67] is shown above[67]. Additional results have been reported by Söll *et al.*[64–66]. Synonym codon sets were determined by stimulating the binding of purified fractions of *E. coli*, yeast, or guinea pig liver aminoacyl-tRNA fractions to *E. coli* ribosomes with trinucleotide codons. The joined symbols adjacent to the codons represent synonym codons recognized by one purified aminoacyl-tRNA fraction from ●, *E. coli*; ▲, yeast; or ■, guinea pig liver. Numerals between symbols represents the number of redundant peaks of aminoacyl-tRNA (aminoacyl-tRNA fractions with the same specificity for codons).

Additional information has been reported by Söll *et al.*[64–66]. Many «universal» species of AA-tRNA were found. However, 7 species of mammalian tRNA were found that were not detected with *E. coli* tRNA and, conversely, 5 species of *E. coli* tRNA were not detected with mammalian tRNA preparations.

Large differences were observed in the concentration of tRNA corresponding to certain codons. Some organisms apparently do not contain AA-tRNA for certain codons. For example, mammalian Ile-tRNA responds well to AUU, AUC, and AUA; whereas *E. coli* Ile-tRNA responds only to AUU and AUC (AUA-deficient). Also, a species of mammalian Arg-tRNA was found responding to ACG but no Arg-tRNA was found corresponding to AGA (AGA-deficient).

Although some variation in codon translation clearly does occur, the remarkable similarity in codon-base sequences recognized by bacterial, amphibian, and mammalian AA-tRNA suggest that most, perhaps all, forms of life on this planet use essentially the same genetic language, and that the language is translated according to universal rules.

Fossil records of microorganisms estimated to be $3.1 \cdot 10^9$ years old have been reported[72]. The first vertebrates appeared approximately $0.5 \cdot 10^9$ years ago; amphibians and mammals appeared 350 and 180 million years ago, respectively. Thus the genetic code probably originated more than $0.6 \cdot 10^9$ years ago. Hinegardner and Engelberg[73] and Sonneborn[74] suggested that the code was frozen after organisms as complex as bacteria had evolved because major alterations in the code would affect the amino acid sequence of most proteins synthesized by the cell and probably would be lethal.

Reliability of Translation

When one considers the number of species of molecules that are required for the synthesis of a single molecule of protein and the fact that the cellular machinery that participates in the assembly process is complex, heterogeneous, and not reliable, the problem of synthesizing protein with precision seems formidable. To synthesize one molecule of protein composed of 400 amino acid residues, 400 AA-tRNA molecules must be selected in the proper sequence. For the synthesis of the corresponding molecule of mRNA, at least 1206 molecules of ribonucleoside triphosphate must be selected in sequence.

One must distinguish between serial operations, that is, successive steps, and parallel, *i.e.*, simultaneous steps. Usually the overall precision of a multistep process deteriorates rapidly as the number of serial steps increases. Two or more serial steps are required for the synthesis of each molecule of AA-tRNA because an AA-tRNA ligase first catalyzes the synthesis of an aminoacyl-adenylate and then catalyzes the transfer of the aminoacyl moiety to an appropriate species of tRNA, yielding AA-tRNA. Many molecules of AA-tRNA can be synthesized in parallel. Although hundreds of sequential selections are required for the synthesis of one molecule of protein, the process of protein synthesis is organized within the cell so that each amino acid usually is selected *independently* of other amino acids. Thus, one translational error usually does not influence the accuracy of other codon translations, and errors usually are not cumulative. However, if an error in translation alters the phase of reading or results in premature termination, subsequent selections obviously will be affected.

Baldwin and Berg[80] have shown that Ile-tRNA ligase from *E. coli* catalyzes the synthesis of AA-tRNA only if both amino acid and tRNA species are selected correctly. If an erroneous aminoacyl-adenylate is synthesized, the enzyme corrects the error by catalyzing the hydrolysis of the aminoacyl-adenylate.

In 1960 Yanofsky and St. Lawrence[81] suggested that certain mutations might result in the production of structurally modified tRNA or AA-tRNA synthetases with altered specificity for amino acid incorporation into protein. Much information is now available concerning suppressor mutations that affect components required for protein synthesis[82]. In addition, factors that influence the precision of protein synthesis have been studied extensively with synthetic polynucleotide templates and *in vitro* protein-synthesizing systems and by determining the binding of AA-tRNA to ribosomes in response to tri- or poly-nucleotide templates. The results show that the precision of codon recognition is affected by the temperature of incubation, pH, concentration of various species of tRNA, concentration of Mg^{2+}, aliphatic amines such as putrescine, spermidine, spermine, streptomycin and related antibiotics, and other compounds.

Most codons probably are translated with relatively little error (0.1–0.01% error or less); however, the level of error can be as high as 50% with certain codons. Hence, the precision of translation can vary from one codon to another at least 5000-fold.

Most errors in codon translation do not result in random amino acid replacements in protein because two out of three bases per codon usually are recognized correctly (*i.e.*, when the precision of translation deteriorates, a codon such as UUU may be translated 80% of the time as phenylalanine, 15% as isoleucine, and 5% as leucine). One codon then is translated by relatively few species of AA-tRNA.

One can only speculate about the biological significance of a flexible, easily modifiable codon-translation apparatus. One extremely interesting possibility is that the codon-recognition apparatus is modified in an orderly, predictable way at certain times during cell growth and differentiation and that such modifications selectively regulate the rate of synthesis of certain species of protein.

Rate of Translation

The *E. coli* chromosome is composed of approximately $3 \cdot 10^6$ base pairs; sufficient information is present to determine the sequence of $1 \cdot 10^6$ amino acids in protein (equivalent to approximately 2500–3000 species of protein or less since duplicate copies of the same gene may be present).

Approximately 20–80 mRNA triplets are translated per second per ribosome at 37^0. One cell may contain 1000–15 000 ribosomes per chromosome, depending upon the rate of growth; therefore, proteins are synthesized at many sites simultaneously. Parallel operations greatly enhance the efficiency of the cell in synthesizing protein.

Concluding Remarks

The genetic code is now essentially deciphered. I have been fortunate in having the collaboration of many enthusiastic associates during the course of our studies. To do justice to the years of effort and the important contributions made by associates and numerous colleagues throughout the world is virtually impossible in the available time. One has only to refer to the comprehensive reviews in the *Cold Spring Harbor Symposium on Quantitative Biology* of 1963 and 1966 to view the breadth of the field and the extent of information now available. Additional information can be found in the recent books by Woese[9] and by Jukes[84].

1. O.T.Avery, C.M.MacLeod and M.McCarty, *J.Exptl.Med.*, 79 (1944) 137.
2. G.W.Beadle and E.L.Tatum, *Proc.Natl.Acad.Sci.* (*U.S.*), 27 (1941) 499.
3. J.Brachet, *Arch.Biol.* (*Liège*), 53 (1942) 207.
4. T.Caspersson, *Naturwiss.*, 29 (1941) 33.
5. P.C.Caldwell and C.Hinshelwood, *J.Chem.Soc.*, Pt.4 (1950) 3156.
6. A.L.Dounce, *Enzymologia*, 15 (1952) 251.

7. G.Gamow, *Nature*, 173 (1954) 318.
8. J.D.Watson and F.H.C.Crick, *Nature*, 171 (1953) 737.
9. C.R.Woese, *The Genetic Code*, Chapter 2, Harper and Row, New York, 1967.
10. A.D.Hershey, J.Dixon and M.Chase, *J.Gen.Physiol.*, 36 (1953) 777.
11. E.Volkin and L.Astrachan, *Virology*, 2 (1956) 149.
12. M.R.Pollock, *Proc.Roy.Soc. (London), Ser.B*, 148 (1958) 340.
13. M.R.Pollock, in I.C.Gunsalus and R.Y.Stanier (Eds.), *The Bacteria*, Vol.4, 1962, p.121.
14. M.R.Lamborg and P.C.Zamecnik, *Biochim.Biophys.Acta*, 42 (1960) 206.
15. A.Tissieres, D.Schlessinger and F.Gros, *Proc.Natl.Acad.Sci. (U.S.)*, 46 (1960) 1450.
16. T.Kameyama and G.D.Novelli, *Biochem.Biophys.Res.Commun.*, 2 (1960) 393.
17. B.Nisman and H.Fukuhara, *Compt.Rend.*, 249 (1959) 2240.
18. M.W.Nirenberg and J.H.Matthaei, *Proc.Natl.Acad.Sci. (U.S.)*, 47 (1961) 1588.
19. J.H.Matthaei and M.W.Nirenberg, *Proc.Natl.Acad.Sci. (U.S.)*, 47 (1961) 1580.
20. M.W.Nirenberg, J.H.Matthaei and O.W.Jones, *Proc.Natl.Acad.Sci. (U.S.)*, 48 (1962) 104.
21. M.Grunberg-Manago, P.J.Ortiz and S.Ochoa, *Biochim.Biophys.Acta*, 20 (1956) 269.
22. M.Singer, O.W.Jones and M.W.Nirenberg, *Proc.Natl.Acad.Sci. (U.S.)*, 49 (1963) 392.
23. J.F.Speyer, P.Lengyel, C.Basilio, A.J.Wahba, R.S.Gardner and S.Ochoa, *Cold Spring Harbor Symp.Quant.Biol.*, 28 (1963) 559.
24. M.W.Nirenberg, O.W.Jones, P.Leder, B.F.C.Clark, W.S.Sly and S.Pestka, *Cold Spring Harbor Symp.Quant.Biol.*, 28 (1963) 549.
25. F.H.C.Crick, L.Barnett, S.Brenner and R.J.Watts-Tobin, *Nature*, 192 (1961) 1227.
26. H.G.Wittmann and B.Wittmann-Leibold, *Cold Spring Harbor Symp.Quant.Biol.*, 28 (1963) 589.
27. R.Arlinghaus, G.Favelukes and R.Schweet, *Biochem.Biophys.Res.Commun.*, 11 (1963) 92.
28. A.Kaji and H.Kaji, *Biochem.Biophys.Res.Commun.*, 13 (1963) 186.
29. M.W.Nirenberg and P.Leder, *Science*, 145 (1964) 1399.
30. P.Leder, M.F.Singer and R.Brimacombe, *Biochemistry*, 4 (1965) 1561.
31. R.E.Thach and P.Doty, *Science*, 147 (1965) 1310.
32. L.A.Heppel, P.R.Whitfield and R.Markham, *Biochem.J.*, 60 (1955) 8.
33. M.Bernfield, *J.Biol.Chem.*, 240 (1965) 4753.
34. M.Bernfield, *J.Biol.Chem.*, 241 (1966) 2014.
35. M.Bernfield and F.M.Rottman, *J.Biol.Chem.*, 242 (1967) 4134.
36. D.Söll, J.Cherayil, D.S.Jones, R.D.Faulkner, H.Hampel, R.M.Bock and H.G. Khorana, *Cold Spring Harbor Symp.Quant.Biol.*, 31 (1966) 51.
37. M.W.Nirenberg, T.Caskey, R.Marshall, R.Brimacombe, D.Kellog, B.Doctor, D.Hatfield, J.Levin, F.Rottman, S.Pestka, M.Wilcox and F.Anderson, *Cold Spring Harbor Symp.Quant.Biol.*, 31 (1966) 11; B.P.Doctor, J.E.Loebel and D.A.Kellogg, *ibid.*, 31 (1966) 543; D.Hatfield, *ibid.*, 31 (1966) 619; S.Pestka and M.Nirenberg, *ibid.*, 31 (1966) 641.

38. H.G.Khorana, H.Buchi, H.Ghosh, N.Gupta, T.M.Jacob, H.Kössel, R.Morgan, S.A.Narang, E.Ohtsuka and R.O.Wells, *Cold Spring Harbor Symp.Quant.Biol.*, 31 (1966) 39.
39. K.Marcker and F.Sanger, *J.Mol.Biol.*, 8 (1964) 835.
40. S.Benzer and S.P.Champe, *Proc.Natl.Acad.Sci.* (*U.S.*), 48 (1962) 1114.
41. A.S.Sarabhai, A.O.W.Stretton, S.Brenner and A.Bolle, *Nature*, 201 (1964) 13.
42. S.Brenner, A.O.W.Stretton and S.Kaplan, *Nature*, 206 (1965) 994.
43. S.Brenner, L.Barnet, E.R.Katz and F.H.C.Crick, *Nature*, 213 (1967) 449.
44. M.G.Weigert and A.Garen, *Nature*, 206 (1965) 992.
45. M.G.Weigert, E.Lanka and A.Garen, *J.Mol.Biol.*, 23 (1967) 391.
46. A.Garen, *Science*, 160 (1968) 149.
47. M.Takanami and Y.Yan, *Proc.Natl.Acad.Sci.* (*U.S.*), 54 (1965) 1450.
48. M.S.Bretscher, H.M.Goodman, J.R.Menninger and J.D. Smith, *J.Mol. Biol.*, 14 (1965) 634.
49. M.C.Ganoza and T.Nakamoto, *Proc. Natl. Acad. Sci. (U.S.)*, 55 (1966) 162.
50. J.A.Last, W.M.Stanley Jr., M.Salas, M.B.Hille, A.J.Wahba and S.Ochoa, *Proc. Natl. Acad. Sci.* (*U.S.*), 57 (1967) 1062.
51. A.R.Morgan, R.D.Wells and H.G.Khorana, *Proc.Natl.Acad.Sci.* (*U.S.*), 56 (1966) 1899.
52. H.Kössel, *Biochim.Biophys.Acta*, 157 (1968) 91.
53. M.R.Capecchi, *Proc.Natl.Acad.Sci.* (*U.S.*), 58 (1967) 1144.
54. M.S.Bretscher, *J.Mol.Biol.*, 34 (1968) 131.
55. C.T.Caskey, R.Tompkins, E.Scolnick, T.Caryk and M.Nirenberg, *Science*, 162 (1968) 135.
56. E.Scolnick, R.Tompkins, T.Caskey and M.Nirenberg, *Proc.Natl.Acad.Sci.* (*U.S.*), 61, (1968) 768.
57. C.Woese, *Nature*, 194 (1962) 1114.
58. R.Eck, *Science*, 140 (1963) 477.
59. B.Weisblum, S.Benzer and R.W.Holley, *Proc.Natl.Acad.Sci.* (*U.S.*), 48 (1962) 1449.
60. M.W.Nirenberg, P.Leder, M.Bernfield, R.Brimacombe, J.Trupin, F.Rottman and C.O'Neal, *Proc.Natl.Acad.Sci.* (*U.S.*), 53 (1965) 1161; J.Trupin, F.Rottman, R.Brimacombe, P.Leder, M.Bernfield and M.Nirenberg, *ibid.*, 53 (1965) 807; R. Brimacombe, J.Trupin, M.Nirenberg, P.Leder, M.Bernfield and T.Jaouni, *ibid.*, 54 (1965) 954.
61. M.R.Bernfield and M.W.Nirenberg, *Science*, 147 (1965) 479.
62. B.F.C.Clark and K.A.Marcker, *J.Mol.Biol.*, 17 (1966) 394.
63. D.A.Kellogg, B.P.Doctor, J.E.Loebel and M.W.Nirenberg, *Proc.Natl.Acad.Sci.* (*U.S.*), 55 (1966) 912.
64. D.Söll, D.S.Jones, E.Ohtsuka, R.D.Faulkner, R.Lohrmann, H.Hayatsu, H.G. Khorana, J.D.Cherayil, A.Hampel and R.M.Bock, *J.Mol.Biol.*, 19 (1966) 556.
65. D.Söll, J.D.Cherayil and R.M.Bock, *J.Mol.Biol.*, 29 (1967) 97; D.Söll, E.Ohtsuka, D.S.Jones, R.Lohrmann, H.Hayatsu, S.Nishimura and H.G.Khorana, *Proc.Natl. Acad.Sci.* (*U.S.*), 54 (1965) 1378.
66. D.Söll and U.L.RajBhandary, *J.Mol.Biol.*, 29 (1967) 113.

67. C. T. Caskey, A. Beaudet and M. Nirenberg, *J. Mol. Biol.*, 37 (1968) 99.
68. F. H. C. Crick, *J. Mol. Biol.*, 19 (1966) 548.
69. R. W. Holley, J. Apgar, G. A. Everett, J. T. Madison, M. Marquisee, S. H. Merrill, J. R. Penswick and A. Zamir, *Science*, 147 (1965) 1462.
70. M. Ikehara and E. Ohtsuka, *Biochem. Biophys. Res. Commun.*, 21 (1965) 257.
71. R. E. Marshall, C. T. Caskey and M. W. Nirenberg, *Science*, 155 (1967) 820.
72. E. Barghoorn and J. Schopf, *Science*, 152 (1966) 758.
73. R. Hinegardner and J. Engelberg, *Science*, 144 (1964) 1031.
74. T. M. Sonneborn, in V. Bryson and H. J. Vogel (Eds.), *Evolving Genes and Proteins*, Academic Press, New York, 1965, p. 377.
75. P. Leder and M. W. Nirenberg, *Proc. Natl. Acad. Sci. (U. S.)*, 51 (1964) 420, 1521.
76. H. Bremer and M. W. Konrad, *Proc. Natl. Acad. Sci. (U. S.)*, 51 (1964) 801.
77. G. S. Stent, *Science*, 144 (1964) 816.
78. R. Byrne, J. G. Levin, H. A. Bladen and M. W. Nirenberg, *Proc. Natl. Acad. Sci. (U. S.)*, 52 (1964) 140.
79. H. A. Bladen, R. Byrne, J. G. Levin and M. W. Nirenberg, *J. Mol. Biol.*, 11 (1965) 78.
80. A. N. Baldwin and P. Berg, *J. Biol. Chem.*, 241 (1966) 839.
81. C. Yanofsky and P. St. Lawrence, *Ann. Rev. Microbiol.*, 14 (1960) 311.
82. L. Gorini and J. R. Beckwith, *Ann. Rev. Microbiol.*, 20 (1966) 401.
83. F. Rottman and M. Nirenberg, *J. Mol. Biol.*, 21 (1966) 555.
84. T. Jukes, *Molecules and Evolution*, Columbia University Press, New York, 1967.

Biography

Marshall Warren Nirenberg was born in New York City on April 10th, 1927, the son of Harry and Minerva Nirenberg. The family moved to Orlando, Florida in 1939. He early developed an interest in biology. In 1948 he received a B. Sc. degree, and in 1952, a M. Sc. degree in Zoology from the University of Florida at Gainesville. His dissertation for the Master's thesis was an ecological and taxonomic study of caddis flies (Trichoptera).

During this period he became interested in biochemistry. He continued studies in this field at the University of Michigan, Ann Arbor, and in 1957 received the Ph. D. degree from the Department of Biological Chemistry. Nirenberg's thesis, performed under the guidance of Dr. James Hogg, was a study of a permease for hexose transport in ascites tumor cells.

From 1957 to 1959 he obtained postdoctoral training with De Witt Stetten Jr., and with William Jakoby at the National Institutes of Health as a fellow of the American Cancer society. During the next year he held a Public Health Service Fellowship and in 1960 became a research biochemist in the Section of Metabolic Enzymes, headed by Dr. Gordon Tompkins, at the National Institutes of Health.

In 1959 he began to study the steps that relate DNA, RNA and protein. These investigations led to the demonstration with H. Matthaei that messenger RNA is required for protein synthesis and that synthetic messenger RNA preparations can be used to decipher various aspects of the genetic code.

In 1962 he became head of the Section of Biochemical Genetics at the National Institutes of Health.

Nirenberg holds honorary degrees from the University of Michigan, Yale University, University of Chicago, University of Windsor (Ontario) and Harvard University. Other honours include: The Molecular Biology Award, National Academy of Sciences, 1962; Paul Lewis Award in Enzyme Chemistry, American Chemical Society, 1964; The National Medal of Science, 1965; The Research Corporation Award, 1966; the Hildebrand Award, 1966; the Gairdner Foundation Award of Merit, 1967; The Prix Charles Leopold Meyer, French Academy of Sciences, 1967; the Joseph Priestly

Award, 1968; and the Franklin Medal, 1968. The Louisa Gross Horwitz Prize, Columbia University, and the Lasker Award were shared with H. G. Khorana in 1968. He is a member of the American Academy of Arts and Sciences and the National Academy of Sciences.

He was married in 1961 to Perola Zaltzman, a chemist from the University of Brazil, Rio de Janeiro. She is now a biochemist at the National Institutes of Health.

1969

M. Delbrück

A.D. Hershey

S.E. Luria

M. Delbrück

A Physicist's Renewed Look at Biology —Twenty Years Later

December 10, 1969

Physics and Biology

At the very beginnings of science the striking dissimilarities between the behavior of living and nonliving things became obvious. Two tendencies can be discerned in the attempts to arrive at a unified view of our world. One tendency is to use the living organism as the model system. This tendency is exemplified by Aristotle. For him, the son of a physician and the keen observer of many forms of life, it was obvious that things develop according to plans. Every animal and plant is generated in some definite way, runs through a cycle of development in which it unfolds its inherent plan, and succumbs to death and decay. For Aristotle, this very obvious feature of the world which surrounds us is *the* model for understanding our (sublunar) world. Astronomy is the exception and offers the contrast of an eternal periodic system subject to neither generation nor decay.

With the ascendance in the Renaissance of the science of physics in our modern sense of the word there seemed to develop at first a peculiar break between the living and the nonliving parts of the world. Life seemed to have unique properties quite irreducible to the world of physics and chemistry: ‹motion generated from within›, ‹chemistry of a very distinct kind›, ‹replication›, ‹development›, ‹consciousness› – each of these aspects of life turned into elements that became more and more foreign to the physicist to the extent that many physicists even today look upon biology as something outside their domain.

A partial reversal of this bizarre partition of the world into the living and the nonliving came with the many proofs that living forms are not, in fact, constant, but over the long range have evolved and that the family tree of this evolution can be traced. The interpretation of evolution in terms of natural selection, expecially after the latter had been put into clearer perspective with the establishment of the science of genetics, suggested a unified view of life, but still left uncertain the connection of life with the nonliving world. The insights of chemistry and its first inroads into biochemistry made it clear that the break between the nonliving world and the living world might not be absolute.

Molecular genetics, our latest wonder, has taught us to spell out the connectivity of the tree of life in such palpable detail that we may say in plain words, «This riddle of life has been solved». The ideas of information storage, of the replication of the stored information and of its programmed readout have become commonplace and have filtered down into the popular magazines and grade school textbooks. The marvel that the mechanical and chemical machinery underlying all these affairs can in fact be worked out is keeping a host of scientists very happy and very busy. With one exception I feel that there is no need to go into the historical aspects of these developments since they have been very adequately treated in the book «*Phage and the Origins of Molecular Biology*» (ref. 4).

The exception is due to the fact that the contribution to this book by N.W. Timofeeff-Ressovsky, although written, for technical reasons could not be included in the book. I hope very much that the time will not be far off when this omission can be rectified. At this moment I would like to describe briefly to what I refer. During the years 1932–1937, while I was assistant to Professor Lise Meitner in Berlin, a small group of theoretical physicists held informal private meetings, at first devoted to theoretical physics but soon turning to biology. Our principal teacher in the latter area was the geneticist, Timofeeff-Ressovsky, who together with the physicist, K. G. Zimmer, at that time was doing by far the best work in the area of quantitative mutation research. A few years earlier H.J. Muller had discovered that ionizing radiations produce mutations and the work of the Berlin group showed very clearly that these mutations were caused either by single pairs of ions or by small clusters of them. Discussions of these findings within our little group strengthened the notion that genes had a kind of stability similar to that of the molecules of chemistry. From the hindsight of our present knowledge one might consider this a trivial statement: what else could genes be but molecules? However, in the mid-thirties, this was not a trivial statement. Genes at that time were algebraic units of the combinatorial science of genetics and it was anything but clear that these units were molecules analyzable in terms of structural chemistry. They could have turned out to be submicroscopic steady state systems, or they could have turned out to be something unanalyzable in terms of chemistry, as first suggested by Bohr[3] and discussed by me in a lecture twenty years ago [reprinted in Cairns *et al.*, 1966[4]]. It is true that our hope at that time to get at the chemical nature of the gene by means of radiation genetics never materialized. The road to success effectively bypassed radiation genetics. Nevertheless, radiation genetics has been through all these decades and is now more than ever a field of great importance, most recently and depressingly so because of the possibilities of large-scale military applications entailing exposure to ionizing radiations.

To illustrate our state of mind at that time I will append to this lecture* a memorandum on the «Riddle of Life», written to clarify my own thinking in the fall of 1937, just before leaving Germany to go to the United States. I found this note a few years ago among my papers. This memorandum would appear to be a summary of discussions at a little meeting in Copenhagen, arranged by Niels Bohr, to which Timofeeff-Ressovsky, H.J. Muller and I had travelled from Berlin. These discussions occurred very much under the impact of the W.M. Stanley findings reporting the crystallization of tobacco mosaic virus[8].

Neurobiology

While molecular genetics has taught us the proper way to reconcile the characteristics of the living world, generation, development towards a goal, and decay, with the contrasting incorruptibility and planlessness of the physical world, it has not resolved our uncertainty about the proper way to relate this language to the notions of ‹consciousness›, ‹mind›, ‹cognition›, ‹logical thought›, ‹truth› – all these notions, too, elements of our ‹world›.

What is language? How does a child come to associate ‹meaning› with a ‹word›? The ability to form abstractions is undoubtedly inherent in our brain, this marvel of a computer. The study of the brain's connectivity, the study of the development of this network in the growing animal, the study of its function and potencies – all of these studies are aspects of the neurobiology of the next decade and they are very appealing ones to many of my colleagues and to many of the new generation of graduate students.

Transducer Physiology

I have two reservations concerning neurobiology. The first reservation is that we are not yet ready to tackle it in a decisive way. I believe that there is a widespread underestimation of the things we do not know and do not understand

* See Appendix (p. 410).

about cell biology and cell–cell interaction. It simply is not enough to know that nerve fibers conduct, that synapses are inhibitory or excitatory, chemical or electrical, that sensory inputs can be transduced, that they result in trains of spikes which measure intensities of stimuli or the time derivatives of these intensities, that all kinds of accommodations occur, etc., etc. I believe that we need a much more basic and detailed understanding of these stimulus response systems, be the stimulus an outside one or a presynaptic signal.

Sensory physiology in a broad sense contains hidden as its kernel an as yet totally undeveloped but absolutely central science: transducer physiology, the study of the conversion of the outside signal to its first ‹interesting› output. I use the word ‹interesting› advisedly because I wish to exclude from the area of study I intend to delimit, for instance, the primary photochemical reactions of the visual systems. I look upon the primary photochemical processes as something ‹uninteresting› because they concern the conversion of a light stimulus into what might be called an olfactory stimulus. A light quantum, in order to be effective as a sensory stimulant, naturally must, in the first instance, create within the cell a primary photoproduct which carries the business further. In thus excluding the photochemistry of the visual process from transducer physiology proper I am excluding the beautiful work on the photochemistry of rhodopsin for which Georg Wald received the Nobel Prize two years ago. Transducer physiology proper comes after this first step, where we are dealing with devices of the cell unparalleled in anything the physicists have produced so far with respect to sensitivity, adaptability and miniaturization. Which biological material will turn out to be the most suitable for bringing us decisive insights in this field? For a number of years I have studied an organelle of the fungus *Phycomyces*, the sporangiophore, in the belief that in the field of transducer physiology, as in genetics, essential progress will require the use of a suitable microorganism. I need not detail this work here since it has very recently been critically reviewed by a group effort of those involved in this work[2]. Let me say here only that this organelle is exquisitely sensitive to light, to gravity, to stretch, and to a stimulus which we believe to be olfactory, and illustrate it with a few slides. Others have proposed and demonstrated the suitability of other systems: chemotaxis of bacteria[1]; olfaction in insects[7]; mechanosensitivity of motoric cilia[10]. We may hope that each of these systems, as well as the lipid-bilayer systems, which can be made to simulate most of the astounding feats of living membranes[5], will contribute to the great discoveries in cell physiology which, in my opinion, are prerequisite for a truly successful venture into neurobiology.

My second reservation regarding the hopes of neurobiology is more disturbing to me and also more nebulous; the eagerness with which we plunge into neurobiology overlooks an essential limitation – the *a priori* aspect of the concept of truth. It is well understood that a computer can be constructed so as to operate with certain axioms and formalized rules of logic, deriving in this way any number of ‹proved declarative sentences›. We may call these sentences true if we have faith in the axioms and the rules of logic and we may be tempted to consider the logical sum of provable sentences as the computer's definition of truth. However, our friends, the logicians, have made it clear to us long ago that in any but the simplest languages we must distinguish between an ‹object language› and a ‹metalanguage›. The word ‹truth›, and thus all discussion of truth, must be excluded from the object language if the language is to be kept free of antinomies. There then follows the strange result that there must be sentences that are true but not probable[9]. Thus the notion of truth, if it is to be meaningful at all, must be distinct and prior to the system of provable sentences, and thus distinct from and prior to the computer which should be looked upon as the embodiment of the system of probable sentences.

Thus, even if we learn to speak about consciousness as an emergent property of nerve nets, even if we learn to understand the processes that lead to abstraction, reasoning and language, still any such development presupposes a notion of truth that is prior to all these efforts and that cannot be conceived as an emergent property of it, an emergent property of a biological evolution. Our conviction of the truth of the sentence, «The number of prime numbers is infinite», must be independent of nerve nets and of evolution, if truth is to be a meaningful word at all.

Artist versus Scientist

Twenty years ago the Connecticut Academy of the Arts and Sciences had a jubilee meeting and on that occasion invited a poet, a composer, and two scientists to ‹create› and to ‹perform›. It was a very fine affair. Hindemith, conducting a composition for trumpet and percussion, and Wallace Stevens, reading a set of poems entitled «An Ordinary Evening in New Haven», were enjoyed by everybody, perhaps most by the scientists. In contrast, the scientists' performances were attended by scientists only. To my feeling this irreciprocity was fitting, although perhaps not intended by the organizers. It is quite rare that scientists are asked to meet with artists and are challenged to match the others' creativeness. Such an experience may well humble the scientist. The medium in which he works does not lend itself to the delight of the listener's ear. When he designs his experiments or executes them with devoted attention to the details he may say to himself. «This is my composition; the pipette is my clarinet». And the orchestra may include instruments of the most subtle design. To others, however, his music is as silent as the music of the spheres. He may say to himself, «My story is an everlasting possession, not a prize composition which is heard and forgotten», but he fools only himself. The books of the great scientists are gathering dust on the shelves of learned libraries. And rightly so. The scientist addresses an infinitesimal audience of fellow composers. His message is not devoid of universality but its universality is disembodied and anonymous. While the artist's communication is linked forever with its original form, that of the scientist is modified, amplified, fused with the ideas and results of others and melts into the stream of knowledge and ideas which forms our culture. The scientist has in common with the artist only this: that he can find no better retreat from the world than his work and also no stronger link with the world than his work.

The Nobel ceremonies are of a nature similar to the one I referred to. Here too scientists are brought together with a writer. Again the scientists can look back on a life during which their work addressed a diminutive audience while the writer, in the present instance Samuel Beckett, has had the deepest impact on men in all walks of life. We find, however, a strange inversion when we come to talking about our work. While the scientists seem elated to the point of garrulousness at the chance of talking about themselves and their work, Samuel Beckett, for good and valid reasons, finds it necessary to maintain a total silence with respect to himself, his work, and his critics. Even though I was more thrilled by the award of the Nobel Prize to him than about the award to me and momentarily looked forward with intense anticipation to hearing his lecture, I now realize that he is acting in accordance with the rules laid down by the old witch at the end of a marionette play entitled «The Revenge of Truth»[6].

«The truth, my children, is that we are all of us acting in a marionette comedy. What is more important than anything else in a marionette comedy is keeping the ideas of the author clear. This is the real happiness in life and now that I have at last come into a marionette play, I will never go out of it again. But you, my fellow actors, keep the ideas of the author clear. Aye, drive them to the utmost consequences.»

Appendix

Preliminary write-up on the topic «Riddle of Life»
(Berlin, August 1937)

We inquire into the relevance of the recent results of virus research for a general assessment of the phenomena peculiar to life.

These recent results all agree in showing a remarkable uniformity in the behavior of individuals belonging to one species of virus in preparations employing physical or chemical treatments mild enough not to impair infective specificity. Such a collection of individuals migrates with uniform velocity in the electrophoresis apparatus. It crystallizes uniformly from solutions such that the specific infectivity is not altered by recrystallization, not even under conditions of extremely fractionated recrystallization. Elementary analysis gives reproducible results, such as might be expected for proteins, with perhaps the peculiarity that the phosphorus and sulfur contents appear to be abnormally small.

These results force us to the view that the viruses are things whose atomic constitution is as well defined as that of the large molecules of organic chemistry. True, with these latter we also cannot speak of unique spatial configurations, since most of the chemical bonds involve free rotation around the bond. We cannot even decide unambiguously which atoms do or do not belong to the molecule, since the degree of hydration and of dissociation depends not only on external conditions, but even when these are fixed, fluctuates statistically from molecule to molecule. Nevertheless, there can be no doubt that such large molecules constitute a legitimate generalization of the standard concept of the chemical molecule. The similarity between virus and molecule is particularly apparent from the fact that virus crystals can be stored indefinitely without losing either their physico-chemical or infectious properties.

Therefore we will view viruses as molecules.

If we now turn to that property of a virus which defines it as a living organism, namely, its ability to multiply within living plants, then we will ask ourselves first whether this accomplishment is that of the host, as a living organism, or whether the host is merely the provider and protector of the virus, offering it suitable nutrients under suitable physical and chemical conditions. In other words, we are asking whether we should view the injection of a virus as a stimulus which modifies the metabolism of the host in such a way as to produce the foreign virus protein instead of its own normal protein, or whether we should view the replication as an essentially autonomous accomplishment of the virus and the host as a nutrient medium which might be replaced by a suitably offered synthetic medium.

Now it appears to me, that upon close analysis the first view can be completely excluded. If we consider that the replication of the virus requires the accurate synthesis of an enormously complicated molecule which is unknown to the host, though not as to general type, yet in all the details of its pattern and therefore of the synthetic steps involved, and if we consider further what extraordinary production an organism puts on to perform in an orderly way the most minute oxidation or synthesis in all those cases that do not involve the copying of a particular pattern – setting aside serology, which is a thing by itself – then it seems impossible to assume that the enzyme system for the host could be modified in such a far-reaching way by the injection of a virus. There can be no doubt that the replication of a virus must take place with the most direct participation of the original pattern and even without the participation of any enzymes specifically produced for this purpose.

Therefore we will look on virus replication as an autonomous accomplishment of the virus, for the general discussion of which we can ignore the host.

We next ask whether we should view virus replication as a particularly pure case of replication or whether it is, from the point of view of genetics, a complex phenomenon. Here we must first point out that with higher animals and plants which reproduce bisexually replication is certainly a very complex phenomenon. This has been shown in a thousand details by genetics, based on Mendel's laws and on modern cytology, and must be so, in order to arrive at any kind of order for the infinitely varied details of inheritance. Specifically, the close cytological analysis of the details of meiosis (reduction division) has shown that it is a specialization of the simpler mitotic division. It can easily be shown that the teleological point of this specialization lies in the possibility of trying out new hereditary factors in ever-new combinations with genes already present, and thus to increase enormously the diversity of the genotypes present at any one time, in spite of low mutation rates.

However, even the simpler mitotic cell division cannot be viewed as a pure case. If we look first at somatic divisions of higher animals and plants, then we find here that an originally simple process has been modified in the most various ways to adapt it to diverse purposes of form and function, such that one cannot speak of an undifferentiated replication. The ability to differentiate is certainly a highly important step in the transition from the protists to the multicellular organisms, but it can probably be related in a natural way to the general property of protists that they can adapt themselves to their environment and change phenotypically without changing genotypically. This phenotypic variability implies that with simple algae like Chlorella we can speak of simple replication only so long as the physical conditions are kept constant. If they are not kept constant, then, strictly speaking, we can only talk of a replication of the genomes which are embedded in a more or less well-nourished, more or less mistreated, specific protoplasma, and which, in extreme cases, may even replicate without cell division.

There can be no doubt, further, that the replication of the genome in its turn is a highly complex affair, susceptible to perturbation in its details without impairing the replication of pieces of chromosomes or of genes. Certainly the crucial element in cell replication lies in the coordination of the replication of a whole set of genes with the division of the cell. With equal certainty this coordination is not a primitive phenomenon. Rather, it requires that particular modification of a simple replication system which accomplishes constancy of supply of its own nutrient. By this modification it initiates the chain of development which until now has been subsumed under the title ‹life›.

In view of what has been said, we want to look upon the replication of viruses as a particular form of a primitive replication of genes, the segregation of which from the nourishment supplied by the host should in principle be possible. In this sense, one should view replication not as complementary to atomic physics but as a particular trick of organic chemistry.

Such a view would mean a great simplification of the question of the origin of the many highly complicated and specific molecules found in every organism in varying quantities and indispensable for carrying out its most elementary metabolism. One would assume that these, too, can replicate autonomously and that their replication is tied only loosely to the replication of the cell. It is clear that such a view in connection with the usual arguments of the theory of natural selection would let us understand the enormous variety and complexity of these molecules, which from a purely chemical point of view appears so exaggerated.

1. J. Adler, Chemoreceptors in bacteria. Studies of chemotaxis reveal systems that detect attractants independently of their metabolism, *Science*, 166 (1969) 1588–1597.
2. K. Bergman, Patricia V. Burke, E. Cerdá-Olmedo, C. N. David, M. Delbrück, K. W. Foster, E. W. Goodell, M. Heisenberg, G. Meissner, M. Zalokar, D. S. Dennison and W. Shropshire Jr., Phycomyces, *Bacteriol, Rev.*, 33 (1969) 99.

3. N. Bohr, Light and Life, *Nature*, 131 (1933) 421, 457; Licht und Leben, *Naturwissenschaften*, 21 (1933) 245; Licht und Leben–noch einmal, *Naturwissenschaften*, 50 (1965) 725.
4. J. Cairns, G. S. Stent and J. D. Watson (Eds.), *Phage and the Origins of Molecular Biology*, Cold Spring Harbor Laboratory of Quantitative Biology, Cold Spring Harbor, N. Y., 1966.
5. M. Delbrück, Lipid bilayers as models of biological membranes, in *The Neurosciences –A Study Program*, Rockefeller University Press, New York, 1970.
6. Isak Dinesen, The roads round Pisa, in *Seven Gothic Tales*, Modern Library, New York, 1934.
7. K. E. Kaissling and E. Priesner, Die Riechschwelle des Seidenspinners, *Naturwissenschaften*, 57 (1970) 23–28.
8. W. M. Stanley, Isolation of a crystalline protein possessing the properties of tobacco-mosaic virus, *Science*, 81 (1935) 644.
9. A. Tarski, Truth and proof, *Sci. Am.*, 220 (1969) 63.
10. U. Thurm, Steps in the transducer process of mechanoreceptors, in J. D. Carthy and G. E. Newell (Eds.), *Invertebrate Receptors*, Academic Press, New York, 1968, p. 199.

Biography

Max Delbrück was born on September 4th, 1906, in Berlin, Germany, the youngest of seven children. His father, Hans Delbrück, Professor of History at the University of Berlin, was for many years editor and political columnist of the *Preussische Jahrbücher*. His mother was a granddaughter of the chemist, Justus von Liebig.

Max Delbrück grew up in a suburb of Berlin (Grunewald) populated by moderately affluent members of the academic, professional, and merchant community, many of them with large families. The period of affluence and lively hospitality before 1914 was followed by the war years with hunger, cold, and death, and the postwar period of revolution, inflation, and impoverishment.

His interest in science dates back to boyhood and was directed first towards astronomy, which he seized upon as a means of finding an identity in an environment of strong personalities. All senior to him, many with high accomplishments, none was in the sciences. The one exception, the oldest boy in the Bonhoeffer family, Karl Friedrich (his father Karl Bonhoeffer was Professor of Psychiatry), eight years older than Max Delbrück, was a physical chemist of high distinction, and became the mentor and lifelong friend of Delbrück. The shift to theoretical physics during the latter part of his graduate studies in Göttingen was an easy one from astrophysics and a natural one in the late twenties, just after the breakthrough of quantum mechanics, for which Göttingen was one of the centers.

Among his friendships during the later student years, the most intense and influential one was with Werner Brock, now emeritus Professor of Philosophy, Freiburg.

There followed three postdoctoral years (1929–1932) abroad, in England, Switzerland, and Denmark. The stay in England, with its immersion into a new language and a new culture, had a vast effect on widening his outlook on life. In Switzerland and Denmark the associations with Wolfgang Pauli and Niels Bohr shaped his attitude toward the pursuit of truth in science.

Delbrück's interest in biology was first aroused by Bohr, in connection with his speculations that the complementarity argument of quantum mechanics might have wide applications to other fields of scientific endeavor and especially in regard to the relations between physics and biology. A move to Berlin in 1932, as assistant to Lise Meither, was largely motivated by the hope that the proximity of the various Kaiser Wilhelm Institutes to each other would facilitate the beginning of an acquaintance with the problems of biology. Paradoxically, this good intention was helped by the rise of Nazism which made official seminars less interesting. A small group of physicists and biologists began to meet privately beginning about 1934. To this group belonged N. W. Timofeeff-Ressovsky (genetics). Out of these meetings grew a paper by Timofeeff, Zimmer, and Delbrück on mutagenesis. A popularization of this paper of 1935 in Schroedinger's little book «*What is Life?*» (1945) had a curiously strong influence on the development of molecular biology in the late 1940's.

The move to the United States in 1937 was made possible by a second fellowship of the Rockefeller Foundation, permitting Delbrück to pursue with greater freedom and effectiveness his interests in biology. He chose Caltech because of its strength in *Drosophila* genetics, and to some extent because of its distance from the impending perils at home. Although his job in Germany seemed reasonably secure, it was clear that political reasons would bar him from advancement.

At Caltech he soon teamed up with E. L. Ellis doing phage research. The fellowship of the Rockefeller Foundation ran out in September 1939. World War II had started and Delbrück elected to stay in the United States. He accepted an instructorship in the Physics Department at Vanderbilt University in Nashville, Tennessee. The years at Vanderbilt were the war years. Both Luria (at Bloomington, Indiana) and Delbrück (at Vanderbilt) were technically enemy aliens, a status affording them the privacy to concentrate on science.

In 1941 Delbrück married Mary Bruce. She has given his life the harmony needed for its fulfillment. They have four children–a first set, Jonathan and Nicola, born in 1947 and 1949, and, since these turned out so happily, a second set, Tobias and Ludina, born in 1960 and 1962.

Since the early 1950's Delbrück's research interests have shifted from molecular genetics to sensory physiology and especially to the idea of introducing here, too, a microorganism of suitable simplicity. He turned to the sporangiophores of *Phycomyces* as a model system for the study of stimulus transductions. The goal of clarifying the molecular nature of the primary transducer processes of sense organs in general and of *Phycomyces* in particular has held his attention since then, with one interruption.

This interruption was the setting up of an institute of molecular genetics at the University of Cologne. Delbrück's goal was to demonstrate the feasibility of modern interdisciplinary research and of the ‹department system› (with several professors in one institute) within a German university setup, and to boost molecular genetics in Germany.

The Institut für Genetik der Universität Köln was formally dedicated on June 22nd, 1962, with Niels Bohr as the principal speaker. His lecture entitled «Light and Life – Revisited» commented on his original one of 1933, which had been the starting point of Delbrück's interest in biology. It was to be Bohr's last formal lecture. He died before completing the preparation of the manuscript of this lecture for publication.

Since 1964 the experimental work on *Phycomyces* is once more being pursued with full force, together with theoretical studies on related systems.

A. D. HERSHEY

Idiosyncrasies of DNA Structure

December 12, 1969

In 1958, what had been learned about the genetic structure of phage particles presented a paradox. On the one hand, genetic crosses revealed only one linkage group[1]. On the other hand, physical evidence suggested that phage particles contained more than one DNA molecule and, probably, more than one species of DNA molecule. The paradox need not be dwelt on here, for it turned out that the physical evidence was mistaken: phage particles contain single DNA molecules that are species specific.

Seeking to resolve the paradox of 1958, my colleagues and I had to start at the beginning by learning how to extract, purify, and characterize DNA molecules. As it happened, our inexperience was not a severe handicap because the existing techniques were still primitive. They were primitive for good reason: until virus particles could be taken apart, nobody had ever seen a solution of uniform DNA molecules. Without knowing it we were entering one of those happy periods during which each techhnical advance yields new information.

Joseph Mandell and I began by attempting to make chromatography of DNA work[2]. We succeeded, as had many chromatographers before us, more by art than by theory[3]. The first application of our method, by Elizabeth Burgi and me, yielded the following results[4].

(*1*) DNA extracted from phage T2 proved to be chromatographically homogeneous.

(*2*) Subjected to a critical speed of stirring, the DNA went over by a single-step process to a second chromatographic species. The second species formed a single band that was not chromatographically homogeneous. We guessed that it consisted of half-length fragments produced by single breaks occurring preferentially near the centers of the original molecules.

(*3*) A single chromatographic fraction of the half-length fragments, when subjected to a higher critical speed of stirring, went over in a single step to a third chromatographic species that we called quarter-length fragments.

(*4*) Unfractionated half-length fragments subjected to stirring could be altered in a gradual but not a stepwise manner, presumably because the fragments of various lengths broke at various characteristic speeds of stirring. (5) These results showed that chromatographic behavior and fragility under shear depended on molecular length, and that our starting material was uniform with respect to length by both criteria.

Burgi and I verified the above results by sedimentation analysis and took pains to isolate precise molecular halves and quarters[5].

At this point we believed we had characteristic DNA molecules in our hands but lacked any method of weighing or measuring them. Fortunately, Irwin Rubenstein and C. A. Thomas Jr. had a method of measurement but were experiencing difficulties in preparing materials. We joined forces to measure by radiographic methods the phosphorus content of T2 DNA molecules and their halves and quarters[6]. We found a molecular weight of 130 million for the intact DNA. Moreover, since the DNA molecule and the phage particle contained equal amounts of phosphorus, there could be only one molecule per particle. Evidently T2 possessed a unimolecular chromosome.

With the materials derived from T2 available as standards, Burgi and I worked out conditions under which sedimentation rates in sucrose could be used as measures of molecular weight[7]. We found the useful relation

$$D_2/D_1 = (M_2/M_1)^{0.35} \qquad (1)$$

in which D means distance sedimented, M means molecular weight, and the subscripts refer to two DNA species. The relation serves to measure the molecular weight of an unknown DNA from that of a known DNA when the two are sedimented in mixture. By this method the DNA of phage λ, for instance, shows a molecular weight of 31 million.

Of course, equation (1) holds only for typical bihelical DNA molecules. As a check for equivalent structures, we measured fragility under hydrodynamic shear, which also depends on molecular weight and molecular structure[8].

While the rudiments of T2 DNA structure were being worked out as described above, genetic analysis of the chromosome was generating its own paradox. By this time T4 had largely replaced T2 for experimental purposes, but the two phages are so closely related that information gained from either one usually applies to both.

The paradox appeared in the work of Doermann and Boehner[9] who found, in effect, that T4 heterozygotes could replicate without segregating and were somehow polarized in structure. These properties were incompatible with the heteroduplex model for heterozygotes and eventually led Streisinger and his colleagues to postulate two radical features of T4 DNA structure: circular permutation and terminal repetition[10]. I shall come back to these features presently, and note here only that both have been confirmed by physical analysis[11]. I turn now to a rather different DNA, that of phage λ.

From the first, our preparations of λ DNA proved refractory in that they refused to pass through our chromatographic column and failed to yield reasonable boundaries in the ultracentrifuge. Only after we achieved sedimentation patterns of high resolution in sucrose did our results begin to make sense[12]. Then we could see in suitable preparations four distinct components. Three of these sedimented at the proper rates according to equation (1) for linear structures with length ratios 1 : 2 : 3. We provisionally called them monomers, dimers, and trimers. The fourth component sedimented faster than the monomer but slower than the dimer. We called it a closed or folded monomer.

Further analysis was possible when we found that heating to about 75° (insufficient to cause denaturation) converted everything into linear monomers. Then we could show that heating dilute solutions at 55° converted monomers entirely into the closed form. Alternatively, heating concentrated solutions at 55° yielded dimers, trimers, and larger aggregates. Moreover, the fragility of the various structures under shear decreased in the proper order: trimers, dimers, closed monomers, then linear monomers. Finally, closure and aggregation were clearly competitive processes, suggesting that each molecule possessed two cohesive sites that were responsible for both processes.

We tested our model by examining molecular halves, which by hypothesis should carry one cohesive site each, and should be able to join only in pairs (Fig. 1). This proved to be correct: by thermal treatment we could reversibly convert molecular halves into structures sedimenting at the rate of unbroken

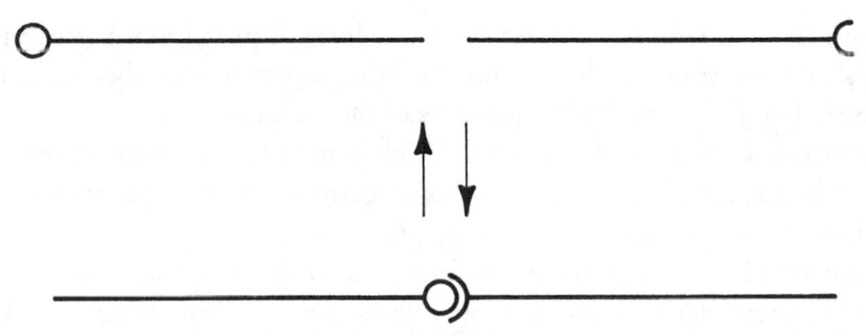

Fig. 1. Reversible joining of molecular halves of λ DNA. The upper part of the figure shows a schematic version of a single molecule cut in two. The lower part of the figure shows the halves rejoined through their terminal cohesive sites.

linear molecules. Furthermore, the rejoined halves exhibited a buoyant density appropriate to paired right and left molecular halves, not to paired right halves or paired left halves[13]. Therefore the cohesive sites were complementary in structure, not just sticky spots. Our results were also consistent with terminally situated cohesive sites, though clear proof of this came from the electron microscopists. The structures corresponding to closed and open forms of λ DNA are diagrammed in Fig. 2.

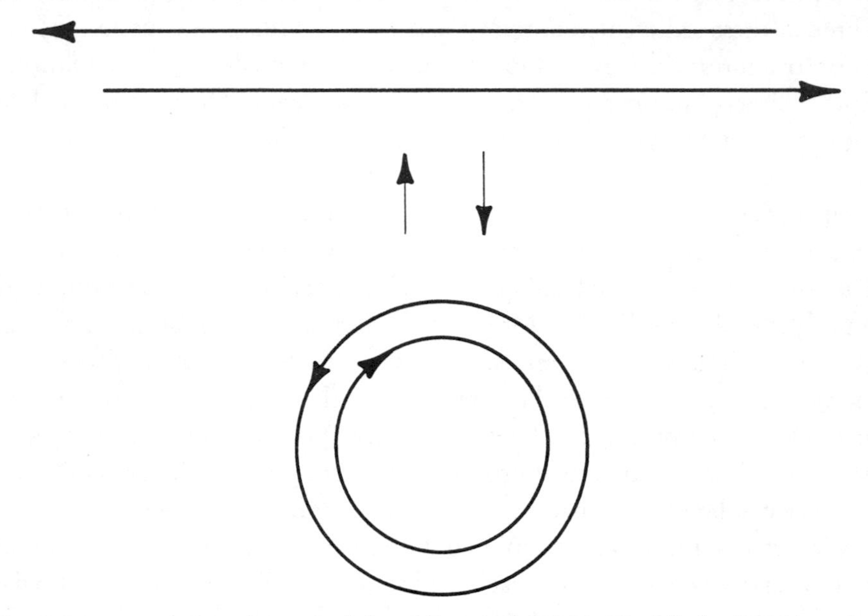

Fig. 2. Interconversion of open and closed forms of λ DNA. The bihelical DNA molecule is indicated by parallel lines of opposite polarity.

Phages T4 and λ do not exhaust the modalities of phage DNA structure[14] but they cover much of the ground. To bring together what they teach us I show in Fig. 3 idealized DNA molecules of three types.

Structure I represents the Watson–Crick double helix in its simplest form. Actually, structure I is not known to exist, perhaps because replication of the molecular ends would be mechanically precarious.

Structure II represents the linear form of λ DNA, as we have seen. Note that on paper it derives from structure I without gain or loss of nucleotides. In nature, structure II cannot derive from I because it contains the sequence za not present in I. In structure II, A and a may be called joining sequences, and they need only be long enough to permit specific base pairing. In λ, the joining

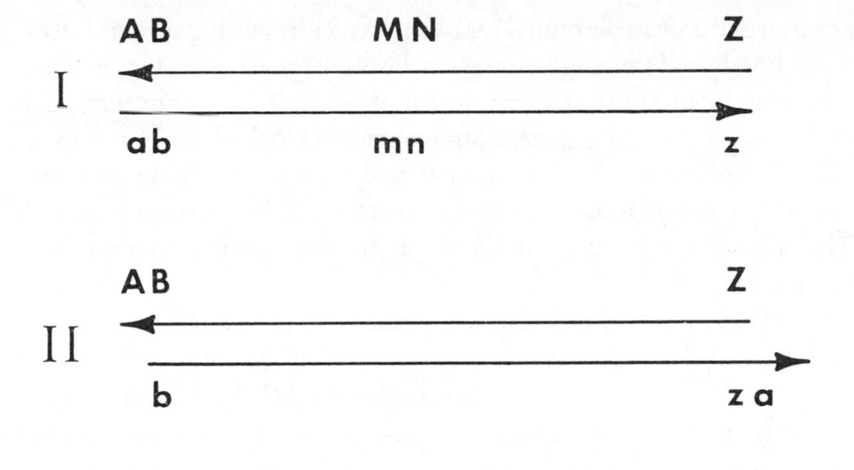

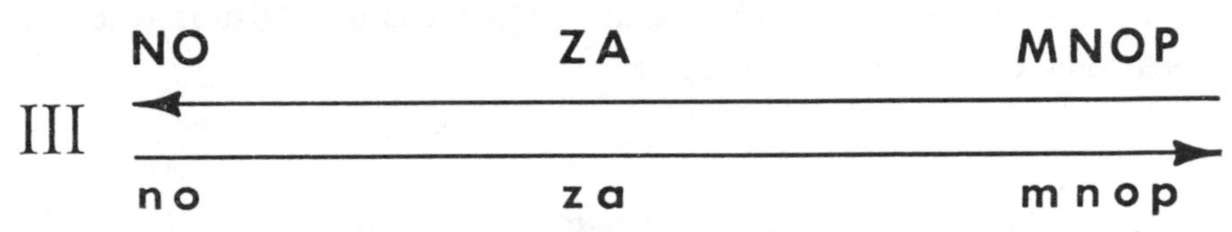

Fig. 3. Three bihelical DNA molecules represented as information diagrams. Each capital letter signifies a nucleotide sequence of arbitrary length; each small letter the corresponding complementary sequence.

sequences each contain about 20 nucleotides[15]. Presumably phage λ circumvents the difficulty of replicating ends by abolishing them (Fig. 2).

Structure III represents T4 DNA. It derives from structure I, on paper, by cyclic permutation followed by the addition of repeats at one end. T4 DNA molecules are said to be circularly permuted, meaning that the 200000 possible cyclic permutations occur with equal frequency. The terminal repetitions in T4 DNA are relatively long: about one per cent of the molecular length containing about two genes. Thus T4 DNA solves the problem of replication of ends by making them dispensable. The same feature protects gene function during permutation, because the severed genes in a particular DNA molecule are always present in a second intact copy.

Structure III has three genetic consequences. First, molecules arising by recombination are heterozygous for markers situated in the terminal repetitions[16]. Second, such heterozygotes are likely to appear one-ended and can replicate before segregation[9]. Third, the circular permutation gives rise to a circular genetic map[10].

T4 cannot generate its permuted DNA molecules by cutting specific internucleotide bonds, and seemingly must somehow cut them to size. In principle, it could do this by measuring either total length or the lengths of the terminal repetitions. Many years ago, George Streisinger and I looked for a shortening of T4 DNA molecules as a result of genetic deletions. We didn't find any shortening. The reason is that T4 cuts its chromosome by measuring overall size. Thus a reduction of the genomic length just increases the lengths of the terminal repetitions[17].

Structure II as seen in λ DNA also has genetic consequences. Here the chromosome ends are clearly generated by cuts at specific internucleotide bonds, with the result that deletions necessarily shorten the DNA molecule[18]. The cohesive sites in λ DNA, though they do not give rise to a circular genetic map, do show up as a joint in the center of the prophage, thus giving to λ a genetic map with two cyclic permutations. In fact Campbell[19] had foreseen the need for potential circularity of the λ chromosome at a time when the DNA could only be described as goo.

Summary

The study of two phage species led to three generalizations probably valid for all viruses.

(*1*) Virus particles contain single molecules of nucleic acid.

(*2*) The molecules are species specific and, with interesting exceptions, are identical in virus particles of a single species.

(*3*) Different viral species contain nucleic acids that differ not only in length and nucleotide sequence but in many unexpected ways as well. I have described only two examples: λ DNA, characterized by terminal joining sequences, and T4 DNA, exhibiting circular permutation and terminal repetition of nucleotide sequences.

Epilogue

In the foregoing account I have deliberately pursued a single line of thought, neglecting both parallel developments in other laboratories and work performed in my own laboratory in which I didn't directly participate. A few of these omissions I feel obliged to repair.

Several people studied breakage of DNA by shear before we did. Davison[20] first noticed the extreme fragility of very long DNA molecules, and he and Levinthal[21] pursued the theory of breakage. However, we first noticed stepwise breakage at critical rates of shear. That observation was needed both to substantiate theory and to complete our evidence for molecular homogeneity.

Davison *et al.*[22] also showed that particles of phage T2 contain single DNA molecules, though they didn't attempt direct measurements of molecular weight.

Physical studies of DNA had of course been under way for some years before analysis of virus particles began. For instance, Doty, McGill and Rice[23] had observed a relation equivalent to our equation (1) containing the exponent 0.37. Their data covered a range of molecular weights below seven million. Larger molecules were not known at the time of their work, and could not have been studied by the existing methods anyway.

Our work on sedimentation of DNA in sucrose would have been considerably eased if we had known of the earlier work on sedimentation of enzymes by Martin and Ames[24].

The first example of a circular DNA[25], as well as the first evidence for one DNA molecule per phage particle[26], came from Sinsheimer's work with phage φX174. Its DNA comes in single strands that weigh only 1.7 million daltons, which is small enough to permit light-scattering measurements.

Elizabeth Burgi[18] demonstrated reductions of molecular weight of DNA in deletion mutants of phage λ already shown by G. Kellenberger *et al.*[27] to contain reduced amounts of DNA per phage particle.

1. A.D. Kaiser, *Virology*, 1 (1955) 424; G. Streisinger and V. Bruce, *Genetics*, 45 (1960) 1289.
2. J.D. Mandell and A.D. Hershey, *Anal. Biochem.*, 1 (1960) 66.
3. I recall a discouraging conversation with Paul Doty, in which he pointed out and I agreed that large DNA molecules could hardly be expected to reach equilibrium with the bed material during passage through a column. But at least Mandell and I were dealing with ionic interactions. A couple of years later Doty and Marmur and their colleagues were to face a real theoretical crisis: having discovered renaturation of DNA, how explain the remarkable speed of the process?
4. A.D. Hershey and E. Burgi, *J. Mol. Biol.*, 2 (1960) 143.
5. E. Burgi and A.D. Hershey, *J. Mol. Biol.*, 3 (1961) 458.

6. I. Rubenstein, C. A. Thomas Jr. and A. D. Hershey, *Proc. Natl. Acad. Sci. (U. S.)*, 47 (1961) 1113.
7. E. Burgi and A. D. Hershey, *Biophys. J.*, 3 (1963) 309.
8. A. D. Hershey, E. Burgi and L. Ingraham, *Biophys. J.*, 2 (1962) 424.
9. A. H. Doermann and L. Boehner, *Virology*, 21 (1963) 551.
10. G. Streisinger, R. S. Edgar and G. Harrar Denhardt, *Proc. Natl. Acad. Sci. (U. S.)*, 51 (1964) 775.
11. C. A. Thomas Jr. and I. Rubenstein, *Biophys. J.*, 4 (1964) 94; L. A. MacHattie, D. A. Ritchie, C. A. Thomas Jr. and C. C. Richardson, *J. Mol. Biol.*, 23 (1967) 355.
12. A. D. Hershey, E. Burgi and L. Ingraham, *Proc. Natl. Acad. Sci. (U. S.)*, 49 (1963) 748.
13. A. D. Hershey and E. Burgi, *Proc. Natl. Acad. Sci. (U. S.)*, 53 (1965) 325.
14. C. A. Thomas Jr. and L. A. MacHattie, *Ann. Rev. Biochem.*, 36 (1967) 485.
15. R. Wu and A. D. Kaiser, *J. Mol. Biol.*, 35 (1968) 523.
16. J. Séchaud, G. Streisinger, J. Emrich, J. Newton, H. Lanford, H. Reinhold and M. Morgan Stahl, *Proc. Natl. Acad. Sci. (U. S.)*, 54 (1965) 1333.
17. G. Streisinger, J. Emrich and M. Morgan Stahl, *Proc. Natl. Acad. Sci. (U. S.)*, 57 (1967) 292.
18. E. Burgi, *Proc. Natl. Acad. Sci. (U. S.)*, 49 (1963) 151.
19. A. Campbell, *Advan. Genet.*, 11 (1962) 101.
20. P. F. Davison, *Proc. Natl. Acad. Sci. (U. S.)*, 45 (1959) 1560.
21. C. Levinthal and P. F. Davison, *J. Mol. Biol.*, 3 (1961) 674.
22. P. F. Davison, D. Freifelder, R. Hede and C. Levinthal, *Proc. Natl. Acad. Sci. (U. S.)*, 47 (1961) 1123.
23. P. Doty, B. McGill and S. Rice, *Proc. Natl. Acad. Sci. (U. S.)*, 44 (1958) 432.
24. R. G. Martin and B. N. Ames, *J. Biol. Chem.*, 236 (1961) 1372.
25. W. Fiers and R. L. Sinsheimer, *J. Mol. Biol.*, 5 (1962) 408.
26. R. L. Sinsheimer, *J. Mol. Biol.*, 1 (1959) 43.
27. G. Kellenberger, M. L. Zichichi and J. Weigle, *J. Mol. Biol.*, 3 (1961) 399.

Biography

Alfred Day Hershey was born on December 4th, 1908, in Owosso, Michigan. He studied at the Michigan State College, where he obtained B.S. in 1930, and Ph.D. in 1934. In 1967 he got an honorary D.Sc. at the University of Chicago.

From 1934 till 1950 he was engaged in teaching and research, at the Department of Bacteriology, Washington University School of Medicine. In 1950 he became a Staff Member, at the Department of Genetics, Carnegie Institution of Washington, Cold Spring Harbor, New York; in 1962 he was appointed Director of the Genetics Research Unit of the same institution.

Alfred Hershey married Harriet Davidson in 1945, they have one son, Peter.

Alfred Hershey is a Member of the American Society for Microbiology, the National Academy of Sciences, and the American Academy of Arts and Sciences. Hershey is Recipient of the Kimber Genetics Award of the National Academy of Sciences, 1965. Michigan State University honored him with an M.D.h.c. in 1970.

S. E. Luria

Phage; Colicins and Macroregulatory Phenomena

December 10, 1969

The early work on bacteriophage growth, mutation, and recombination had the good fortune to serve as one avenue in the growth of molecular biology to its present state of an intellectually satisfying construction. It is unnecessary to recount today the story of that early work, in which it was my fortune to be engaged in friendly and exciting cooperation with Max Delbrück and Alfred Hershey. Even more difficult would be an attempt to trace here the series of developments that led from early phage work to the modern knowledge of virus reproduction, gene replication, gene function and its regulation. My greatest satisfaction derives from the role that has been played by my students and coworkers in these developments and from the personal experience of association with many of the protagonists of this great intellectual adventure.

Phage research has branched off in many directions, each of which has contributed in some measure to the edifice of molecular biology. One of the most notable directions was that of gene function and its regulation. The main contributions of phage research in this area were made in the study of lysogeny by André Lwoff and François Jacob, leading to the formulation of the operon theory by François Jacob and Jacques Monod. The regulatory phenomena considered by this theory concerned the functions of individual genes or groups of genes. In this lecture I wish to deal with approaches to certain aspects of cellular regulation that involve «macroregulatory phenomena». By this I mean those phenomena in which the functional changes observed affect some of the major processes of the living cell, such as the synthesis of DNA, or RNA, or protein, or the energy metabolism, or the selective permeability function of cellular membranes.

The study of antibiotics like penicillin or streptomycin, agents that act in a ‹molar› way on cellular processes, had played an important role in elucidating such processes as the organization of the biosynthesis of the bacterial cell wall or the mechanism of protein synthesis. When an alteration of a major cellular function is produced by the action of an agent such as a bacteriophage or some other macromolecular agent acting in a ‹quantal›, single particle fashion, the situation is even more challenging since some mechanism of amplification must intervene between the individual unit agent and the affected elements of the responding cell. For a viral agent, the amplification mechanism may be the replication of the agent or the expression of its genetic potentials. For a protein agent, for example a bacteriocin, the amplification mechanism must be a change in the integrity of some cellular structure or of the functioning of some cellular control system. In either case, an understanding of the mode of action of such agents on major cellular processes is likely to reveal some interesting aspects of the functional organization of the cellular machinery.

In my laboratory, we are currently using bacteriophages and bacteriocins as probes into macroregulatory phenomena of the bacterial cell. There has not yet been much progress in this field, except in the study of the regulation of genetic transcription. Even a description of current efforts should be of value at least in illustrating what we are after.

Bacteriophage and Macroregulatory Phenomena

An early indication of the potential role of phage as a controller of cellular functions was the observation that irradiated phage T2 retained its host-killing and interfering abilities after losing its reproductive capacity[1]. The bacteria were not grossly disrupted but died. It took years and the development of the biochemical approaches to the study of phage infection before the killing action of phage could be interpreted in terms of physiological mechanisms, that is, of specific inhibitions at the level of macromolecular syntheses. We now know that certain virulent phages, including the T-even coliphages, produce a rapid arrest of synthesis of the protein RNA and DNA of their host cells. Other phages have less drastic or more transient effects on these processes. But our knowledge of the mechanisms of these inhibitions has progressed surprisingly slowly.

Phage infection and host DNA synthesis

Let us take, for example, the effect of phage infection on host DNA. The case of the T-even phages would appear to be the simplest. These phages contain hydroxymethyl cytosine (HMC) instead of cytosine in their DNA[2] and determine among other things the production of an enzyme, deoxycytidine triphosphatase, that destroys dCTP, a specific precursor of host DNA (see summary by Cohen[3]). The bacterial DNA is broken down rather rapidly after infection and is converted to acid-soluble fragments and ultimately to single nucleotides. That double-strand breaks in the bacterial DNA should stop its replication is understandable[4]; but the action of phage in inducing such breaks remains unexplained. Certain mutants of phage T4 fail to convert host DNA to acid-soluble products[5], but the primary breaks still occur and the host DNA is broken into large fragments. The nuclease responsible for these breaks must be specific for cytosine-containing DNA; but no phage gene has yet been found whose mutations prevent the breaking of host DNA.

An even more intriguing situation is that of phage Øe of *Bacillus subtilis*, which has been studied in our laboratory by David Roscoe and Menashe Marcus. This is one of several phages that contain hydroxymethyluracil (HMU) instead of thymine in their DNA and, upon infection, determine a series of enzymatic changes directed at converting the path of DNA synthesis from bacterial type to phage type DNA[6]: a dUMP hydroxymethylase, a thymidylate triphosphate nucleotidohydrolase (dTTPase), an inhibitor of thymidylate synthetase, a dTMP nucleotidase, a dCMP deaminase, and possibly also a deoxynucleotide kinase. Host-DNA synthesis stops a few minutes after infection with phage Øe. Roscoe[7] was able to show that the host DNA remains intact, or at least that double-strand breaks do not occur in detectable numbers. The enzymatic interference with the synthesis of dTTP can be bypassed by using thymine-requiring host bacteria in the presence of thymine and by using phage mutants defective in dTTPase. Under these conditions, the phage produced contains at least 10 and possibly 20% thymine in place of HMU; and yet, synthesis of host DNA is still arrested. Hence we must postulate the existence of some more specific mechanism responsible for the arrest. This mechanism is probably not an inhibition of host-DNA synthesizing enzymes by HMU nucleotides since the arrest of bacterial DNA synthesis is produced also by a phage mutant that lacks the ability to determine either dUMP hydroxymethylase or dTTPase. Yet, the arrest of host-DNA synthesis requires protein synthesis after phage infection; hence there is some specific phage function that inhibits bacterial DNA synthesis. We are currently trying to identify this phage function, which may be exerted at the level of the replication process itself or at some still unrecognized regulatory level.

Let me now turn to the effect of phage infection on the synthesis of RNA and proteins. At least in the case of the T-even phages the arrest in host-protein synthesis appears to be secondary to the arrest of mRNA synthesis[8]. Direct effects on translation of existing messengers may also be present, as they certainly are in some animal virus infections.

The mode of arrest of RNA synthesis remained obscure until about a year ago, when the major discovery was made[9] that at least some phages, including the T-even and T7 coliphages[10], cause an alteration in RNA polymerase that changes its specificity. A factor σ, a component of the polymerase needed for transcription of the «very early» set of phage genes (those transcribed immediately after infection[11]) and presumably also for the transcription of bacterial genes, is altered or destroyed after phage infection. The transcription of other genes of the phages is then made possible by the appearance of some new factor(s) which confer different specificity to a persistent ‹core› portion of the host polymerase[12]. The reasonable assumption is made that σ confers to the polymerase a promoter-recognizing specificity that causes it to initiate mRNA synthesis at specific DNA sites.

In this case, the ‹macroregulatory› phenomenon is brought about not at the level of some purely regulatory mechanism, but at the level of the operational machinery itself. The phage arrests the expression of a whole set of genes by changing the specificity of an enzyme–RNA polymerase.

That this kind of regulation is not peculiar to phage infection has been shown by R. Losick of Harvard together with my student A. L. Sonenshein. Starting from the observation by Sonenshein and Roscoe[13] that the *subtilis* phage Øe fails to grow and to express its functions when it infects bacteria in course of sporulation, Losick and Sonenshein hypothesized that since sporulation involves an arrest of synthesis of many proteins and the appearance of several new ones, the critical step may be a change in specificity in RNA polymerase analogous to the one observed in *E. coli* after T-even phage infection[9]. They succeeded in fact demonstrating that this was the case[14]: a σ-like factor, part of the RNA polymerase of vegetative bacterial cells, is altered or eliminated during sporulation, and this brings about a change in the template specificity of the bacterial polymerase. Remarkably enough, *in vitro* addition of the σ factor from *E. coli* to the core of the *B. subtilis* polymerase restores its original activity!

Note that in this study of sporulation the phage was used, not to investigate some phage-induced change in the cell, but as a probe to reveal a ‹regulatory› phenomenon responsible for a major differentiation in the cell cycle of a bacterium – the change from vegetative to sporulative syntheses. The possible relevance of changes in RNA polymerase, and more generally of macroregulatory changes, to problems of differentiation in higher organisms raises interesting speculations[12,15], and is likely to stimulate new approaches to the study of cellular differentiation.

Macroregulation and Colicins

Next I would like to consider another approach to macroregulation, to which we have recently turned in order to gain further insights into the functional organization of bacterial cells. This involves the study of the mode of action of certain colicins; and, although the history of colicin research is closely interwoven with the history of phage research, it may be instructive to recount the circuitous way in which my present interest in colicins came about. Again it started from a phage problem, the conversion of *Salmonella* somatic antigens by temperate phages discovered by Iséki and Sakai[16]. Dr. Hisao Uetake came to my laboratory in 1956 and together we studied[17] the conversion of antigen 10 to antigen 15 by phage ε^{15}. This collaboration continued when Dr. Takahiro Uchida came from Uetake's laboratory to join me at M.I.T. in 1960, and we were fortunate to bring the problem of antigen conversion to the attention of my colleague Dr. Phillips Robbins. The story of how Robbins and his coworkers[18] solved the problem at the biochemical level, and in the process discovered and elucidated the role of carrier lipids in polysaccharide synthesis, need not be recounted here. My association with this work, however, roused my interest in problems of membranes, particularly in certain remarkable features of the cytoplasmic membrane of bacteria.

In bacterial cells this membrane is the only organelle. It contains enzymes and other constituents that play roles, not only in permeation and active transport, but also in the biosynthesis of the macromolecular components of the bacterial cell wall, such as peptidoglycan and other polysaccharides, including the lipopolysaccharide of the enteric bacteria. In addition, the cytoplasmic membrane is the site of the machinery of terminal respiration and may also play a crucial role in the process of DNA replication and in the segregation of DNA copies at cell division[19]. And yet, the functional organization of this remarkable structure remains obscure. Each group of enzymes and carrier molecules involved in a given biochemical process must presumably be positioned in precise fashion next to each other for efficiency of function. We do not know whether such «supramolecular structures» are solely determined by the intrinsic properties of the individual components, which might be able to reform the functional structures *in vitro* (as in the assembly of viral shells or of bacterial flagella from monomeric proteins) or if the preexisting pattern of molecular organization plays some role in the orderly accretion of new functional elements in the membrane of a growing cell – a priming role or even a catalytic role, for example, a conversion of inactive precursors into active components. There is suggestive evidence for the occurrence of some such enzymatic steps in the assembly of the protein shells of certain complex viruses.[20] An even more intriguing possibility is that the structure of the membrane may play a role not only in the positioning, but also in the functioning of its active constituents, for example, by transmitting conformational signals. This might provide an additional level of regulation of cellular function.

This is where colicins come into the picture. They are protein antibiotics lethal for susceptible strains of coliform bacteria and are produced by other strains of such bacteria that harbor the corresponding genetic determinants or «colicinogenic factors». It has long been known that some colicins arrest the synthesis of macromolecular components of susceptible cells[21]. A major advance was the discovery that different colicins cause different biochemical changes[22] and that the ‹killing› action of some colicins can be reversed by digesting away with trypsin the colicin from the cell receptors[22,23]. This action from the outside, together with the one-hit kinetics of killing by colicins, suggested that a single colicin molecule sitting on some surface component of the cell envelope could exert a bacteriostatic or bactericidal effect through ‹amplification› mechanisms residing in the cell envelope itself. Nomura[22] postulated, therefore, that a colicin attached to a suitable receptor acts on a specific «biochemical target» by bringing about a functional alteration of some specific element of the cytoplasmic membrane. Nomura[22] and I[24] have considered the intriguing possibility that the amplification mechanism may be mediated by conformational changes of the cell membrane as a whole. Changeux and Thiery[25] put forward the same idea in a more specific way, based on consideration of allosteric interactions among membrane proteins.

The three types of actions recognized for colicins by Nomura[22] were: (*1*) arrest of DNA synthesis and breakdown of DNA, typical of colicin E2 action; (*2*) inhibition of protein synthesis, characteristic of colicin E3, which could be traced[26] to a specific alteration on some component of the 30 S ribosomal subunit; and (*3*) overall arrest of macromolecular syntheses, a mechanism common to many colicins (E1, K, A, I). In the cases of colicins E2 and E3 the magnitude of the biochemical effects is strongly dependent on multiplicity, whereas the killing action (defined by inability to grow) is strictly one-hit. Hence there is some question as to whether the effects observed, however specific, are primary or secondary. For colicin K and E1, however, the correlation between killing and inhibitory multiplicities is very good and the biochemical phenomena observed may be more directly related to the primary effects.

How does one molecule of colicin inhibit the synthesis of all macromolecules? An important finding (F. and C. Levinthal, personal communication) was that the inhibition of protein or nucleic acid synthesis was absent when colicin E1 reacted with *E. coli* cells growing in strict anaerobiosis; admission of air brought about a prompt but reversible inhibition. This observation, and the fact that the inhibition of RNA and protein synthesis were simultaneous rather than sequential, led the Levinthals to suggest that the primary action of colicin E1 was on oxidative phosphorylation – a function of the cytoplasmic membrane. ATP levels were drastically decreased although not to zero level.

Starting from this background and from our interest in macroregulatory mechanisms located in the bacterial membrane, my coworkers and I undertook attempts to correlate colicin action with changes in membrane properties such as permeability and transport. I shall refer only to work on colicin E1 and K, where we have had some measure of success.

Kay Fields and I looked first into the possible alterations of the transport and accumulation of β-D-galactosides by colicin-treated *E. coli* cells[27]. Our results indicated that the energy-dependent accumulation process was drastically inhibited, whereas the rate of transport of *ortho*-nitrophenylgalactoside (ONPG), measured by its rate of hydrolysis by the galactosidase of intact cells, was hardly affected. Thus, the cells had not become ‹leaky› to ONPG. The accumulation of α-methylglucoside, which is driven by phosphoenol pyruvate rather than ATP[28], was insensitive to colicin E1, or K, an indication that glycolysis did proceed in colicin-inhibited cells.

When we proceeded to study the fate of glucose used by colicin-treated cells we found, unexpectedly, an indication of what we were looking for: a specific alteration of membrane permeability[29]. The treated cells excreted into the medium almost one-third of the glucose-derived carbon as glucose 6-phosphate, fructose 1,6-diphosphate, dihydroacetone phosphate, and 3-phosphoglycerate. Other intermediates were not excreted in measurable amounts. In addition, pyruvate rather than acetate and CO_2 became the major short-term product of glucose catabolism. This was not due to a leakage of pyruvate since this substance could be converted to lactate if the colicin-treated cells had significant levels of lactic dehydrogenase. The production of pyruvate instead of acetate reflected a specific inhibition, direct or indirect, of pyruvate oxidation.

Also, the effect on energy metabolism turned out to be more complicated than just an inhibition of oxidative phosphorylation. If *E. coli* cells are growing fermentatively on glucose under conditions of adequate but not strict anaerobiosis, the synthesis of protein and nucleic acid is almost as sensitive to inhibition by colicin as in aerobic cells. Even hemin-deficient mutants, which are strongly inhibited by air, prove to be sensitive to the colicins if anaerobiosis is not complete.

These observations suggest that an early effect of these colicins may be a (reversible) alteration of the cytoplasmic membrane, requiring the presence of some oxygen and leading to a block in ATP-dependent processes by limiting ATP availability. This may result either from a reduced ATP production or by an increase in ATP destruction. The fact that biosynthetic processes are blocked despite the significant residual levels of ATP may be due in part to accumulation of AMP and the resulting rise in AMP/ATP ratios[30]. In fact, an *E. coli* mutant with a heat-sensitive AMP kinase behaves at high temperatures very much like colicin-inhibited cells[31].

In a search for further effects of colicins on the membrane, David Feingold, who spent last year as a guest in our laboratory, investigated the effect of colicin E1 on proton uptake by bacteria in the presence of carbonylcyanide *m*-chlorophenylhydrazone (CCCP), a powerful uncoupler of oxidative phosphorylation which promotes H^+ permeation[32]. By itself the colicin produced no increase in proton permeability; in fact, it prevented the slow pH rise observed with normal washed cells. But colicin treatment, even at low multiplicities, sensitized the bacteria to CCCP so that equilibration occurred almost instantly upon the addition of as little as 10^{-6} *M* CCCP to the cell suspension. Thus the action of colicin E1 in *E. coli* mimicked the effects of valinomycin on gram-positive bacteria[32]. Similar findings were made independently by Hirata *et al.*[33]. Experiments are in progress in Feingold's laboratory to decide whether this effect of colicin is secondary to the inhibition of energy metabolism or represents a specific effect on permeability, for example to K^+ ions, permitting exchange with H^+ ions when these gain access through the action of CCCP.

Colicin-tolerant ‹membrane› mutants

Another set of observations has made it possible to tie the response to colicins with the functional properties of the bacterial envelope. Rosa Nagel de Zwaig and I[34] have studied bacterial mutants of a class that is ‹tolerant› to certain colicins; they adsorb the colicins without being inhibited. Similar *tol* or *ref* (‹refractory›) mutants have also been studied in several other laboratories. In line with expectations as to the role of the membrane in the response to colicins, we were gratified to discover that all the *tol* mutants we examined exhibited some membrane defect. Some classes of mutants are fragile so that many cells lyse spontaneously during growth, as though the synthesis of the cell envelope were defective. Like other envelope-defective mutants of enteric bacteria, these *tol* mutants are very sensitive to deoxycholate, possibly because the membrane has become accessible to this surface-active agent. More interesting still, one class of *tol* mutants proves to be very sensitive to a whole series of organic dyes, mostly cations such as acridines, ethidium bromide, and methylene blue. We could show that the dye sensitivity was due to a rapid uptake of the dye by the mutant cells, while the normal cells are almost impermeable. Thus this mutation to colicin-tolerance was correlated with a specific change in membrane permeability*.

Some preliminary analytical studies of the envelopes of normal bacteria and tolerant mutants have not revealed any significant differences between them. The chemistry of the cell envelope of enteric bacteria is extremely complex and remains poorly known. Even when chemical changes are found it is not easy to decide whether they are directly relevant to the phenomena under study. This is true, for example, of the changes in phospholipid composition reported in colicin-treated bacteria[35]. It is encouraging, however, that both the study of response to certain colicins and those of colicin-tolerant mutants have converged to focus our attention on the relation between sites of colicin action and the functions of the bacterial membrane. For the time being the relation is tenuous and inferential. But the observations are encouraging enough to reinforce our hope that the study of colicins may reveal, within the membrane, levels of organization at which some of the essential functions of bacterial cells are masterminded.

* The naive idea of an amplification mechanism of colicin action by over-all conformational changes of the bacterial membrane is not supported by some recent findings with temperature-sensitive *tol* mutants, which indicate that the cell envelope behaves as a mosaic of sensitive and tolerant sites, depending on the temperature at which each site has been synthesized[36].

There are interesting analogies between the present state of colicin research and the state of bacteriophage research in the early 1940's. In both situations, phenomenologies described by pioneer investigators are reexamined by a small group of workers concerned with a new goal. In phage research the goal was to get at elementary phenomena of reproduction, hoping that virus reproduction would help elucidate the replication of genetic materials. In colicin research the goal is to explore the functions of the cytoplasmic membrane of bacteria, with the implicit assumption that the findings may throw light on the general problem of the functional organization of cellular membranes. In both situations, the use of simple bacterial systems represents a departure from the traditional materials of the respective disciplines, genetics and ‹membranology›.

As in bacteriophage research 25 years ago, the practitioners of colicin research today are few, cooperative, and moderately confident of success – and somewhat fearful that success may again transform a quiet area of research into «an elephantine academic discipline»[37]. Again as in phage research, we know that full answers will come only when the problems we are exploring will be ready for a rigorous biochemical approach. It may turn out to be a kind of biochemistry as novel as that of gene function and replication was in its own time. Maybe we will again turn up something meaningful and exciting.

1. S.E.Luria and M.Delbrück, *Arch.Biochem.*, 1 (1942) 207.
2. G.R.Wyatt and S.S.Cohen, *Nature*, 170 (1952) 1072.
3. S.S.Cohen, *Virus-induced Enzymes*, Columbia University Press, New York, 1961.
4. J.Cairns and C.I.Davern, *J.Mol.Biol.*, 17 (1966) 418.
5. E.M.Kutter and J.S.Wiberg. *J.Mol.Biol.*, 38 (1968) 395.
6. H.V.Aposhian, in H.Fraenkel-Conrat (Ed.), *Molecular Basis of Virology*, Reinhold, New York, 1968, p.497.
7. D.H.Roscoe, *Virology*, 38 (1959) 527.
8. R.O.R.Kaempfer and B.Magasanik. *J.Mol.Biol.*, 27 (1967) 453.
9. R.R.Burgess, A.A.Travers, J.J.Dunn and E.K.F.Bautz, *Nature*, 221 (1969) 43.
10. W.C.Summers and R.B.Siegel, *Nature*, 223 (1969) 1111.
11. J.Hosoda and C.Levinthal, *Virology*, 34 (1968) 709.
12. A.A.Travers, *Nature*, 223 (1969) 1107.
13. A.L.Sonenshein and D.H.Roscoe, *Virology*, 39 (1969) 205.

14. R. Losick and A. L. Sonenshein, *Nature*, 224 (1969) 35.
15. R. J. Britten and E. H. Davidson, *Science*, 165 (1969) 349.
16. S. Iseki and T. Sakai, *Proc. Japan Acad.*, 29 (1953) 121.
17. H. Uetake, S. E. Luria and J. W. Burrous, *Virology*, 5 (1958) 68.
18. A. Wright, M. Dankert and P. W. Robbins, *Proc. Natl. Acad. Sci.* (*U. S.*) 54 (1965) 235.
19. A. Ryter, Y. Hirota and F. Jacob, *Cold Spring Harbor Symposia Quant. Biol.*, 33 (1968) 669.
20. W. B. Wood, R. S. Edgar, J. King, I. Lielausis and M. Henninger, *Federation Proc.*, 27 (1968) 1160.
21. F. Jacob, L. Siminovitch and E. Wollman, *Ann. Inst. Pasteur*, 83 (1952) 295.
22. M. Nomura, *Cold Spring Harbor Symposia Quant. Biol.*, 28 (1963) 315.
23. B. L. Reynolds and P. R. Reeves, *Biochem. Biophys. Res. Communs.*, 11 (1963) 140.
24. S. E. Luria, *Ann. Inst. Pasteur*, 107 (1964) 67.
25. J. P. Changeux and J. Thiery, *J. Theoret. Biol.*, 17 (1967) 315.
26. J. Koniskey and M. Nomura, *J. Mol. Biol.*, 26 (1967) 181.
27. K. L. Fields and S. E. Luria, *J. Bacteriol.*, 97 (1969) 57.
28. W. Kundig, S. Ghosh and S. Roseman, *Proc. Natl. Acad. Sci.* (*U. S.*), 52 (1964) 1067.
29. K. L. Fields and S. E. Luria, *J. Bacteriol.*, 97 (1969) 64.
30. D. E. Atkinson, *Ann. Rev. Biochem.*, 35 (1966) 85.
31. D. Cousin, *Ann. Inst. Pasteur*, 113 (1967) 309.
32. F. M. Harold and J. R. Baarda, *J. Bacteriol.*, 96 (1969) 2025.
33. H. Hirata, S. Fukui and S. Ishikawa, *J. Biochem.*, 65 (1969) 843.
34. R. Nagel de Zwaig and S. E. Luria. *J. Bacteriol.*, 94 (1967) 1112.
35. D. Cavard, C. Rampini, E. Barba and J. Polonovski, *Bull. Soc. Chim. Biol.*, 50 (1968) 1455.
36. R. Nagel de Zwaig and S. E. Luria, *J. Bacteriol.*, 99 (1969) 78.
37. G. S. Stent, *Science*, 166 (1969) 479.

Biography

Salvador Edward Luria was born on August 13th, 1912, in Torino, Italy. He has been a naturalized citizen of the U. S. A. since January 1947.

In 1929 he started his studies in Medicine at the University of Torino, where he obtained his M.D. *summa cum laude* in 1935. From 1938 to 1940 he was Research Fellow at the Institute of Radium in Paris; 1940–1942, Research Assistant in Surgical Bacteriology at Columbia University; from 1943 to 1950 he was Instructor, Assistant Professor, and Associate Professor of Bacteriology at Indiana University; in 1950 he was appointed Professor of Microbiology at the University of Illinois; from 1959–1964 he has been Professor of Microbiology at the Massachusetts Institute of Technology; in 1964 he became Sedgwick Professor of Biology at the M.I.T. and in 1965, non-resident Fellow at the Salk Institute for Biological Studies. In 1970 Luria was appointed Institute Professor at the Department of Biology of the M.I.T.

Professor Luria was honoured with the following awards: 1935, Lepetit Prize; 1965, Lenghi Prize, Accademia dei Lincei; 1969, Louisa Gross Horwitz Prize, Columbia University.

He was Guggenheim Fellow, 1942–1943 at Vanderbilt and Princeton; during the year 1963–1964 he worked again in Paris, this time at the Institut Pasteur. He is, or has been, Editor or Member of the Editorial Board of the following journals: *Journal of Bacteriology*, *Virology*, *Experimental Cell Research*, *Journal of Molecular Biology*, *Photochemistry and Photobiology*, *American Naturalist*, *Proceedings of the National Academy of Sciences*, *Annual Review of Genetics*.

Professor Luria is a Member of the National Academy of Sciences, American Academy of Arts and Sciences, American Philosophical Society, American Academy of Microbiology, American Society for Microbiology (President, 1967–1968), American Society of Biological Chemists, Society for General Microbiology, Genetics Society, American Naturalists, Society for the Study of Development and Growth, A. A. A. S., Sigma Xi, A. A. U. P.

Salvador Edward Luria was, in 1945, married to Zella Hurwitz, they have one son, Daniel, who is studying economics. His wife, Zella Hurwitz Luria, Ph. D., is a Professor of Psychology at Tufts University.

1972

Christian B. Anfinsen

Stanford Moore and William H. Stein

Gerald M. Edelman

R.R. Porter

CHRISTIAN B. ANFINSEN

Studies on the Principles that Govern the Folding of Protein Chains

December 11, 1972

The telegram that I received from the Swedish Royal Academy of Sciences specifically cites "... studies on ribonuclease, in particular the relationship between the amino acid sequence and the biologically active conformation ..." The work that my colleagues and I have carried out on the nature of the process that controls the folding of polypeptide chains into the unique three-dimensional structures of proteins was, indeed, strongly influenced by observations on the ribonuclease molecule. Many others, including Anson and Mirsky (1) in the '30s and Lumry and Eyring (2) in the '50s, had observed and discussed the reversibility of denaturation of proteins. However, the true elegance of this consequence of natural selection was dramatized by the ribonuclease work, since the refolding of this molecule, after full denaturation by reductive cleavage of its four disulfide bonds (Figure 1), required that only one of the 105

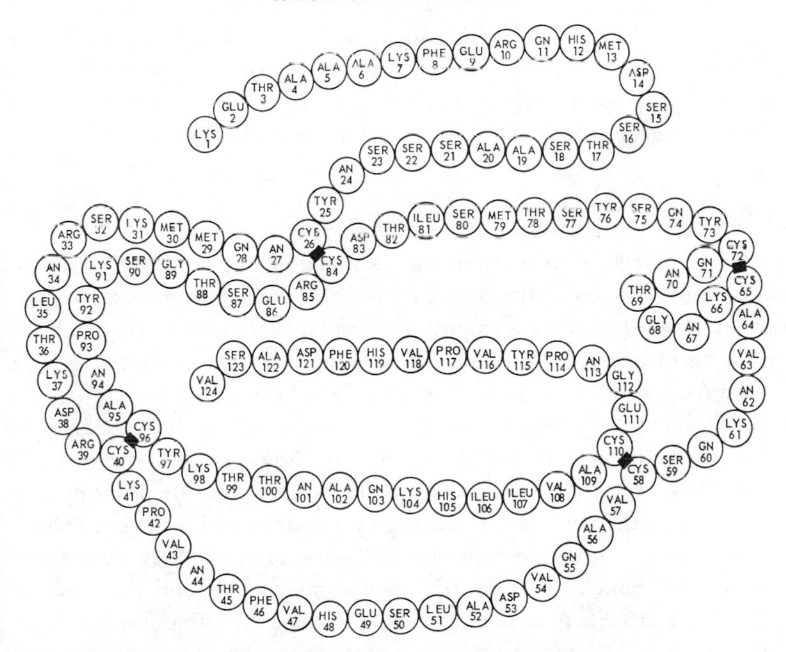

Fig. 1.
The amino acid sequence of bovine pancreatic ribonuclease (3, 4, 5).

possible pairings of eight sulfhydryl groups to form four disulfide linkages take place. The original observations that led to this conclusion were made together with my colleagues, Michael Sela and Fred White, in 1956–1957 (6). These were in actuality, the beginnings of a long series of studies that rather vaguely aimed at the eventual total synthesis of the protein. As we all know, Gutte and Merrifield (7) at the Rockefeller Institute, and Ralph Hirschman and his colleagues at the Merck Research Institute (8), have now accomplished this monumental task.

The studies on the renaturation of fully denatured ribonuclease required many supporting investigations (9, 10, 11, 12) to establish, finally, the generality which we have occasionally called (13) the "thermodynamic hypothesis". This hypothesis—states that the three-dimensional structure of a native protein in its normal physiological milieu (solvent, pH, ionic strength, presence of other components such as metal ions or prosthetic groups, temperature, etc.) is the one in which the Gibbs free energy of the *whole system* is lowest; that is, that the native conformation is determined by the totality of interatomic interactions and hence by the amino acid sequence, in a *given environment*. In terms of natural selection through the "design" of macromolecules during evolution, this idea emphasized the fact that a protein molecule only makes stable, structural sense when it exists under conditions similar to those for which it was selected—the so-called physiological state.

After several years of study on the ribonuclease molecule it became clear to us, and to many others in the field of protein conformation, that proteins devoid of restrictive disulfide bonds or other covalent cross linkages would make more convenient models for the study of the thermodynamic and kinetic aspects of the nucleation, and subsequent pathways, of polypeptide chain folding. Much of what I will review will deal with studies on the flexible and convenient staphylococcal nuclease molecule, but I will first summarize some of the older, background experiments on bovine pancreatic ribonuclease itself.

Support for the "thermodynamic hypothesis."

An experiment that gave us a particular satisfaction in connection with the translation of information in the linear amino acid sequence into native conformation involved the rearrangement of so-called "scrambled" ribonuclease (12). When the fully reduced protein, with 8 SH groups, is allowed to reoxidize under denaturing conditions such as exist in a solution of 8 molar urea, a mixture of products is obtained containing many or all of the possible 105 isomeric disulfide bonded forms (Schematically shown at the bottom right of Figure 2). This mixture is essentially inactive—having on the order of 1 % the activity of the native enzyme. If the urea is removed and the "scrambled" protein is exposed to a small amount of a sulfhydryl group-containing reagent such as mercaptoethanol, disulfide interchange takes place and the mixture eventually is converted into a homogeneous product, indistinguishable from native ribonuclease. This process is driven entirely by the free energy of conformation that is gained in going to the stable, native structure. These experi-

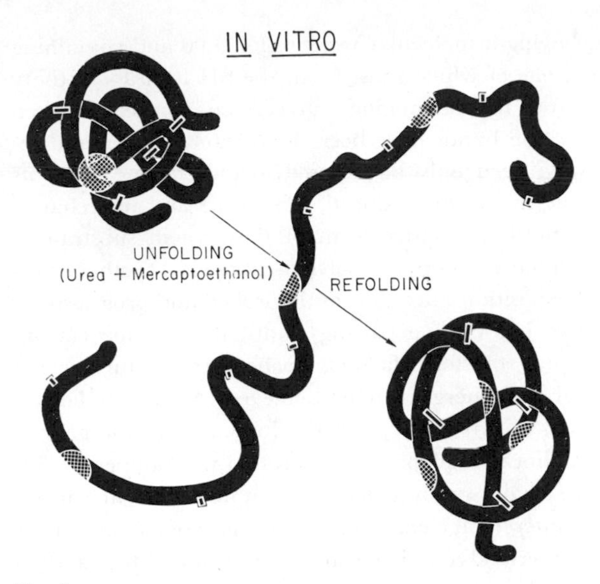

Fig. 2.
Schematic representation of the reductive denaturation, in 8 molar urea solution containing 2-mercaptoethanol, of a disulfide-cross linked protein. The conversion of the extended, denatured form to a randomly cross linked, "scrambled" set of isomers is depicted in the lower right portion of the figure.

ments, incidentally, also make unlikely a process of obligatory, progressive folding during the elongation of the polypeptide chain, during biosynthesis, from the NH_2- to the COOH-terminus. The "scrambled" protein appears to be essentially devoid of the various aspects of structural regularity that characterize the native molecule.

A disturbing factor in the kinetics of the process of renaturation of reduced ribonuclease, or of the "unscrambling" experiments described above, was the slowness of these processes, frequently hours in duration (11). It had been established that the time required to synthesize the chain of a protein like ribonuclease, containing 124 amino acid residues, in the tissues of a higher organism would be approximately 2 minutes (14, 15). The discrepancy between the *in vitro* and *in vivo* rates led to the discovery of an enzyme system in the endoplasmic reticulum of cells (particularly in those concerned with the secretion of extracellular, SS-bonded proteins) which catalyzes the disulfide interchange reaction and which, when added to solutions of reduced ribonuclease or to protein containing randomized SS bonds, catalyzed the rapid formation of the correct, native disulfide pairing in a period less than the requisite two minutes (16, 17). The above discrepancy in rates would not have been observed in the case of the folding of non-crosslinked structures and, as discussed below, such motile proteins as staphylococcal nuclease or myoglobin can undergo virtually complete renaturation in a few a seconds or less.

The disulfide interchange enzyme subsequently served as a useful tool for the examination of the thermodynamic stability of disulfide-bonded protein structures. This enzyme, having a molecular weight of 42,000 and containing three half-cystine residues, one of which must be in the SH form for activity (18, 19), appears to carry out its rearranging activities on a purely random basis. Thus, a protein whose SS bonds have been deliberately broken and reformed in an incorrect way, need only be exposed to the enzyme (with its essential half-cystine residue in the pre-reduced, SH form) and interchange of disulfide bonds occurs until the native form of the protein substrate is reached. Presumably, SS bonds occupying solvent-exposed, or other thermodynamically unfavorable positions, are constantly probed and progressively replaced by more favorable half-cystine pairings, until the enzyme can no longer contact bonds because of steric factors, or because no further net decrease in conformational free energy can be achieved. Model studies on ribonuclease derivatives had shown that, when the intactness of the genetic message represented by the linear sequence of the protein was tampered with by certain cleavages of the chain, or by deletions of amino acids at various points, the added disulfide interchange enzyme, in the course of its "probing", discovered this situation of thermodynamic instability and caused the random reshuffling of SS bonds with the formation of an inactive cross-linked network of chains and chain fragments (e.g., (20)). With two naturally occurring proteins, insulin and chymotrypsin, the interchange enzyme did, indeed, induce such a randomizing phenomenon (21). Chymotrypsin, containing three SS-bonded chains, is known to be derived from a single-chained precursor, chymotrypsinogen, by excision of two internal bits of sequence. The elegant studies of Steiner and his colleagues subsequently showed that insulin was also derived from a single-chained precursor, proinsulin (Figure 3), which is converted to

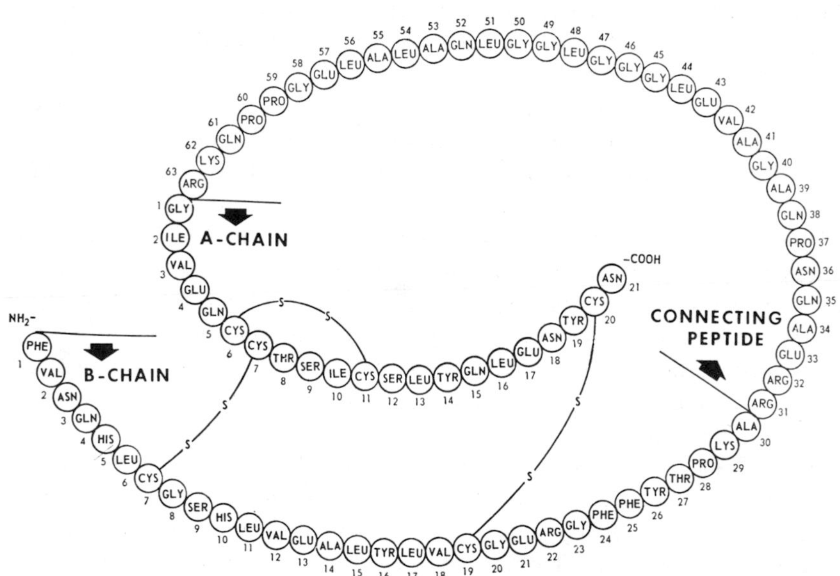

Fig. 3.
The structure of porcine proinsulin (R. E. Chance, R. M. Ellis and W. W. Bromer, Science, *161*, 165 (1968).

the two-chained form, in which we normally find the active hormone, by removal of a segment from the middle of the precursor strand after formation of the 3 SS bonds (22). In contrast, the multichained immune globulins are *not* scrambled and inactivated by the enzyme, reflecting the fact that they are normal products of the disulfide bonding of 4 preformed polypeptide chains.

Factors contributing to the correct folding of polypeptide chains. The results with the disulfide interchange enzyme discussed above suggested that the correct and unique translation of the genetic message for a particular protein backbone is no longer possible when the linear information has been tampered with by deletion of amino acid residues. As with most rules, however, this one is susceptible to many excpetions. First, a number of proteins have been shown to undergo reversible denaturation, including disulfide bond rupture and reformation, after being shortened at either the NH_2- or COOH-terminus (23). Others may be cleaved into two (24, 25, 26), or even three, fragments which, although devoid of detectable structure alone in solution, recombine through noncovalent forces to yield biologically active structures with physical properties very similar to those of the parent protein molecules. Richards and his colleagues (24) discovered the first of these recombining systems, ribonuclease-S (RNase-S), which consists of a 20 residue fragment from the NH_2-terminal end held by a large number of noncovalent interactions to the rest of the molecule, which consists of 104 residues and all four of the disulfide bridges. The work by Wyckoff, Richards and their associates on the three-dimensional structure of this two-fragment complex (27) and on the identification of many of the amino acid side chains that are essential for complementation is classical, as are studies by Hofmann (28) and Scoffone (29) and their colleagues on semi-synthetic analogues of this enzyme derivative. Studies in our own laboratory (30) showed that the 20 residue "RNase-S-peptide"

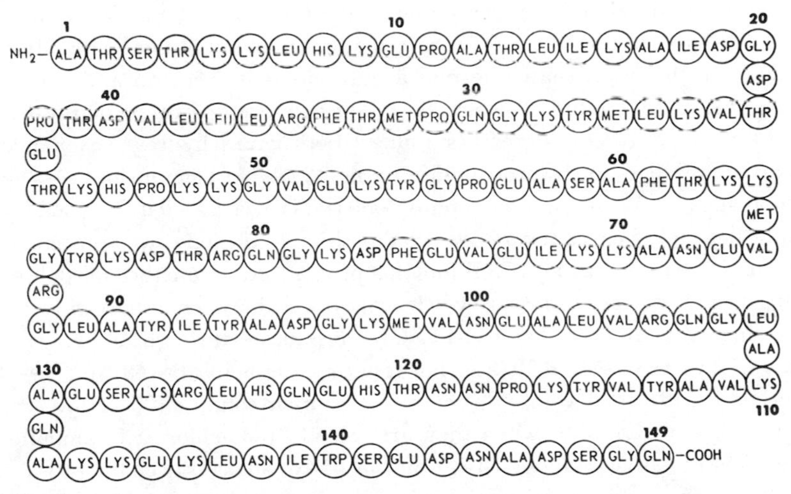

Fig. 4.
Covalent structure of the major extracellular nuclease of *Staphylococcus aureus* (32, 33).

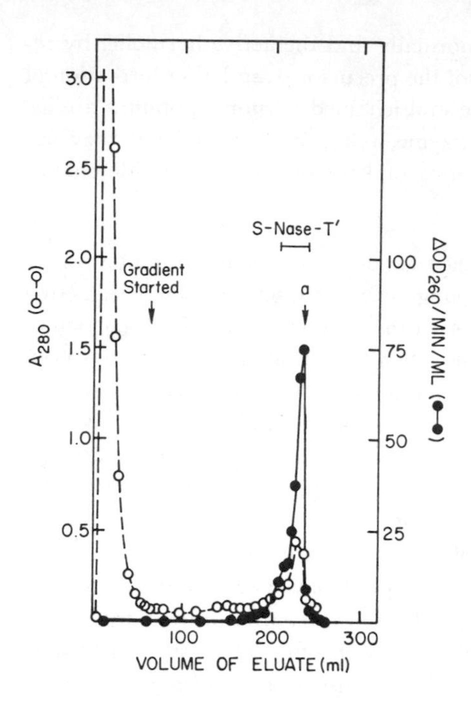

Fig. 5.
Isolation of semisynthetic nuclease-T on a phosphocellulose column following "functional" purification by trypsin digestion in the presence of calcium ions and thymidine-3′5′-diphosphate (41).

fragment could be reduced by 5 residues at its COOH-terminus without loss of enzymic activity in the complex, or of its intrinsic stability in solution.

Other examples of retention of native structural "memory" have been found with complexing fragments of the staphylococcal nuclease molecule (25, 31). This calcium-dependent, RNA and DNA cleaving enzyme (Figure 4) consists of 149 amino acids and is devoid of disulfide bridges and SH groups (32, 33). Although it exhibits considerable flexibility in solution as evidenced by the ready exchange of labile hydrogen atoms in the interior of the molecule with solvent hydrogen atoms (34), only a very small fraction of the total population deviates from the intact, native format at any moment. Spectral and hydrodynamic measurements indicate marked stability up to temperatures ofápproximately 55°. The protein is greatly stabilized, both against hydrogen exchange (34) and against digestion by proteolytic enzymes (35) when calcium ions and the inhibitory ligand, 3′5′-thymidine diphosphate (pdTp), are added. Trypsin, for example, then only cleaves at very restricted positions—the loose aminoterminal portion of the chain and a loop of residues that protrudes out from the molecule as visualized by X-ray crystallography. Cleavage occurs between lysine residues 5 and 6 and, in the sequence -Pro-Lys-Lys-Gly- (residues 47 through 50), between residues 48 and 49 or 49 and 50 (25). The resulting fragments (6—48) and (49—149) or (50—149), are devoid of detectable structure in solution (36). However, as in the case of RNase-S, when they are mixed in stoichiometric amounts regeneration of activity (about 10 %) and of native structural characteristics occurs (the complex is called nuclease-T). Nuclease-T has now been shown (37a) to be closely isomorphous with native nuclease (37b). Thus the cleavages

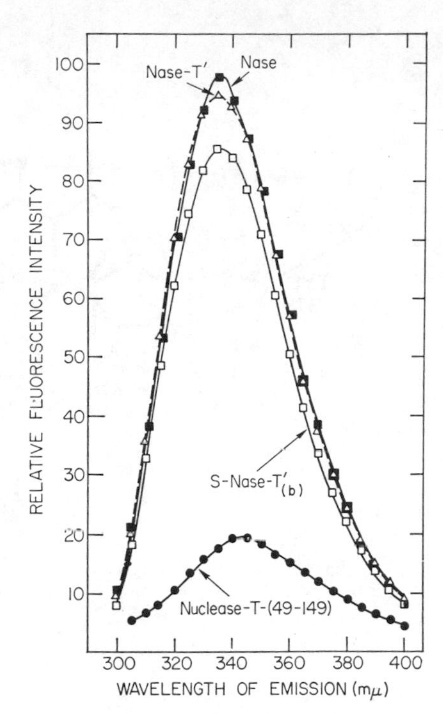

Fig. 6.
Use of fluorescence measurements to determine the relative hydrophobicity (presumably reflecting "nativeness' in the case of nuclease) of the molecular environment of the single tryptophan residue in this protein (39, 41).

and deletions do not destroy the geometric "sense" of the chain. Recently it was shown that residue 149 may be removed by carboxypeptidase treatment of nuclease, and that residues 45 through 49 are dispensible, the latter conclusion the result of solid phase-synthetic studies (38) on analogues of the fragment, (6—47).

Earlier studies by David Ontjes (39) had established that the rapid and convenient solid-phase method developed by Merrifield (40) for peptide synthesis could be applied to the synthesis of analogues of the (6—47) fragment of nuclease-T. The products, although contaminated by sizeable amounts of "mistake sequences" which lack an amino acid residues due to slight incompleteness of reaction during coupling, could be purified by ordinary chromatographic methods to a stage that permitted one to make definite conclusions about the relative importance of various components in the chain. Taking advantage of the limited proteolysis that occurs when nuclease is treated with trypsin in the presence of the stabilizing ligands, calcium and pdTp, Chaiken (41) was able to digest away those aberrant synthetic molecules of (6—47) that did not form a stable complex with the large, native fragment, (49—149). After digestion of the complex, chromatography on columns of phosphocellulose (Figure 5) yielded samples of semisynthetic nuclease-T that were essentially indistinguishable from native nuclease-T. For example, the large enhancement of fluorescence of the single tryptophan residue in nuclease (located at position 140 in the (50—149) fragment) upon addition of the native (6—49) fragment was also shown when, instead, synthetic (6—47) peptide isolated from semisynthetic nuclease-T that had been purified as described above was added (Figure 6).

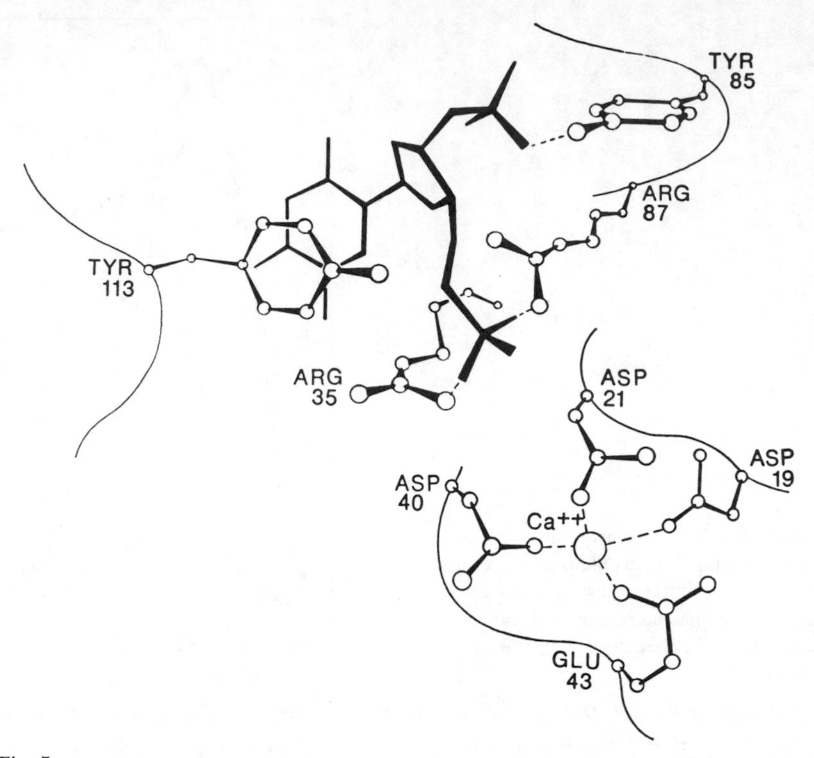

Fig. 7.
Amino acid residues in the sequence of nuclease that are of particular importance in the catalytic activity and binding of substrate and calcium ions (42).

The dispensability, or replaceability, of a number of residues to the stability of the nuclease-T complex was established by examining the fluorescence, activity, and stability to enzymatic digestion, of a large number of semisynthetic analogues (42). As illustrated in Figure 7, interaction with the calcium atom required for nuclease activity normally requires the participation of four dicarboxylic amino acids. Although the *activities* of complexes containing synthetic (6—47) fragments in which one of these had been replaced with an asparagine or glutamine residue were abolished (with one partial exception—asparagine at position 40), three dimensional structure and complex stability was retained for the most part. Similarly, replacement of arginine residue 35 with lysine yielded an inactive complex, but nevertheless one with strong three dimensional similarity to native nuclease-T.

A second kind of complementing system of nuclease fragments (31) consists of tryptic fragment (1—126) and a partially overlapping section of the sequence, (99—149), prepared by cyanogen bromide treatment of the native molecule (shown schematically in Figure 8). These two peptides form a complex with about 15 % the activity of nuclease itself which is sufficiently stable in the presence of pdTp and calcium ions to exhibit remarkable resistance to digestion by trypsin. Thus, many of the overlapping residues in the complex,

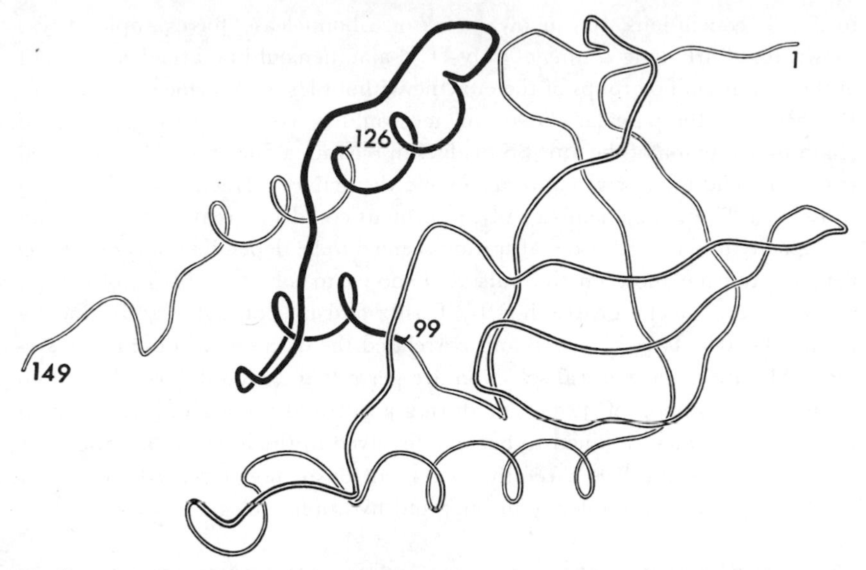

Fig. 8.
A schematic view of the three dimensional structure of staphylococcal nuclease (37b, 53).

(1—126): (99—149), may be "trimmed" away with the production of a derivative, (1—126): (111—149). Further degradation of each of the two components, the former with carboxypeptidases A and B and the latter with leucine aminopeptidase, permits the preparation of (1—124): (114—149) which is as active, and as structurally similar to native nuclease, (as evidenced by estimates of hydrodynamic, spectral, and helical properties) as the parent, undegraded complex. A number of synthetic analogues of the (114—149) sequence have been prepared (43), which also exhibit activity and "native" physical properties when added to (1—126). I will discuss below the manner in which these complexing fragments have been useful in devising experiments to study the processes of nucleation and folding of polypeptide chains.

Mutability of information for chain folding.

Biological function appears to be more a correlate of macromolecular geometry than of chemical detail. The classic chemical and crystallographic work on the large number of abnormal human hemoglobins, the species variants of cytochrome *c* and other proteins from a very large variety of sources, and the isolation of numerous bacterial proteins after mutation of the corresponding genes have made it quite clear that considerable modification of protein sequence may be made without loss of function. In those cases where crystallographic studies of three-dimensional structure have been made, the results indicate that the geometric problem of "designing," through natural selection, molecules that can subserve a particular functional need can be solved in many ways. Only the *geometry* of the protein and its active site need be conserved, except, of course, for such residues as actually participate in a unique way in a catalytic or regulatory mechanism (44). Studies of model systems have led to similar conclusions. In our own work on ribonuclease, for example, it was shown that fairly long chains of poly-D, L-alanine could be attached to eight of the eleven amino groups of the enzyme without loss of enzyme activity (45). Furthermore, the polyalanylated enzyme could be converted to an extended chain by reduction of the four SS bridges in 8 M urea and this fully denatured material could then be reoxidized to yield the active, correctly folded starting substance. Thus, the chemistry of the protein could be greatly modified, and its capacity to refold after denaturation seemed to be dependent only on *internal* residues and not those on the outside, exposed to solvent. This is, of course, precisely the conclusion reached by Perutz and his colleagues (46), and by others (47) who have reviewed and correlated the data on various protein systems. Mutation and natural selection are permitted a high degree of freedom during the evolution of species, or during accidental mutation, but a limited number of residues, destined to become involved in the internal, hydrophobic core of proteins, must be carefully conserved (or at most replaced with other residues with a close similarity in bulk and hydrophobicity).

THE COOPERATIVITY REQUIRED FOR FOLDING AND STABILITY OF PROTEINS.
The examples of non-covalent interaction of complementing fragments of proteins quoted above give strong support to the idea of the essentiality of cooperative interactions in the stability of protein structure. As in the basic rules of languages, an incomplete sentence frequently conveys only gibberish. There appears to exist a very fine balance between stable, native protein structure and random, biologically meaningless polypeptide chains.

A very good example of the inadequacy of an incomplete sequence comes from our observations on the nuclease fragment, (1—126). This fragment contains all of the residues that make up the active center of nuclease. Nevertheless, this fragment, representing about 85 % of the total sequence of nuclease, exhibits only about 0.12 % the activity of the native enzyme (48). The further

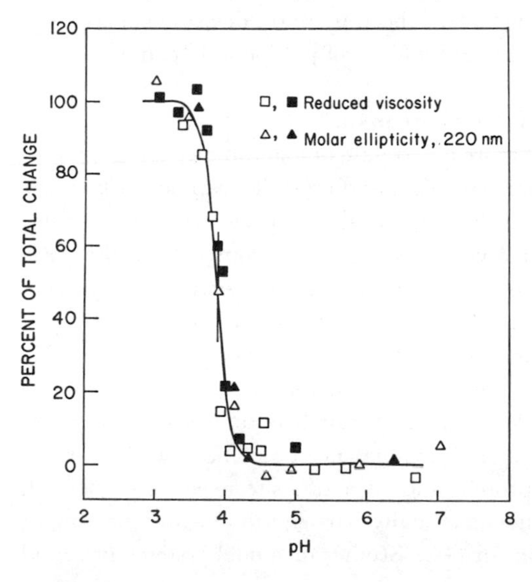

Fig. 9.
Changes in reduced viscocity and molar ellipticity at 220 nm during the acid-induced transition from native to denatured nuclease.
□ and ■, Reduced viscosity; △ and ▲, molar ellipticity at 220 nm. □ and △, Measurements made during the addition of acid; ■ and ▲, measurements made during the addition of base.
A. N. Schechter, H. F. Epstein and C. B. Anfinsen, unpublished results.

addition of 23 residues during biosynthesis, or the addition, *in vitro*, of residues 99—149 as a complementing fragment (31), restores the stability required for activity to this unfinished gene translation.

The transition from incomplete, inactive enzyme, with random structure, to competent enzyme, with unique and stable structure, is clearly a delicately balanced one. The sharpness of this transition may be emphasized by experiments of the sort illustrated in Figure 9. Nuclease undergoes a dramatic change from native globular structure to random disoriented polypeptide over a very narrow range of pH, centered at pH 3.9. The transition has the appearance of a "two-stage" process—either all native or all denatured—and, indeed two-state mathematical treatment has classically been employed to describe such data. In actuality, it has been possible to show, by NMR and spectrophotometric-experiments (49), that one of the 4 histidines and one tyrosine residue of the 7 in nuclease become disoriented before the general and sudden disintegration of organized structure. However, such evidences of a stepwise denaturation and renaturation process are certainly not typical of the bulk of the cooperatively stabilized molecule.

The experiments in Figure 9, involving measurements of intrinsic vicosity and helix-dependent circular dichroism, are typical of those obtained with most proteins. In the case of nuclease, not only is the transition from native to denatured molecule during transfer from solution at pH 3.2 to 6.7 very abrupt, but the process of renaturation occurs over a very short time period. I will not discuss these stop-flow kinetic experiments (50) in detail in this lecture. In brief, the process can be shown to take place in at least two phases; an initial rapid nucleation and folding with a half-time of about 50 milliseconds and a second, somewhat slower transformation with a half time of about 200 milliseconds. The first phase is essentially temperature independent (and therefore possibly entropically driven) and the second temperature dependent.

NUCLEATION OF FOLDING.

A chain of 149 amino acid residues with two rotatable bonds per residue, each bond probably having 2 or 3 permissible or favored orientations, would be able to assume on the order of 4^{149} to 9^{149} different conformations in solution. The extreme rapidity of the refolding makes it essential that the process take place along a limited number of "pathways", even when the statistics are severely restricted by the kinds of stereochemical ground rules that are implicit in a so-called Ramachandran plot. It becomes necessary to postulate the existence of a limited number of allowable initiating events in the folding process. Such events, generally referred to as nucleations, are most likely to occur in parts of the polypeptide chain that can participate in conformational equilibria between random and cooperatively stabilized arrangements. The likelihood of a requirement for cooperative stabilization is high because, in aqueous solution, ionic or hydrogen bonded interactions would not be expected to compete effectively with interactions with solvent molecules and anything less than a sizeable nucleus of interacting amino acid side chains would probably have a very short lifetime. Furthermore, it is important to stress that the

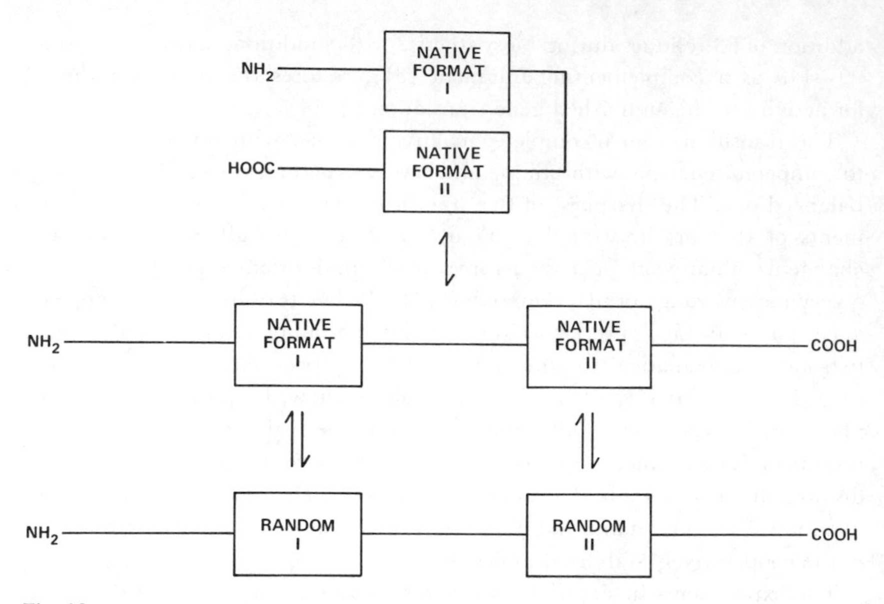

Fig. 10.
How protein chains might fold (see the text for a discussion of this fairly reasonable, but subjective proposal).

amino acid sequences of polypeptide chains designed to be the fabric of protein molecules only make functional sense when they are in the three dimensional arrangement that characterizes them *in the native protein structure.* It seems reasonable to suggest that portions of a protein chain that can serve as nucleation sites for folding will be those that can "flicker" in and out of the conformation that they occupy in the final protein, and that they will form a relatively rigid structure, stabilized by a set of cooperative interactions. These nucleation centers, in what we have termed their "native format", (Figure 10) might be expected to involve such potentially self-dependent substructures as helices, pleated sheets or beta-bends.

Unfortunately, the methods that depend upon hydrodynamic or spectral measurements are not able to detect the presence of these infrequent and transient nucleations. To detect the postulated "flickering equilibria" and to determine their probable lifetimes in solution requires indirect methods that will record the brief appearance of individual "native format" molecules in the population under study. One such method, recently used in our laboratory in a study of the folding of staphylococcal nuclease and its fragments, employs specific antibodies against restricted portions of the amino acid sequence (51).

Figure 8 depicts the three dimensional pattern assumed by staphylococcal nuclease in solution. Major features involving organized structure are the three-stranded antiparallel pleated sheet approximately located between residues 12 and 35, and the three alpha-helical regions between residues 54—67, 99—106 and 121—134. Antibodies against specific regions of the nuclease molecule were prepared by immunization of goats with either polypeptide

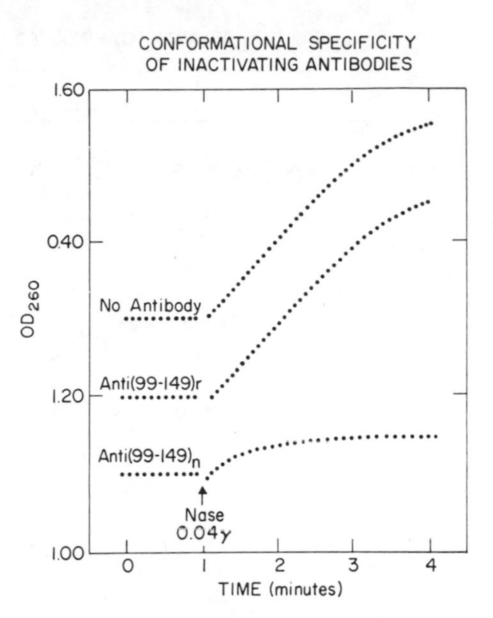

Fig. 11.
Inhibition of nuclease activity by anti-[$(99-149)_n$], and lack of inhibition by anti-[$(99-149)_r$] made against the peptide (99—149), presumably in a random conformation (51).

fragments of the enzyme or by injection of the intact, native protein with subsequent fractionation of the resulting antibody population on affinity chromatography columns consisting of agarose bearing the covalently attached peptide fragment of interest (51, 52). In the former manner there was prepared, for example, an antibody directed against the polypeptide, residues 99—149, known to exist in solution as a random chain without the extensive helicity that characterizes this portion of the nuclease chain when present as part of the intact enzyme. Such an antibody preparation is referred to as anti-$(99-149)_r$, the subscript indicating the disordered state of the antigen.

When, on the other hand, a fraction of anti-native nuclease serum, isolated on an agarose-nuclease column, was further fractionated on agarose-(99—149), a fraction was obtained which was specific for the sequence (99—149) but presumably only when this bit of sequence occupied the "native format". This latter conclusion is based on the observation that the latter fraction, termed anti-$(99-149)_n$ (the subscript "n" referring to the native format) exhibited a strong inhibitory effect on the enzymic activity of nuclease whereas anti-$(99-126)_r$ or anti-$(99-149)_r$ were devoid of such an effect. (see Figure 11). This conclusion was further supported by the observation that the conformation--stabilizing ligands, pdTp and calcium ions, showed a market inhibitory effect on the precipitability of nuclease by anti-$(1-126)_r$ and anti-$(99-149)_r$ but had little effect, if any on such precipitability by anti-$(1-149)_n$ (51). This finding reinforced the idea that many of the antigenic determinants recognized by the anti-*fragment* antibodies are present only in the "unfolded" or "non-native" conformation of nuclease. Analysis of the reaction between anti-(99—149) and nuclease could be shown by measurements of changes in the

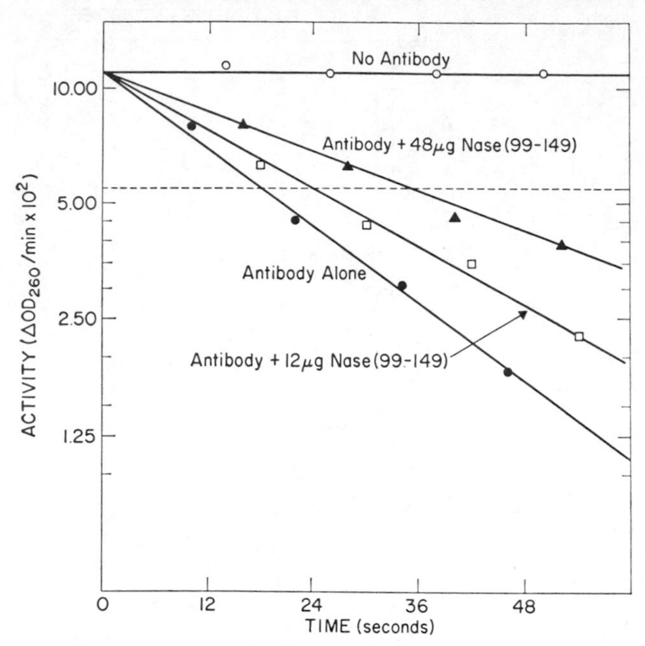

Fig. 12.
Semilogarithmic plot of activity *vs.* time for assays of 0.05 μg of nuclease in the presence of: ○—○ no antibody, ●—● 6 μg of anti-(99—126)$_n$, □—□ 6 μg of anti-(99—126)$_n$ plus 12 μg of (99—149) and ▲—▲ 6 μg of anti-(99—126)$_n$ plus 48 μg of (99—149). The dotted line represents one-half of the initial activity.

kinetics of inhibition of enzyme activity (Fig. 12), to be extremely rapid: k_{off}, on the other hand, is negligibly small.

The system may be described by two simultaneous equilibria, the first concerned with the "flickering" of fragment (99—149), which we shall term "P", from random to native "format", and the second with the association of anti-(99—149)$_n$, which we shall term, simply, "Ab" with fragment P in its native format: i.e. P_n.

$$P_r \rightleftharpoons P_n \quad K_{conf.} = \frac{[P_n]}{[P_r]}$$

$$Ab + P_n \rightleftharpoons AbP_n \quad K_{assoc.} = \frac{[AbP_n]}{[Ab] \cdot [P_n]}$$

$$K_{conf.} = \frac{[AbP_n]}{K_{assoc.} \cdot [Ab] \cdot [P_r]}$$

Two equilibria involving fragment (99—149) of nuclease with the corresponding equilibrium-constant expressions.

The amount of unbound antibody in the second equilibrium may be estimated from measurements of the kinetics of inactivation of the digestion of denatured

Table 1

Studies of the equilibrium between the peptide fragment (99—149) in its random form [99—149)$_r$] and in the form this fragment assumes in the native structure of nuclease (99—149)$_n$]

Abbreviations: P, fragment (99—149); Ab, antibody.

$$K_{conf.} = \frac{[AbP_n]}{K_{assoc.} \cdot [Ab] \cdot [P_T]}$$

$[Ab]_{total\ sites}$ (μM)	$[P_T]$ (μM)	$t_{1/2}$(s)	$[Ab]_{free\ sites}$ (μM)	$[Ab]_{bound\ sites}$ (μM)	$K_{conf.}$	% of P_T as P_n
0.076	0	18	0.076	0	—	—
0.076	0.65	20	0.068	0.0080	2.20×10^{-4}	0.022
0.076	2.0	24	0.057	0.019	2.02×10^{-4}	0.020
0.076	2.6	27	0.051	0.025	2.29×10^{-4}	0.023
0.076	7.8	35	0.039	0.037	1.47×10^{-4}	0.015
0.076	6.5	33	0.042	0.034	1.51×10^{-4}	0.015

$$K_{conf.} = (2.0 \pm 0.4) \times 10^{-4}$$

DNA substrate by a standard amount of nuclease added to the preincubated mixture of fragment (99—149) and anti-(99—149)$_n$. Making the assumption that the affinity of anti-(99—149)$_n$ for (99—149)$_n$ in its folded (P) form is the same as that determined for this antigenic determinant in native nuclease, the value for the term $K_{conf.}$ may be calculated from measureable parameters. A series of typical values shown in Table 1, suggests that approximately 0.02 % of fragment (99—149) exists in the native format at any moment. Such a value, although low, is very large relative to the likelihood of a peptide fragment of a protein being found in its native format on the basis of chance alone.

Empirical considerations of the large amount of data now available on correlations between sequence and three dimensional structure (54), together with an increasing sophistication in the theoretical treatment of the energetics of polypeptide chain folding (55) are beginning to make more realistic the idea of the *a priori* prediction of protein conformation. It is certain that major advances in the understanding of cellular organization, and of the causes and control of abnormalities in such organization, will occur when we can predict, in advance, the three dimensional, phenotypic consequences of a genetic message.

Bibliography

1. Anson, M. L., Advan. Protein Chem. *2*, 361 (1945).
2. Lumry, R. and Eyring, H., J. Phys. Chem. *58*, 110 (1954).
3. Hirs, C. H. W., Moore, S. and Stein, W. H., J. Biol. Chem. *235*, 633 (1960).
4. Potts, J. T., Berger, A., Cooke, J. and Anfinsen, C. B., J. Biol. Chem. *237*, 1851 (1962).
5. Smyth, D. G., Stein, W. H. and Moore, S., J. Biol. Chem. *238*, 227 (1963).
6. Sela, M., White, F. H. and Anfinsen, C. B., Science *125*, 691 (1957).
7. Gutte, B. and Merrifield, R. B., J. Biol. Chem. *246*, 1922 (1971).
8. Hirschmann, R., Nutt, R. F., Veber, D. F., Vitali, R. A., Varga, S. L., Jacob, T. A., Holly, F. W. and Denkewalter, R. G., J. Am. Chem. Soc. *91*, 507 (1969).
9. White, F. H., Jr. and Anfinsen, C. B., Ann. N. Y. Acad. Sci. *81*, 515 (1959).

10. White, F. H., Jr., J. Biol. Chem. *236*, 1353 (1961).
11. Anfinsen, C. B., Haber, E., Sela, M. and White, F. H., Jr., Proc. Nat. Acad. Sci. U. S. *47*, 1309 (1961).
12. Haber, E. and Anfinsen, C. B., J. Biol. Chem. *237*, 1839 (1962).
13. Epstein, C. J., Goldberger, R. F. and Anfinsen, C. B., Cold Spring Harbor Symp. Quant. Biol. *28*, 439 (1963).
14. Dintzis, H. M., Proc. Nat. Acad. Sci. U. S. *47*, 247 (1961).
15. Canfield, R. E. and Anfinsen, C. B., Biochemistry *2*, 1073 (1963).
16. Goldberger, R. F., Epstein, C. J. and Anfinsen, C. B., J. Biol. Chem. *238*, 628 (1963).
17. Venetianer, P. and Straub, F. B., Biochim. Biophys. Acta *67*, 166 (1963).
18. Fuchs, S., DeLorenzo, F. and Anfinsen, C. B., J. Biol. Chem. *242*, 398 (1967).
19. DeLorenzo, F., Goldberger, R. F., Steers, E., Givol, D. and Anfinsen, C. B., J. Biol. Chem. *241*, 1562 (1966).
20. Kato, I. and Anfinsen, C. B., J. Biol. Chem. *244*, 5849 (1969).
21. Givol, D., DeLorenzo, F., Goldberger, R. F. and Anfinsen, C. B., Proc. Nat. Acad. Sci. U. S. *53*, 766 (1965).
22. Steiner, D. F., Trans. N. Y. Acad. Sci. Ser. II *30*, 60 (1967).
23. Anfinsen, C. B., Developmental Biology Supplement *2*, 1 (1968), Academic Press Inc. U.S.A., 1968.
24. Richards, F. M., Proc. Nat. Acad. Sci. U. S. *44*, 162 (1958).
25. Taniuchi, H., Anfinsen, C. B. and Sodja, A., Proc. Nat. Acad. Sci. U. S. *58*, 1235 (1967).
26. Kato, I. and Tominaga, N., FEBS Letters *10*, 313 (1970).
27. Wyckoff, H. W., Tsernoglou, D., Hanson, A. W., Knox, J. R., Lee, B. and Richards, F. M., J. Biol. Chem. *245*, 305 (1970).
28. Hofmann, K., Finn, F. M., Linetti, M., Montibeller, J. and Zanetti, G., J. Amer. Chem. Soc. *88*, 3633 (1966).
29. Scoffone, E., Rocchi, R., Marchiori, F., Moroder, L., Marzotto, A. and Tamburro, A. M., J. Amer. Chem. Soc. *89*, 5450 (1967).
30. Potts, J. T., Jr., Young, D. M. and Anfinsen, C. B., J. Biol. Chem. *238*, 2593 (1963).
31. Taniuchi, H. and Anfinsen, C. B., J. Biol. Chem. *246*, 2291 (1971).
32. Cone, J. L., Cusumano, C. L., Taniuchi, H. and Anfinsen, C. B., J. Biol. Chem. *246*, 3103 (1971).
33. Bohnert, J. L. and Taniuchi, H., J. Biol. Chem. *247*, 4557 (1972).
34. Schechter, A. N., Moravek. L. and Anfinsen, C. B., J. Biol. Chem. *244*, 4981 (1969).
35. Taniuchi, H., Moravek, L. and Anfinsen, C. B., J. Biol. Chem. *244*, 4600 (1969).
36. Taniuchi, H. and Anfinsen, C. B., J. Biol. Chem. *244*, 3864 (1969).
37a. Taniuchi, H., Davies, D. and Anfinsen, C. B., J. Biol. Chem. *247*, 3362 (1972).
b. Arnone, A., Bier, C. J., Cotton, F. A., Hazen, E. E., Jr., Richardson, D. C., Richardson, J. S. and Yonath, A., J. Biol. Chem. *246*, 2302 (1971).
38. Sanchez, G. R., Chaiken, I. M. and Anfinsen, C. B., J. Biol. Chem., 1973, in press.
39. Ontjes, D. and Anfinsen. C. B., J. Biol. Chem. *244*, 6316 (1969).
40. Merrifield, R. B., Science *150*, 178 (1965).
41. Chaiken, I. M., J. Biol. Chem. *246*, 2948 (1971).
42. Chaiken, I. M. and Anfinsen, C. B., J. Biol. Chem. *246*, 2285 (1971).
43. Parikh, I., Corley, L. and Anfinsen, C. B., J. Biol. Chem. *246*, 7392 (1971).
44. Fitch, W. M. and Margoliash, E., Evolutionary Biology, *4*, 67 (1970), Edited by Th. Dobzhansky, M. K. Hecht and W. C. Steere, Appleton-Century-Crofts, New York, 1970.
45. Cooke, J. P., Anfinsen, C. B. and Sela, M., J. Biol. Chem. *238*, 2034 (1963).
46. Perutz, M. F., Kendrew, J. C. and Watson, H. C., J. Mol. Biol. *13*, 669 (1965).
47. Epstein, C. J., Nature *210*, 25 (1966).
48. D. Sachs, H. Taniuchi, A. N. Schechter and A. Eastlake, unpublished work.
49. Epstein, H. F., Schechter, A. N., and Cohen, J. S. *Proc. Nat. Acad. Sci. U. S., 68,* 2042 (1971).

50. Epstein, H. F., Schechter, A. N., Chen, R. F. and Anfinsen, C. B., J. Mol. Biol. *60*, 499 (1971).
51. Sachs, D. H., Schechter, A. N., Eastlake, A. and Anfinsen, C. B., Proc. Nat. Acad. Sci. U. S., *69,* 3790, (1972).
52. Sachs, D. H., Schechter, A. N., Eastlake, A. and Anfinsen, C. B., J. Immunol. *109,* 1300 (1972).
53. Sachs, D. H., Schechter, A. N., Eastlake, A. and Anfinsen, C. B., Biochemistry, *11,* 4268 (1972).
54. Anfinsen, C. B. and Scheraga, H., Adv. in Prot. Chem. ,Vol. 27, 1973, in preparation.
55. H. A. Scheraga, Chemical Reviews, 71, 195 (1971).

CHRISTIAN B. ANFINSEN

Born in Monessen, Pennsylvania, March 26, 1916 Dr. Anfinsen obtained a B.A. degree from Swarthmore College in 1937 and an M.S. in organic chemistry in 1939 from the University of Pennsylvania. He spent the year 1939—40 as a Visiting Investigator at the Carlsberg Laboratory in Copenhagen. In 1943, he received a Ph.D. from Harvard Medical School in biochemistry and spent the next seven years at Harvard Medical School; first as Instructor and then as Assistant Professor of Biological Chemistry. During this time, he spent a year (1947—48) as a Senior Fellow of the American Cancer Society working with Dr. Hugo Theorell at the Medical Nobel Institute. Dr. Anfinsen left Harvard in 1950 to become Chief of the Laboratory of Cellular Physiology and Metabolism in the National Heart Institute of the National Institutes of Health. He was again at Harvard Medical School as Professor of Biological Chemistry in 1962—63 and then returned to the National Institutes of Health to assume his present position.

In Anfinsen's early work, he and Steinberg studied the non-uniform labelling in newly synthesized proteins—a technique with later permitted Dintzis, Canfield and other to determine that proteins are synthesized sequentially from the amino-terminal and *in vivo,* and to calculate the rate at which amino acids are polymerized.

In the mid 1950's Anfinsen began to concentrate on the problem of the relationship between structure and function in enzymes. On the basis of studies on ribonuclease with Sela and White, he proposed that the information determining the tertiary structure of a protein resides in the chemistry of its amino acid sequence. Investigations on reversible denaturation of several proteins served to verify this proposal experimentally. It was demonstrated that, after cleavage of disulfide bonds and disruption of tertiary structure, many proteins could spontaneously refold to their native forms. This work resulted in general acceptance of the "thermodynamic hypothesis". Studies on the rate and extent of renaturation *in vitro* led to the discovery of a microsomal enzyme which catalyzes sulfhydryl-disulfide interchange and thereby accelerates, *in vitro,* the refolding of denatured proteins containing disulfide bonds. In the presence of this enzyme the rate of renaturation approaches that sufficient to account for folding of newly completed polypeptide chains during protein biosynthesis. These findings have given important impetus to studies on the organic synthesis of proteins, since they demonstrate that, under physiological conditions of environment, attainment of the native structure rests solely upon the correct sequential polymerization of the amino acids.

In addition to his research activities, Dr. Anfinsen is an editor of Advances in Protein Chemistry, served on the Editorial Board of the Journal of Biological Chemistry and wrote "The Molecular Basis of Evolution" which was published in 1959. He is active as a member of the Board of Governors of the Weizmann Institute of Science in Rehovot, Israel, and was elected President of the American Society of Biological Chemists for the Academic Year 1971—72. His honors include a Rockefeller Foundation Public Service Award in 1954, a Guggenheim Fellowship in 1958, election to the National Academy of Sciences in 1963 and he Royal Danish Academy in 1964, and Honorary Doctor of Science degrees from Swarthmore College (1965), Georgetown University (1967), and New York Medical College (1969).

In recent years, Anfinsen has devoted himself primarily to comprehensive investigations of an extracellular nuclease of *Staphylococcus aureus*. He and his colleagues have determined the sequence of its 149 amino acids and have described its fundamental enzymological, physical, and immunological properties. They have used an extensive range of spectroscopic and chemical techniques, including new methods of affinity labeling and cross-linking, to delineate the identity and relationship of amino acids in its active site. Dr. Anfinsen has collaborated closely with a crystallographic group at M.I.T., under Professor F. A. Cotton, which has determined the three-dimensional structure of nuclease at high resolution.

Membership and activities:

Board of Governors, Weizmann Institute of Science, Rehovot, Israel
American Society of Biological Chemists (President, 1971—72)
National Academy of Sciences
Royal Danish Academy

Honors:

1954 Rockefeller Foundation Public Service Award
1957 International Union of Pure and Applied Chemistry
Travel Grant to attend Symposium on Proteins, Paris France
1958 Guggenheim Fellowship for Travel to do research at the Weizmann Institute of Science
1959 National Science Foundation Travel Award to attend Conference on Genetic Specificity of Proteins, Copenhagen
1964 NIH Lecture
Kelly Lecture (Purdue University)
1965 D.Sc. (Hon.)—Swarthmore College
1966 Harvey Lecture
1967 D.Sc. (Hon.)—Georgetown University
1969 D.Sc. (Hon.)—New York Medical College
Honorary Fellow, Weizmann Institute of Science
Leon Lecture (University of Pennsylvania)
1970 EMBO Lecturer for Sweden
Visiting Fellow, All Souls College, Oxford
1972 Jubilee Lecture

Stanford Moore and William H. Stein

The Chemical Structures of Pancreatic Ribonuclease and Deoxyribonuclease

December 11, 1972

Introduction

In introducing this summary of experiments on two enzymes, we wish to indicate that the information is representative of what biochemists are obtaining about many proteins. An understanding of the host of reactions in which proteins participate in living cells requires information on the molecular architectures of a wide variety of proteins of different origins and different functions. Such information is coming from laboratories all over the world and draws upon a rich heritage of experience from many investigators. And such knowledge is fundamental to progress in medical research; the Nobel awards this year in Chemistry (concerning ribonuclease) and in Physiology or Medicine (concerning antibodies) both concern basic researches on the chemistry and the biology of proteins.

Occasionally (1) it has been educational to write the structural formula for ribonuclease in full, in terms of its 1,876 atoms of C, H, N, O, and S. Portrayal of the complete molecule with all of the atoms of the amino groups, carboxyl groups, hydroxyl groups, guanido groups, imidazole rings, phenolic groups, indole rings, aromatic, aliphatic, and thiother side chains, sulfhydryl groups, and disulfide bonds, helps in the visualization of the almost infinite number of ways in which such groups could be arranged. This characteristic of proteins makes it possible for nature to design catalysts for such a variety of specific reactions. There is no law that says that a nucleic acid or a polysaccharide could not be an enzyme. But it is understandable that the enzymes so far isolated have turned out to be proteins; a protein is equipped to participate, sometimes through cooperation with coenzymes, in the whole lexicon of organic reactions that require catalysis in the living cell.

Purification of Ribonuclease

The first step in the study of the structure of ribonuclease was, of course, its purification. Ribonuclease was first described in 1920 by Jones (2), who showed that there was present in beef pancreas a relatively heat-stable enzyme capable of digesting yeast nucleic acid. Dubos and Thompson (3) partially purified the enzyme some eighteen years later and in 1940 Kunitz (4) described the isolation of bovine ribonuclease in crystalline form after fractionation by ammonium sulfate precipitation. In order to be as certain as possible that we were beginning the structural study with a single molecular species, we under-

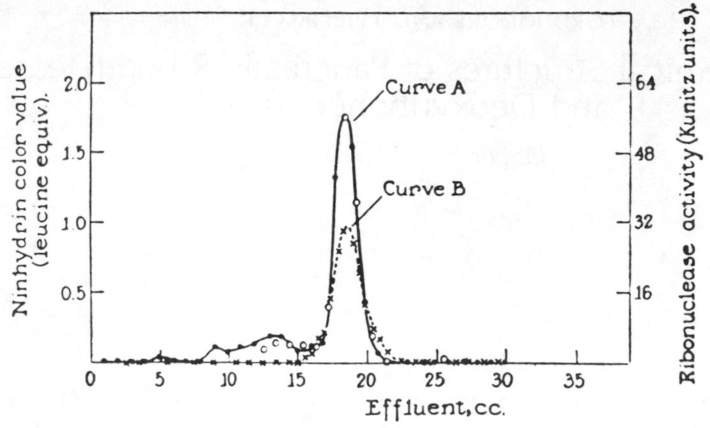

Fig. 1.
Chromatography of crystalline bovine pancreatic ribonuclease (Curve A) on the polymethacrylic acid resin Amberlite IRC-50. Elution was with 0.2 M sodium phosphate buffer at pH 6.45. Curve B was obtained upon rechromatography of material from Peak A. (From (5).

took to apply the potential resolving power of ion exchange chromatography to ribonuclease (Fig. 1). While Werner Hirs, in our laboratory, was exploring the chromatographic purification of ribonuclease on the polymethacrylic resin Amberlite IRC-50 (5,6), Paléus and Neilands (7), in Stockholm, were studying cytochrome C on the same exchanger. These two proteins were the first molecules of their size to be thus purified. The best resolution for ribonuclease (Fig. 2) is now obtained (8) with an exchanger invented in Uppsala, a sulfoethyl cross-linked dextran, which was a development that grew from Porath and Flodin's (9) experiments on gel filtration and drew upon Sober and Peterson's (10) emphasis on the advantages of a carbohydrate matrix for the exchanger.

When pancreatic extracts were analyzed without prior fractionation, two peaks of enzymatic activity were observed by us (6) by ion exchange chromatography and by Martin and Porter (11) by partition chromatography. The major component, ribonuclease A, was selected for the first structural studies. (In later independent experiments, Plummer and Hirs (12, 13) isolated ribonuclease B in pure form from pancreatic juice and showed it to be the same as A but with the addition of a carbohydrate side chain attached to one asparagine residue.)

Amino Acid Analysis

The second step in the structural study of ribonuclease A was the determination of the empirical formula of the chromatographically homogeneous protein in terms of the constituent amino acids. Our appreciation of the importance of quantitative amino acid analysis began in the late 1930's when we had the special privilege of starting our postdoctoral studies in apprenticeship to Max

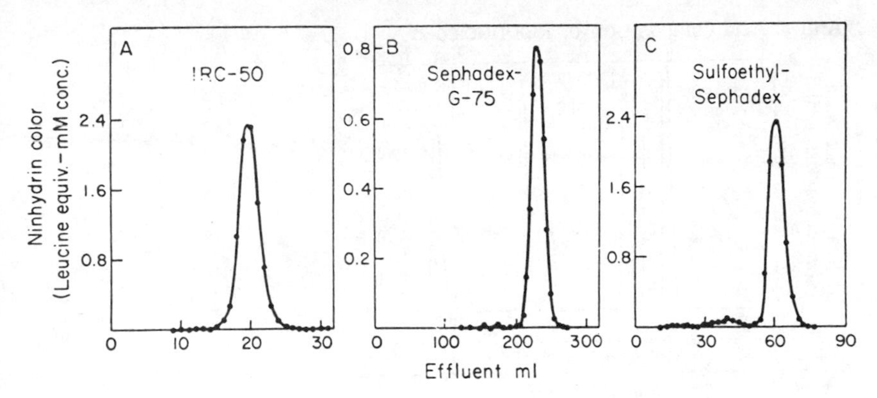

Fig. 2.
Behavior of IRC-50 purified ribonuclease on IRC-50 (A), Sephadex G-75 (B), and sulfoethyl-Sephadex (C) at pH 6.5. From (8).

Bergmann (14). In 1945 it was possible to take a new look at the subject in the light of the renaissance in chromatography stimulated by Martin and Synge in the early 1940's (15—17). In 1949, by combining a quantitative photometric ninhydrin method (18) with elution of amino acids from starch columns by alcohol: water eluents (19, 20) on an automatic fraction collector (21), we were able to analyze a protein hydrolysate in about two weeks by running three such chromatograms to resolve all overlaps. In the early 1950's, the process was speeded up to one week (Fig. 3) by turning to ion exchange chromatography on a sulfonated polystyrene resin (22, 23). In 1958, in co-operation with Darrel Spackman (25, 26) the process was automated (Fig. 4) to give recorded curves (Fig. 5) and the speed was increased to give an over-night run. Shorter columns and faster flow rates (27) permitted an analysis time of about 6 hours. Results from many academic and industrial laboratories have helped to make the procedures simpler and more rapid. Some recent users have adopted 2-hour systems (cf. (28)) and the ninhydrin reagent has been improved (29). In the 1970's, a number of industrially designed analyzers with increased automation have reduced the time for a complete analysis to about 1 hour and increased the sensitivity to the nanomole range. The sharing of knowledge among academic scientists and industrial designers of instruments and manufacturers of ion exchangers has played an important role in progress of biomedical research in this field.

In 1972 there are continuing developments that may make amino acid chromatography more ultramicro and more expeditious. These contributions include the introduction by Udenfriend and his colleagues of an analog of ninhydrin (30—32) that yields, at room temperature, a fluorescent product that can be detected at extremely low concentrations; there is also the continuing possibility that gas chromatography can give fully satisfactory results with amino acid derivatives.

The precision and the sensitivity of current procedures for amino acid analysis have been recently reviewed (33). The developmental research on

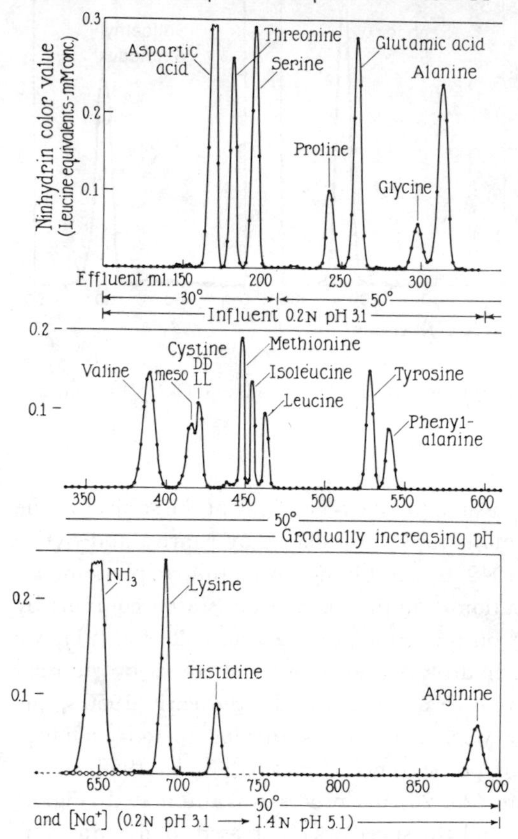

Fig. 3.
The amino acids in an acid hydrolysate of ribonuclease A. The separation of the amino acids was obtained in a five-day run (23) from a 150 × 0.9 cm column of the sulfonated polystyrene resin Dowex 50-X4. From (24).

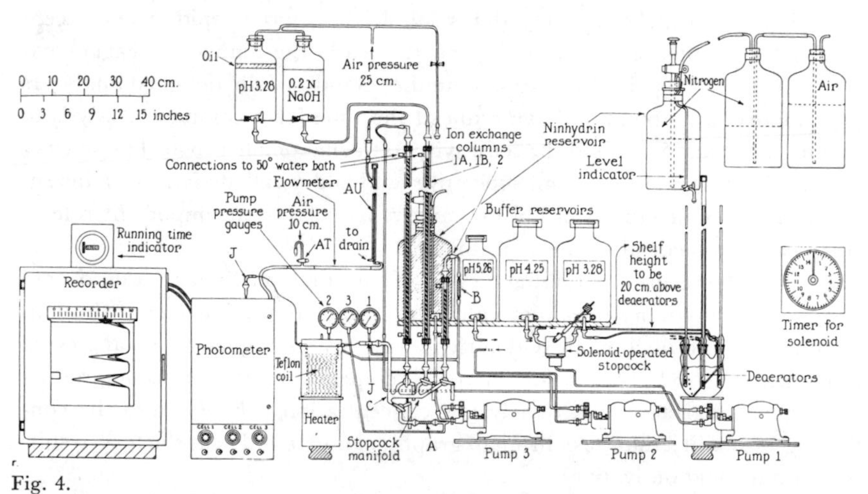

Fig. 4.
Schematic diagram of automatic recording apparatus for the chromatographic analysis of mixtures of amino acids. From (26).

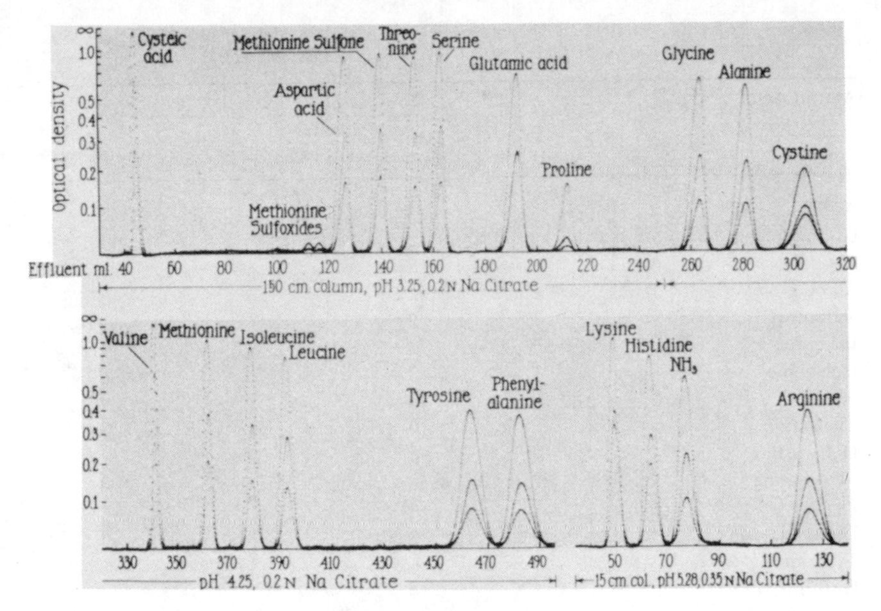

Fig. 5.
Chromatographic analysis of a mixture of amino acids automatically recorded in 22 hours by the equipment shown in Fig. 4. From (26).

quantitative amino acid analysis has also yielded procedures for the isolation of amino acids on a preparative scale (34, 35), for the determination of D- and L-amino acids (36, 37), for the analysis of hydrolysates of foods (38, 39), and for the determination of free amino acids in blood plasma (40), urine (41), mammalian tissues (42), topics that extend beyond the scope of this lecture. Specific discoveries from such studies include the findings by Harris Tallan of 3-methylhistidine (43) and tyrosine-O-sulfate (44) in human urine, acetyl-aspartic acid in brain (45), and cystathionine in human brain (46).

Structure of Ribonuclease

The empirical formula of bovine pancreatic ribonuclease (Table I), determined by the chromatographic methods applied during the structural study, turned out to be that of a molecule containing 124 amino acid residues. From the known mechanisms of protein biosynthesis, coupled with the susceptibility of the peptide bonds to enzymatic hydrolysis, all of the residues are almost certainly of the L-configuration. The calculated molecular weight is 13,683.

The experience with amino acid chromatography led us to try to develop column methods with sufficient resolving power for the separation of the peptides formed by the enzymatic hydrolysis of performic acid-oxidized ribonuclease (47), as in the chromatogram illustrated in Fig. 6 from the experiments of Werner Hirs, who was the first postdoctoral associate to join our laboratory. Fifteen young scholars began their postgraduate careers on the researches summarized in this lecture. Each citation of their contributions connotes our

TABLE I

Amino Acid Composition of Ribonuclease A (47)

Amino Acid	Number of Residues per Molecule (mol. wt. 13,683)
Aspartic acid	15
Glutamic acid	12
Glycine	3
Alanine	12
Valine	9
Leucine	2
Isoleucine	3
Serine	15
Threonine	10
Half-cystine	8
Methionine	4
Proline	4
Phenylalanine	3
Tyrosine	6
Histidine	4
Lysine	10
Arginine	4
Total number of residues	124
Amide NH_3	17

recognition of the ideas, the hard work, and the enthusiasm that facilitate productive research; each of these biochemists shares the credit for the results reported on this occasion and we continue to be stimulated by their current independent accomplishments.

Werner Hirs, through gradient elution from Dowex 50-X2, obtained 100 % yields of the peptides that were completely liberated by tryptic hydrolysis. The elucidation of the sequences of amino acid residues in the peptides and the crossword puzzle-like ordering of the peptides followed many of the principles established by Sanger (48) in his pioneering determination of the structure of insulin, but with the larger molecule of 124 amino acid residues quantitative methods were particularly helpful in the interpretation of the results. A key chemical method in studies of molecules of this size has been the sequential degradation method developed by Pehr Edman (49) with phenylisothiocyanate as the reagent. Instead of determining the resulting phenylthiohydantoins, we have generally used a subtractive procedure in which we utilize the amino acid analyzer to tell us which amino acid has been removed in each step.

The formula for ribonuclease (Fig. 7), largely developed by Werner Hirs (50, 51), Darrel Spackman (52) and Derek Smyth (53, 54), but drawing importantly upon the results of several key experiments by Christian B. Anfinsen and his associates (55—57), in Bethesda, is here written with the customary abbreviations. Ribonuclease was the first enzyme for which the

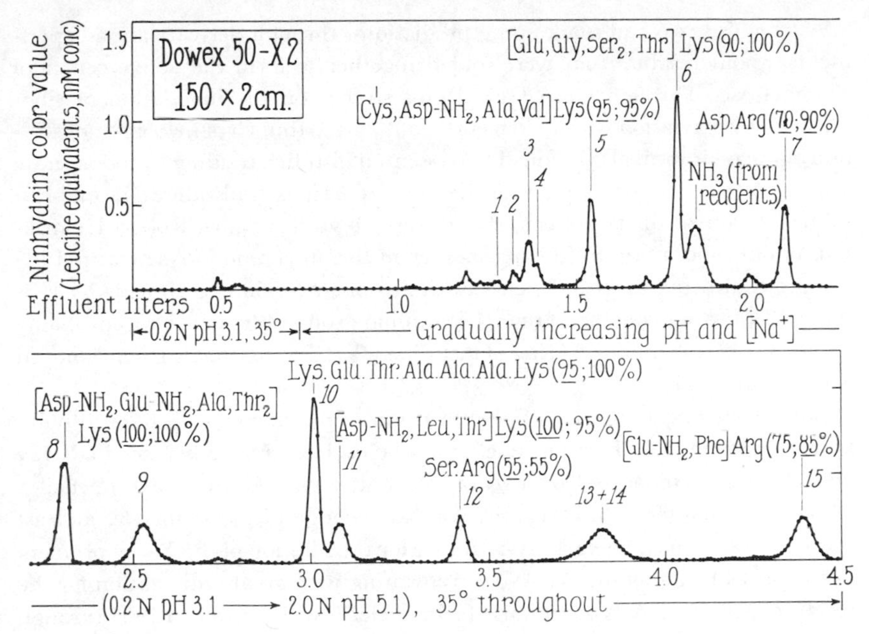

Fig. 6.
Chromatographic separation of the peptides in a tryptic hydrolysate of oxidized ribonuclease A. From (47).

sequence could be written and the determination of its structure was a logical sequel to Sanger's success with the hormone insulin.

The writing of such a two-dimensional formula is only the first step. Linderstrøm-Lang (58) referred to such a sequence as the primary structure of the protein. Catalysis is a three-dimensional operation which involves what Lang termed the secondary and tertiary structures of the chain.

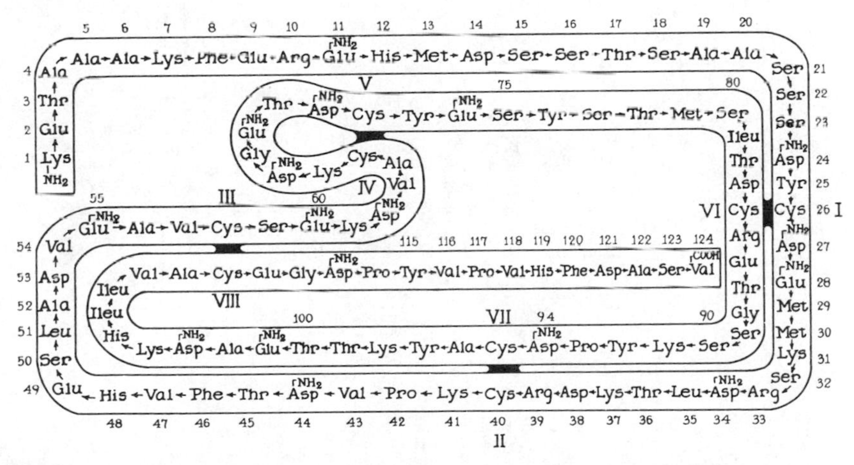

Fig. 7.
The sequence of amino acid residues in bovine pancreatic ribonuclease A. From (54), based upon (50–57).

As chemists, we had made some predictions, through derivatization experiments, about residues that were folded together to form the active center of ribonuclease. Through Gerd Gundlach's studies on the inactiation of ribonuclease by alkylation with iodoacetate (59) and Arthur Crestfield's demonstration of the reciprocal alkylation of two essential histidine residues by iodoacetate at pH 5.5 (60), we concluded that the imidazole rings of histidine-119 and histidine-12 were at the active center and were about 5 Å apart. Robert Heinrikson, in our laboratory, from experiments on the alkylation of lysine at pH 8.5 (61), and drawing upon independent dinitrophenylation experiments by Hirs *et al.* (62), further concluded that the ε-amino group of lysine-41 was probably 7—10 Å from the imidazole ring of histidine-12 and somewhat further removed from histidine-119.

But the chemical approach does not begin to provide enough data to build an adequate model of an enzyme as a whole. The great advances in X-ray crystallography pioneered by Perutz (63) and Kendrew (64) have opened a whole new chapter in this regard, with knowledge of the sequence, at least in considerable part, being a pre-requisite for the solution of the X-ray problem in the present state of the art. We were waiting with great anticipation for the results of X-ray analysis of crystals of ribonuclease which came in 1967 through the researches of Kartha, Bello, and Harker (65) on RNase A and Wyckoff and Richards and their associates (66) on RNase S. In the S-form of the enzyme (67, 68), which is fully active, the chain has been cleaved primarily between the 20th and 21st residues by controlled proteolysis with subtilisin.

Examination of the model shows the approximate positions of the imidazole rings of histidines-12 and -119 and the ε-amino group of lysine-41 to be compatible with the chemical predictions. The substrate for RNase (Fig. 8) is ribonucleic acid, which, from the results of experiments in the laboratories of

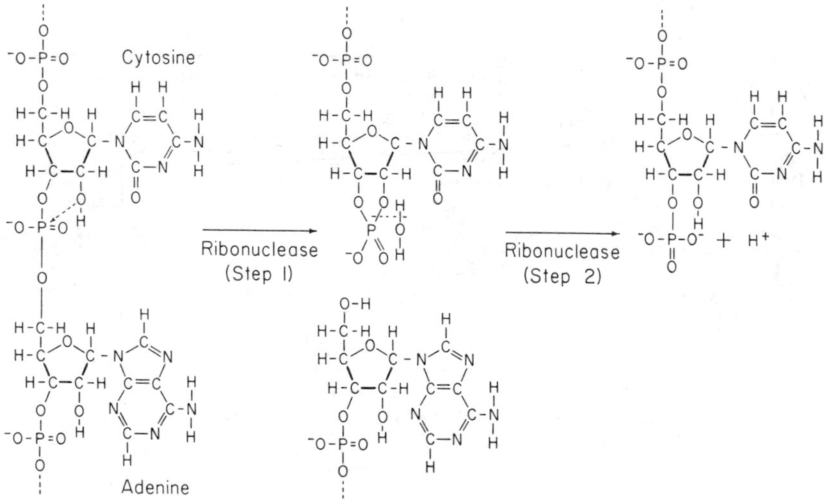

Fig. 8.
The action of ribonuclease on ribonucleic acid (reviewed in (69)).

Todd, of Cohn, and of Markham (reviewed in (69)) is cleaved at the 5′-phosphate ester bond following a pyrimidine-containing nucleotide to give, by transphosphorylation, the 2′, 3′-cyclic phosphate which, in a second step, is hydrolyzed to the 3′-ester. The X-ray data show that the substrate fits in a trough on the surface of the protein with the phosphate moiety near the two imidazole rings of histidines-12 and -119 and with the pyrimidine ring tucked into a hydrophobic pocket close to the aromatic ring of phenylalanine-120

From this picture of the active site (reviewed in (70)), chemical and physical experimentation is progressing in a number of laboratories toward definition of the catalytic process in as explicit terms as possible with primary roles for one charged imidazole and one uncharged imidazole participating in the push-pull which results in transphosphorylation or hydrolysis of the phosphate ester bond.

These further experiments carry the subject into the third chapter in the history of ribonuclease, its chemical synthesis, which has grown from the many innovations in the methods for peptide synthesis in recent years. A preparation with 70 % of the activity of native ribonuclease has been synthesized through a major effort by Gutte and Merrifield (71, 72). An active RNase S-protein has been synthesized by Hirschmann, Denkewalter and associates (73). The yields in the syntheses are dependent upon a very important property of the disulfide bonds of RNase studied by White (74) and by Anfinsen and his associates (75) and reviewed particularly in terms of its special biological significance by Anfinsen (76). The reduced chain with 8 -SH groups folds to give the proper pairs of S-S bonds for the active conformation of the protein. The intramolecular forces that guide such a folding, and the similar forces that contribute to the specific aggregation of the chains of a protein with multiple subunits, such as hemoglobin, form a continuing subject of research.

The ability to synthesize RNase, or major parts thereof, opens new avenues for the identification of residues that may be essential for activity through the preparation of analogs of ribonuclease with substituitions at specific positions. Hofmann and Scoffone and their respective associates (reviewed in (70)) have done this at the amino end of the chain by synthetic variations in Richards' S-peptide. We have cooperated with Merrifield and his associates in the following recent series of experiments which illustrate the application of chemical surgery to the COOH end of the molecule and the use of synthetic replacements.

How can we examine the question of whether the proximity of the aromatic ring of phenylalanine-120 and the pyrimidine ring of the substrate leads to specific interaction between two six-membered rings in a way which is important in the binding of substrate to enzyme? Michael Lin, in our laboratory, was able to cut off, enzymatically (by pepsin according to Anfinsen (77) plus carboxypeptidase A) the last 6 residues from ribonuclease which include phenylalanine-120 and histidine-119. The resulting molecule (78) is completely inactive and does not bind substrate. Concurrently, Gutte and Merrifield had synthesized the 14-residue L-peptide from glutamic acid-111 through valine-124. When this synthetic fragment is mixed with the molecule missing residues

119 to 124, the added peptide is adsorbed and 90 % of the activity of the native enzyme is regained (79). The missing histidine is thus supplied for the active center; other residues in the peptide adsorb on the protein core in such a way as to re-form the binding site and the catalytic site. This result parallels in principle the earlier experiments of Richards and Vithayathil (68) on the removal and adsorption of the S-peptide at the NH_2 end.

When leucine or isoleucine is substituted for phenylalanine at position 120 in the synthetic peptide, the combination of peptide and protein has 10 % of the activity of ribonuclease and binds substrate as effectively as the native enzyme (80). In this way we conclude that the aromatic ring is not essential for binding of the pyrimidine ring or for activity. But the lower activity when leucine or isoleucine is substituted indicates that the aromatic ring of phenylalanine fits into the hydrophobic pocket more specifically than the aliphatic side chains and probably serves to orient histidine-119 more exactly in the delicate balance with histidine-12 that gives the active site its full catalytic power.

Another way to learn what residues can be varied without loss of activity is to study the changes in pancreatic ribonucleases from different species, as has been done for the enzymes from sheep (81), rat (82), and pig (83, 84).

Deoxyribonuclease

A further way to gain insight into what makes ribonuclease so specific for its special substrate is to look at enzymes that hydrolyze similar substrates. In the past few years we have turned our attention to pancreatic deoxyribonuclease. This enzyme, which is about twice the size of ribonuclease, hydrolyzes DNA in the presence of bivalent cations, such as Mn^{++}, to give 5′-mononucleotides and larger fragments (85). Deoxyribonuclease first attracted special attention in the classic work of Avery, MacLeod, and McCarty (86, 87), who showed that the transforming principle of the pneumococcus could be destroyed by the action of the enzyme. McCarty's (88) experiments on the purification of the enzyme from pancreas were followed by those of Kunitz (89) and of

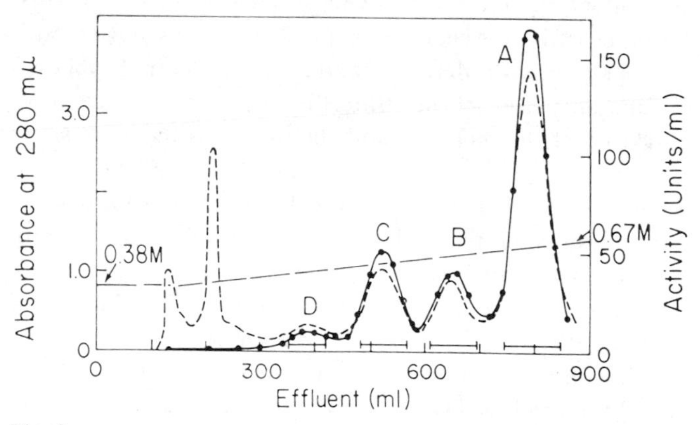

Fig. 9.
Chromatography of bovine pancreatic deoxyribonuclease, prepared by ammonium sulfate precipitation (89), on phosphocellulose at pH 4.7 with a sodium acetate buffer of increasing molarity. From (92).

```
                                          10                                  Carb.
Leu-Lys-Ile-Ala-Ala-Phe-Asn-Ile-Arg-Thr-Phe-Gly-Glu-Thr-Lys-Met-Ser-Asn-
 20                                30
Ala-Thr-Leu-Ala-Ser-Tyr-Ile-Val-Arg-Arg-Tyr-Asp-Ile-Val-Leu-Ile-Glu-Gln-Val-
    40                                    50
Arg-Asp-Ser-His-Leu-Val-Ala-Val-Gly-Lys-Leu-Leu-Asp-Tyr-Leu-Asn-Gln-Asp-Asp-
         60                                    70
Pro-Asn-Thr-Tyr-His-Tyr-Val-Val-Ser-Glu-Pro-Leu-Gly-Arg-Asn-Ser-Tyr-Lys-Glu-
                 80                                    90
Arg-Tyr-Leu-Phe-Leu-Phe-Arg-Pro-Asn-Lys-Val-Ser-Val-Leu-Asp-Thr-Tyr-Gln-Tyr-
                      100                                        110
Asp-Asp-Gly-Cys-Glu-Ser-Cys-Gly-Asn-Asp-Ser-Phe-Ser-Arg-Glu-Pro-Ala-Val-Val-
            |_____________|
                                                                130
Lys-Phe-Ser-Ser-His-Ser-Thr-Lys-Val-Lys-Glu-Phe-Ala-Ile-Val-Ala-Leu-His-Ser-
                                 140                                    150
Ala-Pro-Ser-Asp-Ala-Val-Ala Glu-Ile-Asn-Ser-Leu-Tyr-Asp-Val-Tyr-Leu-Asp-Val-
                                      160                                    170
Gln-Gln-Lys-Trp-His-Leu-Asn-Asp-Val-Met-Leu-Met-Gly-Asp-Phe-Asn-Ala-Asp-Cys-
                                        180                                   |
Ser-Tyr-Val-Thr-Ser-Ser-Gln-Trp-Ser-Ser-Ile-Arg-Leu-Arg-Thr-Ser-Ser-Thr-Phe- |
190                                    200                                    |
Gln-Trp-Leu-Ile-Pro-Asp-Ser-Ala-Asp-Thr-Thr-Ala-Thr-Ser-Thr-Asn-Cys-Ala-Tyr-
 210                                      220
Asp-Arg-Ile-Val-Val-Ala-Gly-Ser-Leu-Leu-Gln-Ser-Ser-Val-Val-Gly-Pro-Ser-Ala-
      230                                          240
Ala-Pro-Phe-Asp-Phe-Gln-Ala-Ala-Tyr-Gly-Leu-Ser-Asn-Glu-Met-Ala-Leu-Ala-Ile-
          250                                257
Ser-Asp-His-Tyr-Pro-Val-Glu-Val-Thr-Leu-Thr
```

Fig. 10.
The sequence of amino acid residues in bovine pancreatic deoxyribonuclease A. From (93, 94).

Lindberg (90). Our studies began when Paul Price, as a graduate student, undertook the chromatographic purification of deoxyribonuclease. His initial studies showed that the enzyme, in the absence of bivalent metals, was extremely sensitive to proteolysis. Success in the purification depended upon keeping metals such as Ca^{++} present or adding diisopropyl phosphorofluoridate to inactivate the pancreatic proteases. He succeeded in resolving preparations of deoxyribonuclease into two active components on sulfoethyl-Sephadex (91) and Hans Salnikow subsequently obtained even higher resolving power (Fig. 9) with phosphocellulose (92). There are three main active components: DNase A is a glycoprotein, DNase B is a sialoglycoprotein, and DNase C is similar to A but with a proline residue substituted for one histidine. These three deoxyribonucleases were also present in the pancreatic juice from a single animal.

The determination of the chemical structure of DNase A was undertaken by

Hans Salnikow (93) and carried to completion this year by Ta-hsiu Liao (94). The working hypothesis for the structure of the molecule (Fig. 10) indicates a single chain of 257 residues with two disulfide bonds. The ordering of the tryptic and chymotryptic peptides in the reduced and carboxymethylated chain and the pairing of the half-cystine residues in the native enzyme were greatly facilitated by the cleavage of the molecule at the four methionine residues by the cyanogen bromide by the method of Gross and Witkop (95). Amino acid analyses at the nanomole level made possible sequence determinations on small amounts of peptides isolated by chromatography or paper electrophoresis.

Some of the special features of the structure can be discussed in reference to the diagram in Fig. 11. The carbohydrate side-chain, which contains 2 residues of N-acetylglucosamine and 2 to 6 residues of mannose (91, 92, 96) and which Brian Catley showed was attached via an aspartamidohexosamine linkage to a -Ser-Asn-Ala-Thr- sequence (96), is found at only one position in the chain, at residue 18. Tony Hugli studied the nitration of deoxyribonuclease (97) by tetranitromethane (98); the enzyme is inactivated by the modification of one tyrosine residue which turns out to be residue 62. Paul Price discovered that inactivation of DNase by iodoacetate in the presence of Cu^{++} and Tris buffer (99) is accompanied by carboxymethylation of one residue of histidine; from the sequences of a 3-carboxymethylhistidine-containing peptide and that of the protein, the essential imidazole ring is found to be in residue 131. DNase C (92) is the result of a mutation which causes one histidine to be replaced by a proline without any change in the activity. Hans Salnikow and Dagmar Murphy (100) have shown that this change occurs at position 118; the histidine at this position in DNase A is thus not essential for enzymatic activity.

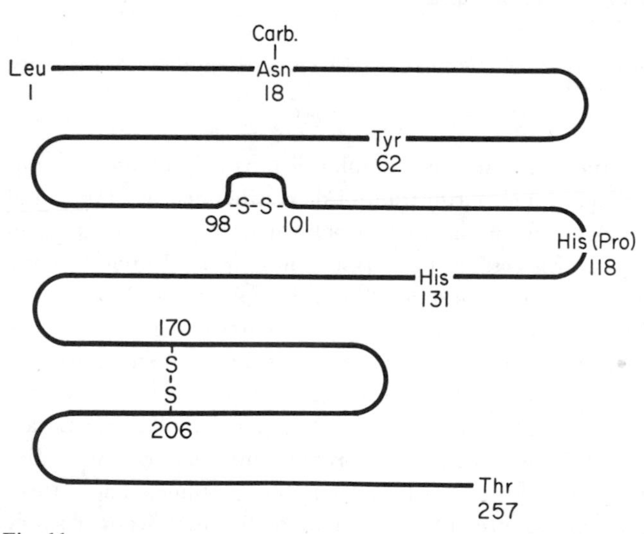

Fig. 11.
Diagram of special features of deoxyribonuclease A (94) and the substitution of Pro for His in deoxyribonuclease C (100).

The two disulfide bonds of deoxyribonuclease possess some unusual properties. Paul Price showed that even without the use of a denaturing agent both bonds are very easily reduced by mercaptans in the absence of calcium to give an inactive product. In the presence of calcium, one bond is stable and one bond is reduced (101) and the product is active. Ta-hsiu Liao has identified the non-essential disulfide bond as the one forming the small loop between residues 98 and 101 (94). When the larger loop, formed by half-cystines 170 and 206, is opened, the activity is lost.

The next step will be the correlation of the chemical evidence with the three-dimensional structure of the enzyme, if X-ray analysis of crystalline DNase A can be successfully accomplished.

Conclusion

In the course of studying enzymes of different functions, we have had the pleasure of cooperation with Kenji Takahashi in the identification of a carboxyl group of glutamic acid as part of the active site of ribonuclease T_1 (102). The essential -SH group and histidine residue of streptococcal proteinase have have been studied in collaboration with Stuart Elliott and Teh-yung Liu (103, 104). The esterification of carboxyl groups at the active center of pepsin was explored with T. G. Rajagopalan (105). There is a vast amount of basic information needed on various enzymes before biochemists can explan catalytic action in full detail. Enzyme chemistry today is in a stage of development that bears some similarity to that of organic chemistry at the beginning of this century. At that time there was great activity in documenting the properties of the myriad small organic compounds conceivable by man and nature. Today, in the polypeptide field, the list of determined structures is relatively small. The enzymes that have been studied first are those that can be prepared in gram quantities, such as ribonucleaase, trypsin, lysozyme, carboxypeptidase, and subtilisin. The experience in the determination of such structures is leading to ultramicromethods which will extend the range of structural studies to tissue enzymes that are present in very small amounts.

From the knowledge of the structures of a large series of enzymes, underlying principles of how nature designs catalysts for given purposes will evolve. And there will be practical dividends from such research on proteins. One example of research-in-progress can illustrate this possibility. The project developed in the following way: In the course of examining the importance of the three-dimensional configuration of RNase to its activity, George Stark had occasion to dissolve the enzyme in 8 M urea at 40° (106). In one of those experiments the RNase was not active after the urea was removed by dialysis, and it turned out that traces of cyanate in the urea solution had carbamylated the ε-NH_2 groups of the enzyme. The chemistry of the subject carries us back to Wöhler's (107) observations on the relationship between ammonium cyanate cyanate and urea in 1828. In 1970, Anthony Cerami and James Manning (108), two young investigators at the Rockefeller University, undertook to explore, fully on their own initiative, whether traces of cyanate in urea might have a role in the reported beneficial effect of urea on the sickling of erythrocytes of individuals carrying hemoglobin S. They have discovered that there is such an effect of cyanate on human erythrocytes, and that it is accompanied by carbamylation of the α-NH_2 groups of the valine residues of the α- and β-chains of hemoglobin S. The knowledge that a relatively simple chemical modification of hemoglobin S can restore nearly normal function to the deficient molecule opens the possibility that a genetic defect in man might be remedied, not by having to change the gene, but by redivatizing the protein.

Such results afford an example of the manner in which one finding leads to another in basic research and ultimately to possible benefits to man. When we consider biochemistry in 1972, it is important to realize how fragmentary is our knowledge of the molecular basis of life. Very few macromolecules can be discussed in the detail with which ribonuclease or hemoglobin can be defined. Such knowledge of structure-function relationships is basic to the rational approach to the intricate synergisms of living systems.

ACKNOWLEDGEMENTS

The researches from our laboratory on ribonuclease and deoxyribonuclease summarized in this lecture have been possible through financial backing by The Rockefeller University, The United States Public Health Service, and The National Science Foundation.

REFERENCES

1. Stein, W. H., and Moore, S. Scientific American 204, 81–92, 1961.
2. Jones, W., Am. J. Physiol. 52, 203–207, 1920.
3. Dubos, R. J., and Thompson, R. H. S., J. Biol. Chem. *124*, 501–510, 1938.
4. Kunitz, M., J. Gen. Physiol. *24*, 15–32, 1940.
5. Hirs, C. H. W., Stein, W. H., and Moore, S., J. Amer. Chem. Soc. *73*, 1893, 1951.
6. Hirs, C. H. W., Moore, S., and Stein, W. H., J. Biol. Chem. *200*, 493–506, 1953.
7. Paléus, S., and Neilands, J. B., Acta Chem. Scand. *4*, 1024–1030, 1950.
8. Crestfield, A. M., Stein, W. H., and Moore, S., J. Biol. Chem. *238*, 618–621, 1963.
9. Porath, J., and Flodin, P., Nature *183*, 1657–1659, 1959.
10. Sober, H. A., and Peterson, E. A., J. Amer. Chem. Soc. *76*, 1711–1712, 1954.
11. Martin, A. J. P., and Porter, R. R., Biochem. J. *49*, 215–218, 1951.
12. Plummer, T. H., Jr., and Hirs, C. H. W., J. Biol. Chem. *238*, 1396–1401, 1963.
13. Plummer, T. H., Jr., and Hirs, C. H. W., J. Biol. Chem. *239*, 2530–2538, 1964.
14. Moore, S., Stein, W. H., and Bergmann, M., Chem. Rev. *30*, 423–432, 1942.
15. Martin, A. J. P., and Synge, R. L. M., Biochem. J. *35*, 1358–1368, 1941.
16. Martin, A. J. P., Ann. Rev. Biochem. *19*, 517–535, 1950.
17. Synge, R. L. M., Analyst *71*, 256–258, 1946.
18. Moore, S., and Stein, W. H., J. Biol. Chem. *176*, 367–388, 1948.
19. Moore, S., and Stein, W. H., J. Biol. Chem. *178*, 53–77, 1949.
20. Stein, W. H., and Moore, S., J. Biol. Chem. *178*, 79–91, 1949.
21. Stein, W. H., and Moore, S., J. Biol. Chem. *176*, 337–365, 1948.
22. Moore, S., and Stein, W. H., J. Biol. Chem. *192*, 663–681, 1951.
23. Moore, S., and Stein, W. H., J. Biol. Chem. *211*, 893–906; *211*, 907–913, 1954.
24. Hirs, C. H. W., Stein, W. H., and Moore, S., J. Biol. Chem. *211*, 941–950, 1954.
25. Moore, S., Spackman, D. H., and Stein, W. H., Anal. Chem. *30*, 1185–1190, 1958.
26. Spackman, D. H., Stein, W. H., and Moore, S., Anal Chem. *30*, 1190–1206, 1958.
27. Spackman, D. H., Federation Proc. *22*, 244, 1963.

28. Spackman, D. H., In Methods in Enzymology (C. H. W. Hirs, Editor), Vol. 11, Academic Press Inc., New York, pp. 3–15, 1967.
29. Moore, S., J. Biol. Chem. *243*, 6281–6283, 1968.
30. Samejima, K., Dairman, W., and Udenfriend, S., Anal. Biochem. *42*, 222–236, 1971.
31. Samejima, K., Dairman, W., Stone, J., and Udenfriend, S., Anal. Biochem. *42*, 237–247, 1971.
32. Weigele, M., De Bernardo, S. L., Tengi, J. P., and Leimgruber, W., J. Amer. Chem. Soc. *94*, 5927–5928, 1972.
33. Moore, S., In Chemistry and Biology of Peptides (J. Meienhofer, Editor), Ann Arbor Science Publications, Inc., Ann Arbor, Mich., pp. 629—653, 1972.
34. Hirs, C. H. W., Moore, S., and Stein, W. H., J. Biol. Chem. *195*, 669–683, 1952.
35. Hirs, C. H. W., Moore, S., and Stein, W. H., J. Amer. Chem. Soc. *76*, 6063–6065, 1954.
36. Manning, J. M., and Moore, S., J. Biol. Chem. *243*, 5591–5597, 1968.
37. Manning, J. M., J. Amer. Chem. Soc. *92*, 7449–7454, 1970.
38. Schram, E., Dustin, J. P., Moore, S., and Bigwood, E. J., Anal. Chim. Acta *9*, 149–161, 1953.
39. Dustin, J. P., Czakowska, C., Moore, S., and Bigwood, E. J., Anal. Chim. Acta *9*, 256–262, 1953.
40. Stein, W. H., and Moore, S. J. Biol. Chem. *211*, 915–926, 1954.
41. Stein, W. H., J. Biol. Chem. *201*, 45–58, 1953.
42. Tallan, H. H., Moore S., and Stein, W. H., J. Biol. Chem. *211*, 927–939, 1954.
43. Tallan, H. H., Moore, S., and Stein, W. H., J. Biol. Chem. *206*, 825–834, 1954.
44. Tallan, H. H., Bella, S. T., Stein, W. H., and Moore, S., J. Biol. Chem. *217*, 703–708, 1955.
45. Tallan, H. H., Moore, S., and Stein, W. H., J. Biol. Chem. *219*, 257–264, 1956
46. Tallan, H. H., Moore, S., and Stein, W. H., J. Biol. Chem. *230*, 707–716, 1958.
47. Hirs, C. H. W., Moore, S., and Stein, W. H., J. Biol. Chem. *219*, 623–642, 1956.
48. Sanger, F., Science *129*, 1340–1344, 1959.
49. Edman, P., Acta Chem. Scand. *10*, 761–768, 1956.
50. Hirs, C. H. W., J. Biol. Chem. *235*, 625–632, 1960.
51. Hirs, C. H. W., Moore, S., and Stein, W. H., J. Biol. Chem. *235*, 633–647, 1960.
52. Spackman, D. H., Stein, W. H., and Moore, S. ,J. Biol. Chem. *235*, 648–659, 1960.
53. Smyth, D. G., Stein, W. H., and Moore, S., J. Biol. Chem. *237*, 1845–1850, 1962.
54. Smyth, D. G., Stein, W. H., and Moore, S., J. Biol. Chem. *238*, 227–234, 1963.
55. Anfinsen, C. B., Redfield, R. R., Choate, W. L., Page, J., and Carroll, W. R., J. Biol. Chem. *207*, 201–210, 1954.
56. Redfield, R. R., and Anfinsen, C. B., J. Biol. Chem. *221*, 385–404, 1956.
57. Potts, J. T., Berger, A., Cooke, J., and Anfinsen, C. B., J. Biol. Chem. *237*, 1851–1855, 1962.
58. Linderstrøm-Lang, K., Lane Medical Lectures, Stanford University Publications, Vol. 6, 1–115, 1952.
59. Gundlach, H. G., Stein, W. H., and Moore, S., J. Biol. Chem. *234*, 1754–1760, 1959.
60. Crestfield, A. M., Stein, W. H., and Moore, S., J. Biol. Chem. *238*, 2413–2420; *238*, 2421–2428, 1963.
61. Heinrikson, R. L., J. Biol. Chem. *241*, 1393–1405. 1966.
62. Hirs, C. H. W., Halmann, M., and Kycia, J. H., In T. W. Goodwin and I. Lindberg, (Editors), Biological Structure and Function, Vol. 1, Academic Press, Inc., New York, p. 41, 1961.
63. Perutz, M. F., Science *140*, 863–869, 1963.
64. Kendrew, J. C., Science *139*, 1259–1266, 1963.
65. Kartha, G., Bello, J., and Harker, D., Nature *213*, 862–865, 1967.
66. Wyckoff, H. W., Hardman, K. D., Allewell, N. M., Inagami, T., Johnson, L. N., and Richards, F. M., J. Biol. Chem. *242*, 3984–3988, 1967.
67. Richards, F. M., C. R. Trav. Lab. Carlsberg *29*, 322–328, 329–346, 1955.

68. Richards, F. M., and Vithayathil, P. J., J. Biol. Chem. *234*, 1459–1465, 1959.
69. Brown, D. M., and Todd, A. R., In The Nucleic Acids: Chemistry and Biology (E. Chargaff and J. N. Davidson, Editors), Vol. 1, Academic Press, Inc., New York, p. 409, 1955.
70. Richards, F. M., and Wyckoff, H. G., In The Enzymes (P. D. Boyer, Editor), Third Ed., Vol. 4, Academic Press, Inc., New York, pp. 647–806, 1971.
71. Gutte, B., and Merrifield, R. B., J. Amer. Chem. Soc. *91*, 501–502, 1969.
72. Gutte, B., and Merrifield, R. B., J. Biol. Chem. *246*, 1922–1941, 1971.
73. Hirschmann, R., Nutt, R. F., Veber, D. F., Vitali, R. A., Varga, S. L., Jacob, T.A., Holly, F. W., and Denkewalter, R. G., J. Amer. Chem. Soc. *91*, 507–508, 1969.
74. White, F. H., J. Biol. Chem. *235*, 383–389, 1960.
75. Goldberger, R. F., Epstein. C. J., and Anfinsen, C. B., J. Biol. Chem. *328*, 628–635, 1963.
76. Anfinsen, C. B., Les Prix Nobel en 1972, Stockholm, 1973.
77. Anfinsen, C. B., J. Biol. Chem. *221*, 405–412, 1956.
78. Lin, M. C., J. Biol. Chem. *245*, 6726–6731, 1970.
79. Lin, M. C., Gutte, B., Moore, S., and Merrifield, R. B., J. Biol. Chem. *245*, 5169–5170, 1970.
80. Lin, M. C., Gutte, B., Caldi. D. G., Moore, S., and Merrifield, R. B., J. Biol. Chem. *247*, 4768–4774, 1972.
81. Anfinsen, C. B., Aqvist, S. E. G., Cooke, J. P., and Jonsson, B., J. Biol. Chem. *234*, 1118–1123, 1959.
82. Beintema, J. J., and Gruber, M., Biochim. Biophys. Acta *147*, 612–614, 1967.
83. Jackson, R. L., and Hirs, C. H. W., J. Biol. Chem. *245*, 637–653, 1970.
84. Phelan, J. J., and Hirs, C. H. W., J. Biol. Chem. *245*, 654–661, 1970.
85. Laskowski, M., Sr., In The Enzymes (P. D. Boyer, Editor), 3. Ed., Vol. 4, Academic Press, Inc., New York, pp. 289–311, 1971.
86. Avery, O. T., MacLeod, C. M., and McCarty, M., J. Exp. Med. *79*, 137–157, 1944.
87. McCarty, M., and Avery, O. T., J. Exp. Med. *83*, 89–96, 1946.
88. McCarty, M., J. Gen. Physiol. *29*, 123–139, 1946.
89. Kunitz, M., J. Gen. Physiol. *33*, 349–362, 1950.
90. Lindberg, U., Biochemistry *6*, 335–342, 1967.
91. Price, P. A., Liu, T.-Y., Stein, W. H., and Moore, S., J. Biol. Chem. *244*, 917–923, 1969.
92. Salnikow, J., Moore, S., and Stein, W. H., J. Biol. Chem. *245*, 5685–5690, 1970.
93. Salnikow, J., Liao, T.-H., Moore, S., and Stein, W. H., J. Biol. Chem., *248*, 1480—1488, 1973.
94. Liao, T.-H., Salnikow, J., Moore, S., and Stein, W. H., J. Biol. Chem., *248* 1489—1495, 1973.
95. Gross, E., and Witkop, B., J. Biol. Chem. *257*, 1856–1860, 1962.
96. Catley, B. J., Moore, S., and Stein, W. H., J. Biol. Chem. *244*, 933–936, 1969.
97. Hugli, T. E., and Stein, W. H., J. Biol. Chem. *246*, 7191–7200, 1971.
98. Riordan, J. F., Sokolovsky, M., and Vallee, B. L., J. Amer. Chem. Soc. *88*, 4104–4105, 1966.
99. Price, P. A., Moore, S., and Stein, W. H., J. Biol. Chem. *244*, 924–928, 1969.
100. Salnikow, J., and Murphy, D., J. Biol. Chem., *248*, 1499—1501, 1973.
101. Price, P. A., Stein, W. H., and Moore, S., J. Biol. Chem. *244*, 929–932, 1969.
102. Takahashi, K., Stein, W. H., and Moore, S., J. Biol. Chem. *242*, 4682–4690, 1967.
103. Liu, T.-Y., Stein, W. H., Moore, S., and Elliott, S. D., J. Biol. Chem. *240*, 1143–1149, 1965.
104. Liu, T.-Y., J. Biol. Chem. *242*, 4029–4032, 1967.
105. Rajagopalan, T. G., Stein, W. H., and Moore, S., J. Biol. Chem. *241*, 4295–3297, 1966.
106. Stark, G. R., Stein, W. H., and Moore, S., J. Biol. Chem. *235*, 3177–3181, 1960.
107. Wöhler, F., Pogg. Ann. *12*, 253–256, 1828.
108. Cerami, A., and Manning, J. M., Proc. Nat. Acad. Sci. USA *68*, 1180–1183, 1971.

STANFORD MOORE

Stanford Moore was born in 1913 in Chicago, Illinois, and grew up in Nashville, Tennessee, where his father was a member of the faculty of the School of Law of Vanderbilt University. His developmental years were in a home environment which made the pursuit of knowledge an eagerly adopted undertaking. He had the opportunity to attend a high school administered by the George Peabody College for Teachers in Nashville. A skilled teacher of science, R. O. Beauchamp, kindled an interest in chemistry. Moore entered Vanderbilt University undecided between a career in chemistry or aeronautical engineering. The courses which he took in the engineering school presaged a concern for instrumentation. But a gifted professor of organic chemistry, Arthur Ingersoll, succeeded in presenting the study of molecular architecture as an even more appealing discipline. Moore graduated from Vanderbilt (B.A. 1935, *Summa cum laude*) with a major in chemistry. The faculty recommended him for a Wisconsin Alumni Research Foundation Fellowship which took him to the University of Wisconsin where he received his Ph.D. in organic chemistry in 1938.

His thesis research was in biochemistry in the laboratory of Karl Paul Link. The first lessons that the young graduate student received from the skilled hands of his professor were in the microanalytical methods of Pregl for the determination of C, H, and N; Link had recently returned from Europe where he had studied in the laboratory of Fritz Pregl in Graz. This training from Link in microchemistry was especially valuable for a student who was later to be concerned with the quantitative analysis of proteins. Moore's thesis was on the characterization of carbohydrates as benzimidazole derivatives. The experience of bringing that work from the bench to the printed page under Link's guidance marked Moore's transition from a student to a productive scholar.

Karl Paul Link was a friend of Max Bergmann, who had recently arrived from Germany to lead a laboratory at the Rockefeller Institute for Medical Research in New York. Through that friendship, Moore was encouraged to join the Bergmann Laboratory in 1939, which was an internationally renowned center of research on the chemistry of proteins and enzymes. During Emil Fischer's last years Max Bergmann had been his senior research associate, and Bergmann had attracted to Rockefeller a group of versatile chemists who maintained a tradition of innovative research and high productivity. After nearly three valuable years in such company, which included William H. Stein, the advent of World War II drew Moore out of the laboratory to serve as a junior administrative officer in Washington for academic and industrial chemical projects administered by the Office of Scientific Research and Development. At the close of the war, he was on duty with the Operational Research Section attached to the Headquarters of the United States Armed Forces in the Pacific Ocean Area, Hawaii.

During the war years, the situation at the Rockefeller Institute had changed. The untimely death of Max Bergmann in 1944 had brought to a close the major chapter in biochemistry which the contributions of his laboratory comprised. Moore's decision to return to Rockefeller was influenced by Herbert Gasser, then the Director of The Rockefeller Institute, who offered to give modest space to Moore and Stein to pursue the theme of research which they had begun with Bergmann or any new lines of investigation that appealed to them. Thus began the collaboration that led to the development of quantitative chromatographic methods for amino acid analysis, their automation, and the utilization of such techniques, in cooperation with younger associates, in the researches in protein chemistry summarized in the Nobel Lecture by Moore and Stein in this volume.

The investigations were conducted in an atmosphere at Rockefeller that encouraged interdepartmental cooperation and international consultation that would expedite research. Interludes included Moore's tenure of the Francqui Chair at the University of Brussels in 1950, where, at the generous invitation of E. J. Bigwood, a laboratory of amino acid analysis was organized in the School of Medicine. Moore had the opportunity to round out the year in Europe with six months in England at the University of Cambridge where he shared part of a laboratory with Frederick Sanger during the time of the pioneering studies on insulin. In 1968, Moore was a Visiting Professor of Health Sciences at the Vanderbilt University School of Medicine.

Memberships and Activities: American Society of Biological Chemists (Treasurer, 1956—59; Editorial Board, 1950—60; President, 1966), American Chemical Society, hon. member Belgian Biochemical Society, foreign correspondent Belgian Royal Academy of Medicine, Biochemical Society (Great Britain), U.S. National Academy of Sciences (Chairman, Section of Biochemistry, 1970), American Academy of Arts and Sciences, Harvey Society, Chairman of Panel on Proteins of the Committee on Growth of the National Research Council (1947—49), Secretary of the Commission on Proteins of the International Union of Pure and Applied Chemistry (1953—57), Chairman of the Organizing Committee for the Sixth International Congress of Biochemistry (1964), President of the Federation of American Societies for Experimental Biology (1970).

Honors:

Docteur *honoris causa* from the Faculty of Medicine of the University of Brussels (1954) and from the University of Paris (1964). Award shared with William H. Stein: American Chemical Society Award in Chromatography and Electrophoresis, 1964; Richards Medal of the American Chemical Society, 1972; Linderstrøm-Lang Medal, Copenhagen, 1972.

WILLIAM H. STEIN

I was born June 25, 1911 in New York City, the second of three children, to Freed M. and Beatrice Borg Stein. My father was a business man who was greatly interested in communal affairs, particularly those dealing with health, and he retired quite early in life in order to devote his full time to such matters as the New York Tuberculosis and Health Association, Montefiore Hospital and others. My mother, too, was greatly interested in communal affairs and devoted most of her life to bettering the lot of the children of New York City. During my childhood, I received much encouragement from both of my parents to enter into medicine or a fundamental science.

My early education was at the Lincoln School of Teachers College of Columbia University in New York City, a school which was considered progressive for that time and which fostered in me an active interest in creative arts, music, and writing. There I had my first course in chemistry which proved to be an extremely valuable and interesting one. I left this school when I was about sixteen and went to an excellent preparatory school in New England, namely Phillips Exeter Academy, which was at the time, although it has changed since, a much more rigid and much more demanding educational experience than I had had at Lincoln. It was at Exeter that I was introduced to standards of precision of writing, and of work generally which I think has stood me in very good stead, and I believe that the combination of a progressive school and a more demanding school such as I enjoyed was an ideal preparation. From Exeter I went to Harvard where I had a very enjoyable, although not a very academically distinguished career, and graduated from the college in 1933 at the depths of the economic depression. I had majored in chemistry at college and decided to continue on at Harvard as a graduate student in that subject. This proved to be a rather unfortunate experience because my first graduate year was undistinguished, to say the very least. I was almost ready to abandon a career in science when it was suggested to me that I might enjoy biochemistry much more than straight organic chemistry.

The next year, I transferred to the Department of Biochemistry, then headed by the late Hans Clarke at the College of Physicians and Surgeons, Columbia University in New York. The department at Columbia was an eye-opener for me. Professor Clarke had succeeded in surrounding himself with a fascinating and active faculty and an almost equally stimulating group of graduate students. From both of these I learned a tremendous amount in a short time. My thesis involved the amino acid analysis of the protein elastin, which was then thought to play a role in coronary artery disease and I completed the requirements for my degree at Columbia late in 1937 and went directly to the laboratory of Max Bergmann at the Rockefeller Institute.

While still a graduate student, I had the extreme good fortune to marry, in 1936, Phoebe Hockstader who has been of enormous support to me ever since. We have three sons, William H. Jr., 35; David F., 33; Robert J., 28.

Bergmann was, I still feel, one of the very great protein chemists of this century and he, too, had the ability to surround himself with a most talented group of postdoctoral colleagues. In the laboratory at the time that I was there were, of course, Dr. Moore, and, in addition, Dr Joseph S. Fruton, Dr. Emil L. Smith, Dr. Klaus Hofmann, Dr. Paul Zamecnik, and many others. It was impossible not to learn a great deal about the business of research in protein chemistry from Bergmann, himself, and from the outstanding group he had around him.

The task of Moore and myself was to devise accurate analytical methods for the determination of the amino acid composition of proteins, because Bergmann firmly believed, as did we, that the amino acid analysis of proteins bore the same relationship to these macromolecules that elementary analysis bore to the chemistry of simpler organic substances. It was during this period in the mid-thirties that Bergmann and Fruton and their colleagues were working out the specificity of proteolytic enzymes, work which has had a profound effect upon our knowledge of how enzymes function and has made it possible to use these proteolytic enzymes as tools for the degradation and subsequent derivation of structure of protein molecules ever since.

Work on proteins was suspended during the war for other more pressing matters and Dr. Moore left the laboratory in order to be of assistance in Washington and elsewhere. Our entire group was engaged in working for the Office of Scientific Research and Development. Bergmann's death in 1944 robbed the world of a distinguished chemist and, of course, left the laboratory without a chief. The group continued to function until the end of the war at which time Moore and I had the very great good fortune to be asked by Dr. Herbert Gasser, Director of the Institute, to stay on at Rockefeller with the freedom to do anything we pleased in the biochemical field.

In the meantime, had come the remarkable developments in England on the separation of amino acids by paper chromatography by Martin and Synge and Sanger had started his classical work on the derivation of the structure of insulin. It was then, perhaps, not surprising that Moore and I resumed our collaboration, and following a suggestion of Synge began to try to separate amino acids on columns of potato starch. We were very fortunate in hitting upon a type of potato starch which was well-suited to our needs almost immediately, and from that day on began to work first on the amino acid analysis, and then on the structural analysis of proteins. From columns of potato starch, we progressed to columns of ion exchange resins, developed the automatic amino acid analyzer, and together with a group of very devoted and extremely skillful collaborators, began work on the structure of ribonuclease. These columns were also used for other purposes. In the course of the early work, wc developed a drop-counting automatic fraction collector which is now a common instrument in most biochemical laboratories throughout the world.

I should like to emphasize that the development of methods grew out of a need rather than a particular desire to develop methods as ends in themselves. We needed to know the amino acid composition of proteins, we needed to be able to separate and analyze peptides in good yield, and we needed to be able to purify proteins chromatographically. Since there were no methods for doing any of these things at the time that we started, we had to devise them ourselves. We not only wanted to know what the amino acid sequence of an enzyme such as ribonuclease was, but we tried to find out as much as we could about what made it an enzyme and after we had taken that particular enzyme about as far as we thought we could profitably go, we turned to a number of others which have been listed in the Nobel Lecture.

During all of this time, we had the undeviating support of an enlightened administration at Rockefeller who believed in allowing us to do those things which we thought to be important, and, during the last years of this work, we also have had great financial assistance from the NIH. For this and particularly for the very large number of devoted and talented colleagues which we have had in the laboratory we shall be forever grateful.

During all of this time, each of us, naturally, developed interests outside of the laboratory. I, for example, became greatly concerned about the promulgation of scientific information and have been attached, in one way or another, to the Journal of Biological Chemistry for a matter of over fifteen years. During this time it has been my privilege to work with a knowledgeable and dedicated group of biochemists who have devoted themselves unselfishly to serving the interests of their fellow biochemists throughout the world.

Scientific Societies—National Academy of Sciences, American Academy of Arts and Sciences, American Society of Biological Chemists, Biochemical Society of London, American Chemical Society, American Association for the Advancement of Science, Harvey Society of New York.

I was a member of the Editorial Committee of the Journal of Biological Chemistry, which is an elective office, for six years and Chairman of this Committee for three, 1958—61. After the conclusion of my work on the Editorial Committee, I became a member of the Editorial Board of the Journal of Biological Chemistry in 1962, and then an Associate Editor from 1964 until 1968. I assumed the Editorship, succeeding John T. Edsall, in 1968, a post I was forced to relinquish by illness in 1971.

Other Activities—Member of the Council of the Institute of Neurological Diseases and Blindness of the NIH, 1961—66; Chairman of the U.S. National Committee for Biochemistry, 1968—69; Philip Schaffer Lecturer at Washington University at St. Louis, 1965; Harvey Lecturer, 1956; Phillips Lecturer at Haverford College, 1962; Visiting Professor at the University of Chicago, 1961; Visiting Professor at Harvard University, 1964; Member of Medical Advisory Board, Hebrew University-Hadassah Medical School, 1957—1970; Trustee, Montefiore Hospital, 1948—.

Awards (shared with Stanford Moore): American Chemical Society Award in Chromatography and Electrophoresis, 1964; Richards Medal of the American Chemical Society, 1972; Kaj Linderstrøm-Lang Award, Copenhagen, 1972.

GERALD M. EDELMAN

Antibody Structure and Molecular Immunology

December 12, 1972

Some sciences are exciting because of their generality and some because of their predictive power. Immunology is particularly exciting, however, because it provokes unusual ideas, some of which are not easily come upon through other fields of study. Indeed, many immunologists believe that for this reason, immunology will have a great impact on other branches of biology and medicine. On an occasion such as this in which a very great honor is being bestowed, I feel all the more privileged to be able to talk about some of the fundamental ideas in immunology and particularly about their relationship to the structure of antibodies.

Work on the structure of antibodies has allied immunology to molecular biology in much the same way as previous work on hapten antigens allied immunology to chemistry. This structural work can be considered the first of the projects of molecular immunology, the task of which is to interpret the properties of the immune system in terms of molecular structures. In this lecture, I should like to discuss some of the implications of the structural analysis of antibodies. Rather than review the subject, which has been amply done (1—4), I shall emphasize several ideas that have emerged from the structural approach. Within the context of these ideas, I shall then consider the related but less well explored subject of antibodies on the surfaces of lymphoid cells, and describe some recently developed experimental efforts of my colleagues and myself to understand the molecular mechanisms by which the binding of antigens induces clonal proliferation of these cells.

Antibodies occupy a central place in the science of immunology for an obvious reason: they are the protein molecules responsible for the recognition of foreign molecules or antigens. It is, therefore, perhaps not a very penetrating insight to suppose that a study of their structure would be valuable to an understanding of immunity. But what has emerged from that study has resulted in both surprises and conceptual reformulations.

These reformulations provided a molecular basis for the selective theories of immunity first expounded by Niels Jerne (5) and MacFarlane Burnet (6) and therefore helped to bring about a virtual revolution of immunological thought. The fundamental idea of these theories is now the central dogma of modern immunology: molecular recognition of antigens occurs by selection among clones of cells already committed to producing the appropriate antibodies, each of different specificity (Figure 1). The results of many studies by

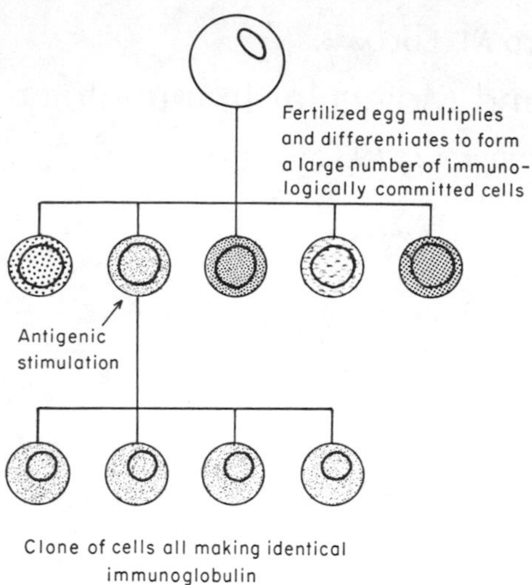
Fertilized egg multiplies
and differentiates to form
a large number of immuno-
logically committed cells
Antigenic
stimulation
Clone of cells all making identical
immunoglobulin

Fig. 1.
A diagram illustrating the basic features of the clonal selection theory. The stippling and shading indicate that different cells have antibody receptors of different specificities, although the specificity of all receptors on a given cell is the same. Stimulation by an antigen results in clonal expansion (maturation, mitosis and antibody production) of those cells having receptors complementary to the antigen.

cellular immunologists (see references 1 and 2) strongly suggest that each cell makes antibodies of only one kind, that stimulation of cell division and antibody synthesis occurs after interaction of an antigen with receptor antibodies at the cell surface, and that the specificity of these antibodies is the same as that of the antibodies produced by daughter cells. Several fundamental questions are raised by these conclusions and by the theory of clonal selection. How can a sufficient diversity of antibodies be synthesized by the lymphoid system? What is the mechanism by which the lymphocyte is stimulated after interaction with an antigen?

In the late 1950's, at the beginnings of the intensive work on antibody structure, these questions were not so well defined. The classic work of Landsteiner on hapten antigens (7) had provided strong evidence that immunological specificity resulted from molecular complementarity between the determinant groups of the antigen molecule and the antigen-combining site of the antibody molecule. In addition, there was good evidence that most antibodies were bivalent (8) as well as some indication that antibodies of different classes existed (9). The physico-chemical studies of Tiselius (10) had established that antibodies were proteins that were extraordinarily heterogeneous in charge. Moreover, a number of workers had shown the existence of heterogeneity in the binding constants of antibodies capable of binding a single hapten antigen (11). Despite the value of all of this information, however, little was known of the detailed chemical structure of antibodies or of what are now called the immunoglobulins.

The Multichain Structure of Antibodies: Problems of Size and Heterogeneity

If the need for a structural analysis of antibodies was great, so were the experimental difficulties: antibodies are very large proteins (mol. wt. 150,000 or greater) and they are extraordinarily heterogeneous. Two means were adopted around 1958 in an effort to avoid the first difficulty. Following the work of Petermann (12) and others, Rodney Porter (13) applied proteolytic enzymes, notably papain, to achieve a limited cleavage of the gamma globulin fraction of serum into fragments. He then successfully fractionated the digest, obtaining antigen binding (Fab) and crystallizable (Fc) fragments. Subsequently, other enzymes such as pepsin were used in a similar fashion by Nisonoff and his colleagues (14). I took another approach, in an attempt to cleave molecules of immunoglobulin G and immunoglobulin M into polypeptide chains by reduction of their disulfide bonds and exposure to dissociating solvents such as 6 M urea (15). This procedure resulted in a significant drop in molecular weight, demonstrating that the immunoglobulin G molecule was a multichain structure rather than a single chain as had been believed before. Moreover, corresponding chains obtained from both immunoglobulins had about the same size. The polypeptide chains (16) were of two kinds (now called light and heavy chains) but were obviously not the same as the fragments obtained by proteolytic cleavage and therefore the results of the two cleavage procedures complemented each other. Ultracentrifugal analyses indicated that one of the polypeptide chains had a molecular weight in the vicinity of 20,000, a reasonable size for determination of the amino acid sequence by the methods available int he early 1960's.

Nevertheless, the main obstruction to a direct analysis of antibody structure was the chemical heterogeneity of antibodies and their antigen binding fragments. Two challenging questions confronted those attempting chemical analyses of antibody molecules at that time. First, did the observed heterogeneity of antibodies reside only in the conformation of their polypeptide chains as was then widely assumed, or did this heterogeneity reflect differences in the primary structures of these chains, as required implicitly by the clonal selection theory? Second, if the heterogeneity did imply a large population of molecules with different primary structures, how could one obtain the homogeneous material needed for carrying out a detailed structural analysis?

These challenges were met simultaneously by taking advantage of an accident of nature rather than by direct physicochemical assault. It had been known for some time that tumors of lymphoid cells called myelomas produced homogeneous serum proteins that resembled the normal heterogeneous immunoglobulins. In 1961, M. D. Poulik and I showed that the homogeneity of these proteins was reflected in the starch gel electrophoretic patterns of their dissociated chains (16). Some patients with multiple myeloma excrete urinary proteins which are antigenically related to immunoglobulins but whose nature remained obscure since their first decsription by Henry Bence Jones in 1847. These Bence Jones proteins were most interesting, for they could be readily obtained from the urine in large quantities, were homogeneous, and had low molecular weights. It seemed reasonable to suggest (16) that Bence Jones proteins represented one of the chains of the immunoglobulin molecule that was synthesized by the myeloma tumor but not incorporated into the homogeneous myeloma protein and therefore excreted into the urine.

This hypothesis was corroborated one exciting afternoon when my student

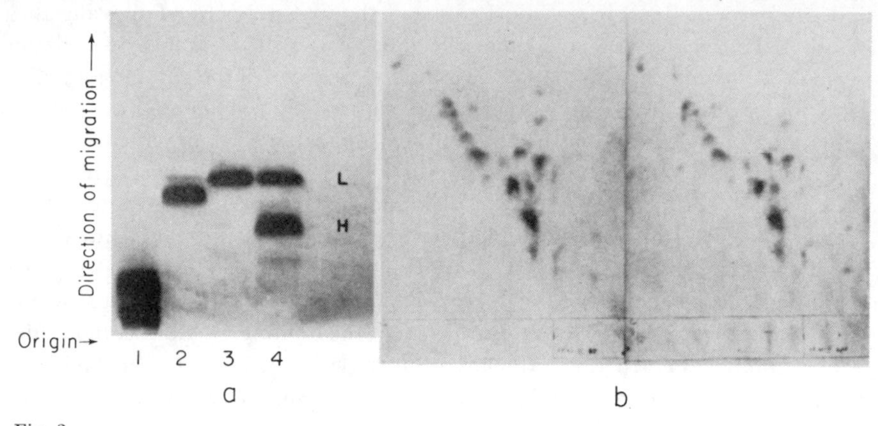

Fig. 2.
Comparisons of light chains isolated from serum IgG myeloma proteins with urinary Bence Jones proteins from the same patient. (a) Starch gel electrophoresis in urea. 1) serum myeloma globulin, 2) urinary Bence Jones protein, 3) Bence Jones protein reduced and alkylated, 4) myeloma protein reduced and alkylated. L – light chain; H – heavy chain. (b) Two-dimensional high voltage electrophoresis of tryptic hydrolysates. Pattern on left is of urinary Bence Jones protein; that on right is of light chain isolated from the serum myeloma protein of the same patient.

Joseph Gally and I (17) heated solutions of light chains isolated from our own serum immunoglobulins in the classical test for Bence Jones proteinuria. They behaved as Bence Jones proteins, the solution first becoming turbid, then clearing upon further heating. A comparison of light chains of myeloma proteins with Bence Jones proteins by starch gel electrophoresis in urea (17) and by peptide mapping (18) confirmed the hypothesis (Figure 2). Indeed, Berggård and I later found (19) that in normal urine there were counterparts to Bence Jones proteins that shared their properties but were chemically heterogeneous.

No physical means was known at the time that was capable of fractionating antibodies to yield homogeneous proteins. It was possible, however, to prepare specifically reactive antibodies by using the antigen to form antigen-antibody aggregates and then dissociating the complex with free hapten. Although we knew that these specifically prepared antibodies were still heterogeneous in their electrophoretic properties, it seemed possible that antibodies to different haptens might show differences in their polypeptide chains. Baruj Benacerraf had prepared a collection of these antibodies, and together with our colleagues (20) we decided to compare their chains, using the same methods that we had used for Bence Jones proteins. The results were striking: purified antibodies showed from 3 to 5 sharp bands in the Bence Jones or light chain region and antibodies of different specificities showed different patterns. In sharp contrast, normal immunoglobulin showed a diffuse zone extending over the entire range of mobilities of these bands. These experiments showed not only that antibodies of different specificities were structurally different but also that their heterogeneity was limited.

The results of these experiments on Bence Jones proteins and purified antibodies had a number of significant implications. Because different Bence Jones proteins had different amino acid compositions, it was clear that immunoglobulins must vary in their primary structures. This deduction, confirmed later by Koshland (21) for specifically purified antibodies, lent strong support to selective theories of antibody formation. Moreover, it opened the possibility of beginning a direct analysis of the primary structure of an immunoglobulin molecule, for not only were the Bence Jones proteins composed of homogeneous light chains, but their subunit molecular weight was only 23,000. The first report by Hilschmann and Craig (22) on partial sequences of several different Bence Jones proteins indicated that the structural heterogeneity of the light chains was confined to the amino terminal (variable) region, whereas the carboxyl terminal half of the chain (the constant region) was the same in all chains of the same type. This finding was soon extended by studies of other Bence Jones proteins (23).

Although some work had also been done on the heavy chains of immunoglobulins, there was much less information on their structure. For instance, it was suspected but not known that they also had variable regions resembling those of light chains. Comparisons of heavy chains and light chains even at this early stage did, however, clarify the nature of another source of antibody heterogeneity: the existence of immunoglobulin classes (24).

Antibodies within a particular class have similar molecular weights, carbohydrate content, amino acid compositions and physiological functions (Table 1) but still possess heterogeneity in their net charge and antigen binding affinities. Studies of classes in various animal species indicated that both the multichain structure and the heterogeneity are ubiquitous properties of immunoglobulins. The different classes apparently emerged during evolution (25) to carry out various physiologically important activities that have been named effector functions in order to distinguish them from the antigen-binding or recognition function. The various manifestations of humoral immune responses as well as their prophylactic, therapeutic and pathological consequences can now be generally explained in terms of the properties of the particular class of antibody mediating that response. As a result of comparing their chain structure, it became clear that although immunoglobulins of all classes contain similar kinds of light chains (Table 1), the distinctive class character (24) is conferred by structural differences in the heavy chains, specifically in their constant regions, as I shall discuss later.

With the clarification of the nature of the heterogeneity of immunoglobulin chains and classes, attention could be turned to the problem of relating the structure and evolution of antibodies within a given class to their antigen-binding and effector functions. We chose to concentrate on immunoglobulin G, for this was the most prevalent class in mammals and the work on chain structure suggested that it would be sufficiently representative.

Table 1. *Human Immunoglobulin Classes.*

Class	Physiological Properties	Heavy Chain[1]	Light Chains	Molecular[2] Formula	Molecular weight ($\times 10^{-3}$) and sedimentation constant	Carbohydrate Content
IgG	Complement fixation; placental transfer	γ	$\varkappa$ or λ	$(\gamma_2\varkappa_2)$ or $(\gamma_2\lambda_2)$	143–149; 6.7S	2.5 %
IgA	Localized protection in external secretions	α	$\varkappa$ or λ	$(\alpha_2\varkappa_2)$ or $(\alpha_2\lambda_2)$	158–162; 6.8–11.4S	5–10 %
IgM	Complement fixation; early immune response	μ	$\varkappa$ or λ	$(\mu_2\varkappa_2)$ or $(\mu_2\lambda_2)$	800–950; 19.0S	5–10 %
IgD	Unknown	δ	$\varkappa$ or λ	$(\delta_2\varkappa_2)$ or $(\delta_2\lambda_2)$	175–180; 6.6S	10 %
IgE	Reagin activity; mast cell fixation,	ε	$\varkappa$ or λ	$(\varepsilon_2\varkappa_2)$ or $(\varepsilon_2\lambda_2)$	185–190; 8.0S	12 %

(1) The class distinctive features of these chains are in their constant regions.
(2) IgA can have additional unrelated chains called SC and J; J chains are also found in IgM.

The Complete Covalent Structure and the Domain Hypothesis

An understanding of the chain structure and its relation to the proteolytic fragments (26, 27) made feasible an attempt to determine the complete structure of an immunoglobulin G molecule. My colleagues and I started this project in 1965 and before it was completed in 1969 (28) seven of us had spent a good portion of our waking hours on the technical details. One of our main objectives was to provide a complete and definitive reference structure against which partial structures of other immunoglobulins could be compared. In particular, we wished to compare the detailed structure of a heavy chain and a light chain from the same molecule.

Another objective was to examine in detail the regional differentiation of the structure that had been evolved to carry out different physiological functions in the immune response. The work of Porter (13) had shown that the so-called Fab fragment of immunoglobulin G was univalent and bound antigens whereas the Fc fragment did not. This provided an early hint that immunoglobulin molecules were organized into separate regions, each mediating different functions. In accord with selective theories of immunity, it was logical to suppose that V regions from both the light and the heavy chains mediated the antigen binding functions. Early evidence that some of the C regions were concerned with physiologically significant effector functions was obtained by showing that Fc fragments would bind components of the complement system (29), a complex group of proteins responsible for immunologically induced cell lysis. A more detailed assignment of structure to function required a knowledge of the total structure.

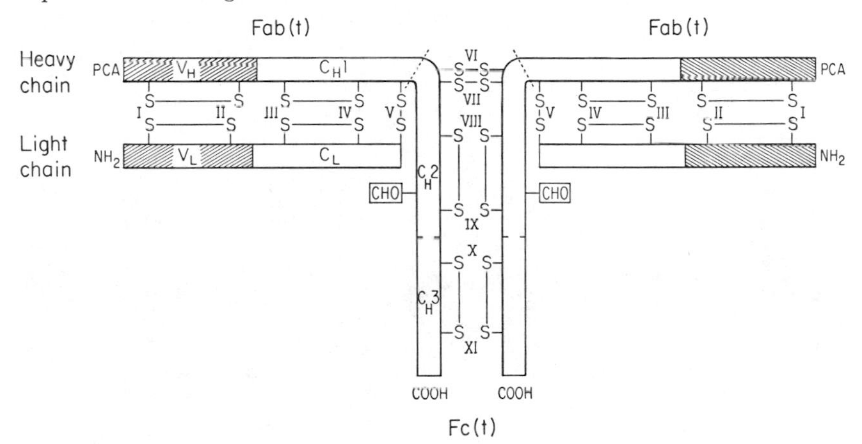

Fig. 3.
Overall arrangement of chains and disulfide bonds of the human γG_1 immunoglobulin, *Eu.* Half-cystinyl residues are I – XI; I – V designates corresponding half-cystinyl residues in light and heavy chains. PCA, pyrrolidonecarboxylic acid; CHO, carbohydrate. Fab(t) and Fc(t) refer to fragments produced by trypsin, which cleaves the heavy chain as indicated by dashed lines above half-cystinyl residues VI. Variable regions, V_H and V_L, are homologous. The constant region of the heavy chain (C_H) is divided into three regions, C_H1, C_H2 and C_H3, that are homologous to each other and to the C region of the light chain. The variable regions carry out antigen-binding functions and the constant regions the effector functions of the molecule.

Amino acid sequence analysis of the Fc region of normal rabbit γ chains by Hill and his colleagues (30) demonstrated that the carboxyl terminal portion of heavy chains was homogeneous. On the basis of internal homologies in this region, Hill (30) and Singer and Doolittle (31) proposed the hypothesis that the genes for immunoglobulin chains evolved by duplication of a gene of sufficient size to specify a precursor protein of about 100 amino acids in length. Although direct confirmation of this hypothesis is obviously not possible, it was strongly supported by the results of our analysis (28) of the complete amino acid sequence and arrangement of the disulfide bonds of an entire IgG myeloma protein.

Comparisons of the amino acid sequences of the heavy chain of this protein with others studied in Porter's laboratory (32) and by Bruce Cunningham and his colleagues in our laboratory (33) showed that heavy chains had variable (V_H) regions, i.e., regions that differed from one another in the sequences of the 110—120 residues beginning with the amino terminus (Figure 3).

Examination of the amino acid sequences (Figures 4 and 5) allowed us to draw the following additional conclusions:

1) The variable (V) regions of light and heavy chains are homologous to each other, but they are not obviously homologous to the constant regions of these chains. V regions from the same molecule appear to be no more closely related than V regions from different molecules.

2) The constant (C) region of γ chains consists of three homology regions,

```
                    SEQUENCE HOMOLOGY IN EU VARIABLE REGIONS

                                                    1                                   10
     EU V_L  (RESIDUES 1-208)                      ASP ILE GLN MET THR GLN SER PRO SER THR
     EU V_H  (RESIDUES 1-114)                      PCA VAL GLN LEU VAL GLN SER GLY  -  ALA

                                        20                                              30
LEU SER ALA SER VAL GLY ASP ARG VAL THR ILE THR CYS ARG ALA SER GLN SER ILE ASN
GLU VAL LYS LYS PRO GLY SER SER VAL LYS VAL SER CYS LYS ALA SER GLY GLY THR PHE

                                            40
THR  -   -  TRP LEU ALA TRP TYR GLN GLN LYS PRO GLY LYS ALA PRO LYS LEU LEU MET
SER ARG SER ALA ILE ILE TRP VAL ARG GLN ALA PRO GLY GLN GLY LEU GLU TRP MET GLY

     50                                         60
TYR LYS ALA SER SER  -  LEU GLU SER GLY VAL PRO SER ARG PHE ILE GLY SER GLY SER
GLY ILE VAL PRO MET PHE GLY PRO PRO ASN TYR ALA GLN LYS PHE GLN GLY  -  ARG VAL

         70                                                                     80
GLY THR GLU PHE THR  -   -   -   -   -   -   -  LEU THR ILE SER SER LEU GLN PRO
THR ILE THR ALA ASP GLU SER THR ASN THR ALA TYR MET GLU LEU SER SER LEU ARG SER

                                        90
ASP ASP PHE ALA THR TYR TYR CYS GLN GLN  -  TYR ASN SER ASP SER LYS MET PHE GLY
GLU ASP THR ALA PHE TYR PHE CYS ALA GLY GLY TYR GLY ILE TYR SER PRO GLU GLU TYR

100
GLN GLY THR LYS VAL GLU VAL LYS GLY
ASN GLY GLY LEU VAL THR
```

Fig. 4.
Comparison of the amino acid sequences of the V_H and V_L regions of protein Eu. Identical residues are shaded. Deletions indicated by dashes are introduced to maximize the homology.

SEQUENCE HOMOLOGY IN EU CONSTANT REGIONS

		110										120	
EU C_L (RESIDUES 109-214)	THR	VAL	ALA	ALA	PRO	SER	VAL	PHE	ILE	PHE	PRO	PRO	SER
EU C_H1 (RESIDUES 119-220)	SER	THR	LYS	GLY	PRO	SER	VAL	PHE	PRO	LEU	ALA	PRO	SER
EU C_H2 (RESIDUES 234-341)	LEU	LEU	GLY	GLY	PRO	SER	VAL	PHE	LEU	PHE	PRO	PRO	LYS
EU C_H3 (RESIDUES 342-446)	GLN	PRO	ARG	GLU	PRO	GLN	VAL	TYR	THR	LEU	PRO	PRO	SER

									130										
ASP	GLU	GLN	-	-	LEU	LYS	SER	GLY	THR	ALA	SER	VAL	VAL	CYS	LEU	LEU	ASN	ASN	PHE
SER	LYS	SER	-	-	THR	SER	GLY	GLY	THR	ALA	ALA	LEU	GLY	CYS	LEU	VAL	LYS	ASP	TYR
PRO	LYS	ASP	THR	LEU	MET	ILE	SER	ARG	THR	PRO	GLU	VAL	THR	CYS	VAL	VAL	VAL	ASP	VAL
ARG	GLU	GLU	-	-	MET	THR	LYS	ASN	GLN	VAL	SER	LEU	THR	CYS	LEU	VAL	LYS	GLY	PHE

140												150							
TYR	PRO	ARG	GLU	ALA	LYS	VAL	-	-	GLN	TRP	LYS	VAL	ASP	ASN	ALA	LEU	GLN	SER	GLY
PHE	PRO	GLU	PRO	VAL	THR	VAL	-	-	SER	TRP	ASN	SER	-	GLY	ALA	LEU	THR	SER	GLY
SER	HIS	GLU	ASP	PRO	GLN	VAL	LYS	PHE	ASN	TRP	TYR	VAL	ASP	GLY	-	VAL	GLN	VAL	HIS
TYR	PRO	SER	ASP	ILE	ALA	VAL	-	-	GLU	TRP	GLU	SER	ASN	ASP	-	GLY	GLU	PRO	GLU

		160										170							
ASN	SER	GLN	GLU	SER	VAL	THR	GLU	GLN	ASP	SER	LYS	ASP	SER	THR	TYR	SER	LEU	SER	SER
-	VAL	HIS	THR	PHE	PRO	ALA	VAL	LEU	GLN	SER	-	SER	GLY	LEU	TYR	SER	LEU	SER	SER
ASN	ALA	LYS	THR	LYS	PRO	ARG	GLU	GLN	GLN	TYR	-	ASP	SER	THR	TYR	ARG	VAL	VAL	SER
ASN	TYR	LYS	THR	THR	PRO	PRO	VAL	LEU	ASP	SER	-	ASP	GLY	SER	PHE	PHE	LEU	TYR	SER

		180										190							
THR	LEU	THR	LEU	SER	LYS	ALA	ASP	TYR	GLU	LYS	HIS	LYS	VAL	TYR	ALA	CYS	GLU	VAL	THR
VAL	VAL	THR	VAL	PRO	SER	SER	SER	LEU	GLY	THR	GLN	-	THR	TYR	ILE	CYS	ASN	VAL	ASN
VAL	LEU	THR	VAL	LEU	HIS	GLN	ASN	TRP	LEU	ASP	GLY	LYS	GLU	TYR	LYS	CYS	LYS	VAL	SER
LYS	LEU	THR	VAL	ASP	LYS	SER	ARG	TRP	GLN	GLU	GLY	ASN	VAL	PHE	SER	CYS	SER	VAL	MET

		200													210				
HIS	GLN	GLY	LEU	SER	SER	PRO	VAL	THR	-	LYS	SER	PHE	-	-	ASN	ARG	GLY	GLU	CYS
HIS	LYS	PRO	SER	ASN	THR	LYS	VAL	-	ASP	LYS	ARG	VAL	-	-	GLU	PRO	LYS	SER	CYS
ASN	LYS	ALA	LEU	PRO	ALA	PRO	ILE	-	GLU	LYS	THR	ILE	SER	LYS	ALA	LYS	GLY		
HIS	GLU	ALA	LEU	HIS	ASN	HIS	TYR	THR	GLN	LYS	SER	LEU	SER	LEU	SER	PRO	GLY		

Fig. 5.
Comparison of the amino acid sequences of C_L, C_H1, C_L2 and C_L3 regions. Deletions, indicated by dashes, have been introduced to maximize homologies. Identical residues are darkly shaded; both light and dark shadings are used to indicate identities which occur in pairs in the same position.

C_H1, C_H2 and C_H3, each of which is closely homologous to the others and to the constant regions of the light chains.

3) Each variable region and each constant homology region contains one disulfide bond, with the result that the intrachain disulfide bonds are linearly and periodically distributed in the structure.

4) The region containing all of the interchain disulfide bonds is at the center of the linear sequence of the heavy chain and has no homologous counterpart in other portions of the heavy or light chains.

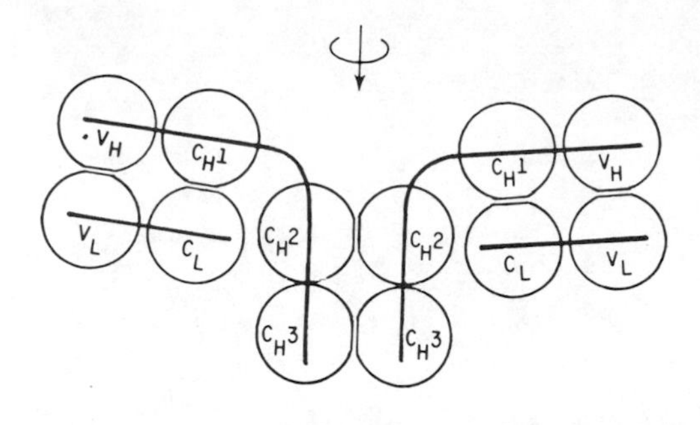

Fig. 6.
The domain hypothesis. Diagrammatic arrangement of domains in the free immunoglobulin G molecule. The arrow refers to a dyad axis of symmetry. Homology regions (see Figures 3, 4 and 5) which constitute each domain are indicated: V_L, V_H — domains made up of variable homology regions; C_L, C_H1, C_H2, and C_H3 — domains made up of constant homology regions. Within each of these groups, domains are assumed to have similar three-dimensional structures and each is assumed to contributed to an active site. The V domain sites contribute to antigen recognition functions and the C domain sites to effector functions.

These conclusions prompted us to suggest that the molecule is folded in a congeries of compact domains (28, 33) each formed by separate V homology regions or C homology regions (Figure 6). In such an arrangement, each domain is stabilized by a single intrachain disulfide bond and is linked to neighboring domains by less tightly folded stretches of the polypeptide chains. A twofold pseudosymmetry axis relates the V_LC_L to the V_HC_H1 domains and a true dyad axis through the disulfide bonds connecting the heavy chains relates the C_H2-C_H3 domains. The tertiary structure within each of the homologous domains is assumed to be quite similar. Moreover, each domain is assumed to contribute to at least one active site mediating a function of the immunoglobulin molecule.

This last supposition is nicely demonstrated by the interaction of V region domains. The reconstitution of active antibody molecules by recombining their isolated heavy and light chains (34, 35, 36) as well as affinity labelling experiments (31) confirmed our early hypothesis that the V regions of both heavy and light chains contributed to the antigen-combining sites. Moreover, the experiments of Haber (37) provided the first indication that Fab fragments of specific antibodies could be unfolded after reduction of their disulfide bonds and refolded in the absence of antigen to regain most of their antigen binding activity. This clearly indicated that the information for the combining site was contained entirely in the amino acid sequences of the chains. That this information is contained completely in the variable regions is strikingly shown by the recent isolation of antigen-binding fragments consisting only of V_L and V_H (38). The chain recombination experiments suggested an hypothesis to account in part for antibody diversity: the various combinations of different heavy and light chains expressed in different lymphocytes allow the formation of a large number of different antigen-combining sites from a relatively small number of V regions.

One of the remaining structural tasks of molecular immunology is to obtain a direct picture of antigen-binding sites by X-ray crystallography of V domains at atomic resolution. Although crystals of the appropriate molecule or fragment yielding diffraction patterns that extend beyond Bragg spacings of 3.0 Å have not yet been obtained, it is likely that continued searching will provide them. The details of a particular antigen-antibody interaction revealed by such a study will be of enormous interest. For example, certain sequence positions of V regions are hypervariable (39) and are very good candidates for direct contribution to the site. It will be particularly important to understand how the basic three-dimensional structure can accomodate so many amino acid substitutions. X-ray crystallographic work may also show in detail how the disulfide bonds in each of the V domains provide essential stability to the site (28, 33, 40).

The proposed similarities in tertiary structures among C domains have not been established nor have the functions of the various C domains been fully determined. There is a suggestion that C_H2 may play a role in complement fixation (41). A good candidate for binding to the lymphocyte cell membrane is C_H3, the function of which may be concerned with the mechanism of lymphocyte triggering following the binding of antigen by V domains. The C_H3 domain has already been shown to bind to macrophage membranes (42) and there is now some evidence that lymphocytes can synthesize isolated domains (43, 44, 45) similar to C_H3 as separate molecules.

Although many details are still lacking, the gross structural aspects of the domain hypothesis have received direct support from X-ray crystallographic analyses of Fab fragments (46) and whole molecules (47) in which separate domains were clearly discerned. Indirect support for the hypothesis has also come from experiments (38, 48) on proteolytic cleavage of regions between domains.

It is not completely obvious why the domain structure was so strictly preserved during evolution. One reasonable hypothesis is that although there was a functional need for association of V and C domains in the same molecule, there was also a need to prevent allosteric interactions among these domains. Whatever the selective advantages of this arrangement, it is clear that immunoglobulin evolution by gene duplication permitted the possibility of modular alteration of immunological function by addition or deletion of domains.

Translocons: Proposed Units of Evolution and Genetic Function

The evolution by gene duplication of both the domain structure and the immunoglobulin classes raises several questions about the number and arrangement of the structural genes specifying immunoglobulins. Although time does not permit me to discuss this complex subject in detail, I should like to suggest how structural work has sharpened these questions.

According to the theory of clonal selection, it is necessary that there preexist in each individual a large number of different antibodies with the capacity to bind different antigens. One of the most satisfying conclusions that emerged from structural analysis is that the diversity of the V regions of antibody chains is sufficient to satisfy this requirement. This diversity arises at three levels of structural or genetic organization, two of which are now reasonably well understood:

1) V regions from both heavy and light chains contribute to the antigen-

binding site and therefore the number of possible antibodies may be as great as the product of the number of different V_L and V_H regions.

2) Analyses of the amino acid sequences of V regions of light chains by Hood (49) and Milstein (50) and later of heavy chains from myeloma proteins (32, 33) indicated that V regions fall into subgroups of sequences which must be specified by separate genes or groups of genes. Within a subgroup, the amino acid replacements at a particular position are of a conservative type consistent with single base changes in codons of the structural genes. Variable regions of different subgroups differ much more from each other than do variable regions within a subgroup.

Although different V region subgroups are specified by a number of non-allelic genes (50), the analysis of genetic or allotypic markers suggests that C regions of a given immunoglobulin class are specified by no more than one or two genes. These allotypic markers, first described by Grubb (51) and Oudin (52) provide a means in addition to sequence analysis for understanding the genetic basis of immunoglobulin synthesis (4). V regions specified by a number of different genes can occur in chains each of which may have the same C region specified by a single gene. It therefore appears that each immunoglobulin chain is specified by two genes, a V gene and a C gene (4, 49, 50).

Work in a number of laboratories (reviewed in reference 4) has shown that the genetic markers on the two types of light chains are not linked to those of the heavy chains or to each other. These findings and the conclusion that there are separate V and C genes led Gally and me to suggest (4) that immunoglobulins are specified by three unlinked gene clusters (Figure 7). The clusters have been named translocons (4) to emphasize the fact that some mechanism must be provided to combine genetic information from V region loci with information from C region loci to make complete V–C structural genes. According to this hypothesis, the translocon is the basic unit of immuno-

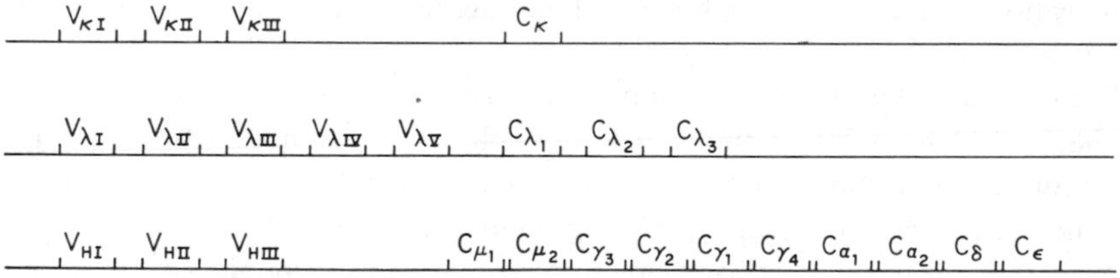

$V_{\kappa I}$ $V_{\kappa II}$ $V_{\kappa III}$ C_{κ}

$V_{\lambda I}$ $V_{\lambda II}$ $V_{\lambda III}$ $V_{\lambda IV}$ $V_{\lambda V}$ C_{λ_1} C_{λ_2} C_{λ_3}

V_{HI} V_{HII} V_{HIII} C_{μ_1} C_{μ_2} C_{γ_3} C_{γ_2} C_{γ_1} C_{γ_4} C_{α_1} C_{α_2} C_{δ} C_{ϵ}

Fig. 7.
A diagrammatic representation of the proposed arrangement in mammalian germ cells of antibody genes in three unlinked clusters termed translocons. κ and λ chains are each specified by different translocons and heavy chains are specified by a third translocon. The exact number and arrangement of V and C genes within a translocon is not known. Each variable region subgroup (designed by a subscript corresponding to chain group and subgroup) must be coded by at least one separate germ line V gene. The number of V genes within each subgroup is unknown, however, as is the origin of intrasubgroup diversity of V regions. A special event is required to link the information from a particular V gene to that of a given C gene. The properties of the classes and subclasses (see Table 1) are conferred on the constant regions by C genes.

globulin evolution, different groups of immunoglobulin chains having arisen by duplication and various chromosomal rearrangements of a precursor gene cluster. Presumably, gene duplication during evolution also led to the appearance of V region subgroups within each translocon.

The key problem of the generation of immunoglobulin diversity has been converted by the work on chains and subgroups to the problem of the origin of sequence variations within each V region subgroup. It is still not known whether there is a germ line gene for each V region within a subgroup or whether each subgroup contains only a few genes (see Figure 7) and intra-subgroup variation arises by somatic genetic rearrangements of translocons within precursors of antibody forming cells. At this time, therefore, we can conclude that only the basis but not the origin of diversity has been adequately explained by the work on structure. Although structural analysis of various immunoglobulin classes will continue to be important, it does not in itself seem likely to lead to an explanation of the origin of antibody diversity. What will probably be required are imaginative experiments on DNA, RNA and their associated enzymes obtained from lymphoid cells at the proper stage of development.

In this abbreviated and necessarily incomplete account, I have attempted to show how structural work on immunoglobulins has provided a molecular basis for a number of central features of the theory of clonal selection. The work on humoral antibodies is just a beginning, however, for two great problems of molecular and cellular immunology remain to be solved. The first problem, the origin of intrasubgroup diversity, will undoubtedly receive great attention in the next few years. The second problem is concerned with the triggering of the clonal expansion of lymphocytes after combination of their receptor antibodies with antigens and the quantitative description of the population dynamics of the responding cells. An adequate solution to this problem must also account for the phenomenon of specific immune tolerance as described by the original work of Medawar and his associates (53).

For the remainder of this lecture, I shall turn my attention to some recent attempts that my colleagues and I have made to see whether these problems can be profitably studied using molecular approaches.

Lymphocyte Stimulation by Means of Lectins

The mechanisms of the cellular events underlying immune responses and immune tolerance remain a major challenge to theoretical and practical immunology (53, 60). How does a given antigen induce clonal proliferation or immune tolerance in certain subpopulations of cells?

Cells reactive to a given antigen constitute a very small portion of the lymphocyte population and are difficult to study directly. Two means have been used to circumvent this difficulty: the application of molecules that can stimulate lymphocytes independent of their antigen binding specificity, and fractionation of specific antigen binding lymphocytes for studies of stimulation by antigens of known structure. Although the problem of lymphocyte stimulation is far from being solved, both of these approaches are valuable particularly when used together.

Antigens are not the only means by which lymphocytes may be stimulated. It has been found that certain plant proteins called lectins can bind to glycoprotein receptors on the lymphocyte surface and induce blast transformation, mitosis and immunoglobulin production (see reference 54 for a review). Different lectins have different specificities for cell surface glycoproteins and different molecular structures although their mitogenic properties can be quite similar. In addition, they have a variety of effects on cell metabolism and transport. Such effects are independent of the antigen binding specificity of the cell and they may therefore be studied prior to specific cell fractionation.

The fact that antigens and lectins of different specificity and structure may stimulate lymphocytes suggests that the induction of mitosis is a property of membrane-associated structures that can respond to a variety of receptors. Triggering appears to be independent of the specificity of these receptors for their various ligands. To understand mitogenesis, it is therefore necessary to solve two problems. The first is to determine in molecular detail how the lectin binds to the cell surface and to compare it to the binding of antigens. The second is to determine how the binding induces metabolic changes necessary for the initiation of cell division. These changes are likely to include the production or release of a messenger which is a final common pathway for the stimulation of the cell by a particular lectin or antigen.

One of the important requirements for solving these problems is to know the complete structure of several different mitogenic lectins. This structural information is particularly useful in trying to understand the molecular transformation at the lymphocyte surface required for stimulation. With the knowledge of the three-dimensional structure of a lectin, various amino acid side chains at the surface of the molecule may be modified by group reagents which also may be used to change the valence of the molecule. The activities of the modified lectin derivatives may then be observed in various assays of their effects on cell surfaces and cell functions.

My colleagues and I (55) have recently determined both the amino acid sequence and three-dimensional structure of the lectin, concanvalin A (Con A) (Figure 8). This lectin has specificity for glucopyranosides, mannopyranosides and fructofuranosides and binds to glycoproteins and possibly glycolipids at a variety of cell surfaces. The purpose of our studies was to know the exact size and shape of the molecule, its valence and the structure and distribution of its binding sites.

With this knowledge in hand, we have been attempting to modify the structure and determine the effects of that modification on various biological activities of the lymphocyte. So far, there are several findings suggesting that such alterations of the structure have distinct effects. Con A in free solution stimulates thymus-derived lymphocytes (T cells) but not bone marrow-derived lymphocytes (B cells), leading to increased uptake of thymidine and blast transformation. The curve of stimulation of T cells by native Con A shows a rising limb representing stimulation and a falling limb (Figure 9)

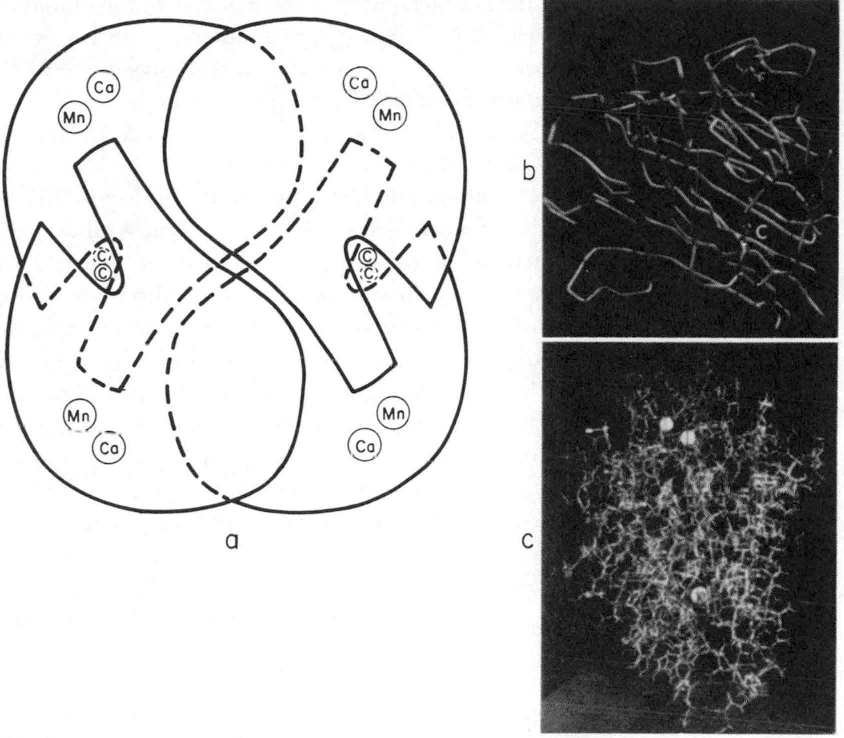

Fig. 8.
Three-dimensional structure of concanavalin A, a lectin mitogenic for lymphocytes. (a) Schematic representation of the tetrameric structure of Con A viewed down the z axis. The proposed binding sites for transition metals, calcium, and saccharides are indicated by Mn, Ca and C respectively. The monomers on top (solid lines) are related by a twofold axis, as are these below. The two dimers are paired across an axis of D_2 symmetry to form the tetramer. (b) Wire model of the polypeptide backbone of the concanavalin A monomer oriented approximately to correspond to the monomer on the upper right of the diagram in (a). The two balls at the top represent the Ca and Mn atoms and the ball in the center is the position of an iodine atom in the sugar derivative, b-iodophenylglucoside, which is bound to the active site. Four such monomers are joined to form the tetramer as shown in (a). (c) A view of the Kendrew model of the Con A monomer rotated to show the deep pocket formed by the carbohydrate binding site. (White ball at the bottom of the figure is at the position of the iodine of b-iodophenylglucoside). The two white balls at the top represent the metal atoms.

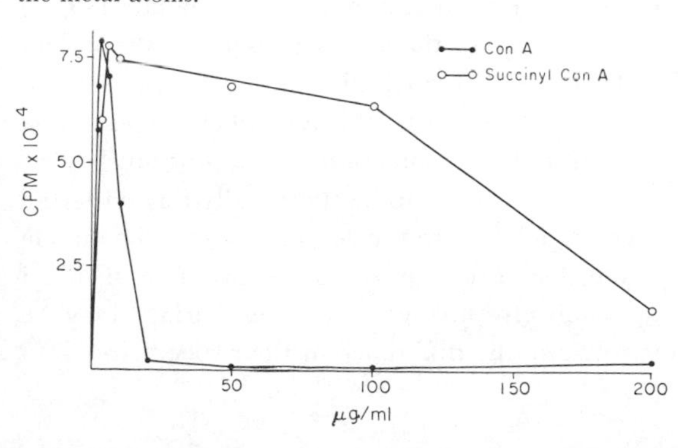

Fig. 9.
Stimulation of uptake of radioactive thymidine by mouse spleen cells after addition of concanavalin A and succinylated concanavalin A in increasing doses (μg/ml). CPM — counts per minute.

probably the result of cell death. The fact that the mitogenic effect and inhibition effect are dose dependent suggests an analogy to stimulation and tolerance induction by antigens. When Con A is succinylated, it dissociates from a tetramer to a dimer without alteration of its carbohydrate binding specificity. Although succinylated Con A is just as mitogenic as native Con A, the falling limb is not seen until much higher doses are reached.

Succinylation of Con A also alters another property of the lectin. It has been shown that, at certain concentrations, the binding of Con A to the cell surface restricts the movement of immunoglobulin receptors (56, 57). This suggests that it somehow changes the fluidity of the cell membrane resulting in reduction of the relative mobility of these receptors. In contrast, succinylated Con A has no such effect although it binds to lymphocytes to the same extent as the native molecule. Both the abolition of the killing effect in mitogenic assays and the failure to alter immunoglobulin receptor mobility in B cells after succinylation of Con A may be the result of change in valence or o alteration in the surface charge of the molecule. Examination of other derivatives and localization of the substituted side chains in the three-dimensional structure will help to establish which is the major factor. Recent experiments suggest that the valence is probably the major factor, for addition of divalent antibodies against Con A to cells that had bound succinylated Con A resulted again in restriction of immunoglobulin receptor mobility.

Con A may also be modified by cross-linking several molecules. A very striking effect is seen if the surface density of the Con A molecules presented to the lymphocyte is increased by cross-linking it at solid surfaces (58). Con A in free solution stimulates mouse T cells to an increased incorporation o radioactive thymidine but has no effect on B cells. When cross-linked at a solid surface, however, it stimulates mainly mouse B cells, although both T and B cells have approximately the same number of Con A receptors (58). Similar results have been obtained with other lectins (59). A reasonable interpretation of these phenomena (although not the only one) is that the lectin acts at the cell surface rather than inside the cell, that the presence of a high surface density of the mitogen is an important variable in exceeding the threshold for the lymphocyte stimulation, and that the threshold differs in the two kinds of lymphocytes.

Alteration of the structure and function of various lectins appears to be a promising means of analyzing the mechanism of lymphocyte stimulation. One intriguing hypothesis is that cross-linkage of the proper subsets of glycoprotein receptors by lectins is essentially equivalent in inducing cell transformation to cross-linkage of immunoglobulin receptors in the lymphocyte membrane by multivalent antigens. The central effector function of receptor antibodies, triggering of clonal proliferation, may turn out to be specifically related to the mode of anchorage of the antibody molecule to the cell membrane. The mode of attachment of antibody and lectin receptors to membrane-associated structures and their perturbation by crosslinkage at the cell surface may be similar and have similar effects despite the difference in their specificities and molecular structures.

Antibodies on the Surfaces of Antigen-Binding Cells

The most direct attack on the problem of lymphocyte stimulation is to explore the effects of antigens of known molecular geometry on specifically purified populations of lymphocytes. For this and other reasons, it is necessary to develop methods for the specific fractionation of antigenbinding cells.

In carrying out this task it is important both theoretically and operationally to discriminate between antigen-binding and antigen-reactive cells. In clonal selection, the phenotypic expression of the immunoglobulin genes is mediated in the animal by somatic division of precommitted cells (Figure 10). The pioneering work of Nossal and Mäkela and later of Ada and Nossal (see reference 60) clearly showed that each cell makes antibodies of a single specificity and that there are different populations of specific antigen-binding cells. An animal is capable of responding specifically to an enormous number of antigens to which it is usually never exposed, and it therefore must contain genetic information for synthesizing a much larger number of different immunoglobulin molecules on cells than actually appear in detectable amounts in the bloodstream. In other words, the immunoglobulin molecules whose properties we can examine may represent only a minor fraction of those for which genetic information is available.

One may distinguish two levels of expression in the synthesis of immunoglobulins that I have termed for convenience the *primotype* and the *clonotype* (4). The primotype consists of the sum total of structurally different immunoglobulin molecules or receptor antibodies generated within an organism

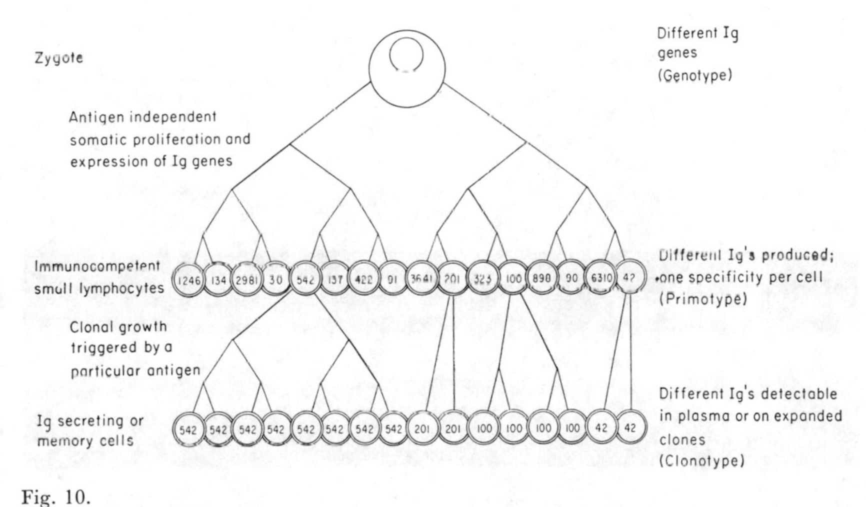

Fig. 10.
A model of the somatic differentiation of antibody-producing cells according to the clonal selection theory. The number of immunoglobulin genes may increase during somatic growth so that, in the immunologically mature animal, different lymphoid cells are formed each committed to the synthesis of a structurally distinct receptor antibody (indicated by an arabic number). A small proportion of these cells proliferate upon antigenic stimulation to form different clones of cells, each clone producing a different antibody. This model represents bone marrow-derived (B) cells but with minor modifications it is also applicable to thymus-derived (T) cells.

during its lifetime. The number of different molecules in the primotype is probably orders of magnitude greater than the number of different effective antigenic determinants to which the animal is ever exposed (Figure 10). The clonotype consists of those different immunoglobulin molecules synthesized as a result of antigenic stimulation and clonal expansion. These molecules can be detected and classified according to antigen-binding specificity, class, antigenic determinants, primary structure, allotype, or a variety of other experimentally measurable molecular properties. As a class, the clonotype is smaller than the primotype and is wholly contained within it (Figure 10).

Although a view of the clonotype is afforded by the analysis of humoral antibodies, we know very little about the primotype. It is therefore important

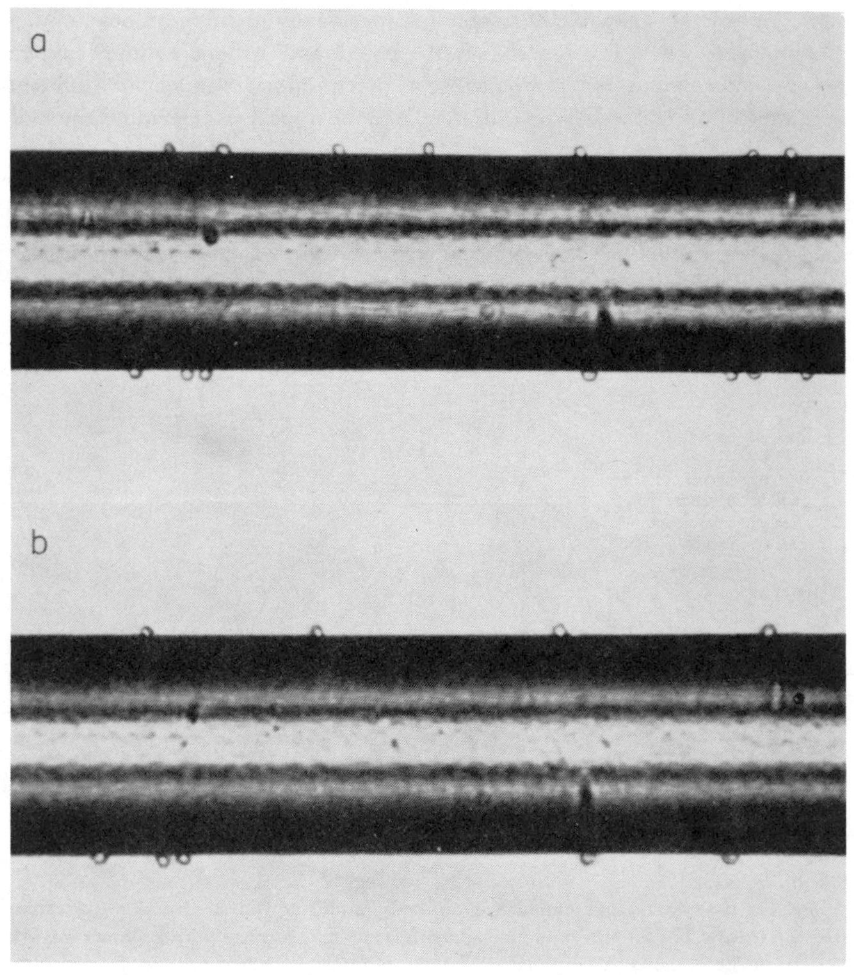

Fig. 11.
Lymphoid cells from the mouse spleen bound by their antigen-specific receptors to a nylon fiber to which dinitrophenyl bovine serum albumin has been coupled. Treatment of bound cells in (a) with antiserum to the T cell surface antigen θ and with serum complement destroys the T cells leaving B cells still viable and attached (b). See Table 2. Magnification: X235.

to attempt to fractionate the cells of the immune system according to the specificity of their antigen-binding receptors (61). We have been attempting to approach this problem of the specific fractionation of lymphocytes using nylon fibers to which antigens have been covalently coupled (62, 63). The derivatized fibers are strung tautly in a tissue culture dish so that cells in suspension may be shaken in such a way as to collide with them. Some of the cells colliding with the fibers are specifically bound to the covalently coupled antigens by means of their surface receptors. Bound cells may be counted microscopically *in situ* by focusing on the edge of the fiber (Figure 11). After washing away unbound cells, the specifically bound cells may be removed by plucking the fibers and shearing the cells quantitatively from their sites of attachment. The removed cells retain their viability provided that the tissue culture medium contains serum.

Derivatized nylon fibers have the ability to bind both thymus-derived lymphocytes (T cells) and bone marrow-derived (B cells) (64) according to the specificity of their receptors for a given antigen (65) (Figure 11, Table 2).

Table 2.
Characterization of mouse lymphoid cells fractionated according to their antigen-binding specificities. Nylon fibers were derivatized with hapten conjugates of bovine serum albumin and mice were immunized with each of the designated haptens coupled to hemocyanin. Inhibition of binding was achieved by addition of hapten-protein conjugates (250 μg/ml) or rabbit anti-mouse immunoglobulin (Ig) (250 μg/ml) to the cell suspension. High avidity cells are defined as those which are prevented from binding by concentrations of Dnp-bovine serum albumin of less than 4 μg/ml in the cell suspensions. Cells inhibited by higher concentrations are defined as low avidity cells. Virtually complete inhibition occurs at levels of homologous hapten greater than 100 μg/ml.

Antigen on Fiber	Dnp (NO_2-phenyl-NO_2)	Dnp	Tosyl (CH_3-phenyl-SO_2)	Tosyl
Immunization	none	Dnp	none	Tosyl
Cells Bound to Fiber (per cm)	1200	4000	800	2000
% Inhibition of Binding by:				
Dnp	90	95	5	10
Tosyl	1	2	75	87
Anti-Ig	85	93	73	90
High Avidity Cells (per cm)	100	2800	—	—
Low Avidity Cells (per cm)	1200	1200	—	—
% T Cells	41	39	43	—
% B Cells	59	56	54	—

About 60 % of spleen cells specifically isolated are B cells and the remainder are T cells. By the use of appropriate antisera to cell surface receptors (Table 2), the cells of each type can be counted on the fibers and most of the cells of one type or the other may then be destroyed by the subsequent addition of serum complement. In this way, one can obtain populations of either T or B cells that are highly enriched in their capacity to bind a given antigen (Figure 11).

Cells of either kind may be further fractionated according to the relative affinity of their receptors. This can be accomplished by prior addition of a chosen concentration of free antigen, which serves to inhibit specific attachment of subpopulations of cells to the antigen-derivatized fibers by binding to their receptors. As defined by this technique, cells capable of binding specifically to a particular antigen constitute as much as 1 % of a mouse spleen cell population. Very few of these original antigen-binding cells appear to increase in number after immunization, however, and the cells that do respond are those having receptors of higher relative affinities (62) (Table 2).

Whether these populations correspond to the primotype and clonotype remains to be determined. It is significant, however, that fiber-binding cells do not include plaque forming (66) cells, and it is therefore possible to fractionate antigen-binding cells from cells that are already actively secreting antibodies. Recent experiments indicate that the antigen-binding cells isolated by this method may be transferred to irradiated animals to reconstitute a response to the antigen used to isolate them. This suggests that the antigen-specific population of cells removed from the fibers contains precursors of plaque-forming cells.

We have been rather encouraged by these findings, for the various methods of cell fractionation appear to have promise not only in determining the specificity and range of T and B cell receptors for antigens but also in analyzing the population dynamics of T and B cells in both adult and developing animals. Now that fractionated populations of lymphocytes specific for particular antigens are available, it should be possible to determine the connection between lectin-induced and antigen-induced changes by comparing responses to both agents on the same cells.

Although many experiments remain to be done in this area of the molecular immunology of the cell surface, continued analysis of the mitogenic mechanism should undoubtedly clarify the problems of immune induction and tolerance. The results obtained using lymphocytes may also have general significance, however, and bear upon the nature of cell division in normal and tumor cells as well as upon growth control and cell-cell interactions in developmentaf biology. Immunology can be expected to play a double role in these areas ol study, for it will be a tool as well as a model system of central importance.

Conclusion

Immunology has been and is a curiously reflexive science, generating its own tools for understanding, such as antibodies to antibody molecules themselves. While this approach is a powerful one, a fundamental understanding of immunological problems requires chemical analysis. The determination of the molecular structure of antibodies is a persuasive example and its virtual completion has allied immunology to molecular biology in a very satisfying way:

1) The heterogeneity of antibodies and complexity of immunoglobulin classes have been rationalized in a fashion consistent with selective theories of immunity.

2) The structural basis for differentiation of the biological activity of antibodies into antigen-binding and effector functions has been made clear.

3) The detailed analysis of antibody primary structure has provided a basis for studying the molecular genetics of the immune response, particularly the origin of diversity and the commitment of each cell to the synthesis of one kind of antibody.

4) A general framework has been provided for studying antibodies at the cell surface, opening several molecular approaches for analyzing stimulation and cell triggering.

5) Finally, it is perhaps not too extravagant to suggest that the extensions of the ideas and methods of molecular immunology to fields such as developmental biology has been facilitated. In this sense, immunology provides an essential tool as well as a model with distinct advantages: dissociable cells with unique gene products of known structure; the capacity to induce specific cloned cell lines for *in vitro* analysis; the means to fractionate cells according to their state of differentiation and binding specificity, allowing quantitative studies of their selection, interaction and population dynamics.

Whether or not the immune response turns out to be a uniquely useful model, we can expect that continued work by molecular and cellular immunologists will solve the major problems of the origin of diversity and the induction of antibody synthesis and tolerance. In view of the intimate connection of these problems with problems of gene expression and cellular regulation, their solution should bring valuable insights to other important areas of eukaryotic biology and again transform immunology both as a discipline and as an increasingly important branch of medicine.

ACKNOWLEDGEMENT

By its very nature, science is a communal enterprise. I am deeply aware of the essential contributions to this work made by my many colleagues and friends throughout the last fifteen years. This occasion recalls the daily life we have shared with warmth and affection as well as the personal debt of gratitude that I owe them. I am equally cognizant of the fact that the knowledge of antibody structure was developed by many laboratories and researchers throughout the world. Not all of this work has been cited, for specific recognition here runs the risk of an unintentional omission; reference may be made to the reviews cited in the bibliography.

In addition to the fundamental support of the Rockefeller University, the work of my colleagues and myself was supported by grants from the National Institutes of Health and the National Science Foundation.

References

1. Cold Spring Harbor Symposia on Quantitative Biology, "Antibodies," *32*, 1967.
2. Nobel Symposium, 3, Gamma Globulins, Structure and Control of Biosynthesis (Killander, J. editor), Almqvist and Wiksell, Stockholm, 1967.
3. Edelman, G. M. and Gall, W. E., Ann. Rev. Biochem., *38*, 415, 1969.
4. Gally, J. A. and Edelman, G. M., Ann. Rev. Genet., *6*, 1, 1972.
5. Jerne, N. K., Proc. Natl. Acad. Sci. U.S., *41*, 849, 1955.
6. Burnet, F. M., The Clonal Selection Theory of Acquired Immunity, Vanderbilt University Press, Nashville, Tennessee, 1959.
7. Landsteiner, K., The Specificity of Serological Reactions, 2nd ed., Harvard University Press, Cambridge, Massachusetts, 1945.
8. Marrack, J. R., The Chemistry of Antigens and Antibodies, 2nd ed., (Medical Research Council Special Report Series, No. 230), London, His Majesty's Stationery Office, 1938.
9. Pedersen, K. O., Ultracentrifugal Studies on Serum and Serum Fractions, Uppsala, Almqvist and Wiksell, 1945.
10. Tiselius, A., Biochem. J., *31*, 313; 1464, 1937.
11. Karush, F., Advan. Immunol., *2*, 1, 1962.
12. Petermann, M. L., J. Biol. Chem., *144*, 607, 1942.
13. Porter, R. R., Biochem., *73*, 119, 1959.
14. Nisonoff, A., Wissler, F. C., Lipman, L. N. and Woernley, D. L. Arch. Biochem. Biophys., 89, 230, 1960.
15. Edelman, G. M., J. Am. Chem. Soc., *81*, 3155, 1959.
16. Edelman, G. M. and Poulik, M. D., J. Exp. Med., *113*, 861, 1961.
17. Edelman, G. M. and Gally, J. A., J. Exp, Med., *116*, 207, 1962.
18. Schwartz, J. and Edelman, G. M., J. Exp. Med., *118*, 41, 1963.
19. Berggård, I. and Edelman, G. M. Proc. Natl. Acad. Sci. U.S., *49*, 330, 1963.
20. Edelman, G. M., Benacerraf, B., Ovary, Z., and Poulik, M. D., Proc. Natl. Acad. Sci. U.S., *47*, 1751, 1961.
21. Koshland, M. E. and Englberger, F. M., Proc. Natl. Acad. Sci. U.S., *50*, 61, 1963.
22. Hilschmann, N. and Craig, L. C., Proc. Natl. Acad. Sci. U.S., *53*, 1403, 1965.
23. Titani, K., Whitley, E., Jr., Avogardo, L. and Putnam, F. W., Science, *149*, 1090, 1965.
24. Bull. World Health Org., *30*, 447, 1964.
25. Marchalonis, J. and Edelman, G. M., J. Exp. Med., *122*, 601, 1965; *124*, 901, 1966.
26. Fleishman, J. B., Pain, R. H., and Porter, R. R., Arch. Biochem. Biophys., Supplement 1, 174, 1962; Fleishman, J. B., Porter, R. R. and Press, E. M., Biochem. J., *88*, 220, 1963.
27. Fougereau, M. and Edelman, G. M., J. Exp. Med., *121*, 373, 1965.
28. Edelman, G. M., Gall, W. E., Waxdal, M. J. and Konigsberg, W. H., Biochemistry, *7*, 1950, 1968;

   Waxdal, M. J., Konigsberg, W. H., Henley, W. L. and Edelman, G. M., Biochemistry, *7*, 1959, 1968;

   Gall, W. E., Cunningham, B. A., Waxdal, M. J., Konigsberg, W. H. and Edelman, G. M., Biochemistry, *7*, 1973, 1968;

   Cunningham, B. A., Gottlieb, P. D., Konigsberg, W. H. and Edelman, G. M., Biochemistry, *7*, 1983, 1968;

   Edelman, G. M., Cunningham, B. A., Gall, W. E., Gottlieb, P. D., Rutishauser, U. and Waxdal, M. J., Proc. Natl. Acad. Sci. U.S., *63*, 78, 1969.

   Gottlieb, P. D., Cunningham, B. A., Rutishauser, U. and Edelman, G. M., Biochemistry, *9*, 3155, 1970;

   Cunningham, B. A., Rutishauser, U., Gall, W. E., Gottlieb, P. D., Waxdal, M. J. and Edelman, G. M., Biochemistry, *9*, 3161, 1970;

   Rutishauser, U., Cunningham, B. A., Bennett, C., Konigsberg, W. H. and Edelman, G. M., Biochemistry, *9*, 3171, 1970;

Bennett, C., Konigsberg, W. H., and Edelman, G. M., Biochemistry, *9*, 3181, 1970; Gall, W. E., and Edelman, G. M., Biochemistry, *9*, 3188, 1970; Edelman, G. M., Biochemistry, *9*, 3197, 1970.
29. Amiraian, K. and Leikhim, E. J., Proc. Soc. Exptl. Biol. Medl, *108*, 454, 1961; Taranta, A. and Franklin, E. C., Science, *134*, 1981, 1961.
30. Hill, R. L., Delaney, R., Fellows, R. R., Jr., and Lebovitz, H. E., Proc. Natl. Acad. Sci. U.S., *56*, 1762, 1966.
31. Singer, S. J. and Doolittle, R. E., Science, *153*, 13, 1966.
32. Press, E. M. and Hogg, N. M., Nature, *223*, 807. 1969.
33. Cunningham, B. A., Gottlieb, P. D., Pflumm, M. N. and Edelman, G. N. in *Progress in Immunology*, (B. Amos, ed.), Academic Press, Inc., N.Y., pp. 3–24, 1971; Cunningham B. A., Pflumm, M. N., Rutishauser, U., and Edelman, G. M., Proc. Natl. Acad. Sci. U.S., *64*, 997, 1969.
34. Franěk, F. and Nezlin, R. S., Biokhimia, *28*, 193, 1963.
35. Edelman, G. M., Olins, D. E., Gally, J. A. and Zinder, N. D., Proc. Natl. Acad. Sci. U.S., *50*, 753, 1963.
36. Olins, D. E. and Edelman, G. M., J. Exp. Med., *119*, 799, 1964.
37. Haber, E., Proc. Natl. Acad. Sci. U.S., *52*, 1099, 1964.
38. Inbar, D., Hachman, J. and Givol, D. Proc. Natl. Acad. Sci. U.S., *69*, 2659, 1972.
39. Wu, T. T. and Kabat, E. A., J. Exp. Med., *132*, 211, 1970.
40. Edelman, G. M. Ann. N.Y. Acad. Sci., *190*, 5, 1971.
41. Kehoe, J. M. and Fougereau, M., Nature, *224*, 1212, 1970.
42. Yasmeen, D., Ellerson, J. R., Dorrington, K. J., and Pointer, R. H., J. Immunol., in press.
43. Berggård, I. and Bearn, A. G., J. Biol. Chem., *243*, 4095, 1968.
44. Smithies, O. and Poulik, M. D., Science, *175*, 187, 1972.
45. Peterson, P. A., Cunningham, B. A., Berggård, I. and Edelman, G. M. Proc. Natl. Acad. Sci. U.S., *69*, 1967, 1972.
46. Poljak, R. J., Amzel, L. M., Avey, H. P., Becka, L. N. and Nisonoff, A., Nature New Biol., *235*, 137, 1972.
47. Davies, D. R., Sarma, V. R., Labaw, L. W., Silverton, E. W. and Terry, W. D., in *Progress in Immunology*, (B. Amos, ed.), Academic Press, N.Y., pp. 25–32, 1971.
48. Gall, W. E. and D'Eustachio, P. C., Biochemistry, *11*, 4621, 1972.
49. Hood, L., Gray, W. R., Sanders, B. G. and Dreyer, W. J., Cold Spring Harbor Symp. Quant. Biol., *32*, 133, 1967.
50. Milstein, C., Nature, *216*, 330, 1967.
51. Grubb, R., *The Genetic Markers of Human Immunoglobulins*, Springer-Verlag Berlin-Heidelberg-New York, 1970; Acta Path. Microbiol. Scand., *39*, 195, 1956.
52. Oudin, J., Compt. Rend. Acad. Sci., *242*, 2489; 2606, 1956; J. Exp. Med., *112*, 125, 1960.
53. Medawar, P. B. in *Les Prix Nobel en 1960*, Imprimerie Royal, P. A. Norstedt and Söner, Stockholm.
54. Sharon, N. and Lis, H., Science, *177*, 949, 1972.
55. Edelman, G. M., Cunningham, B. A., Reeke, G. N., Jr., Becker, J. W., Waxdal, M. J., and Wang, J. L., Proc. Natl. Acad. Sci. U.S., *69*, 2580, 1972.
56. Yahara, I. and Edelman, G. M. Proc. Natl. Acad. Sci. U.S., *69*, 608, 1972.
57. Taylor, R. B., Duffus, P. H., Raff, M. C. and DePetris, S., Nature New Biol., *233*, 225, 1971; Loor, R., Forni, L. and Pernis, B., Eur. J. Immunol., *2*, 203, 1972.
58. Andersson, J., Edelman, G. M., Möller, G. and Sjöberg, O., Eur. J. Immunol., *2*, 233, 1972.
59. Greaves, M. F. and Bauminger, S., Nature New Biol., *235*, 67, 1972.
60. Nossal, G. J. V. and Ada, G. L., *Antigens, Lymphoid Cells and the Immune Response*, Academic Press, N.Y., 1971.
61. Wigzell, H. and Andersson, B., J. Exp. Med., *129*, 23, 1969.
62. Edelman, G. M., Rutishauser, U. and Millette, C. F. Proc. Natl. Acad. Sci. U.S., *68*, 2153, 1971.

63. Rutishauser, U., Millette, C. F. and Edelman, G. M. Proc. Natl. Acad. Sci. U.S., *69*, 1596, 1972.
64. Gowans, J. L., Humphrey, J. H. and Mitchison, N.A., A discussion on cooperation between lymphocytes in the immune response. Proc. Roy. Soc. London B, *176*, No. 1045, pp. 369–481, 1971.
65. Rutishauser, U. and Edelman, G. M. Proc. Natl. Acad. Sci. U.S., *69*, 3774, 1972.
66. Jerne, N. K., Nordin, A. A. and Henry, C. in "Cell Bound Antibodies," (B. Amos and H. Koprowski, eds.), Wistar Institute Press, pp. 109–125, 1963.

GERALD M. EDELMAN

Dr. Gerald M. Edelman was born on July 1, 1929 in New York City to Edward Edelman and Anna Freedman Edelman. His father is a practicing physician in New York.

After his education in New York public schools, Edelman attended Ursinus College in Pennsylvania and received the B.S. degree, magna cum laude, in 1950. He then attended the Medical School of the University of Pennsylvania where he received the M.D. degree in 1954. In the succeeding year, he was a Medical House Officer at the Massachusetts General Hospital. He became a Captain in the U.S. Army Medical Corps in 1955 and practiced general medicine at a Station Hospital connected with the American Hospital in Paris, France. In 1957, he joined the Rockefeller Institute as a graduate fellow in the laboratory of Dr. Henry G. Kunkel.

After receiving the Ph.D. degree in 1960, he remained at the Rockefeller Institute as Assistant Dean of Graduate Studies and started work in his own laboratory. In 1963, he became Associate Dean of Graduate Studies, a position from which he retired in 1966. From that time to the present, he has been a Professor of the Rockefeller University.

Edelman is a member of the National Academy of Sciences, the American Academy of Arts and Sciences, the American Society of Biological Chemists and the American Association of Immunologists, as well as a number of other scientific societies. He was a member of the Biophysics and Biophysical Chemistry Study Section of the National Institutes of Health from 1964 to 1967. Presently, he is an Associate of the Neurosciences Research Program at Massachusetts Institute of Technology, a member of the Board of Governors of the Weizmann Institute of Science and a member of the Advisory Board of the Basel Institute for Immunology.

He has given the Carter-Wallace Lectures at Princeton University in 1965, the National Institutes of Health Biophysics and Bioorganic Chemistry Lectureship at Cornell University in 1971, and delivered the Darwin Centennial Lectures at the Rockefeller University in 1971. In 1972, he was the first Felton Bequest Visiting Professor at the Walter and Eliza Hall Institute for Medical Research in Melbourne, Australia.

Edelman has received the Spencer Morris Award of the University of Pennsylvania in 1954, the Eli Lilly Award in Biological Chemistry given by the American Chemical Society in 1965, and the Annual Alumni Award of Ursinus College in 1969. In addition to his studies of antibody structure, his research interests have included the application of fluorescence spectroscopy and fluorescent probes to the study of proteins and the development of new methods of fractionation of both molecules and cells. His present research interests include work on the primary and three-dimensional structures of proteins, experiments on the structure and function of plant mitogens and studies of the cell surface.

In 1950, Edelman married Maxine M. Morrison. They have two sons, Eric and David and one daughter, Judith.

R. R. PORTER

Structural Studies of Immunoglobulins

December 12, 1972

In 1946, when I was starting work as a research student under the supervision of Dr. F. Sanger, the second edition of Karl Lansteiner's book 'The Specificity of Seriological Reactions' (1) reached England. In it was summarised the considerable body of information available on the range of antibody specificity and much of it was Landsteiner's own work or by others using his basic technique of preparing antibodies against haptenes and testing their ability to inhibit the precipitation of the antisera and the conjugated protein. Also described in this book was the work in Uppsala of Tiselius and Pederson in collaboration with Heidelberger and Kabat in which they showed that all rabbit antibodies were in the γ globulin fraction of serum proteins and that they had a molecular weight of 150,000. This combination of an apparently infinite range of antibody combining specificity associated with what appeared to be a nearly homogenous group of proteins astonished me and indeed still does.

ACTIVE FRAGMENTS OF ANTIBODIES

The preparation of antibodies by dissociation of specific precipitates with strong salt solutions or in acidic conditions had been described, as had the preparation in fair yield of γ globulin fractions by salting out techniques from whole serum, so an experimental approach to the structural basis of antibody combining specificity was possible. A start had indeed been made by showing that the whole molecule was not required for the combining specificity. Parventjev (2) had introduced pepsin treatment of serum as a commercial method of purification of horse antitoxins and Petermann and Pappenheimer (3) studied the reaction but using purified horse antidiphtheria toxin rather than whole serum for the peptic digestion. They showed that a product able to flocculate with the toxoid or neutralise toxin could be obtained and that it had a molecular weight of 113,000, i.e. substantially less than that of the original molecule. Petermann (4) showed later that human γ globulin could be split by papain to give what she estimated, by using the ultracentrifuge, to be quarter molecules. No antibody activities were, however, investigated in this study.

At about the same time Landsteiner (5) was extending his investigation of the antigenic specificity of protein antigens and had found that crude but apparently low molecular weight peptides from an acid digest of silk fibroin could inhibit the precipitation of soluble fibroin with its rabbit antiserum. This finding in conjunction with the haptene and other studies suggested that antigenic sites and presumably, therefore, antibody combining sites were small, certainly very much smaller than the antibody molecules and further attempts to obtain fragments of an antibody molecule, which retained the power to combine with the antigen seemed worthwhile. Testing for such active fragments was by their ability to inhibit the combination of the antigen and whole antibody. However, although a variety of conditions of hydrolysis by acids or enzymes were investigated (6) only papain gave an active product and this appeared to be from N terminal amino acid assay, the quarter molecules previously described by Petermann (4). There was no doubt that the combining site was in these smaller fragments and hence a substantial reduction in the magnitude of the structural problem had been achieved but protein molecules of molecular weight 40,000 were still a formidable prospect. This work was carried out in Sanger's laboratory in Cambridge and with his guidance N terminal amino acid assay and the terminal sequences were attempted but proved unhelpful in that they suggested that the rabbit γ globulins and antibodies had a single open polypeptide chain and that the biologically active

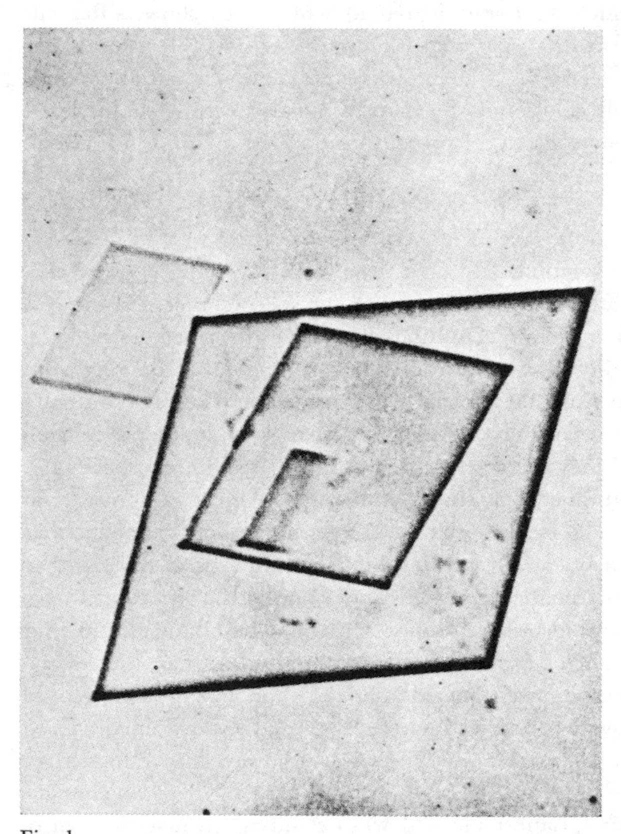

Fig. 1.
Crystals formed from a papain digest of rabbit IgG, the Fc fragment.

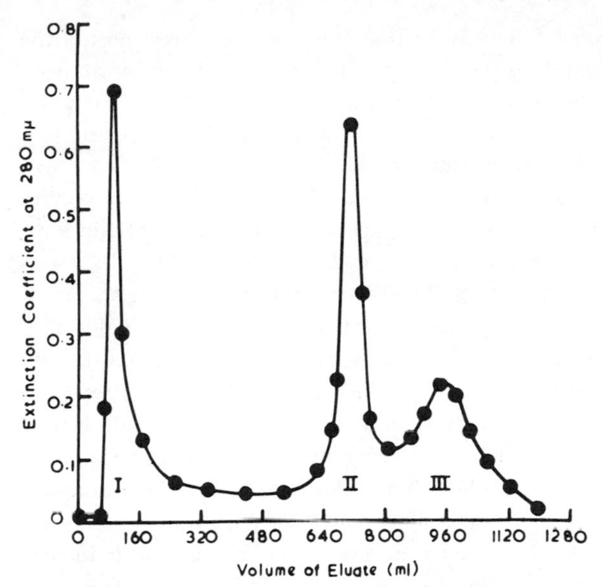

Fig. 2.
Fractionation of a papain digest of rabbit IgG on carboxylmethyl cellulose in sodium acetate buffer pH 5.5 with a gradient from 0.01M to 0.9M. Fractions I and II (Fab) carry the antibody combining sites and Fraction III (Fc) will crystallise easily.

quarter molecules also had the same alanine N terminal acid. The possilbiity that there might be blocked N terminal residues was not considered.

A return to papain digestion of rabbit γ globulin was made seven years later but in place of the crude enzyme preparation used earlier, a crystalline enzyme (7) was used in much lower concentrations, at 1/100 the weight of substrate (8). Under these conditions a number of points missed previously became apparent. First, there was a very high recovery of total protein after dialysis of the digest—very few small peptides were formed in spite of the wide specificity of the enzyme. The products, all of very similar size (sedimentation value 3.5S) were one third of the original rather than one quarter and most surprising one of the products of digestion crystallised very easily in diamond shaped plates (fig. 1) during the dialysis at neutrality in the cold room. This last observation suggesting that a protein which itself would never crystallise could give a fragment which presumably was more homogenous and hence able to crystallise was quite unexpected and indeed was unacceptable. The crystals were dismissed as coming from the less soluble amino acids and discarded without further consideration over several months. Fortunately, my neighbour in the adjoining laboratory at the National Institute of Medical Research in London was the X-ray crystallographer, Dr. Olga Kennard and when I eventually asked her opinion she immediately gave the view that they were protein crystals. They were then identified with the third peak obtained by fractionation of the digest products on CM cellulose (fig. 2). They were named Fraction III and are now known as the Fc fragment. Fractions I and II were the components of the digest which retained the combining specificity of the orginal antibody. Fraction III had no such activity but did carry most of the antigenic specificity of the rabbit γ globulins when tested with antiserum from goats, rats and guinea pigs. More detailed studies (9) showed that fraction I and II were very similar, chemically and antigenically, and later Nisonoff and colleagues (10) showed that the distinction between these two fractions was artificial. If a basic fraction of γ globulin was treated with papain two molecules of II and one of III were obtained and an acidic fraction of γ globulin gave two molecules of I and one of III. The slight differences in charge between fractions I and II reflected the charge heterogeneity of the starting material.

Nisonoff *et al* (11) had also returned to the peptic digestion of γ globulins but using the rabbit protein rather than horse antitoxin and showed that on reduction the 100,000 molecular weight product comparable to that reported earlier would give half molecules very similar to the fractions I and II. The latter are now named Fab and the peptic digest product $(Fab')_2$.

The papain digest studies established that γ globulin, now named immunoglobulin gamma or IgG, was formed from three globular sections which were probably rather tightly folded as they were exceptionally resistant to further degradation by papain. The Fc fragment was apparently common to all molecules while the two identical Fab fragments each carried a combining site and with it the inherent variability associated with the whole antibody. An attempt was made (9) to relate this tripartite structure with the supposedly single open polypeptide structure deduced from end group analysis of rabbit IgG. Of course, it made no sense and progress depended upon the demonstration by Edelman (12) that in fact, human IgG and therefore presumably IgG of all species were multichain proteins. It followed that there must be blocked N terminal amino acids and that estimation of the free N terminal amino acid was of only limited significance.

The four peptide chain structure

The solution of the gross structure of immunoglobulins depended upon establishing the relationship between the peptide chains identified by Edelman and the products of papain digestion. This was achieved easily when the con-

Fig. 3.
Double diffusion of the heavy and light chains of rabbit IgG against goat anti rabbit Fab and goat anti rabbit Fc. Note that light chain reacts only with anti Fab while heavy chain reacts with both anti Fab and anti Fc, i.e. Fab contains parts of heavy and light chains while Fc contains only parts of heavy chain.

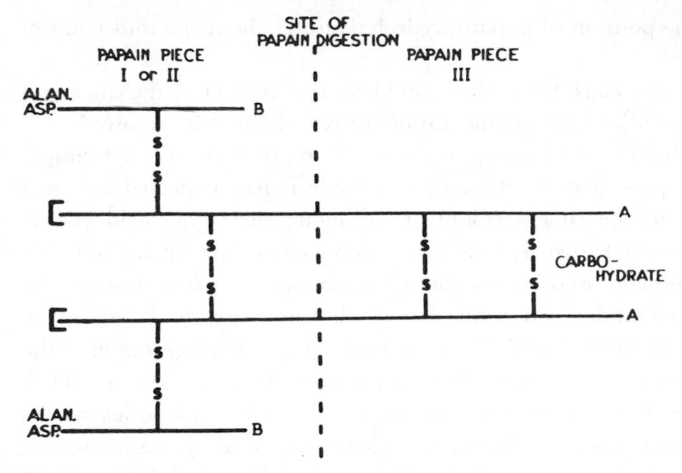

Fig. 4.
The four chain structure of rabbit IgG postulated on the basis of the double diffusion experiment of figure 3 and supporting chemical evidence.

ditions of isolation of the chains were modified using reduction in the absence of denaturing agents—conditions in which predominantly interchain disulphide bonds are broken. No fall in molecular weight followed such reduction but the chains were dissociated and separated when run on Sephadex columns in acetic or propionic acid to give heavy and light chains of molecular weights approximately 50,000 and 20,000 respectively. These chains now remained soluble at neutrality and retained antigenic specificities. A double diffusion plate using antisera specific to Fab or Fc showed that Fab contained antigenic sites common to both heavy and light chains but Fc those common to heavy chains only (fig. 3). This led to the postulated four chain structure (13) (fig. 4). More detailed studies were in agreement with this structure (14, 15 and 16)

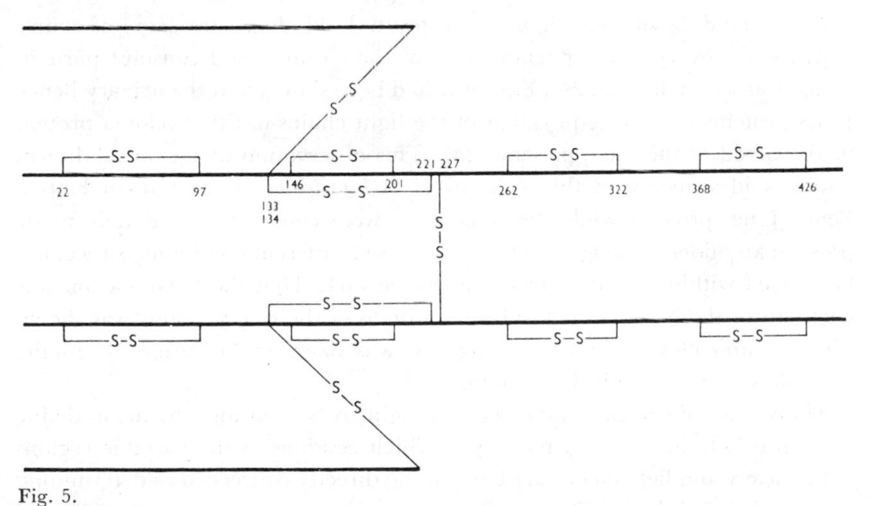

Fig. 5.
The structure of rabbit IgG showing the rather complex arrangement of inter and intra disulphide bonds in the heavy chain.

and confirmed the position of papain hydrolysis to be about the mid-point of the heavy chain.

It was some years before the rather complex arrangement of the inter and intra chain disulphide bonds of the rabbit heavy chains was resolved (17) (fig. 5). The mechanism of blocking of the α amino groups of the N terminal amino acids also proved more difficult to establish than expected as it was found to be due to the ringed residue pyrrolidone carboxylic acid (PCA) (18, 19) which was well known as an artefact arising from N terminal glutamine residues. All attempts to find N terminal glutamine in IgG were unsuccessful even when isolated under conditions in which conversion to PCA during handling appeared to be excluded. Evidence has been given that glutamine is the residue incorporated into the peptide chain during synthesis (20) but PCA appears to be the terminal residue present in the immunoglobulins in the blood and it has been suggested that enzymatically catalysed cyclisation occurs intracellularly. It has been assumed that PCA is the only blocked N terminal residue in immunoglobulin molecules from all species, but there appears to have been little careful study.

Antibody combining sites

While the four peptide chain model clarified many aspects of the structure of antibodies, it made no contribution to our understanding of what features gave the possibility of forming antibodies of innumerable different specificities. It seemed at the time to increase the difficulties as the possibilities of variation were reduced. I made repeated attempts to obtain digest products of Fab which still bound the antigen but without success (21) but in fact progress has just been reported in isolating a peptic digest product of a mouse myeloma protein MOPC 315. This fragment named Fv appears to be formed from the N terminal half of the Fab molecule i.e. of the light and Fd chains and retains its full affinity for a dinitrophenyl hapten (22).

Understanding of the origin of the multiple binding sites came of course with the discovery of the phenomenon of the variable and constant parts in Bence Jones proteins (23, 24). Earlier it had been shown that the urinary Bence Jones proteins were the equivalent of the light chains of the myeloma protein in the blood of the same patient (25). This observation of the variability in amino acid sequence in the N terminal 107 residues of the human Kappa Bence Jones proteins while the remainder were constant immediately made possible an understanding as to how millions of different combining sites could be formed within the same structural framework. That the phenomenon was common to the N terminal 110 residues or so of the heavy chains was shown (26, 27) and clearly the combining site was likely to be formed from the variable sections of both these chains.

Many lines of evidence have been brought to bear in an attempt to define more precisely in chemical terms just which residues in the variable regions of the heavy and light chains are likely to be directly concerned in determining the specificity of the binding site. They have been reviewed recently (28) and so the main conclusions could perhaps be just listed here:

1. The size of the antigenic site appears to be of the order of a hexapeptide or hexasaccharide.

2. This is comparable to the size of the substrate of a hydrolytic enzyme such as lysozyme. In this enzyme 15—20 amino acid residues have been identified as probable 'contact amino acids' i.e. residues forming a bond with the substrate. Hence this appears to be the likely number of residues to be expected to line the antibody combining site and to play a direct role in determining specificity. If any residue could occupy any of these 15—20 positions, the possible number of variants is indeed high.

3. There are at least three hypervariable sections in each of the heavy and light chain variable regions. They have been demonstrated rather clearly in the plots of Kabat (29, 30) of the frequence of occurrence of different residues in myeloma proteins along the 110 or so positions of the variable regions. This hypervariability is also apparent in sequence studies of the heavy chains of rabbit IgG and both agree in suggesting that in most cases the hypervariability is confined to one or two positions but in the region 96—110 of the heavy chains four or five positions may be exceptionally variable. In the rabbit γ chain there is a section in this position across which no satisfactory sequence has been obtained, presumably because of the complexity of the sequences (31).

4. Several pieces of evidence suggest that these six hypervariable sections in the two variable regions may be brought together in the intact molecule to contribute to the structure of the combining site. The most direct evidence comes from affinity labelling studies in which an antibody is allowed to bind a haptene to which a reactive group has been attached. Covalent reaction will follow and after subsequent hydrolysis the labelled peptides can be identified and placed by comparison with the known amino acid sequence. A variety of affinity labelling techniques has been introduced and used both with natural antibodies and with a mouse myeloma protein showing high affinity for the dinitrophenyl group. Though the work is in some cases incomplete, all agree in that the labelled reagents are found attached to residues in or near one or other of the six hypervariable sections.

5. It is likely that there is a hydrophobic region adjacent to the combining site which may not contribute to specificity but could increase the affinity of binding for an appropriate antigenic site.

The precise details of the combining site must await the completion of the crystallographic studies on immunoglobulins and their fragments now being undertaken in several laboratories. There will be considerable interest in seeing how far the above prediction from chemical studies prove to be correct when the full structures become available.

The genetic origin of the multiple forms of antibodies

While the discovery of the variable region and particularly of the hypervariable sections in it, seemed likely to answer the basic question about the structural origin of multiple binding specificities it clearly raised very difficult problems as to the genetic origin of these many different amino acid sequences. This topic has formed the basis of many discussions and reviews and decisive evidences in favour of any of the many theories advanced is lacking. I only wish to make a brief comment here about the sequence work on the variable region of rabbit IgG heavy chains which developed as part of the other structural work discussed above.

There are perhaps two main points under immediate discussion. First, are both heavy and light chains each the products of two genes one coding for the variable region and one for the constant? Second, are the multiple genes which code for the variable regions, germ line genes or are they the product of somatic mutation of a much smaller number of germ line genes?

In each case it would be of obvious value if allelic variants able to act as genetic markers could be found in both the variable and constant regions. Markers in the constant regions have, in fact, been found in many of the different classes and subclasses of immunoglobulins of different species. The phenotypic character followed is the antigenic specificity of the immunoglobulin. This specificity has been correlated with amino acid changes but no direct identification of the specificity and the sequence change by demonstration of, say, the inhibitory power of a small peptide, has proved to be possible. Presumably a much larger section of molecule is necessary for the integrity of the antigenic site. However, in many cases there is no doubt that the genetic markers of the constant region have been found as they can be shown to be present in the Fc fragment. Allocation of genetic markers to the variable region is less clear though it is likely that this is where the rabbit 'a' locus allotypes orginate.

In 1963, Todd (32) made the very surprising observation that the 'a' locus allelic specificities of rabbit immunoglobulins were common to IgM and IgG. As the structural work progressed it became clear that these antigenic specificities were carried by the μ and γ chains respectively though these chains differed obviously in chemical structure and hence in their structural genes. It then seemed possible that the 'a' locus specificities could be determined by the variable regions which might be common to both chains. This observation of Todd's was the first to raise the possibility of two genes being concerned in the structure of the heavy chain, and now two examples of crossovers out of about 400 offspring between allelic specificities undoubtedly determined by structure of the constant parts of the γ chains and the 'a' locus specificities have been reported. If extended, these studies will strengthen the evidence that two genes code for the γ chains. However, the establishment of the structural basis of the 'a' locus specificities depends at present on correlation of amino acid sequences in the variable region with the specificity and lack of any similar correlation in the constant region sequences. This is rather indirect but but as far as it is acceptable, such correlation extending to about 16 positions in the γ chains has been found working with pooled rabbit IgG (33). Work with homogeneous rabbit anti polysaccharide antibodies has confirmed this correlation for some but not all these positions (34, 35).

These 'a' locus markers could clearly be of decisive importance in genetic studies and indeed their existence in the variable region would be taken by some as strong evidence against the likelihood of many million copies of the variable regions being present in the germ line. It is, therefore, worthwhile to attempt to obtain unequivocal evidence that these apparent genetic markers of the variable region are indeed such. Alternatively, however, it should also be possible using the chemical evidence now obtained to follow the inheritance directly of a given allelic peptide rather than an antigenic specificity. One such peptide which occurs in two or three forms among different rabbits can be identified rather easily by autoradiography after reaction of the half cysteine residue at position 92 with ^{14}C iodoacetic acid (36). Early evidence suggests that it is indeed behaving as an heritable character and if confirmed will be direct evidence for a structural gene marker in the variable region. Further work is in progress and this should contribute to knowledge of the genetic origin of the variable region.

Some aspects of the structural studies of immunoglobulins have reached completion in that full chemical structures are now available for several human myeloma proteins and almost complete structures for rabbit immunoglobulins. The solution of the structural basis of the combining specificity of antibodies which seemed to me the central problem also appears to be nearing completion. There is still a role for structural work in the solution of the genetic origins of antibodies and obviously there are many other applications not discussed during lecture. Interaction of immunoglobulins with complement components and cell surfaces are two which are already arousing rapidly increasing interest.

References

1. Landsteiner, K., The Specificity of Serological Reaction. Harvard University Press, Cambridge U.S.A. (1946).
2. Parventjev, I. A., U.S. Patent 2,065,196 (1936).
3. Petermann, M. L., and Pappenheimer, A. M., J. Phys. Chem. *45*, 1 (1941).
4. Petermann, M. L., J. Am. Chem. Soc. *68*, 106 (1946).
5. Landsteiner, K., J. Exptl. Med. *75*, 269 (1942).
6. Porter, R. R., Biochem. J. *46*, 479 (1950).
7. Kimmel, J. R., and Smith, E. L., J. Biol. Chem. *207*, 515 (1954).
8. Porter, R. R., Nature *182*, 670 (1958).
9. Porter, R. R.. Biochem. J. *73*, 119 (1959).
10. Palmer, J. L., Mandy, W. J., and Nisonoff, A., Proc. Nat. Acad. Sci. US *48*, 49 (1962).
11. Nisonoff, A., Wissler, F. C., Lipman, L. N., and Woernley, D. L., Arch. Biochem. Biophys. *89*, 230 (1960).
12. Edelman, G. M., J. Am. Chem. Soc. *81*, 3155 o1959).
13. Porter, R. R., Basic Problems of Neoplastic Disease (A. Gellhorn and E. Hirschberg Eds) Columbia Univ. Press, New York (1962).
14. Fleishman, J. B., Porter, R. R., and Press, E. M., Biochem. J. *88*, 220 (1963).
15. Crumpton, M. J., and Wilkinson, J. M., Biochem. J. *88*, 228 (1963).
16. Pain, R. H., Biochem. J. *88*, 234 (1963).
17. O'Donnell, I. J., Frangione, B., and Porter, R. R., Biochem. J. *116*, 261 (1970).
18. Press, E. M., Piggot, P. J., and Porter, R. R., Biochem. J. *99*, 356 (1966).
19. Wilkinson, J. M., Press, E. M., and Porter, R. R., Biochem. J. *100*, 303 (1966).
20. Stott, D. I., and Munro, A. J., Biochem. J. *128*, 1221 (1972).
21. Porter, R. R., Brookhaven Symposium in Biology *13*, 203 (1960).
22. Inbar, D., Hochman, J., and Givol, D., Proc. Nat. Acad. Sci. US. *69*, (1972).
23. Hilschmann, N., and Craig, L. C., Proc. Nat. Acad. Sci. US *53*, 1403 (1965).
24. Titani, K., Whitley, E., Avogardo, L., and Putnam, F. W., Science *149*, 1090 (1965).

25. Edelman, G. M., and Gally, J. A., J. Exptl. Med. *116*, 207 (1962).
26. Press, E. M., and Hogg, N. M., Nature *223*, 807 (1969).
27. Edelman, G. M., Cunningham, B. A., Gall, W. E., Gottlieb, P. D., Rutishauser, U., and Waxdal, M. J., Proc. Nat. Acad. Sc. US *63*, 78 (1969).
28. Porter, R. R., Contemporary Topics in Immunochemistry (Ed. F. P. Inman) Plenum Press New York Vol *1*, 145 (1972).
29. Wu, T. T., and Kabat, E. A., J. Exptl. Med. *132*, 211 (1970).
30. Kabat, E. A., Wu, T. T., Ann. New York Acad. Sci. *190*, 382 (1971).
31. Fruchter, R. G., Jackson, S. A., Mole, L. E., and Porter, R. R., Biochem. J. *116*, 249 (1970).
32. Todd, C. W., Biochem. Biophys. Res. Commun. *11*, 170 (1963).
33. Mole. L. E., Jackson, S. A., Porter, R. R., and Wilkinson, J. M., Biochem. J. *124*, 301 (1971).
34. Fleischman, J. B., Biochemistry *10*, 2753 (1971).
35. Jaton, J. C., and Braun, D. G., Biochem. J. *130*, 539 (1972).
36. Mole, L. E., Unpublished Work (1972).

RODNEY R. PORTER

Rodney Robert Porter born 8 October 1917 at Newton-le-Willows, Lancashire, England.

He was educated at the Ashton-in-Makerfield Grammar School taking his Hons.B.Sc. (Biochemistry) in 1939 at the University of Liverpool and his Ph.D. at the University of Cambridge in 1948.

After one year's postdoctoral work at Cambridge, Professor Porter joined the scientific staff of the National Institute of Medical Research in 1949 and was there until 1960 when he joined St. Mary's Hospital Medical School, London University as the first Pfizer Professor of Immunology.

In 1967, he was appointed Whitley Professor of Biochemistry in the University of Oxford and a Fellow of Trinity College, Oxford.

Amongst his awards are those of:

Fellow of the Royal Society, 1964

Gairdner Foundation Award of Merit, 1966

Ciba Medal (Biochemical Society), 1967

Karl Landsteiner Memorial Award from the American Association of Blood Banks, 1968

National Academy of Science, U.S.A., Foreign Member 1972

He took his Ph.D. at Cambridge under the supervision of Dr. F. Sanger investigating protein chemistry. In 1948 Professor Porter started to investigate the structure of antibodies, but on moving to Mill Hill he worked on methods of protein fractionation collaborating with Dr. A. J. P. Martin. The particular interest was in chromatographic methods of fractionation.

He returned to the study of the chemical structure of antibodies leading to the finding of the three fragments produced by splitting with papain in 1958—59. He continued on this work at St. Mary's Hospital Medical School and put forward the peptide chain structure of antibodies in 1962.

Since moving to Oxford he has been concerned with the structure of antibody combining site, of the genetic markers of immunoglobulins and recently in the chemical structure of some of the early complement components.

During the war years 1940—46 Professor Porter was in the army serving with the R.A., R.E., and R.A.S.C., finishing with the rank of Major. He was with the First Army in 1942 in the invasion of Algeria and with the 8th Army during the invasion of Silicy and then Italy. He remained with the Central Mediterranean Forces in Italy, Austria, Greece and Crete until January 1946.

1975

Renato Dulbecco
David Baltimore
Howard M. Temin

Renato Dulbecco

From the Molecular Biology of Oncogenic DNA Viruses to Cancer

December 12, 1975

Oncogenic viruses, able to elicit tumour formation in animals have been on the scientific scene for many years. After the early discovery of Ellerman and Bang at the beginnig of this century, Peyton Rous opened up the field in its second decade and in prophetic words gave a good hint of things to come. However, these discoverics were soon forgotten and only after a long eclipse was interest in oncogenic viruses revived in the 'fifties. My involvement in this field began at that time, when Rubin and Temin worked in my laboratory with the Rous Sarcoma Virus. When in 1958 polyoma virus, a new oncogenic virus with different properties, was isolated, I jumped at the new opportunity and started working with it. Within a short time polyoma virus became the main interest of my laboratory, to be joined, a few years later by SV_{40}, another papovavirus. It became clear fairly soon that the molecular biology of these viruses could be worked out, and I set out to find the molecular basis of cancer induction. The results that I and a number of brilliant young collaborators have obtained during the following fifteen years have brought us close to that goal. I will review the most interesting steps of our work and will then ask some questions concerning the nature of cancer and about perspectives for prevention and treatment. I stress the relevance of my work for cancer research because I believe that science must be useful to man.

INTEGRATION: THE PROVIRUS

Let me start with a brief review of our work in the molecular events in transformation. The first results, crucial for future developments, showed that polyoma virus could be assayed in certain cell cultures (1), which we call permissive, and could induce a cancer-like state in other cultures (2, 3) in which the virus does not grow, which we call non-permissive. The induction of the cancer-like state in vitro was called transformation. We were able to show that the virus containes DNA (4), and within a few years we gave the first evidence of its cyclic, or circular, shape (5), which is important for two critical biological events: DNA replication and integration. In integration, which we discovered a few years later with SV_{40} (6), the viral DNA becomes a provirus, i. e. it establishes permanent, covalent bonds with the cellular DNA. The cyclic configuration explains how a complete molecule of the SV_{40} DNA can be integrated without losses.

Integration is one of the key events in virus induced cell transformation. It explained the persistence of the transformed state in the cell clone deriving from a transformed cell, since the provirus replicates with the cellular DNA. It also permitted us to resolve one of the main questions about the role of viruses in transformation. It was known at the time that papova viruses leave their footprints in the cells of the cancers they induce and those they transform *in vitro,* in the form of characteristic antigens. However, it was not known whether the antigens were expressed by viral genes or by derepressed cellular genes. Hence, it was uncertain whether cells were transformed by the expression of viral genes persisting in the cells or alternatively if the virus altered the cells by a hit-and-run mechanism, changing the expression of cellular genes and then leaving. The demonstration that viral DNA is integrated in the cells in conjunction with the finding that the provirus is transcribed into messenger RNA (7) hundreds of generations after the establishment of a transformed clone, made the hit-and-run hypothesis unlikely and supported a continuing role of viral gene functions in determining transformation. This possibility was later supported by observations with abortively transformed cells, which behave as transformed only for several generations after infection, but then return to normality (8). When they are back to normal these cells no longer contain the viral DNA (9).

The viral genes that remain unexpressed in the transformed cells, such as those for capsid protein in SV_{40}-transformed cells, were also interesting, although in a different way. In fact their expression could be renewed in heterokaryons formed by fusing transformed cells with permissive cells (10, 11), a result that gave the first evidence that the viral functions are under the control of cellular functions. The provirus thus became a tool for studying regulation of DNA transcription in animal cells. Subsequently, the presence of giant RNAs containing viral sequences in the nucleus of transformed or lytically infected cells (12) raised the question of the initiation and termination signals for transcription in animal cells, as well as the question of processing of nuclear RNA precursors of messenger RNA, questions that are still largely unresolved.

VIRAL FUNCTIONS IN TRANSFORMATION

Efforts were in the meantime directed at identifying the viral genes transcribed in the transformed cells. It was established that in lytic infection with SV_{40} the whole viral DNA is transcribed in two nearly equal parts, one early, before the inception of replication of the viral DNA, the other late after DNA replication has begun and that the early RNA is also present in transformed cells (7). Subsequently, the early and the late messengers were found to be transcribed from different DNA strands (13), an observation that facilitated the further characterization of the viral transcripts. Later work in other laboratories using specific fragments produced by restriction endonucleases confirmed and refined these findings; and the results were extended to adenoviruses by showing that a segment of the early part of that DNA is always present and transcribed in transformed cells (14, 15).

These facts suggestad that some early viral function is essential for maintaining the transformed state but they could also be interpreted differently: for instance, transformation might be caused by the mere presence of the viral DNA in the cellular DNA, the persistent viral functions being perhaps required for establishing and maintaining integration.

Attempts were made to solve the dilemma by isolating temperature sensitive mutants affecting either initiation or maintenance of transformation. Many transformation mutants were found, all clustered in a segment of the early region of the viral DNA, designated as the A gene, but they were all initiation mutants (16, 17, 18). These mutations prevent the onset of transformation at high but not at low temperature and cells transformed at low temperature remain transformed at high temperature. It was not possible to find clear cut maintenance mutants, i.e. capable of causing a complete reversion of the phenotype when cells transformed at low temperature were shifted to high temperature: however, careful observation later showed that the initiation mutants are also partial maintenance mutants, since the cells they transform undergo a partial reversion of the phenotype at high temperature (19, 20, 21). This result shows that the viral genes play a continuing role in transformation; however, the failure to obtain complete maintenance mutants suggests that the relation between viral gene expression and cell phenotype is complex.

SEARCH FOR THE VIRAL TRANSFORMING PROTEIN

Further progress in this subject has been achieved by studying the proteins specified by the early region of the viral DNA. This work has centred around the so-called T antigen (22) present in the nucleus of cells infected or transformed by SV_{40} and whose synthesis and properties are affected by mutations of the A gene (23, 24, 25). In non-permissive transformed cells the antigen is a protein of molecular weight of about 94,000 daltons (26), which binds firmly to double-stranded DNA, but without much specificity (27, 28, 29). That the T antigen is specified by the viral DNA is strongly suggested by its *in vitro* synthesis using a wheat germ extract primed with various messengers (30), especially since the size of the product depends on the nature of the messenger. Thus when the messenger was viral RNA made *in vitro* by transcribing SV_{40} DNA with *E. coli* RNA polymerase, an antigenic protein of about 62,000 daltons was synthesized; but when mRNA extracted from infected cells was used the protein synthesized was, like the T antigen of transformed cells, of about 94,000 daltons. The discrepancy of the two molecular weights makes it very unlikely that the T antigen is a cellular protein modified by a viral function, because two different proteins would have to be modified in the same extract depending on the messenger used. In contrast, the synthesis of a shorter polypeptide chain with the artificial messenger may be justified by the absence of accessory signals, such as cap, poly A, and possibly other modifications. Further definition of these findings awaits peptide maps of the various products.

Since the early, transforming, part of the SV_{40} genome can specify proteins of a molecular weight of about 100,000 daltons altogether, the T antigen is likely to be its sole product and, therefore, to be the transforming protein. However, the same protein must also initiate viral DNA replication, which cannot initiate at high temperature in cells infected by mutants of the A gene. The different functions in transformation and lytic infections could be performed by different domains of the same protein, or could result from modifications (such as phosphorylation and glycosylation) or from processing. Processing of SV_{40} T antigen seems to occur in lytically infected cells which contain a smaller T antigen of about 84,000 daltons; this smaller size contrasts with the regular size (94,000 daltons) of the antigen specified *in vitro* by mRNA extracted from the same cells (26). Whether the two forms of the antigen have different roles in transformation and DNA replication still remains to be established.

Since the transforming protein should control both initiation and maintenance of transformation the partial reversion of the phenotype of cells transformed by A mutants when shifted to high temperature may be explained by a decreased requirement for the transforming protein once transformation has taken place, which in turn could result from a positive feedback stabilizing the transformed state. For instance, unstable protein monomers specified by the mutated gene might form self-stabilizing oligomers (31), or the transforming protein might generate changes that tend to favour the transformed state. An example of the latter model is the β galactosidase induction in *E. coli* which is maintained by inducer concentrations much smaller than that required for initiating induction, because inducer is pumped into the cells by the induced permease (32). I wonder whether a certain degree of self-stabilization of the state of gene expression is a general property of animal cells, which has developed for maintaining differentiation.

CELLULAR EVENTS IN TRANSFORMATION

I now wish to turn to cellular events participating in transformation, which will be the main problem after the remaining questions on the role of the virus have been answered. Among the cellular events are functional changes and mutations. Some functional changes, which affect many cellular properties, are associated with the shift of resting cells to growing state after infection with polyoma virus or SV_{40} (33, 34); other changes observed in transformed and in cancer cells in general consist of the re-expression of cellular genes normally expressed in a preceding state of differentiation, in foetal life (35). These functional changes might be caused by the binding of transforming proteins to DNA; if so they may be mediated by an alteration of transcription of the cellular DNA. However, we do not know whether the transcription pattern changes because experiments based on competition hybridization have given ambigous results. Perhaps the methodology is not good enough. Cloning of cellular DNA fragments in phages or plasmids may afford the necessary probes for carrying out significant experiments.

In order to understand further how the virus deregulates cellular growth we would need a detailed knowledge of the mechanisms of growth regulation in animal cells, which is now lacking. However, certain useful ideas about growth regulation are now available, and can be used to draw inferences on the action of the virus. Thus it seems clear that with a given cell type growth regulation involves a complex chain of events, beginning with extracellular regulators of many kinds, probably interacting with the cell plasma membrane. Cytoplasmic mediators then appear to transmit regulatory signals from the plasma membrane to the nucleus, where they perhaps control DNA-binding proteins similar to the transforming protein of papova viruses. The complexity of growth regulation increases markedly when different cell types are considered, since they seem to recognize different sets of extracellular regulators and may have different mediators and DNA binding proteins.

Proceeding from this general picture it would be tempting to propose that the viral transforming protein replaces one of the normal nuclear regulatory proteins of the cell, and being unaffected by the mediators that control the normal protein, keeps growth related transcription going, bypassing the signals of the plasma membrane. If so, however, the transformed state should be dominant over the normal state in cell hybrids, whereas the contrary is usually true (36). On the other hand, the dominance of the normal state could be explained if the transformed cells had a changed surface, unable to respond to regulatory signals. Such a change could result from the re-expression of foetal functions to make the transformed cells anachronistic, i. e. belonging to a stage of differentiation inappropriate to that of the organism which contains them. The cells with an anachronistic surface being insensitive to the growth regulators present in the adult organisms, which operate on adult cells, would grow without control. A striking support of the role of cell anachronism in cancer has been obtained with teratoma, a tumour originating when cells from an early embryo are transplanted to an adult environment. When, after many transplants cells of this tumour are introduced back into a blastocyst, i. e. an early embryo, they return to normality (37, 38), presumably because the internal growth control of the cells becomes again matched by the environmental regulators of the recipient embryo. In this model a hybrid cell formed by fusing a transformed and a normal cell may be untransformed, if the normal partner contributes normal surface components which respond to the normal extracellular regulators. For this result to be possible, anachronistic transcription should not be initiated after cell fusion on the DNA deriving from the normal parent. The virological studies suggest that this may well be the case, since the initiation of transformation seems to require much more transforming protein than its maintenance.

It would be important to recognize the developmental period in which the anachronistic genes of transformed or cancer cells are normally expressed, not only for understanding but possibly also for controlling cancer. In fact, if the growth regulators specific for the periods expressed in cancer cells could be identified they could be used for halting the growth of the cancer cells.

ROLE OF CELLULAR MUTATIONS

I will now consider the other cellular events important in viral transformation, i. e. cellular mutations. Several results suggest that cellular mutations may be needed for obtaining the full state of transformation with papova viruses. Thus after infection primary cultures generate clones with various degree of transformation, some of which appear to undergo full transformation in steps (39) which may correspond to the occurrence of cellular mutations. Cells that achieve full transformation inmediately, as are common with permanent lines, may have already undergone similar mutations before infection. Some cellular mutations occuring in transformed cells may even be virus-induced, because in the early stages of transformation by papova viruses cells of primary cultures have frequent chromatid breaks (39). Conversely, cells fully transformed by SV_{40} can revert to a relatively normal phenotype although they still contain normal viral DNA and T antigen (40, 41). It is conceivable that these mutations are reversions of mutations of the former kind, which enhance the transformed state of the cells. Stepwise transformation may not only occur with viruses. Thus I have observed it in primary cultures exposed to a chemical carcinogen. In this experiment fully transformed cells evolved from the normal cells, which have limited life, generating first cells with unlimited life but unable to form colonies in agar, then cells with progressively increasing colony forming efficiency in agar, to finally reach 100 % efficiency.

All these observations show the important role of cellular mutations in cell transformation induced by different agents. This conclusion is reinforced and generalised by additional findings, such as 1) the experimental enhancement of the transforming activity of viruses by mutagenic agents (42), 2) by the elevated cancer frequency in some genetic diseases, and 3) by the evidence that most carcinogens are pro-mutagens, i. e. generate mutagenic substances when acted upon by normal metabolism (43). Most of the carcinogens themselves must be activated by metabolism in a similar way in order to induce cancer.

PROSPECTS FOR CANCER PREVENTION

I will now turn to some general deductions concerning the etiology and possible prevention of human cancer, which derive from the various points I have discussed so far. One deduction, deriving from the persistence of the viral DNA in the cells is that we can test whether a given DNA virus is a possible agent of human cancer by looking for its DNA in the cancer cells. I think that much more extensive surveys than those carried out so far as warranted, but they should have a sensitivity sufficient to detect fragments of viral DNA of about one million daltons, which is within the reach of modern technology, even with the most difficult viruses. A positive finding would be significant because DNA viruses do not appear to exist in widespread endogenous forms.

Another deduction is that somatic mutations are one of the fundamental ingredients of cancer although they appear to require the occurrence of several other events not yet understood. The role of mutations in turn suggest that the incidence of cancer in man could be reduced by identifying as many promutagens as possible, and by eliminating them from the environment. One important feature of this approach to cancer prevention is that it can be started now, since these substances can be identified with simple bacterial tests suitable for mass screening (44). The feasibility of prevention is shown by the fact that the promutagens already identified in a preliminary screening, such as tobacco or some hair dyes, are inessential for human life (45, 46).

However, it is practically difficult to achieve a substantial reduction of the use of these substances, as shown by the example of tobacco. According to epidemiological evidence tobacco smoke is the agent of lung cancer in man, which in Britain is responsible for one in eight of all male deaths (47). Yet only mild sanctions have been imposed on tobacco products, such as a vague health warning on cigarette packets, which sounds rather like an official endorsement. Any limitation on the use of tobacco is left to the individual, although it is clear that the individual cannot easily exercise voluntary restraint in the face of very effective advertisements, especially as he does not usually appreciate the danger of a cumulative action over a long period of time.

The lax attitude of governments towards tobacco probably also derives from the difficulty of appreciating epidemiological evidence, especially since this evidence is contradicted from time to time by single-minded individuals who use incomplete or even erroneous analyses of the data, and whose views are magnified out of all proportion by the media. However, the recent recognition that tobacco smoke contains promutagens contributes direct experimental evidence on the dangers of tobacco smoke, on which there cannot be any equivocation. I, therefore, call on governments to act towards severly discouraging tobacco consumption; and to act now because it will be at least thirty years before their action has its full effect.

Although tobacco smoke is a striking example of an environmental carcinogen, many others are known and probably many more remain to be identified. Identification by conventional tests is difficult because they are costly and laborious; but they can now be replaced by the bacterial tests for promutagens. Since the tests are easy and inexpensive it should be possible to investigate many normal constituents of the environment, and every new compound before it is offered to the public. The feasibility of such a programme is borne out by the finding that most of the commonly available substances are not promutagens (45, 46). Given the strong correlation between mutagenicity and carcinogenicity (43), any promutagen is suspect and, if all possible, should be withdrawn.

In fact, this is precisely the attitude that scientists have taken for themselves concerning the experiments in genetic engineering, which carry the theoretical possibility of creating new virus-like molecules, endowed with carcinogenic activity. Although the danger is only hypothetical, experiments which might be very useful for science and society have been postponed until they can be carried out under the strictest safeguards (48). Governments have accepted this position and are eager to impose severe restrictions on the performance of these experiments. While I fully approve of their concern, I cannot help noticing that they follow a double standard; if there is any doubt—

you must discourage experiments, but if there is any doubt—you cannot discourage cigarettes.

BIOLOGISTS AND SOCIETY

This discussion about cancer prevention is a development of the experimental results obtained in the field of oncogenic viruses, but it is also strongly influenced by the new social conscience of many scientists. Historically, science and society have gone separate ways, although society has provided the funds for science to grow and in return science has given society all the material things it enjoys. In recent years, however, the separation between science and society has become excessive, and the consequences are felt especially by biologists. Thus while we spend our life asking questions about the nature of cancer and ways to prevent or cure it, society merrily produces oncogenic substances and permeates the environment with them. Society does not seem prepared to accept the sacrifices required for an effective prevention of cancer. The situation is clearly unacceptable and we biologists would like to see it corrected. We have ourselves begun to put our house in order, by banning some experiments that may contain a risk for mankind. We would like to see society take a similar attitude, abandoning selfish practices that are dangerous for society itself. We would also like to see a new co-operation of science and society for the benefit of all mankind and hope that the dominant forces in society will recognize that this is a necessity.

REFERENCES:

1. Dulbecco, R., and Freeman, G., *Virology, 8*, 396 (1959).
2. Vogt, M., and Dulbecco, R., *Proc. Nat. Acad. Sci. USA., 46*, 365 (1960).
3. Dulbecco, R., and Vogt, M., *Proc. Nat. Acad. Sci. USA., 46*, 1617 (1960).
4. Smith, J. D., Freeman, G., Vogt, M., and Dulbecco, R., *Virology, 12, 185* (1960).
5. Dulbecco, R., and Vogt, M., *Proc. Nat. Acad. Sci. USA., 50*, 236, (1963)
6. Sambrook, J., Westphal, H., Srinivasan, P. R., and Dulbecco, R., *Proc. Nat. Acad. Sci. USA., 60*, 1288 (1968).
7. Oda, K., and Dulbecco, R., *Proc. Nat. Acad. Sci. USA., 60*, 525 (1968).
8. Stoker, M., *Nature (London), 218*, 234 (1968).
9. Berg, P., and Stoker, M., Personal communication.
10. Koprowski, H., Jensen, F. C., and Steplewski, Z., *Proc. Nat. Acad. Sci. USA., 58*, 127 (1967).
11. Watkins, J. F., and Dulbecco, R., *Proc. Nat Acad. Sci. USA., 58*, 1396 (1967).
12. Tonegawa, S., Walter, G., Bernardini, A., and Dulbecco, R., *Cold Spring Harbor Symp. Quant. Biol., 35*, 823 (1970).
13. Lindstrom, D. M., and Dulbecco, R., *Proc. Nat. Acad. Sci. USA., 69*, 1517, (1969).
14. Sharp. P. A., Petterson, U., and Sambrook, J., *J. Mol. Biol., 86*, 709 (1974).
15. Sharp, P. A., Gallimore, P. H., and Flint, S. J., *Cold Spring Harbor Symp. Quant. Biol., 39*, 457 (1974).
16. Fried, M., *Proc. Nat. Acad. Sci. USA., 53*, 486 (1965).
17. Eckhart, W., *Virology, 38*, 120 (1969).
18. Tegtmeyer, P., and Ozer, H. L., *J. Virol., 8*, 516 (1971).
19. Martin, R. G., Chou, J. Y., Avila, J., and Saral, R., *Cold Spring Harbor Symp. Quant. Biol., 39*, 17 (1974).

20. Butel, J. S., Brugge, J. S., and Noonan, C. A., *Cold Spring Harbor Symp. Quant. Biol., 39,* 25 (1974).
21. Kimura, G., and Itagaki, A., *Proc. Nat. Acad. Sci. USA., 72,* 673 (1974).
22. Black. P. H., Rowe, P. W., Turner, H. C., and Hubner, R. J., *Proc. Nat. Acad. Sci. USA., 50,* 1148 (1963).
23. Tegtmeyer, P., *Cold Spring Harbor Symp. Quant. Biol., 39,* 9 (1974).
24. Oxman, M., Takemoto, K. K., and Eckhart, W., *Virology, 49,* 675 (1972).
25. Paulin, D., and Cuzin, F., *J. Virol., 15,* 393 (1975).
26. Carroll, R. B., Personal communication.
27. Carroll, R. B., Hager, L., and Dulbecco, R., *Proc. Nat. Acad. Sci. USA., 71,* 3754 (1974).
28. Jessel, D., Hudson, J., Landau, T., Tenen, D., and Livingston, D. M., *Proc. Nat. Acad. Sci. USA., 72,* 1960 (1975).
29. Reed, S. I., Ferguson, J., Davis, R. W., Stark, G. R., *Proc. Nat. Acad. Sci. USA., 72,* 1605 (1975)
30. Smith, A. E., Bayley, S. T., Wheeler, T., and Mangel, W. F., In "*In vivo Transcription and Translation of Viral Genomes*"., Eds., Haenni, A., and Beaud, J., *INSERM*, Paris, *47,* 331 (1975).
31. Dulbecco, R., *Proc. Royal Society, Lond. B., 189,* 1—14 (1975).
32. Novick, A., and Weiner, M., *Proc. Nat. Acad. Sci. USA., 43,* 553 (1957).
33. Dulbecco, R., Hartwell, L. H., and Vogt, M., *Proc. Nat. Acad. Sci. USA., 53,* 403 (1965).
34. Hartwell, L. H., Vogt, M., and Dulbecco, R., *Virology, 27,* 262 (1965).
35. Coggin, J. H., *J. Immunol., 105,* 524 (1970).
36. Wiener, F., Klein, G., and Harris, H., *J. Cell Sci, 8,* 681 (1971).
37. Mintz, B., and Illmensee, K., *Proc. Nat. Acad. Sci. USA., 72,* 3585 (1975).
38. Papaionnou, V. E., McBurney, M. W., Gardner, R. L., and Evans, M. J., *Nature (Lond.), 258,* 70 (1975).
39. Vogt, M., and Dulbecco, R., *Cold Spring Harbor Symp. Quant. Biol., 27,* 367 (1962).
40. Pollack, R. E., Green, H., and Todaro, G. J., *Proc. Nat. Acad. Sci. USA., 60,* 126 (1968).
41. Renger, H. C., and Basilico, C., *Proc. Nat. Acad. Sci. USA., 69,* 109 (1972).
42. Stich, H. F., San, R. H. C., and Kawazoe, Y., *Nature (Lond)., 229,* 416, (1971).
43. McCann, J., Choi, E., Yamasaki, E., and Ames, B. N., *Proc. Nat. Acad. Sci. USA.*, In press.
44. Ames, B. N., Lee, F. D., and Durston, W. E., *Proc. Nat. Acad, Sci. USA.*, 70, 782 (1973).
45. Kier, L. D., Yamasaki, E., and Ames, B. N., *Proc. Nat. acad. Sci. USA., 71,* 4159 (1974).
46. Ames, B. N., Kammen, H. O., and Yamasaki, E., *Proc. Nat. Acad. Sci.·USA., 72,* 2423 (1975).
47. Doll, R., *J. Roy Stat. Soc., Series A, 134,* 133 (1971)
48. Berg, P., Baltimore, D., Brenner, S., Roblin, R. O., and Singer, M. F., *Proc. Nat. Acad. Sci. USA., 72,* 1981 (1975).

RENATO DULBECCO

I was born in Catanzaro, Italy, from a Calabrese mother and a Ligurian father. I stayed in that city for a short time; my father was called into the army (World War I) and we moved to the north, Cuneo and Torino. At the end of the war my father, who was in the "Genio Civile", was sent to Imperia, Liguria, where we stayed for many years. The life I remember begins at Imperia, where I went to school, including the Ginnasio-Liceo "De Amicis". What I remember most of that period, besides my family and the few friends, was the rocky beach where I spent most of my time during the summer holiday, and a small meterological observatory, where I used to spend lots of my free time throughout the year. There I developed a strong liking for physics, which I put to good use by building an electronic seismograph, probably one of the first of its kind, which actually worked.

I graduated from high school at 16 (1930) and went to the University in Torino. Although I liked especially physics and mathematics for which I had considerable talent, I decided to study medicine. This profession had for me a strong emotional appeal, which was reinforced by having an uncle who was an excellent surgeon.

In Torino I was a very successful student, but I soon realized that I was interested in biology more than in applied medicine. So I went to work with Giuseppe Levi, the professor of Anatomy, where I learned Hystology and the rudiments of cell culture. For my degree, however, I went to morbid anatomy and pathology. In Levi's laboratory I met two students who later had a strong influence on my life: Salvador Luria and Rita Levi-Montalcini.

All through the student years I was at the top of my class although I was two years younger than everbody else.

After taking my MD degree in 1936 I was called up for military service as a medical officer. In 1938 I was disharged and returned to pathology. A year later however I was called up again because of the Second World War. I was sent briefly to the French front, and a year later to Russia. There I had a narrow escape on the front of the Don during a major Russian offensive in 1942: I was hospitalized for several months and sent home. When Mussolini's governement collapsed and Italy was taken over by the German army I hid in a small village in Piemonte and joined the Resistance, as physician of the local partisan units. I continued to visit the Institute of Morbid Anatomy in Torino where I joined in underground political activities together with Giacomo Mottura, a senior collegue. I was part of the "Committee for National Liberation" of the city of Torino, and became a councillor of that city in the first postwar city council. However, the life of routine politics was not for me and within months I left that position to return to the laboratory. I also went back to school, enrolling in regular courses in physics, which I pursued for the next two years.

I moved back to Levi's Institute and worked together with Levi-Montalcini, who encouraged me to go to USA to work in modern biology. My dream was to work in genetics of some very simple organism, possibly using radiations. This dream became a reality after Luria, who was in USA since the beginning of the war, and was working in this very field, came in the summer of 1946 to Torino. He encouraged me and offered me a small salary for working in his group. I was urged in this direction by Rita Levi-Montalcini, who was herself preparing to go to another laboratory in USA. So in the autumn 1947 we both embarked for the US.

I went to work with Luria in Bloomington, Indiana, where I shared with him a small laboratory under the roof, to be soon joined by Jim Watson. Within a year I had made two good pieces of work, using my mathematical knowledge, and discovered photoreactivation of phage inactivated by ultraviolet light. This attracted the interest of Max Delbrück, who offered me a job in his group at Caltech.

I moved to Caltech in the summer 1949. I remember that memorable trip from Indiana to California with my family in an old car, with our limited possessions in a small trailer behind. I was fascinated by the beauty and immensity of the USA and the kindness of its people. Reaching the Pacific Ocean in Oregon was like arriving at a new world, an impression that continued and increased as we made our way south to Pasadena. I resolved at that time that I would not like to live anywhere else in the world — a resolution that I changed only some twenty-three years later.

At Caltech I continued to work with phages for a few years. One day I was told by Delbrück that a rich citizien had given Caltech a fund for work in the animal virus field. He asked me whether I was interested. My medical background and the experience gained in Levi's laboratory came back to me and I accepted. After visiting the major centers of animal virus work in the US I set out to discover the way to assay animal viruses by a plaque technic, similar to that used for phages, using cell cultures. Whitin less than a year, I worked out such a method, which opened up animal virology to quantitative work. I used the technic for studying the biological properties of poliovirus. These successes brought me an appointment first to associate professor, then full professor at Caltech.

In the late fifties I had as a student Howard Temin, who, together with Harry Rubin, then a postdoctoral fellow in my laboratory, worked on the Rous Sarcoma Virus. Their work started my interest in the tumor virus fields. I myself started working on an oncogenic virus, polyoma virus, in 1958, and continued until now. This work has led to discovering many aspects of the interaction of this virus (and of SV_{40}) with the host cells in lytic infection and transformation.

I moved from Caltech to the Salk Institute in 1962, and in 1972 to the Imperial Cancer Research Fund Laboratories in London. One of the reasons for the latter move was the opportunity to work in the field of human cancer.

My work throughout the years has been strongly influenced by my associates. Giuseppe Levi taught me the essential value of criticism in scientic work, Rita Levi-Montalcini helped me to determine my goals at an early stage; Salvador Luria introduced me to viruses; Herman Muller, at the University of Indiana taught me the significance of Genetics; Max Delbrück helped me understand the scientific method and the goals of biology, and Marguerite Vogt contributed to my knowledge of animal cell cultures. Perhaps more important than all this, the daily interaction through the years with a continuously changing group of young investigators shaped my work. For although I had general goals, the actual path followed by my research was pragmatically determined by what could be done at any given time, and my young collaborators were an essential part of this process. I always did as much as possible of the experimental work with my own hands, but in the later part of my research career this became progressively less feasible, both because the demand on my time increased and because the increasing technical sophistication and complexities of the experiments demanded a great deal of specialized skills.

Since 1962 my scientific life has had the support of my second wife, Maureen, who for some years helped in my experiments. Without her affectionate encouragement and sound advice I doubt whether I would have been able to accomplish what I have done.

DAVID BALTIMORE

Viruses, Polymerases and Cancer

December 12, 1975

The study of biology is partly an exercise in natural esthetics. We derive much of our pleasure as biologists from the continuing realization of how economical, elegant and intelligent are the accidents of evolution that have been maintained by selection. A virologist is among the luckiest of biologists because he can see into his chosen pet down to the details of all of its molecules. The virologist sees how an extreme parasite functions using just the most fundamental aspects of biological behavior.

A virus is a form of life with very simple requirements (1). The basic needs of a virus are a nucleic acid to be transmitted from generation to generation (the genome) and a messenger RNA to direct the synthesis of viral proteins. The critical viral proteins that the messenger RNA must encode are those that coat the genome and those that help replicate the genome. One of the great surprises of modern virology has been the discovery of the variety of genetic systems that viruses have evolved to satisfy their needs. Among the animal viruses, at least 6 totally different solutions to the basic requirements of a virus have been found (2).

If we look back to virology books of 15 years ago, we find no appreciation yet for the variety of viral genetic systems used by RNA viruses (3). Since then, the various systems have come into focus, the last to be recognized being that of the retroviruses ("RNA tumor viruses"). As each new genetic system was discovered, it was often the identification of an RNA or DNA polymerase that could be responsible for the synthesis of virus-specific nucleic acids that gave the most convincing evidence for the existence of the new system.

Now that the life-styles of different types of viruses have been delineated we can ask what relation there is between a virus' multiplication cycle and the disease it causes. In general, this question has no simple answer because disease symptoms do not correlate with the biochemical class of the virus. For instance, both poliovirus and rhinovirus are picornaviruses but one causes an intestinal infection with paralysis while the other causes the common cold. One class of RNA viruses, however, does have a unique symptom associated with it: the biochemically-defined group of viruses called the retroviruses are the only RNA viruses known to cause cancer. For a virologist interested in cancer, the problem is to first understand the molecular biology of retroviruses and then to understand how they cause the disease.

In what follows, I will review my personal involvement in uncovering the different genetic systems of RNA viruses, a story which leads to the recognition of the unique style of retroviruses. I will then consider what is known about the relationship between the biochemistry of retroviruses and their ability to be oncogenic.

As I tell my story I will mention a few of the many co-workers, teachers and students who have influenced my thinking or contributed their labors and ideas to the products of my laboratory. To mention all of the people to whom I am indebted would make too long a list; I can only say that the honors I receive are in large measure testaments to their accomplishments.

PICORNAVIRUSES

My work on the genetic systems of RNA viruses dates back to my graduate school days. As part of my introduction to animal virology during a Cold Spring Harbor course, I heard Richard Franklin describe his then-recent experiments using autoradiography to show that Mengovirus, a picornavirus and a close relative of poliovirus, could shut off the nuclear synthesis of cellular RNA early after infection and could later induce new RNA synthesis in the cytoplasm which appeared to represent synthesis of viral RNA (4). I decided to go to the Rockefeller University as a graduate student with Richard Franklin in order to work on the system he had developed.

Before I began to study how picornavirus RNA was made, it was already known from the work of Simon that picornavirus RNA synthesis was independent of DNA synthesis (5). Furthermore, studies with actinomycin D had shown that neither synthesis nor expression of cellular DNA was involved in viral RNA synthesis (6). These results suggested that Mengovirus might make a cytoplasmic RNA-dependent RNA synthesis system. The concept that viruses induce synthesis of their own enzymes had strong precedents in bacteriophage systems—Seymour Cohen's work had shown decisively that new virus-specified enzymes were found in infected bacteria (7).

We approached the problem of the virus' effect on intracellular RNA synthesis as a question in enzymology. We first showed that the nuclei from Mengovirus-infected cells were greatly reduced in their ability to carry out cell-free DNA-dependent RNA synthesis compared to nuclei of uninfected cells (8). Later, we showed that cytoplasmic extracts of Mengovirus- or poliovirus-infected cells contained an RNA synthesis activity not evident in uninfected cells and not inhibited by actinomycin D (9). When we learned that the system made RNA of the size and structure of virion RNA (10), it became clear that it represented the postulated viral RNA-dependent RNA synthesis system.

While there has been extensive further analysis of crude cytoplasmic systems (11, 12) and impressive enrichment of the RNA synthesis system has been achieved (13), no pure enzyme able to make picornavirus RNA has ever been isolated so the detailed mechanism of viral RNA synthesis still remains obscure. Most of our knowledge of how picornavirus RNA is made has come from studies on the virus-specific RNA molecules in infected cells and their kinetics of labeling by radioactive precursors. Such research has been carried out in many laboratories (11, 12); my work in this area was done in association with James Darnell and Marc Girard. Together we found and studied the relations between the poliovirus double-stranded RNA, the poliovirus replicative intermediate and the poliovirus replication complex (14). Marc Girard's precise *in vitro* analysis of RNA synthesis capped this whole series of experiments (15).

A crucial part of the viral genetic system is the manner of translation of the viral messenger RNA. While working on viral maturation, my first graduate student, Michael Jacobson, and I began to realize that proteolytic cleavage was an important part of the process (16). Our work led us to suggest that the whole 7500-nucleotide viral genome might be translated into a single continuous polypeptide that we have called a polyprotein (17, 18). Recently, this work culminated in the demonstration that poliovirus RNA can be translated into this continuous polyprotein in a cell-free system and that some of the cleavages that make the polyprotein into the functional proteins appear to occur in extracts of uninfected cells (19).

The demonstration that the poliovirus genome RNA is the messenger RNA for the synthesis of viral proteins, coupled with the demonstration of the infectivity of viral RNA (20), implies that the poliovirus genetic system is very simple. The existing evidence confirms this simplicity—as seen diagrammatically in Figure 1, it appears that the incoming "plus" strand of

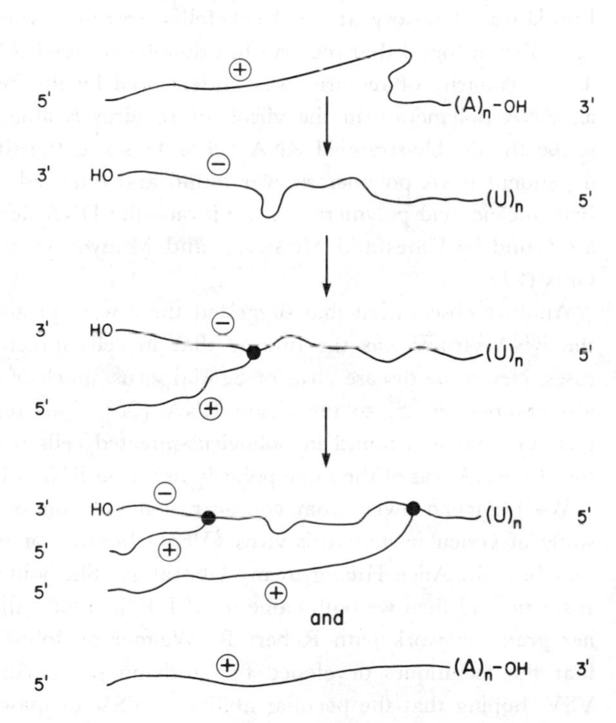

Figure 1. Schematic representation of poliovirus-specific RNA synthesis in the cytoplasm of infected cells.

RNA synthesizes a "minus" strand which in turn synthesizes a series of plus strands. This diagrammatic simplicity of poliovirus replication hides a fair amount of as yet undeciphered complexity as shown by the work of Eckard Wimmer and his colleagues as well as by work in my laboratory. For instance, the 3′-ends of the virion RNA and messenger RNA are both poly(A), the 5′-end of the minus strand is poly(U), so we assume that they are templates for each other (21). But these homopolymer stretches have very variable lengths even in the progeny of a cloned virus; what then determines their length? The poly(A) serves some obscure but necessary function in the life-cycle of the virus (22); what is this function? There is no triphosphate 5′-terminus, either free or capped, on the virion RNA or messenger RNA (23, 24); how then is the synthesis of these molecules initiated? The 5′-end of the virion RNA and messenger RNA are different (24); what does this mean?

VESICULAR STOMATITIS VIRUS

Most of the work in my laboratory until 1969 centered on poliovirus. We had assumed that all RNA viruses would be similar in their basic molecular biology but during the 1960's results emerging from various laboratories implied that poliovirus, rather than being a model for all RNA viruses, used one out of a collection of different viral genetic systems. Probably the first dramatic demonstration of the variety in RNA viruses came from next door to Richard Franklin's laboratory at the Rockefeller Institute where Peter Gomatos and Igor Tamm found that reovirus has double-stranded RNA as its genome (25). The peculiarity of reovirus was underscored by the demonstration later that an RNA polymerase in the virion of reovirus is able to assymetrically transcribe the double-stranded RNA (26). This was the first virion-bound RNA-dependent RNA polymerase ever found and followed after the finding of the first nucleic acid polymerase in a virion—the DNA-dependent RNA polymerase found by Kates and McAuslan and Munyon *et al.* in virions of vaccinia virus (27).

Another observation that suggested there were profound differences among the RNA viruses was the finding that in cells infected by the paramyxoviruses, Newcastle disease virus or Sendai virus, much of the virus-specific RNA was complementary to the virion RNA (28). This result was in sharp contrast to what was found in poliovirus-infected cells where most of the virus-specific RNA was of the same polarity as virion RNA (11).

We branched away from concentration only on poliovirus to include the study of vesicular stomatitis virus (VSV) because of the lucky circumstances that brought Alice Huang to my laboratory. She joined me first at the Salk Institute and then we both came to M.I.T. in 1968. Alice had studied VSV as her graduate work with Robert R. Wagner at Johns Hopkins. We decided that the techniques developed for studying poliovirus should be applied to VSV, hoping that the peculiar ability of VSV to spawn and then be inhibited by short, defective particles could be understood at the molecular level. A graduate student, Martha Stampfer, joined in this work and together we found that we had bitten off an enormous problem because VSV induced synthesis of so many species of RNA. In poliovirus-infected cells, only three species of RNA are seen but we found at least 9 RNA's in VSV-infected cells and one of these RNA's was clearly heterogeneous (29)—later work showed it had 4 components (30, 31). In our description of this work we said that 9 RNA species "seems exorbitant" (29) but we soon realized that each RNA had its place in the cycle of growth and growth inhibition of VSV.

As we were beginning to unravel the multiple VSV RNA's, Schaeffer et al. (32) published a paper showing that the major VSV-induced RNA's in infected cells, like those induced by Sendai and Newcastle disease viruses, were complementary in base sequence to the virion RNA. We confirmed and extended that observation, showing that the virus-specific RNA recovered from the polyribosomes of infected cells (the viral messenger RNA) was all complementary to virion RNA (33). This finding presented a pregnant paradox: if all viruses were like poliovirus and induced a new polymerase in the infected cell how could a virus that carried as its genome the strand of RNA complementary to messenger RNA ever start an infection? There seemed two possible solutions: the RNA came into the cells and was copied by a cellular enzyme to make the messenger RNA to initiate the infection cycle or the RNA came into the cell carrying an RNA polymerase with it.

Because no convincing evidence for RNA to RNA transcription existed (or exists) for any uninfected cell, the possibility of a polymerase with the incoming RNA seemed attractive. This possibility implied that there might be polymerase activity demonstrable in disrupted virions of VSV. Almost as soon as the power of this reasoning was clear to us, we had shown the existence of the virion RNA polymerase (34). The demonstration of this enzyme was the piece of evidence that led to the realization that there is a huge class of viruses, now called negative strand viruses (35), that all carry the strand of RNA complementary to the messenger RNA as their genome and that carry an RNA polymerase able to copy the genome RNA to form multiple messenger RNA's.

RETROVIRUSES

The discovery of a virion polymerase in VSV led us to search for such enzymes in other viruses. Because Newcastle disease virus made a lot of complementary RNA after infection it seemed a logical candidate and after an initial failure (34), we found activity in virions of that virus (36). But a more exciting possibility occurred to me; maybe by looking for a virion polymerase, light could be shed on the puzzle of how RNA tumor viruses multiply.

In his Nobel lecture, Howard Temin has outlined how he was led to postulate a DNA intermediate in the growth of RNA tumor viruses (37). Although his logic was persuasive, and seems in retrospect to have been flawless, in 1970 there were few advocates and many skeptics. Luckily, I had no experience in the field and so no axe to grind—I also had enormous respect for Howard dating back to my high school days when he had been the guru of the Summer School I attended at the Jackson Laboratory in Maine. So I decided to hedge my bets—I would look for either an RNA or a DNA polymerase in virions of RNA tumor viruses.

To make the foray into tumor virology, I needed some virus. Peter Vogt first sent me some Rous sarcoma virus and, although I later used it to good advantage, I initially assayed for an RNA polymerase in this viral preparation and failed to find anything. Then George Todaro helped me utilize the resources of the Special Virus Cancer Program of the National Cancer Institute to get some Rauscher mouse leukemia virus. Using that virus preparation I set out to look for a DNA polymerase activity. With little difficulty, I was able to demonstrate that virions of Rauscher virus contained a ribonuclease-sensitive DNA polymerase activity and, after confirming the results with Rous sarcoma virus, I knew that the machinery for making a DNA copy of the RNA genome was wrapped up inside the virions of RNA tumor viruses (38). Simultaneously with my work, Temin and Mizutani discovered the DNA synthesis activity in Rous sarcoma virus (39).

BIOCHEMISTRY OF REVERSE TRANSCRIPTASE

Once the DNA polymerase activity had been demonstrated in the virions of what we now call retroviruses, many laboratories began to study the enzyme. A correspondent of *Nature* dubbed the enzyme "reverse transcriptase" (40) and this romantic name has become common parlance. About 2 years after its discovery, Howard Temin and I reviewed the literature on the enzyme (41). That review and later compendia (42) make a detailed rehash of the biochemistry of retroviruses superfluous. So, I will only present a brief sketch of the picture we have today of how a retrovirus multiplies and how the reverse transcriptase functions. I will not attempt to credit all of the people who have helped to clarify this picture.

There are two separate time-periods that can be distinguished after infection of a cell by a retrovirus (Fig. 2). The first period, which lasts a few hours, involves reverse transcription and establishment of the DNA provirus as an integrated part of the cellular DNA; the second period involves the expression of the integrated genome and synthesis of progeny virions.

Analysis of the first period of retrovirus growth has focused on the types of virus-specific DNA molecules that are produced. One important type of DNA that has been found is a closed circular duplex DNA of about 5.7×10^6 daltons (43). This DNA can transfect cells with one-hit kinetics (44) and therefore contains the total viral genetic information. Other DNA's that may be on the way to becoming the closed circular form are also evident (45). It has been hard to get definitive evidence as to what DNA form integrates but presumably it is the circular duplex DNA. Whatever the form that integrates, the evidence is quite good for acquisition of proviral DNA by the chromosomes of infected cells (46, 47).

The second period in a productive retrovirus growth cycle starts when the integrated genome begins making viral RNA (48). Synthesis of viral proteins and progeny virions ensues and the cell ever-after continues to make viral products except for variations imposed by the cell cycle (49). Among the

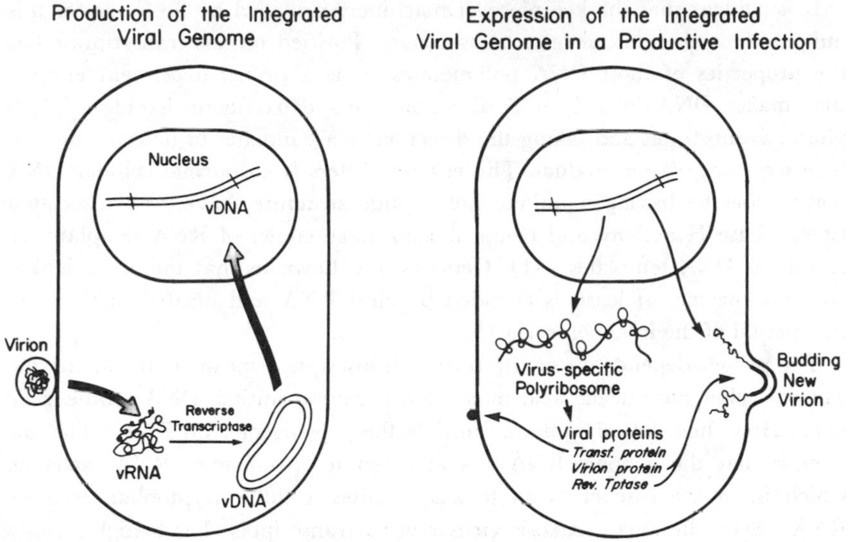

Production of the Integrated Viral Genome

Expression of the Integrated Viral Genome in a Productive Infection

Nucleus

vDNA

Virion

Reverse Transcriptase

vRNA

vDNA

Virus-specific Polyribosome

Viral proteins

Transf. protein

Virion protein

Rev. Tptase

Budding New Virion

Figure 2. The life cycle of an RNA tumor virus.
Based on present knowledge (42), the life cycle of an RNA tumor virus can be separated into two parts. In the first part the virion attaches to the cell and somehow allows its RNA along with reverse transcriptase to get into the cell's cytoplasm. There the reverse transcriptase causes the synthesis of a DNA copy of the viral RNA. A fraction of the DNA can be recovered as closed, circular DNA (43) and it is presumably that form which integrates into the cellular DNA. Once the proviral DNA is integrated into cellular DNA it can then be expressed by the normal process of transcription. The two types of product which have been characterized are new virion RNA and messenger RNA. Much of the messenger RNA which specifies the sequence of viral protein is of the same length as the virion RNA but there may also be shorter messenger RNA's (48). The virus-specific proteins have 2 known functions: one is the transformation of cells which occurs when, for instance, a sarcoma virus infects a fibroblast, the second is to provide the protein for new virion production. The transforming protein is shown here as acting at the cell surface but that is only a hypothesis.

viral proteins made in the infected cell may be a product that changes the growth properties of the cell (50); in such a case the retrovirus becomes a tumor virus.

The second period of the infection cycle can be dissociated from the first in a number of experimental systems. For instance, mammalian cells infected by avian viruses can gain viral DNA but not express it (46). Also, cells can have viral genomes that they inherited from their ancestors and such genomes are generally not being transcribed. Nonexpressed genomes can be activated: bromodeoxyuridine and iododeoxyuridine, for instance, stimulate the expression of inherited, silent viral genomes (51). The exact mechanism of activation of the genome for transcription, initiation of the transcript and termination of transcription are obscure, as are any processing events of the initial transcript which may occur.

It is evident that the key piece of machinery provided by the virus for this unique life cycle is the reverse transcriptase. Purified reverse transcriptase has the properties of most DNA polymerases: it is a primer-dependent enzyme that makes DNA in a $5' \rightarrow 3'$ direction using deoxyribonucleoside triphosphates as substrates and taking the direction of a template for determining the base sequence of the product. The enzyme differs from normal cellular DNA polymerases by having a unique polypeptide structure, having an associated ribonuclease H activity and being able to make copies of RNA templates as readily as DNA templates (41). Genetics has shown us that the avian leukosis viral enzyme, at least, is encoded by viral RNA and needed only in the first period of the infection cycle (52).

The primer-dependence of the reverse transcriptase means that the enzyme can only elongate nucleic acid molecules, it cannot initiate DNA synthesis *de novo*. How then does the enzyme initiate the copying of viral RNA? The answer is that the genome RNA has attached to it a primer RNA molecule which is, in the case of avian leukosis viruses, cellular tryptophan transfer RNA (53). The avian leukosis virus reverse transcriptase has a high-affinity binding site for that transfer RNA which the enzyme presumably uses for precise initiation of reverse transcription (54).

RETROVIRUSES AND CANCER

The last 15 years of research in animal virology has allowed us to see the diversity of genetic systems used by the various types of RNA viruses and has most recently shown us how distinct the retroviruses are from the others. Rather than using an entirely RNA-dependent replication and transcription machinery, the retroviruses have included the DNA provirus in their life-cycle. Having a DNA intermediate does not make their mode of growth especially complicated—the DNA formally takes the place of the "minus" strand in the picornavirus genetic system—but the DNA is probably the clue to why retroviruses are the only ones able to cause cancer. The DNA provides the necessary stability to the virus-cell interaction so that a viral gene product can permanently change the growth properties of an infected cell. Equally significant, the DNA stage is probably important to the ease with which retroviruses carry out genetic recombination (55); it is quite possible that the recombination system can bring together host cell genetic information with viral information and that in this way non-oncogenic retroviruses become oncogenic (56).

So the inclusion of a proviral stage in the retrovirus life-cycle may provide critical capabilities towards the development of an oncogenic potential. But the actual transformation of cells by retroviruses is a highly selective process; each type of oncogenic virus transforms a very limited range of cell types (57). If we assume that all transforming genes of viruses are like those of Rous sarcoma virus, genes that are not necessary to the growth of the virus (50, 58), then we can postulate that each type of transforming virus makes a specific type of transforming protein. Such a protein, by this model, would not be critically involved in the multiplication cycle of the virus. Isolation of such transforming proteins and elucidation of their mechanism of action is the present challenge of cancer virology. Not only will such work help us to understand carcinogenesis, it may also be important to the study of developmental biology because of the intimate relationship between the differentiated state of cells and the type of virus able to transform them.

Another implication of the occurrence of a proviral stage in the life-cycle of retroviruses is that cells can harbor such viruses as genetically silent DNA molecules. In fact, in most, if not all, animal species, the normal cells of the body have DNA related to one or more types of retroviruses (59). They receive that DNA by inheritance, not infection, and in favorable cases it can be as precisely located in the chromosomes as any gene (60). What is the significance of these genes that look like viruses?

There have been three types of explanations for virus-related genes that are inherited in the germ line of so many animal species:

1) They are an aspect of the normal genetic complement of the animal and they are virus-related because they are the progenitors of retroviruses. These genes play some important role in the life of the animal and so are not dispensable. This explanation is basically the protovirus hypothesis put forward by Howard Temin (37).

2) They are genes inserted into the chromosome of some ancestral animal by a retrovirus infection of the germ cells of that animal. Because once the provirus is integrated it remains stably associated with the chromosome, the viral genes are inherited by progeny of the original infected animal. There is one force that can eliminate such genes from a species, the slow but inexorable process of mutation. As part of this explanation of inherited viral genomes, therefore, it has been suggested that the viral genes have some positive influence on the life of the animal and so are maintained by positive selection. This explanation is closely related to the virogene-oncogene hypothesis (61).

3) The previous explanation can be modified by the exclusion of any positive role for viral genes in the life of the infected animal. There are a number of reasons why positive selection may be an unnecessary attribute to postulate. For one thing, mutations are rare events and totally inactivating mutations are much rarer. Also, the virus can be genetically invigorated by becoming a replicating virus in the body of the animal and then reinfecting the germ line. When the virus starts multiplying as an independent entity, the burden of debilitating mutations it might have accumulated could be purged if a sufficient number of generations intervened between the activation of the latent provirus and its reintegration into the germ line. The reintegration might even replace the original provirus (62). If the viral genes were not transcribed in most cells that have the viral genome, as appears to be the case, the negative effect of having one or a few integrated genomes would be slight and probably insufficient to cause a serious negative selection against animals with inherited proviruses.

I would argue that the third explanation above is the one most likely to be correct. It is an explanation that maintains the separation of viral activities and cellular activities and does not require the *ad hoc* postulation of beneficial properties of viral products. It treats retroviruses like any other virus, as an entity with its own life-style and its own accomodation with evolution.

In summary, I have tried here to develop the view of retroviruses as one of a number of solutions to the problem of creating a virus. Each virus directs synthesis of two critical classes of proteins: proteins for replication and proteins for constructing the virus particle. By encoding the reverse transcriptase, retroviruses have evolved the ability to integrate themselves into the cell chromosome as a provirus. This is a very sheltered environment in which to live, only mutation interferes with the continual transmission of the virus to the progeny of an animal that is infected in its germ cells. In this context, the ability of some retroviruses to cause cancer is a gratuitous one. But it is today the most challenging and important attribute of these retroviruses and the one that will dominate future research efforts in this area.

REFERENCES:

1. Diener, T., in *Advances in Virus Research,* vol. 17, Smith, K. M., Lauffer, M. A. and Bang, F. B. Eds. (Academic Press, New York and London, 1972), pp. 295—313.
2. Baltimore, D. *Bacteriological Reviews* 35, 235 (1971).
3. Luria, S. E., in *The Viruses,* vol. I, Burnet, F. M. and Stanley, W. M., Eds. (Academic Press, New York, 1959), pp. 549—568.
4. Franklin, R. M. and Rosner, J. *Biochem. Biophys. Acta* 55, 240 (1962).
5. Simon, E., *Virology* 13, 105 (1961).
6. Reich, E., Franklin, R. M., Shatkin, A. J. and Tatum, E. L., *Science* 134, 556 (1961).
7. Cohen, S. S., *Fed. Proc.* 20, 641 (1961).
8. Baltimore, D. and Franklin, R. M., *Proc. Nat. Acad. Sci. U.S.A.* 48. 1383 (1962). Franklin, R. M. and Baltimore, D., *Cold Spring Harbor Symp. Quant. Biol.* 27, 175 (1962).
9. Baltimore, D. and Franklin, R. M. *Biochem. Biophys. Res. Commun.* 9, 388, (1962). Baltimore, D. and Franklin, R, M. *J. Biol. Chem.* 238, 3395 (1963). Baltimore, D., Franklin, R. M., Eggers, H. J. and Tamm, I. *Proc. Nat. Acad. Sci. U.S.A.* 49, 843 (1963). Baltimore, D. and Franklin, R. M., *Cold Spring Harbor Symp. Quant. Biol.* 28, 105 (1963).
10. Baltimore, D., *Proc. Nat. Acad. Sci. U.S.A.* 51, 450 (1964).
11. Baltimore, D., in *The Biochemistry of Viruses,* Levy, H. B., Ed., (Marcel Dekker Inc., New York, 1969), pp. 101—176.
12. Levintow, L., in *Comprehensive Virology,* vol. 2. Fraenkel-Conrat, H. and Wagner, R. R., Eds. (Plenum Press, New York, 1974), pp. 109—169.
13. Lundquist, R. E., Ehrenfeld, E. and Maizel, J. E., *Proc. Nat. Acad. Sci. U.S.A.* 71, 4773 (1974).
14. Baltimore, D., Becker, Y. and Darnell, J. E., *Science* 143, 1034 (1964). Baltimore, D., *J. Mol. Biol.* 18, 421 (1966). Baltimore, D., Girard, M. and Darnell, J., *Virology* 29, 179 (1966). Baltimore, D. and Girard, M., *Proc. Nat. Acad. Sci. U.S.A.* 56, 741 (1966). Girard, M., Baltimore, D. and Darnell, J. E., *J. Mol. Biol.* 24, 59 (1967). Baltimore, D., *J, Mol. Biol.* 32, 359 (1968).
15. Girard, M., *J. Virol.* 3, 376 (1969).
16. Jacobson, M. and Baltimore, D., *J. Mol. Biol.* 33, 369 (1968).
17. Jacobson, M. F. and Baltimore, D., *Proc. Nat. Acad. Sci. U.S.A.* 61, 77 (1968). Jacobson, M. F., Asso, J. and Baltimore, D., *J. Mol. Biol.* 49, 657 (1970).

18. Baltimore, D., Harvey Lectures (Academic Press, New York, 1975), in press.
19. Villa-Komaroff, L., Guttman, N., Baltimore, D. and Lodish, H. F., *Proc. Nat. Acad. Sci. U.S.A.* 72, 4157 (1975).
20. Colter, J. S., Birel, H. H., Mayer, A. W. and Brown, R. A., *Virology* 4, 522 (1957). Alexander, H. E., Koch, G., Mountain, I. M., Sprunt, K. and Van Damme, O., *Virology* 5, 172 (1958).
21. Yogo, Y. and Wimmer, E., *Proc. Nat. Acad. Sci. U.S.A.* 69, 1877 (1972). Yogo, Y., Teng, M. and Wimmer, E., *Biochem. Biophys. Res. Commun.* 61, 1101 (1974). Spector, D. H. and Baltimore, D., *J. Virol.* 15, 1418 (1975). Spector, D. H. and Baltimore, D., *Virology* 67, 498 (1975).
22. Spector, D. H. and Baltimore, D., *Proc. Nat. Acad. Sci. U.S.A.* 71, 2983 (1974).
23. Wimmer, E., *J. Mol. Biol.* 68, 537 (1972).
24. Nomoto, A., Lee, Y. F. and Wimmer, E., *Proc. Nat. Acad, Sci. U.S.A.,* 73, 375 (1976). Hewlett, M. J., Rose, J. K. and Baltimore, D., *Proc. Nat. Acad. Sci. U.S.A.*, 73, 327 (1976).
25. Gomatos, P. J. and Tamm, I., *Proc. Nat. Acad. Sci. U.S.A.* 49, 707 (1963).
26. Shatkin, A. J. and Sipe, J. D., *Proc. Nat. Acad. Sci. U.S.A.* 61, 1462 (1968). Borsa, J. and Graham, A. F., *Biochem. Biophys. Res. Commun.* 33, 895 (1968).
27. Kates, J. R. and McAuslan, B. R., *Proc. Nat. Acad. Sci. U.S.A.* 58, 1134 (1967). Munyon, W., Paoletti, E. and Grace, J. T., *Prot. Nat. Acad. Sci. U.S.A.* 58, 2280 (1967).
28. Kingsbury, D., *J. Mol. Biol.* 18, 204 (1966). Bratt, M. A. and Robinson, W. S., *J. Mol. Biol.* 23, 1 (1967).
29. Stampfer, M., Baltimore, D. and Huang, A. S., *J. Virol.* 4, 154 (1969).
30. Morrison, T. G., Stampfer, M., Lodish, H. F. and Baltimore, D., in *Negative Strand Viruses,* vol. 1, Mahy, B. W. J. and Barry, R. D., Eds. (Academic Press, London, New York, and San Francisco, 1975), pp. 293—300.
31. Rose, J. K. and Knipe, D., *J. Virol.* 15, 994 (1975).
32. Schaffer, F. L., Hackett, A. J. and Soergel, M. E., *Biochem, Biophys. Res. Commun.* 31, 685 (1965).
33. Huang, A. S., Baltimore, D. and Stampfer, M., *Virology* 42, 946 (1970).
34. Baltimore, D., Huang, A. S. and Stampfer, M., *Proc. Nat. Acad. Sci U.S.A.* 66, 572 (1970).
35. Mahy, B. W. J. and Barry, R.D., Eds., *The Negative Strand Viruses,* Vol. 1 & 2, (Academic Press, London, New York and San Francisco, 1975).
36. Huang, A. S., Baltimore, D. and Bratt, M. A., *J. Virol.* 7, 389 (1971).
37. Temin, H., Nobel address, 1975.
38. Baltimore, D., *Nature* 226, 1209 (1970).
39. Temin, H. and Mizutani, S., *Nature* 226, 1211 (1970).
40. Anonymous, *Nature* 228, 609 (1970).
41. Temin, H. and Baltimore, D., in *Advances in Virus Research,* vol. 17 Smith, K. M., Lauffer, M. A. and Bang, F. B. (eds), (Academic Press, New York and London, 1972), pp. 129—186.
42. Cold Spring Harbor Symp. Quant. Biol., vol. 39 (Cold Spring Harbor Laboratory, New York, 1975). Tooze, J., Ed., *The Molecular Biology of Tumor Viruses*, (Cold Spring Harbor Laboratory, New York, 1973). Bishop, J. M. and Varmus, H. E., in *Cancer,* vol. 2, Becker, F. F., Ed., (Plenum Press, New York, 1975), pp. 3—48.
43. Varmus, H. E., Guntaka, R. V., Fan, W. J. W., Heasley, S. and Bishop, J. M., *Proc. Nat. Acad. Sci. U.S.A.* 71, 3874 (1974). Gianni, A. M., Smotkin, D. and Weinberg, R. A., *Proc. Nat. Acad. Sci. U.S.A.* 72, 447 (1975).
44. Smotkin, D., Gianni, A. M., Rozenblatt, S. and Weinberg, R. A., *Proc. Nat. Acad. Sci. U.S.A.* 72, 4910 (1975).
45. Gianni, A. M. Weinberg, R. A., Nature 255, 646 (1975).
46. Varmus, H. E., Vogt, P. K. and Bishop, J. M., *Proc. Nat. Acad. Sci. U.S.A.* 70, 3067 (1973).

47. Markham, P. D. and Baluda, M. A., *J. Virol.* 12, 721 (1973).
48. Fan, H. and Baltimore, D., *J. Mol. Biol.* 80, 93 (1973).
49. Leong, J. A., Levinson, W. and Bishop, J. M., *Virology* 47, 133 (1972). Paskind, M. P., Weinberg, R. A. and Baltimore, D., *Virology* 67, 242 (1975).
50. Martin, G. S., *Nature* 227, 1021 (1970).
51. Lowy, D. R., Rowe, W. P., Teich, N. and Hartley, J. W., *Science,* 174, 155 (1971).
52. Verma, I. M., Mason, W. S., Drost, S. D. and Baltimore, D., *Nature* 251, 27 (1974).
53. Dahlberg, J. E., Sawyer, R. C., Taylor, J. M., Faras, A. J., Levinson, W. E., Goodman, H. M. and Bishop, J. M., *J. Virol.* 13, 1126 (1974).
54. Panet, A., Haseltine, W. A., Baltimore, D., Peters, G., Harada, F. and Dahlberg, J. E., *Proc. Nat. Acad. Sci. U.S.A.* 72, 2535 (1975).
55. Wyke, J., Bell, J. G. and Beamund, J. A., *Cold Spring Harbor Symp. Quant. Biol.* 39, 897 (1975). Wyke, J., *Reviews on Cancer* 417, 91 (1975).
56. Weiss, R. A., in *Possible Episomes in Eukaryotes, 4th Lepetit Symposium* Sylvestri, L. G., Ed., (North-Holland, Amsterdam, 1973), pp. 130—141. Stehelin, D., Varmus, H. E., and Bishop, J. M., *J. Mol. Biol.,* in press (1975).
57. Gross, L., *Oncogenic Viruses,* 2nd edition (Pergamon Press, New York, 1972).
58. Hanafusa, H., in *Cancer,* vol. 2, Becker, F. F., Ed. (Plenum Press, New York, 1975), pp. 49—90.
59. Todaro, G. J., Beneviste, R. E., Callahan, R., Lieber, R. R. and Sherr, C., *Cold Spring Harbor Symp. Quant. Biol.* 39, 1159 (1975). Beneviste, R. E. and Todaro, G. J., *Proc. Nat. Acad. Sci. U.S.A.* 72, 4090 (1975).
60. Chattopadhyay, S. K., Lowy, D. R., Teich, N. M., Levine, A. S. and Rowe, W. P., *Cold Spring Harbor Symp. Quant. Biol.* 39, 1085 (1975). Chattopadhyay, S. K., Rowe, W. P., Teich, N. M. and Lowy, D. R., *Proc. Nat. Acad. Sci. U.S.A.* 72, 906 (1975).
61. Huebner, R. J. and Todaro, G. J., *Proc. Nat. Acad. Sci. U.S.A.* 64, 1087 (1969). Todaro, G. J. and Huebner, R. J., *Proc. Nat. Acad. Sci. U.S.A.* 60, 1009 (1972).
62. Temin, H. M., *Virology,* 13, 159 (1961).

DAVID BALTIMORE

My interest in Biology began when I was a high school student and spent a summer at the Jackson Memorial Laboratory in Bar Harbor, Maine. There I first experienced research biology and saw research biologists at work; this experience led me to become a biology major in college.

I went on to Swarthmore College where I began as a major in biology but switched to chemistry later so that I could carry out a research thesis. Between my last two years at Swarthmore I spent a summer at the Cold Spring Harbor Laboratories working with Dr. George Streisinger and the experience of working with and watching that great teacher led me to molecular biology.

I started graduate school at Massachusetts Institute of Technology in biophysics but when I decided to work on animal viruses I left M.I.T. to study for a summer with Dr. Philip Marcus at the Albert Einstein Medical College and to take the animal virus course at Cold Spring Harbor, then taught by Dr. Richard Franklin and Dr. Edward Simon. I joined Dr. Franklin at the Rockefeller Institute to do my thesis work and then continued in animal virology as a postdoctoral fellow with Dr. James Darnell. I had already found that much could be learned by studying virus-specific enzymes, so I studied for a while with Dr. Jerard Hurwitz at the Albert Einstein College of Medicine to learn from someone who knew enzymology as a professional.

My first independent position was at the Salk Institute in La Jolla, California where I had the rare opportunity to work in association with Dr. Renato Dulbecco. After 2 1/2 years away from a university setting, I returned to M.I.T. in 1968 and have remained there. In 1974, I joined the staff of the M.I.T. Center for Cancer Research under the directorship of Dr. Salvador Luria because I had found that my research interests, that previously had involved mainly the non-oncogenic RNA viruses, were more and more focused on the problems of cancer.

DATE AND PLACE OF BIRTH:
March 7, 1938 in New York, New York
EDUCATION:

1956—1960 Swarthmore College, Swarthmore, Pennsylvania B.A. with high honors in Chemistry, 1960

1960—1961 Massachusetts Institute of Technology, Cambridge Massachusetts, graduate courses toward Ph.D.

1961—1964 Rockefeller University, New York, New York. Ph.D. received in 1964

POSITIONS HELD:

1963—1964	Postdoctoral Fellow, Massachusetts Institute of Technology, Cambridge, Mass.
1964—1965	Postdoctoral Fellow, Albert Einstein College of Medicine, Bronx, New York
1965—1968	Research Associate, Salk Institute for Biological Studies, La Jolla, California
1968—1972	Associate Professor of Microbiology, Massachusetts Institute of Technology, Cambridge, Mass.
1972—present	Professor of Biology, Massachusetts Institute of Technology, Cambridge, Mass.
1973—present	American Cancer Society Professor of Microbiology

HOWARD M. TEMIN

The DNA Provirus Hypothesis

The Establishment and Implications of RNA-directed DNA Synthesis

December 12, 1975

I. INTRODUCTION

Your Majesty, fellow scientists, ladies and gentlemen: It is a great honor for me to be here today to discuss the DNA provirus and RNA-directed DNA synthesis, and it has been a pleasure for my family and me to be here in Stockholm this week. The Nobel Prize is an honor not only for me but also for all those who have been working with avian RNA tumor viruses. The Nobel Prize is also an honor for the American people, whose tax dollars and private contributions have supported my work.

The genetic information in RNA is transferred to DNA during the replication of some viruses, including some that cause cancer. This transfer of information from the messenger molecule, RNA, to the genome molecule, DNA, apparently contradicted the "central dogma of molecular biology", formulated in the late 1950's. This mode of information transfer was first postulated and established for the replication of Rous sarcoma virus, a strongly transforming avian C-type ribodeoxyvirus. (Ribodeoxyviruses are RNA viruses that replicate through a DNA intermediate.)

In this lecture, I shall discuss the experiments that led to the formulation of the DNA provirus hypothesis; the experiments that established the DNA provirus hypothesis and, therefore, the existence of RNA-directed DNA synthesis; some aspects of the present status of our knowledge of the mechanism of formation of the DNA provirus; and, finally, some implications of this work for the questions of the origin of animal viruses, how cancers may be caused by viruses, and how the majority of cancers, which do not involve infectious viruses, are caused.

The majority of the ideas I shall discuss today came from experiments with Rous sarcoma virus (RSV), the prototype RNA tumor virus. Rous sarcoma virus was originally described by Peyton Rous in 1911. He stated, "A transmissible sarcoma of the chicken has been under observation in this laboratory for the past fourteen months, and it has assumed of late a special interest because of its extreme malignancy and a tendency to wide-spread metastasis. In a careful study of the growth, tests have been made to determine whether it can be transmitted by a filtrate free of the tumor cells . . . Small quantities of a cell-free filtrate have sufficed to transmit the growth to susceptible fowl." (Rous, 1911).

Although Rous and his associates carried out many experiments with RSV, as the virus is now called, and had many prophetic insights into its behavior, they and other biologists of that time did not have the scientific concepts or the technical tools to exploit his discovery. And in about 1915 Rous himself stopped work with RSV.

The major scientific concepts required to understand the behavior of RSV were that genetic information was contained in and transferred from nucleic acids, developed especially by Avery, MacLeod, and McCarthy (1944), and by Watson and Crick (1953), as well as the concept that viral genomes could become part of cell genomes, developed especially by Lwoff (1965). The major technical tools required were those of quantitative virology and of the study of animal viruses in cell culture, developed especially by Delbrück (Cairns, Stent, and Watson, 1966), Enders, Robbins, and Weller (1955), and Dulbecco (1966).

My first contact with RSV was in 1956 when, as a graduate student at the California Institute of Techonology, I was asked by Dr. Harry Rubin, a postdoctoral fellow in Professor Dulbecco's laboratory, to try and make more quantitative the observations of Manaker and Groupé (1956) that discrete foci of altered chicken embryo cells were associated with Rous sarcoma virus in tissue culture (see also Rubin, 1966; Temin, 1971c).

II. ASSAY FOR ROUS SARCOMA VIRUS

I soon found that addition of RSV to cultures of chicken embryo cells in a sparse layer, rather than in a crowded monolayer as then used for the assay of other animal viruses, led to the appearance of foci of transformed cells (Figure 1). The number of these foci was proportional to the concentration of virus, and the foci resulted from altered morphology and altered control of multiplication of the infected cells (Temin and Rubin, 1958). The foci were cell culture analogs of tumors in chickens.

This assay allowed RSV to be studied like other viruses, leading to the demonstration that RSV-infected cells could produce virus and divide (Temin and Rubin, 1959) and to the demonstration by Crawford and Crawford (1961) that the genome of RSV was RNA. The assay for RSV was also a model for the assay of other transforming viruses, such as polyoma virus, as discussed by Dr. Dulbecco (1976).

Further observation of RSV-induced foci revealed that some of the foci contained long fusiform cells rather than the rounded cells seen in the focus in Figure 1 (Temin, 1960). Virus produced by these fusiform foci caused the formation of further foci of long fusiform cells, that is, the virus from these foci was a genetic variant.

These and other observations indicated that viral genes controlled the morphology of transformed cells and led to the hypothesis that transformation is the result of the action of viral genes, that is, transformation is a conversion analogous to lysogenic conversion. This hypothesis has been amply confirmed for RSV by the isolation of variant viruses temperature-sensitive for transformation or defective for transformation (Martin, 1970; Vogt, 1971; Kawai and Hanafusa, 1972).

These observations also led to the study of differences between transformed and normal cells. At least two important results came from these stud-

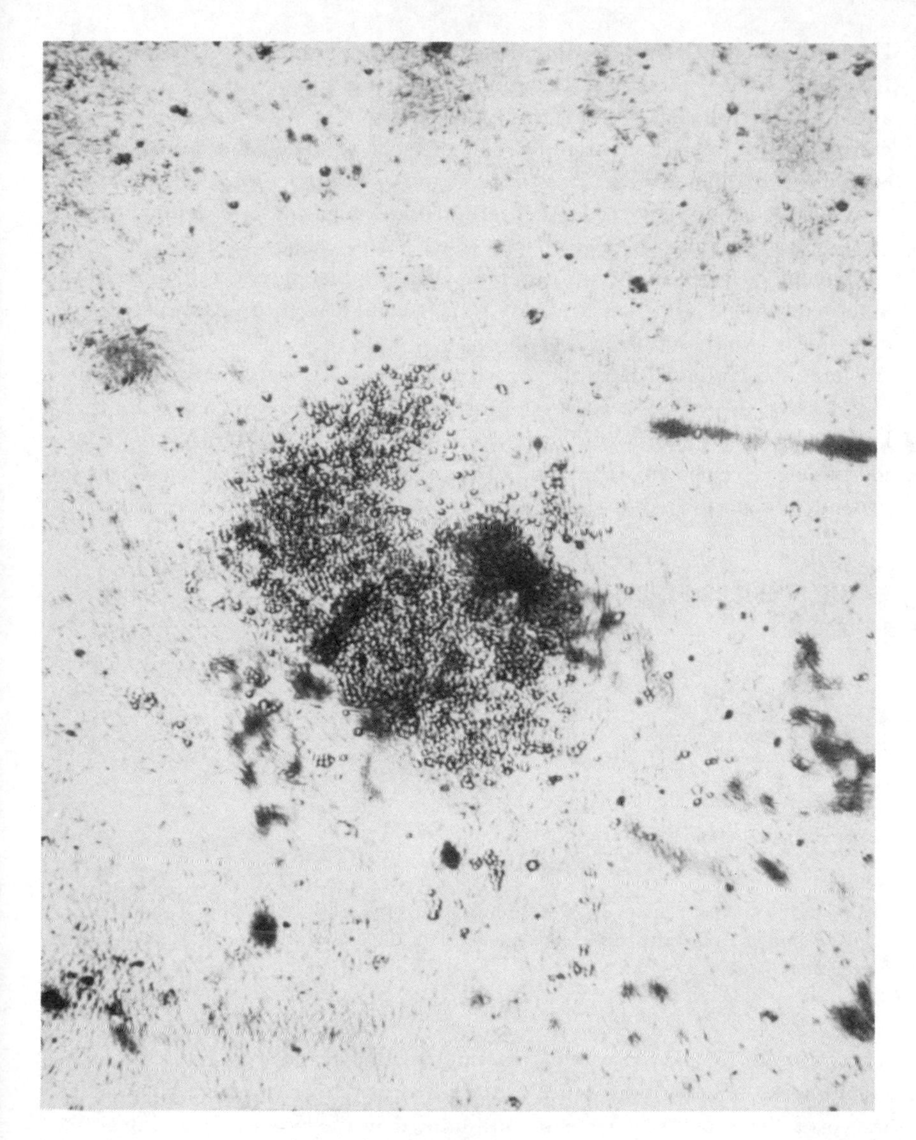

Figure 1. Focus induced by Rous sarcoma virus in chicken cells. A sparse monolayer os chicken embryo fibroblasts was exposed to Bryan standard Rous sarcoma virus. The cellf were overlaid with tissue culture medium and incubated at 38° C for ten days. This photograph of one focus was taken with an inverted microscope at a magnification of 25.

ies: 1. the concept of an altered requirement of transformed cells for specific multiplication-stimulating factors in serum (Temin, 1967b; Pierson and Temin, 1972; Dulak and Temin, 1973); and 2. the discovery by Reich and coworkers of increased production by transformed cells of an activator of a serum protease (Reich, 1975).

III. THE PROVIRUS HYPOTHESIS

In 1960 I studied the kinetics of mutation of the viral genes controlling cell and focus morphology, the effects of mutation in these viral genes on the morphology of infected cells, and the inheritance of these genes in cells infected with two different Rous sarcoma viruses (Temin, 1961). These studies demonstrated that these viral genes mutated at a high rate, that mutation in a viral gene present in an infected cell often led to change in the morphology of that infected cell, that two different viruses infecting one cell were stably inherited, and that the intracellular viral genomes were probably located at only one or two sites in the cell genome.

These observations led to the provirus hypothesis (Figure 2) — infection of chicken cells by RSV leads to the formation of one or two copies of a regularly inherited structure with the information for progeny virus and for cell morphology. (Svoboda *et al.* (1963) from studies of RSV-infected rat cells independently postulated the existence of a provirus in RSV-infected cells.)

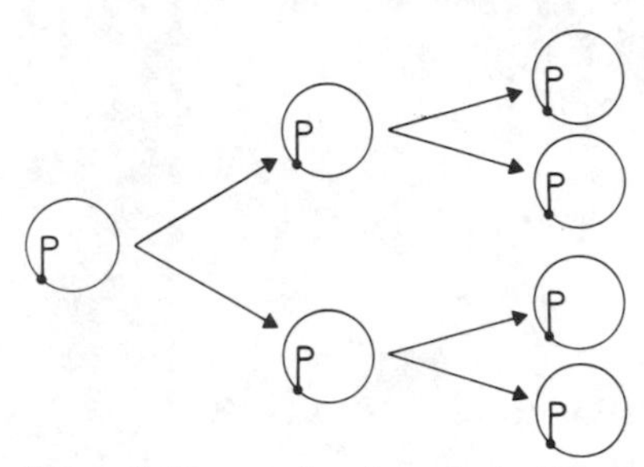

Figure 2. The provirus hypothesis. Virus information (P) is contained in infected cells in one or two copies of a regularly inherited structure with the information for progeny virus and for cell morphology.

The provirus hypothesis was a genetic hypothesis and contained no statement about the molecular nature of the provirus. However, the regular inheritance of the provirus led me to postulate that the provirus was integrated with the cell genome.

The provirus hypothesis was further supported by the behavior of converted RSV-infected chicken cells that were not producing infectious virus (Temin, 1962). (Analysis of similar cells by others led to the concept of defectiveness of some strongly transforming RNA tumor viruses (Hanafusa, Hanafusa, and Rubin, 1963; Hartley and Rowe, 1966).)

IV. THE DNA PROVIRUS HYPOTHESIS

At the time of my formulation of the provirus hypothesis in 1960, the general rules for information transfer in living systems were being clearly established in what was called "the central dogma of molecular biology", that is, genetic information is transferred from DNA to RNA to protein. RNA viruses were an apparent exception to this "dogma". Studies with the newly discovered RNA bacteriophage and with animal RNA viruses, especially using the antibiotic actinomycin D, indicated that RNA viruses transferred their information from RNA to RNA and from RNA to protein and that DNA was not directly involved in the replication of these RNA viruses (Reich *et al.*, 1962).

Although I was unable to reconcile the regular inheritance of the provirus with its being RNA, I still tried in 1962, after I had arrived at the University of Wisconsin—Madison, to use actinomycin D to isolate the provirus of Rous sarcoma virus, just as David Baltimore and others were using actinomycin to study the intermediates in the replication of other animal RNA viruses (Franklin and Baltimore, 1962).

However, when actinomycin D was added to Rous sarcoma virus-producing cells, it inhibited virus production (Figure 3). Control experiments demonstrated that this inhibition was neither of early events in infection (as was found by Barry, Ives, and Cruickshank (1962) with influenza virus) nor of the ability of the treated cells to support replication of other animal RNA viruses. These results indicated to me that the provirus was DNA.

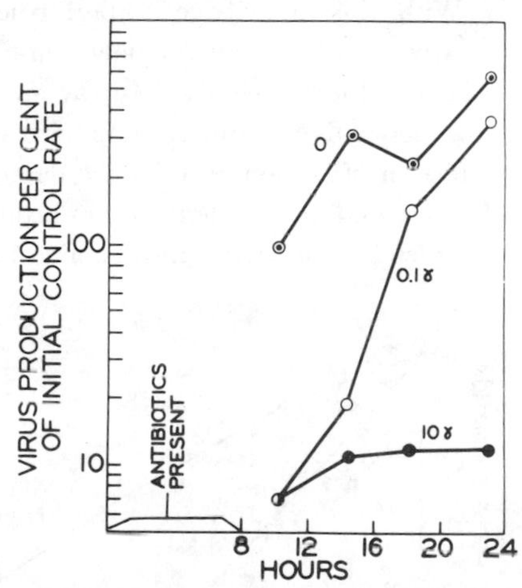

Figure 3. Effects of actinomycin D on production of Rous sarcoma virus. Chicken cells producing RSV were exposed to 0, 0.1 or 10 μg (γ)/ml of actinomycin D. After 8 hours, the medium was removed, the cells were washed, and fresh medium was added. At the indicated times, the medium was harvested and assayed for focus forming units of RSV. (Taken from Temin, 1963.)

I carried out further experiments that indicated that new DNA synthesis was required for RSV infection and that new RSV-specific DNA was found in infected chicken cells (Temin, 1964a,b; see also Bader, 1965).

Based on the results of these experiments, I proposed the DNA provirus hypothesis at a meeting in the Spring of 1964 (Temin, 1964c)—the RNA of infecting RSV acts as a template for the synthesis of viral DNA, the provirus, which acts as a template for the synthesis of progeny RSV RNA (Figure 4).

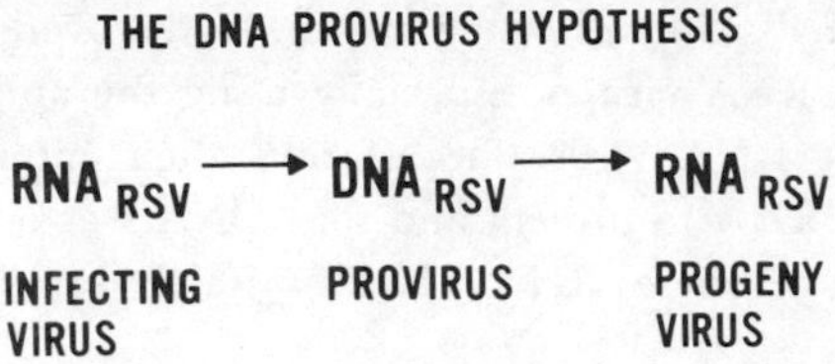

THE DNA PROVIRUS HYPOTHESIS

$RNA_{RSV} \longrightarrow DNA_{RSV} \longrightarrow RNA_{RSV}$

INFECTING VIRUS	PROVIRUS	PROGENY VIRUS

Figure 4. The DNA provirus hypothesis.

At this meeting and for the next 6 years this hypothesis was essentially ignored.

My co-workers and I tried in 1964 and 1965 to obtain direct molecular evidence for the DNA provirus by looking for RNA-directed DNA polymerase activity in cells soon after infection, for infectious DNA in infected cells, and for better systems of nucleic acid hybridization. These initial efforts were unsuccessful (Temin, 1966).

I then developed systems with better controlled cells to study RSV infection—at first, synchronized cells, and later stationary cells (Temin, 1967a, 1968a). Experiments with these cells indicated that a normal replicative cell cycle was needed for initiation of RSV production.

With this knowledge, I performed experiments that demonstrated more clearly a requirement for new non-S phase DNA synthesis for RSV infection (Temin, 1968b; see also Murray and Temin, 1970), and I demonstrated that this new DNA synthesis was virus-specific (Temin, 1970a). Finally, using infection of stationary cells, we demonstrated that the newly synthesized viral DNA could be labeled with 5-bromodeoxyuridine and inactivated by light (Figure 5) (Boettiger and Temin, 1970). However, our attempts at this time

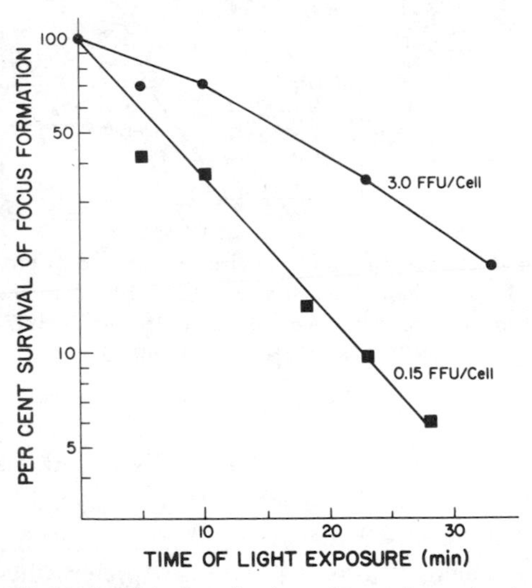

Figure 5. Light inactivation of focus formation by chicken cells infected with RSV in the presence of 5-bromodeoxyuridine. Stationary chicken cells were exposed to RSV at two multiplicities of infection (0.15 or 3.0 focus forming units per cell), incubated in medium containing 5-bromodeoxyuridine, exposed to light, and plated on rat cells to determine the number of focus forming cells surviving. (Taken from Boettiger and Temin, 1970.)

to isolate the bromodeoxyuridine-labeled viral DNA were unsuccessful (Boettiger, 1972).

V. RSV VIRION DNA POLYMERASE

In 1969 Dr. Satoshi Mizutani came to my laboratory. He demonstrated that no new protein synthesis was required for the synthesis of viral DNA during RSV infection of stationary chicken cells (quoted in Temin, 1971a), and, therefore, that the DNA polymerase that synthesized viral DNA existed before the infection of the chicken cells. This work was never published completely for, in December, 1969, we decided that the experiments indicated that RSV virions contain a DNA polymerase, and we decided to look for the virion polymerase first.

There were precedents for virion polymerases. In 1967 Kates and McAuslan, and Munyon, Paoletti, and Grace had found a DNA-directed RNA polymerase in poxvirus virions, and in 1968 Borsa and Graham, and Shatkin and Sipe had found an RNA-directed RNA polymerase in virions of reovirus. (The conclusion that RSV virions contain a DNA polymerase could have been deduced in 1967 or 1968 from the DNA provirus hypothesis and the existence of these virion polymerases, but it was not (but see Baltimore, 1976).)

RSV virions contain an endogenous DNA polymerase activity with the following characteristics (Figure 6). The virion polymerase activity incor-

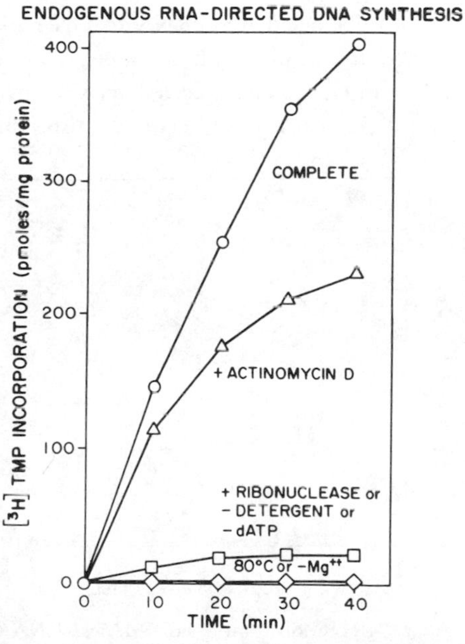

Figure 6. Endogenous RNA-directed DNA synthesis by avian leukosis virus virions. Purified virions (2 μg protein) of an avian leukosis virus were incubated in a complete system (Mizutani, Kang, and Temin, 1973) with the indicated pretreatments, additions, or subtractions, and the incorporation of label was measured.

porates deoxyribonucleoside monophosphates into DNA and requires all four deoxyribonucleoside triphosphates, a divalent cation, and a detergent to disrupt the virion envelope. Furthermore, the polymerase activity is inactivated by heat, which denatures the polymerase, and by ribonuclease, which destroys the template, and it is partially resistant to actinomycin D. (All but one of these characteristics, actinomycin D resistance (McDonnell *et al.*, 1970), were presented in our original paper (Temin and Mizutani, 1970), which was published together with the paper of Dr. Baltimore (Baltimore, 1970).) We call this virion enzyme activity "endogenous RNA-directed DNA polymerase activity".

The avian RNA tumor virus DNA polymerases are stable and easy to solubilize and study (see Temin and Baltimore, 1972). Numerous workers have purified these enzymes, especially from avian myeloblastosis virus, and this DNA polymerase has become a standard reagent for molecular biologists. It is especially useful because it has no deoxyribonuclease activity, but it does have ribonuclease H activity. (Ribonuclease H activity degrades the RNA strand of an RNA · DNA hybrid molecule, but not single-stranded RNA.)

VI. THE ESTABLISHMENT OF THE DNA PROVIRUS HYPOTHESIS

Although the discovery of the RSV virion DNA polymerase immediately provided convincing evidence for the DNA provirus hypothesis, actual proof of the existence of a DNA provirus depended upon later work involving nucleic acid hybridization and infectious DNA experiments.

Neiman (1972) was the first to demonstrate convincingly increased hybridization of labeled RSV RNA to DNA of infected chicken cells. We have confirmed his results with another avian RNA virus that replicates through a DNA intermediate, spleen necrosis virus, which gives a clearer and cleaner result (Figure 7). (Others have also confirmed Neiman's results (for example

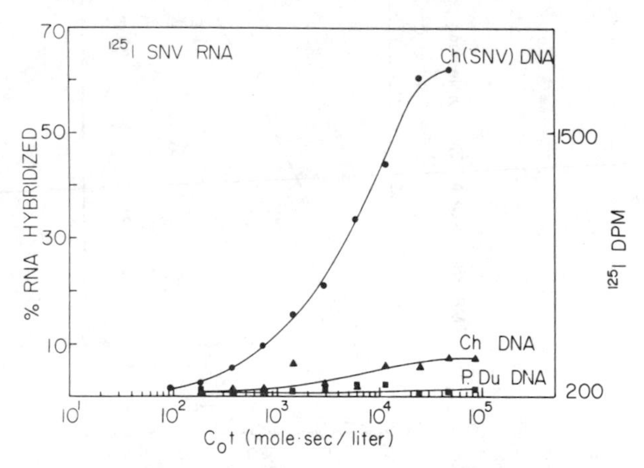

Figure 7. Hybridization of labeled viral RNA to DNA from infected and uninfected cells. ^{125}I-labeled RNA from spleen necrosis virus (SNV) was incubated for different times with a large excess of DNA from uninfected chicken (Ch) or Peking duck (P. Du) cells or from spleen necrosis virus-infected chicken (Ch(SNV)) cells, and the percentage of RNA that was ribonuclease-resistant was determined (Kang and Temin, 1974).

Varmus, Heasley, and Bishop, 1974; Shoyab, Baluda, and Evans, 1974).) Therefore, the DNA of ribodeoxyvirus-infected cells contains new nucleotide sequences complementary to the RNA of the infecting ribodeoxyvirus.

To a virologist an even more satisfying proof for the existence of the DNA provirus was the demonstration, first by Hill and Hillova (1972), of infectious DNA for RSV. We, as well as others, have repeated and extended their work, making it more quantitative (Table 1). Rous sarcoma virus-infected cells, but not uninfected cells, contain a nucleic acid with the information for RSV (the provirus). This information is contained in DNA as shown by its inactivation by deoxyribonuclease, its resistance to alkali, ribonuclease, and Pronase, and its density in equilibrium cesium chloride density gradient centrifugation. A single molecule of about 6×10^6 daltons of double-stranded DNA is sufficient to cause infection, and the efficiency of infection is similar to that of the DNA isolated from animal small DNA viruses (Cooper and Temin, 1974).

Table 1. Infectious Rous sarcoma virus DNA.[a]

DNA	Infectious dose 50 (ID_{50}) (μg)
RSV-infected chicken cell	0.1
RSV-infected chicken cell, deoxyribonuclease	> 10
RSV-infected chicken cell, alkali	1.0
RSV-infected chicken cell, ribonuclease	0.1
RSV-infected chicken cell, Pronase	0.1
RSV-infected chicken cell, cesium chloride density gradient centrifugation	0.1
RSV-infected rat cell	0.1

[a] DNA was isolated from RSV-infected chicken or rat cells, treated as indicated, and assayed for infectivity in chicken fibroblasts. Infectivity is presented as the amount of DNA required to infect half of the assay cultures. (Taken from Cooper and Temin, 1974.) The lower the amount of DNA required for infection, the more infectious the DNA was.

VII. STATUS OF KNOWLEDGE OF THE MECHANISM OF FORMATION OF THE DNA PROVIRUS AT THE PRESENT TIME (NOVEMBER, 1975)

The existence of a DNA provirus for RSV has been established. In addition, some knowledge has been gained of the details of the molecular mechanisms for the formation of the RSV provirus. Especially notable has been the work of Bishop and Varmus and their colleagues at the University of California—San Franscisco Medical School (Bishop and Varmus, 1975).

After infection of susceptible cells by RSV, the virion DNA polymerase synthesizes a DNA copy of the viral RNA, probably using a cellular transfer RNA molecule associated with the viral RNA as a primer for the DNA synthesis. After the formation of the RNA · DNA hybrid molecule, there is synthesis of a second strand of DNA, perhaps after degradation of the viral RNA by the ribonuclease H activity of the virion DNA polymearse. Double-stranded closed circular viral DNA appears. Viral DNA becomes integrated with host DNA. However, neither the mechanism for integration nor whether virion-associated enzymes (Mizutani *et al.*, 1971) are involved in integration is known.

We have been studying the formation of the provirus of spleen necrosis virus (SNV), a cytopathic member of a species of avian ribodeoxyviruses distinct from the avian leukosis viruses like RSV. Some interesting contrasts, as well as similarities, have been found.

Instead of using only a pre-formed primer for DNA synthesis, spleen necrosis virus may at times synthesize an RNA primer *de novo* (Mizutani and Temin, 1975). The virions of spleen necrosis virus contain an RNA polymerase activity as well as a DNA polymerase activity (Mizutani and Temin, 1976) (Figure 8). This RNA polymerase activity can initiate synthesis of new RNA chains, and its product RNA, a small molecule, is hydrogen-bonded to viral RNA. Thus, SNV virions contain both DNA polymerase and RNA polymerase activities—the only virions so far reported to contain both of these enzyme activities.

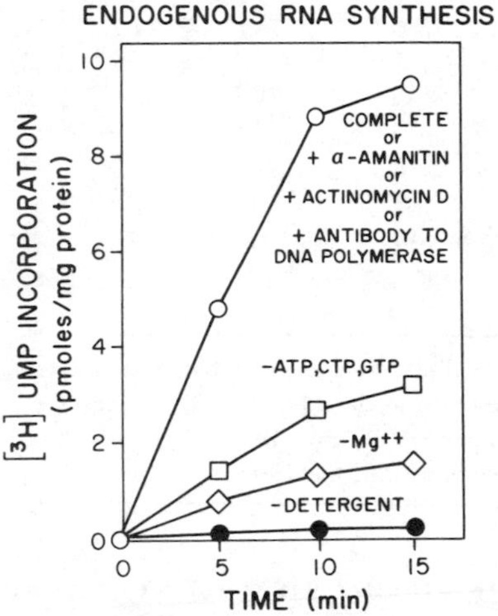

Figure 8. Endogenous RNA synthesis by reticuloendotheliosis virus virions. Purified virions (2 μg protein) of SNV were incubated in a complete system with the indicated additions, subtractions, or pretreatments, and the incorporation of label was measured. (Taken from Mizutani and Temin, 1976.)

We have also studied the kinetics of formation of infectious SNV DNA (Fritsch and Temin, 1976) (Figure 9). After infection of chicken cells by SNV, infectious viral DNA appeared in an unintegrated form, found in the supernatant of a Hirt extract (Hirt, 1967), shortly before it appeared in an integrated form, found in the pellet of a Hirt extract. Surprisingly there were large further increases in the amounts of both unintegrated and integrated viral DNAs, and some unintegrated viral DNA persisted for over a week after infection. In contrast to these results with dividing cells, little infectious vi-

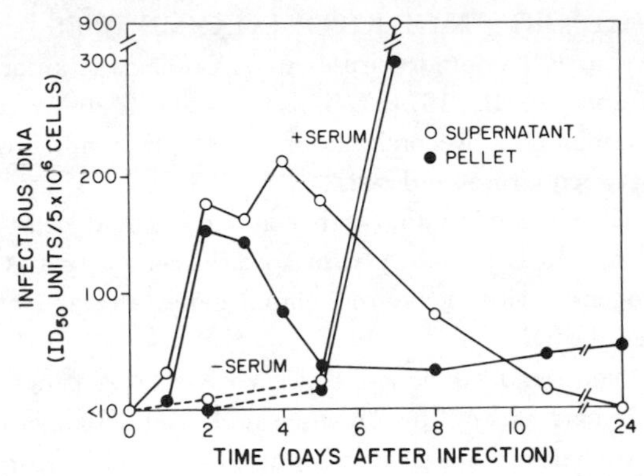

Figure 9. Kinetics of formation of infectious DNA in SNV-infected multiplying and stationary chicken cells. Chicken cells were exposed to SNV at a multiplicity of infection of 5 plaque forming units per cell, and medium with or without serum was added. At different times, the cells were fractionated by Hirt extraction (Hirt, 1967), and the DNAs in the supernatant and pellet fractions were assayed for infectivity. (Taken from Fritsch and Temin, 1976.)

ral DNA was formed in stationary cells exposed to SNV. This result indicates that a normal replicative cell cycle is required for formation of infectious viral DNA (also see Humphries and Temin, 1974).

The forms of unintegrated infectious viral DNA were analyzed by agarose gel electrophoresis (Figure 10). Three forms were found, reminiscent of the three forms of DNA in papovavirus virions (see Tooze, 1973). The majority of the infectious viral DNA was in linear molecules, but there were minor components of closed circular and nicked infectious SNV DNA.

Thus, the early events in ribodeoxyvirus infection are complex, and much remains to be learned before we can describe the formation of the provirus in molecular detail.

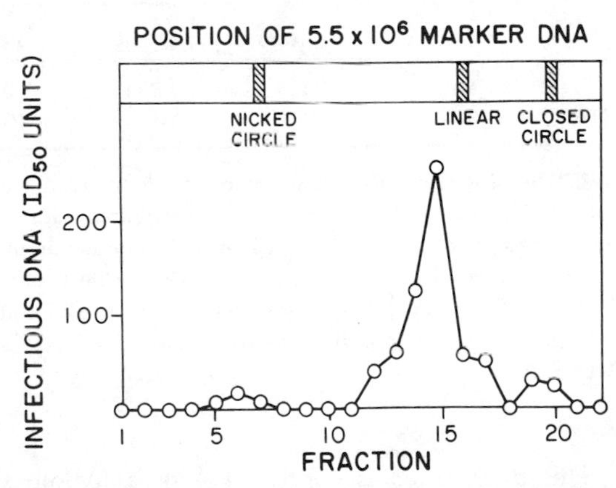

Figure 10. Electrophoresis of unintegrated infectious SNV DNA. The supernatant fraction from Hirt extraction of cells 65 hours after infection by SNV was electrophoresed in a 0.7 % agarose gel in the presence of ethidium bromide with DNA from plasmid RSF 1010 as a marker. The positions of the marker DNAs were established visually, and each fraction was assayed for infectivity. (Taken from Fritsch and Temin, 1976.)

VIII. ORIGIN OF RIBODEOXYVIRUSES

Avian RNA tumor viruses undergo a great amount of genetic variation (see Temin, 1971b; 1974a). This variation is the result of both mutation and recombination. Recombination takes place not only between viruses, but also between viruses and cells.

The recombination between viruses and cells does not appear to be random, but is primarily with specific cellular genes. These genes are called endogenous ribodeoxyvirus-related genes and are, of course, part of the normal cellular DNA.

Endogenous avian leukosis virus-related genes were first recognized about 10 years ago by the presence and Mendelian inheritance of a Rous sarcoma virus virion antigen in some uninfected chicken cells (Dougherty and DiStefano, 1966; Payne and Chubb, 1968). Later an avian leukosis virus virion envelope protein was found in some uninfected chicken cells, and, finally, nucleotide sequences of avian leukosis virus RNA were found in the DNA of all uninfected chicken cells (see Tooze, 1973; Temin, 1974a). (Similar results have been found with mammalian leukemia viruses and cells.)

Study of the phylogenetic distribution of the endogenous avian leukosis virus-related nucleotide sequences revealed (Table 2) a relationship between the amount of these sequences in cell DNA from a particular species of bird and the closeness of the relationship of that species to chickens; for example, more avian leukosis virus nucleotide sequences were found in pheasant DNA than in duck DNA (Kang and Temin, 1974; see also Benveniste and Todaro, 1974).

Table 2. Endogenous avian ribodeoxyvirus-related nucleotide sequences in avian cell DNAs.[a]

Virus	DNA				
	Chicken	Pheasant	Quail	Turkey	Duck
RAV-O	55	20	15	10	< 1
SNV	10	10	10	10	< 2

[a] ^{125}I-labeled RNAs of Rous-associated virus-0, an avian leukosis virus, and of spleen necrosis virus, a reticuloendotheliosis virus, were incubated with an excess of DNA from uninfected cells as described in the legend of Figure 7. The maximum amounts of hybridization from curves like those in Figure 7 are listed. (Taken from Kang and Temin, 1974.) In contrast to RAV-O RNA, SNV RNA hybridized equally to DNA of all the gallinaceous birds. This difference reflects the horizontal transmission of SNV and the vertical transmission of RAV-O.

This distribution is consistent with a hypothesis (the protovirus hypothesis) I originally proposed in 1970 to explain the origin of ribodeoxyviruses—ribodeoxyviruses evolved from normal cellular components (Temin, 1970b, 1974d). The normal cellular components are the endogenous ribodeoxyvirus-related genes. These genes are involved in normal DNA to RNA to DNA information transfer. This normal process of information transfer in cells could not exist only for its ability to give rise to viruses. It must exist as a result of its role in normal cellular processes, for example, cell differentiation, antibody formation, and memory (Temin, 1971d).

One prediction of this protovirus hypothesis is that there are relationships between ribodeoxyvirus and cell DNA polymerases. We have demonstrated such relationships by an antibody blocking test (Mizutani and Temin, 1974). In this test, for example, the activity of an antibody against avian leukosis virus DNA polymerases was blocked by incubation with chicken cell DNA polymerases or a DNA polymerase from an otherwise unrelated avian ribodeoxyvirus.

Therefore, certain predictions of the protovirus hypothesis for the origin of ribodeoxyviruses have been verified. But, obviously, much further work must be done to establish or disprove this hypothesis.

IX. FURTHER IMPLICATIONS OF THESE STUDIES

The protovirus hypothesis can explain the origin of ribodeoxyviruses, but it does not help in understanding the origin of other animal viruses. The presence of an RNA polymerase activity in virions of spleen necrosis virus might, however, present a clue to the origin of the other animal enveloped RNA viruses. As Dr. Baltimore has described, many animal enveloped RNA viruses contain an RNA polymerase activity (Baltimore, 1976). If there were genetic changes so that the SNV RNA polymerase activity synthesized a complete copy of SNV RNA rather than only a small molecule, the first step in the synthesis of a viral RNA intermediate would occur (Figure 11). Further genetic changes leading to copying of the newly synthesized RNA strand would complete the replication of the viral RNA. Therefore, I propose that other animal enveloped RNA viruses evolved from ribodeoxyviruses. (The recent reports of DNA intermediates in carrier cultures of some animal enveloped RNA viruses (Zhdanov, 1975; Simpson and Iinuma, 1975) could indicate a vestige of the origin of these viruses from ribodeoxyviruses.)

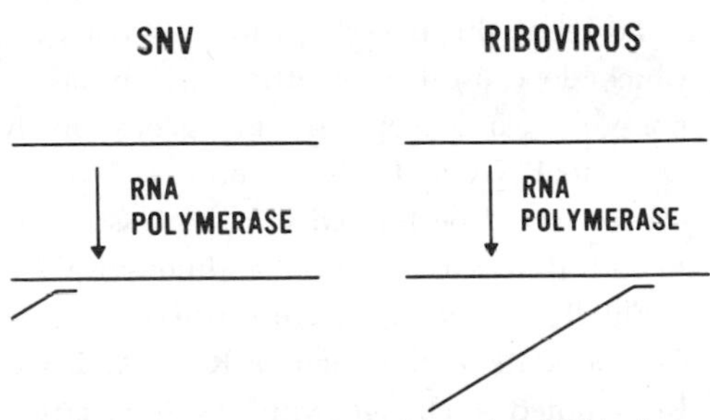

Figure 11. Initial RNA synthesis by SNV and by other RNA viruses.

Animal small DNA viruses might also have originated from ribodeoxyviruses. As discussed above, the unintegrated infectious DNA in SNV-infected cells exists in several forms, and the amount of the unintegrated DNA increases for several days after infection. This unintegrated ribodeoxyvirus DNA could represent a precursor of animal small DNA viruses. Continued replication of unintegrated viral DNA and encapsidation in viral proteins would also be required. Therefore, I propose that animal small DNA viruses also evolved from ribodeoxyviruses.

In most of this discussion of virus replication and virus origins, I have not mentioned cancer. In fact, the absence of such discussion makes an important point: RNA tumor virus replication is not sufficient for cancer formation by RNA tumor viruses. Strongly transforming RNA tumor viruses like RSV cause cancer by introducing genes for cancer into cells. But there are viruses that replicate in much the same way as RSV, for example, SNV or Rous-associated virus-O, that do not cause cancer because they do not contain genes for cancer (Temin, 1974c).

In addition, the majority of human cancers are not caused primarily by infectious viruses like RSV (Temin, 1974b), but by other types of carcinogens, for example, the chemicals in cigarette smoke (Hammond, 1964). These non-viral carcinogens probably act to mutate a special target in the cell DNA to genes for cancer (Figure 12).

THE PROTOVIRUS HYPOTHESIS
FOR THE
ORIGIN OF THE GENES FOR CANCER

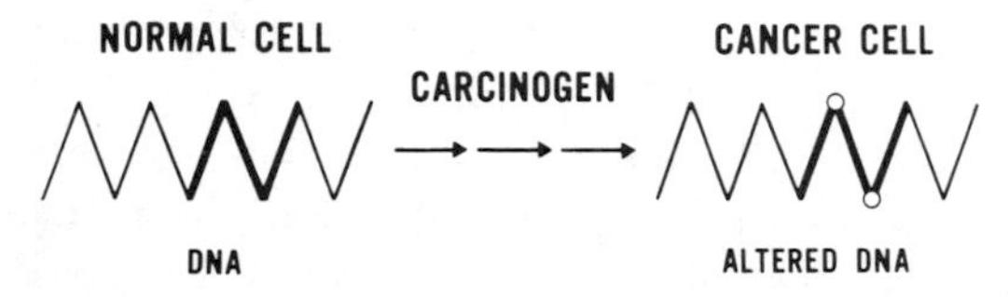

Figure 12. The protovirus hypothesis for the origin of the genes for cancer. The heavy lines indicate DNA involved in DNA to RNA to DNA information transfer.

To relate this hypothesis to the existence of animal RNA viruses like RSV, which do cause cancer efficiently, I have suggested that the targets for the non-viral carcinogens are the genes involved in information transfer from DNA to RNA to DNA (Temin, 1974b,c). Under this hypothesis, genes for cancer would be formed in a process involving RNA-directed DNA synthesis in both RNA virus-induced and non-viral carcinogen-induced cancers.

Finally, to end this lecture where it began with Peyton Rous and RSV, we can speculate on the origin of RSV. As I quoted at the beginning of my lecture, Rous noted a change with transplantation in the behavior of the chicken tumor. This change, I propose, was the result of the formation of RSV, that is, the Rous sarcoma appeared before the Rous sarcoma virus. More specifically, other events not involving a virus led to the formation of genes for cancer and the chicken sarcoma. This sarcoma was infected by an avian leukosis virus, and RSV was formed by a rare recombination (Figure 13).

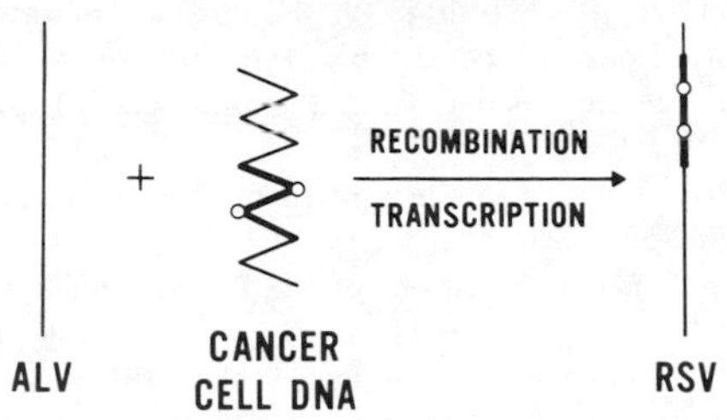

Figure 13. A hypothesis for the origin of Rous sarcoma virus (RSV). A straight line represents RNA, and a zig-zag line represents DNA. ALV is avian leukosis virus.

X. SUMMARY

I have discussed the observations and experiments that led to the formulation and establishment of the provirus hypothesis and the DNA provirus hypothesis, which includes RNA-directed DNA synthesis for the formation of the provirus.

I have also discussed some aspects of the present status of our knowledge of the mechanism of formation of the DNA provirus both to point out the work remaining to be done and to illustrate hypotheses for the origins of ribodeoxyviruses and the origins of other animal enveloped RNA viruses and of animal small DNA viruses.

Finally, I have indicated that I do not believe that infectious viruses cause most human cancers, but I do believe that viruses provide models of the processes involved in the etiology of human cancer.

ACKNOWLEDGEMENTS

I should like to acknowledge three types of support: financial, intellectual, and personal.

The work from my laboratory has been supported by grants from the National Cancer Institute and the American Cancer Society. I held a Research Career Development Award from the National Cancer Institute and am now an American Cancer Society Research Professor.

My work has been supported intellectually by colleagues in my laboratory, especially Satoshi Mizutani, by colleagues at McArdle Laboratory and the University of Wisconsin—Madison, and by co-workers in the field of avian RNA tumor viruses, especially Peter Vogt, Hidesaburo Hanafusa, Marcel Baluda, Jan Svoboda, Peter Duesberg, Robin Weiss, J. Michael Bishop, and Harold Varmus.

I have been supported personally by my family, especially my wife, Rayla.

I thank all these and the others who have supported me.

REFERENCES:

Avery, O. T., C. M. MacLeod, and M. McCarty 1944. Studies on the chemical nature of the substance inducing transformation of pneumococcal types. Induction of transformation by a desoxyribonucleic acid fraction isolated from Pneumococcus Type III. J. Exp. Med. *79*: 137—158.

Bader, J. P. 1965. The requirement for DNA synthesis in the growth of Rous sarcoma and Rous-associated viruses. Virology 26: 253—261.

Baltimore, D. 1970. RNA-dependent DNA polymerase in virions of RNA tumor viruses. Nature *226:* 1209-1211.

Baltimore, D. 1976. Viruses, polymerases and cancer. *In:* Le Prix Nobel 1975, this volume.

Barry, R. D., D. R. Ives, and J. G. Cruickshank. 1962. Participation of deoxyribonucleic acid in the multiplication of influenza virus. Nature *194*: 1139—1140.

Bishop, J. M., and H. E. Varmus. 1975. The molecular biology of RNA tumor viruses. *In:* Cancer, Vol. 2. F. F. Becker (Ed.), pp. 3—48, Plenum Press, N.Y.

Benveniste, R. E., and G. J. Todaro. 1974. Evolution of type C viral genes: I. Nucleic acid from baboon type C virus as a measure of divergence among primate species. Proc. Nat. Acad. Sci. U.S.A., *71:* 4513—4518.

Boettiger, D. 1972. A probable early DNA intermediate in Rous sarcoma virus replication. Ph. D. Dissertation, University of Wisconsin-Madison.

Boettiger, D., and H. M. Temin. 1970. Light inactivation of focus formation by chicken embryo fibroblasts infected with avian sarcoma virus in the presence of 5-bromodeoxyuridine. Nature *228:* 622—624.

Borsa, J., and A. F. Graham. 1968. Reovirus: RNA polymerase activity in purified virions. Biochem. Biophys. Res. Commun. *33*: 895—901.

Cairns, J., G. S. Stent, and J. D. Watson (Eds). 1966. Phage and The Origins of Molecular Biology. Cold Spring Harbor Laboratory.

Cooper, G. M., and H. M. Temin. 1974. Infectious Rous sarcoma and reticuloendotheliosis virus DNAs. J. Virol. *14:* 1132—1141.

Crawford, L. V., and E. M. Crawford. 1961. The properties of Rous sarcoma virus purified by density gradient centrifugation. Virology *13:* 227—232.

Dougherty, R. M., and H. S. Di Stefano. 1966. Lack of relationship between infection with avian leukosis virus and the presence of COFAL antigen in chick embryos. Virology *29:* 586—595.

Dulak, N. C., and H. M. Temin. 1973. A partially purified polypeptide fraction from rat liver cell conditioned medium with multiplication-stimulating activity for embryo fibroblasts. J. Cell. Physiol. *81:* 153—160.

Dulbecco, R. 1966. The plaque technique and the development of quantitative animal virology. *In:* Phage and The Origins of Molecular Biology. J. Cairns, G. S. Stent, and J. D. Watson (Eds.), pp. 287—291. Cold Spring Harbor Laboratory.

Dulbecco, R. 1976. From the molecular biology of oncogenic DNA viruses to cancer. *In:* Le Prix Nobel 1975, this volume.

Enders, J. F., F. C. Robbins, and T. H. Weller. 1954. The cultivation of the poliomyelitis viruses in tissue culture. *In:* Nobel Lectures in Physiology or Medicine, 1942—1962, pp. 448—467. Elsevier Publ. Co., 1964.

Franklin, R. M., and D. Baltimore. 1962. Patterns of macromolecular synthesis in normal and virus-infected mammalian cells. Cold Spring Harbor Symp. Quant. Biol. XXVII: 175—198.

Fritsch, E., and H. M. Temin. 1976. Formation and structure of infectious DNA of spleen necrosis virus. Submitted for publication.

Hammond, E. C. 1964. Smoking in relation to mortality and morbidity. Findings in first thirty-four months of follow-up in a prospective study started in 1959. J. Nat. Cancer Inst. *32:* 1161—1188.

Hanafusa, H., T. Hanafusa, and H. Rubin. 1963. The defectiveness of Rous sarcoma virus. Proc. Nat. Acad. Sci. U.S.A. *49:* 572—580.

Hartley, J. W., and W. P. Rowe. 1966. Production of altered cell foci in tissue culture by defective Moloney sarcoma virus particles. Proc. Nat. Acad. Sci. U.S.A. *55:* 780—786.

Hill, M., and J. Hillova. 1972. Virus recovery in chicken cells tested with Rous sarcoma cell DNA. Nature New Biol. 237: 35—39.

Hirt, B. 1967. Selective extraction of polyoma DNA from infected mouse cell cultures. J. Mol. Biol. *26*: 365—369.

Humphries, E. H., and H. M. Temin. 1974. Requirement for cell division for initiation of transcription of Rous sarcoma virus RNA. J. Virol. *14*: 531—546.

Kang, C.-Y., and H. M. Temin. 1974. Reticuloendotheliosis virus nucleic acid sequences in cellular DNA. J. Virol. *14:* 1179—1188.

Kates, J. R., and B. R. McAuslan. 1967. Poxvirus DNA-dependent RNA polymerase. Proc. Nat. Acad. Sci. U.S.A. *58*: 134—141.

Kawai, S., and H. Hanafusa. 1972. Genetic recombination with avian tumor virus. Virology *49*: 37—44.

Lwoff, A. 1965. Interaction among virus, cell, and organism. *In:* Nobel Lectures in Physiology or Medicine, 1963—1970, pp. 174—185. Elsevier Publ. Co., 1972.

Manaker, R., and V. Groupé. 1956. Discrete foci of altered chicken embryo cells associated with Rous sarcoma virus in tissue culture. Virology *2*: 838—840.

Martin, G. S. 1970. Rous sarcoma virus: a function required for the maintenance of the transformed state. Nature *227:* 1021—1023.

McDonnell, J. P., A.-C. Garapin, W. E. Levinson, N. Quintrell, L. Fanshier, and J. M. Bishop. 1970. DNA polymerase of Rous sarcoma virus: delineation of two reactions with actinomycin. Nature *228*: 433—435.

Mizutani, S., H. M. Temin, M. Kodama, and R. D. Wells. 1971. DNA ligase and exonuclease activities in virions of Rous sarcoma virus. Nature New Biol. *230*: 232—235.

Mizutani, S., and H. M. Temin. 1974. Specific serological relationship among partially purified DNA poymerases of avian leukosis-sarcoma viruses, reticuloendotheliosis viruses, and avian cells. J. Virol. *13:* 1020—1029.

Mizutani, S., C.-Y. Kang, and H. M. Temin. 1974. Endogenous RNA-directed DNA polymerase activity in virions of RNA tumor viruses and in a fraction from normal chicken cells. *In:* Methods of Enzymology. Vol. XXIX. Nucleic Acids and Protein Synthesis, Part E. L. Grossman, and K. Moldave (Eds.), pp. 119—124. Academic Press, N.Y.

Mizutani, S., and H. M. Temin. 1975. Endogenous RNA synthesis is required for endogenous DNA synthesis by reticuloendotheliosis virus virions. *In:* Fundamental Aspects of Neoplasia. A. A. Gottlieb, O. J. Plescia, and D. H. L. Bishop (Eds.), pp. 235—242. Springer-Verlag. N.Y.

Mizutani, S., and H. M. Temin. 1976. An RNA polymerase activity in purified virions of avian reticuloendotheliosis viruses. J. Virol. 19: 610—619.

Munyon, W., E. Paoletti, and J. T. Grace, Jr. 1967. RNA polymerase activity in purified infectious vaccinia virus. Proc. Nat. Acad. Sci. U.S.A. *58:* 2280—2287.

Murray, R. K., and H. M. Temin. 1970. Carcinogenesis by RNA sarcoma virus. XIV. Infection of stationary cultures with murine sarcoma virus (Harvey). Int. J. Cancer *5:* 320—326.

Neiman, P. E. 1972. Rous sarcoma virus nucleotide sequences in cellular DNA: Measurement of RNA-DNA hybridization. Science *178:* 750—753.

Payne, L. N., and R. C. Chubb. 1968. Studies on the nature and genetic control of an antigen in normal chick embryos which reacts in the COFAL test. J. Gen. Virol. *3*: 379—391.

Pierson, R. W. Jr., and H. M. Temin. 1972. The partial purification from calf serum of a fraction with multiplication-stimulating activity for chicken fibroblasts in cell culture and with nonsuppressible, insulin-like activity. J. Cell. Physiol. *79:* 319—330.

Reich, E. 1975. Plasminogen secretion by neoplastic cells and macrophages. *In:* Proteases and Biological Control. E. Reich, D. B. Rifkin, and E. Shaw (Eds.), pp. 333—341. Cold Spring Harbor Laboratory.

Reich, E., R. M. Franklin, A. J. Shatkin, and E. L. Tatum. 1962. Action of actinomycin D on animal cells and viruses. Proc. Nat. Acad. Sci. U.S.A. *48*: 1238—1245.

Rous, P. 1911. A sarcoma of the fowl transmissible by an agent separable from the tumor cells. J. Exp. Med. *13:* 397—411.

Rubin, H. 1966. Quantitative tumor virology. *In:* Phage and the Origin of Molecular Biology. J. Cairns, G. S. Stent, and J. D. Watson (Eds.), pp. 292—300. Cold Spring Harbor Laboratory.

Shatkin, A. J., and J. D. Sipe. 1968. RNA polymerase activity in purified reoviruses. Proc. Nat. Acad. Sci. U.S.A. *61*: 1462—1469.

Shoyab, M., M. A. Baluda, and R. Evans. 1974. Acquisition of new DNA sequences after infection of chicken cells with avian myeloblastosis virus. J. Virol. *13:* 331—339.

Simpson, R. W., and M. Iinuma. 1975. Recovery of infectious proviral DNA from mammalian cells infected with respiratory syncytial virus. Proc. Nat. Acad. Sci. U.S.A. *72:* 3230—3234.

Svoboda, J., P. Chyle, D. Simkovic, and I. Hilbert. 1963. Demonstration of the absence of infectious Rous virus in rat tumor XC, whose structurally intact cells produce Rous sarcoma when transferred to chick. Folia Biol. (Praha) *9*: 77—81.

Temin, H. M. 1960. The control of cellular morphology in embryonic cells infected with Rous sarcoma virus *in vitro.* Virology *10*: 182—197.

Temin, H. M. 1961. Mixed infection with two types of Rous sarcoma virus. Virology *13:* 158—163.

Temin, H. M. 1962. Separation of morphological conversion and virus production in Rous sarcoma virus infection. Cold Spring Harbor Symp. Quant. Biol. *XXVII*: 407—414.

Temin, H. M. 1963. The effects of actinomycin D on growth of Rous sarcoma virus *in vitro.* Virology *20*: 577—582.

Temin, H. M. 1964a. The participation of DNA in Rous sarcoma virus production. Virology *23*: 486—494.

Temin, H. M. 1964b. Homology between RNA from Rous sarcoma virus and DNA from Rous sarcoma virus-infected cells. Proc. Nat. Acad. Sci. U.S.A. *52:* 323—329.

Temin, H. M. 1964c. Nature of the provirus of Rous sarcoma. Nat. Cancer Inst. Monograph *17*: 557—570.

Temin, H. M. 1966. Genetic and possible biochemical mechanisms in viral carcinogenesis. Cancer Res. *26*: 212—216.

Temin, H. M. 1967a. Studies on carcinogenesis by avian sarcoma viruses. V. Requirement for new DNA synthesis and for cell division. J. Cell. Physiol. *69:* 53—64.

Temin, H. M. 1967b. Studies on carcinogenesis by avian sarcoma viruses. VI. Differential multiplication of uninfected and of converted cells in response to insulin. J. Cell. Physiol. *69:* 377—384.

Temin, H. M. 1968a. Carcinogenesis by avian sarcoma viruses. X. The decreased requirement for insulin-replaceable activity in serum for cell multiplication. Int. J. Cancer *3:* 771—787.

Temin, H. M. 1968b. Carcinogenesis by avian sarcoma viruses. Cancer Res. *28:* 1835—1838.

Temin, H. M. 1970a. Formation and activation of the provirus of RNA sarcoma viruses. *In:* The Biology of Large RNA viruses. R. D. Barry and B. W. J. Mahy (Eds.), pp. 233—249. Academic Press, N.Y.

Temin, H. M. 1970b. Malignant transformation of cells by viruses. Perspectives Biol. Med. *14:* 11—26.

Temin, H. M. 1971a. The role of the DNA provirus in carcinogenesis by RNA tumor viruses. *In*: The Biology of Oncogenic Viruses. Lepetit Colloquia in Biology and

Medicine, Vol. 2. L. G. Silvestri (Ed.), pp. 176—187. North Holland Publ. Co., Amsterdam.

Temin, H. M. 1971b. Mechanism of cell transformation by RNA tumor viruses. Ann. Rev. Microbiol. *25*: 610—648.

Temin, H. M. 1971c. RNA tumor viruses and cancer. *In*: Virus Y Cancer—Homenaje a F. Duran-Reynals. V. Congress National Society Esp. Bioq. W. M. Stanley, J. Casals, J. Oro, and R. Segura (Eds.), pp. 331—357, Barcelona, Spain, Imprenta Socitra.

Temin, H. M. 1971d. Guest Editorial. The protovirus hypothesis: Speculations on the significance of RNA-directed DNA synthesis for normal development and carcinogenesis. J. Nat. Cancer Inst. *46:* III—VIII.

Temin, H. M. 1974a. The cellular and molecular biology of RNA tumor viruses, especially avian leukosis-sarcoma viruses, and their relatives. Advan. Cancer Res. *19*: 47—104.

Temin, H. M. 1974b. Introduction to virus-caused cancers. Cancer *34*: 1347—1352.

Temin, H. M. 1974c. On the origin of the genes for neoplasia: G. H. A. Clowes Memorial lecture. Cancer Res. *34:* 2835—3841.

Temin, H. M. 1974d. On the origin of RNA tumor viruses. Ann. Rev. Genetics *8:* 155—177.

Temin, H. M., and H. Rubin. 1958. Characteristics of an assay for Rous sarcoma virus and Rous sarcoma cells in tissue culture. Virology *6*: 669—688.

Temin, H. M., and H. Rubin. 1959. A kinetic study of infection of chick embryo cells *in vitro* by Rous sarcoma virus. Virology *8:* 209—222.

Temin, H. M., and S. Mizutani. 1970. RNA-dependent DNA polymerase in virions of Rous sarcoma virus. Nature *226:* 1211—1213.

Temin, H. M., and D. Baltimore. 1972. RNA-directed DNA synthesis and RNA tumor viruses. Advan. Virus. Res. *17*: 129—186.

Tooze, J. (Ed.). 1973. The Molecular Biology of Tumor Viruses. Cold Spring Harbor Laboratory.

Varmus, H. E., S. Heasley, and J. M. Bishop. 1974. Use of DNA-DNA annealing to detect new virus-specific DNA sequences in chicken embryo fibroblasts after infection by avian sarcoma virus. J. Virol. *14:* 895—903.

Vogt, P. K. 1971. Spontaneous segregation of nontransforming viruses from cloned sarcoma viruses. Virology *46*: 939—946.

Watson, J. D., and F. H. C. Crick. 1953. Genetical implications of the structure of deoxyribonucleic acid. Nature *171:* 964—967.

Zhdanov, V. M. 1975. Integration of viral genomes. Nature *256:* 471—473.

HOWARD M. TEMIN

I was born on December 10, 1934 in Philadelphia, Pennsylvania, United States of America, the second of three sons of Annette and Henry Temin. My father was an attorney, and my mother has been continually active in civic affairs, especially educational ones. My older brother, Michael, is also an attorney in Philadelphia, and my younger brother, Peter, is a Professor of Economics at the Massachusetts Institute of Technology, Cambridge, Mass.

I received my elementary and high school education in the public schools of Philadelphia. My specific interest in biological research was focused by summers (1949—1952) spent in a program for high school students at the Jackson Laboratory in Bar Harbor, Maine, and a summer (1953) spent at the Institute for Cancer Research in Philadelphia. I attended Swarthmore College from 1951 to 1955, majoring and minoring in biology in the honors program. After another summer (1955) at the Jackson Laboratory, I became a graduate student in biology at the California Institute of Technology in Pasadena, California, majoring in experimental embryology. After a year and a half, I changed my major to animal virology, becoming a graduate student in the laboratory of Professor Renato Dulbecco. My doctoral thesis was on Rous sarcoma virus. Much of my early work on this virus was carried out with the close collaboration of Dr. Harry Rubin, then a postdoctoral fellow in Professor Dulbecco's laboratory. At Cal Tech, I was also greatly influenced by Professor Max Delbrück and by Dr. Matthew Meselson. After finishing my Ph.D. degree in 1959, I remained for an additional year in Professor Dulbecco's laboratory as a postdoctoral fellow. In that year, I performed the experiments that led to the formulation in the same year of the provirus hypothesis for Rous sarcoma virus.

In 1960, I moved to Madison as an Assistant Professor in the McArdle Laboratory for Cancer Research, which is also the Department of Oncology, in the Medical School, The University of Wisconsin-Madison. My first laboratory was in the basement, with a sump in my tissue culture lab and with steam pipes for the entire building in my biochemistry lab. Here I performed the experiments that led in 1964 to my formulating the DNA provirus hypothesis. In the fall of 1964, the entire department moved to a new building. I became successively Associate Professor, Full Professor, Wisconsin Alumni Research Foundation Professor of Cancer Research, and, in 1974, American Cancer Society Professor of Viral Oncology and Cell Biology. From 1964 to 1974, I also held a Research Career Development Award from the National Cancer Institute.

During my first years at Wisconsin, I worked with only two technicians.

My first postdoctoral fellow joined me in 1963, and my first graduate student, in 1965. I had no more than two or three postdoctoral fellows and graduate students at one time until about 1968.

During the late 1960's, about half of my time was spent in studying the control of multiplication of uninfected and Rous sarcoma virus-infected cells in culture. This work led to my appreciation of the role of specific serum factors in the control of cell multiplication and the demonstration that a multiplication-stimulating factor in calf serum for chicken fibroblasts was the same as somatomedin.

I serve on the editorial boards of several journals, including the *Journal of Cellular Physiology,* the *Journal of Virology,* and the *Proceedings of the National Academy of Sciences U.S.A.* I have also been a member of the Virology Study Section of the National Institutes of Health. In addition, I do much other paper and grant reviewing.

Since the general acceptance of the DNA provirus hypothesis in 1970, I have received many honors, including the Warren Triennial Prize (with David Baltimore); the Pap Award of the Papanicolaou Institute, Miami, Florida; the Bertner Award, M. D. Anderson Hospital and Tumor Institute, Houston, Texas; the U. S. Steel Foundation Award in Molecular Biology, National Academy of Sciences U.S.A.; the American Chemical Society Award in Enzyme Chemistry; the Griffuel Prize, Association Developpment Recherche Cancer, Villejuif, France; the G.H.A. Clowes Award, American Association for Cancer Research; the Gairdner International Award (with David Baltimore); the Albert Lasker Award in Basic Medical Research; and honorary degrees from Swarthmore College and New York Medical College. I have also presented several honorary lectures. I am a fellow of the American Academy of Arts and Sciences and a member of the National Academy of Sciences, U.S.A.

In 1962 I married Rayla Greenberg of Brooklyn, New York, a population geneticist. She has been a constant source of support and warmth. We have two daughters, Sarah Beth and Miriam.

Index

Numbers in italic type refer to figures or tables